EXPLORING THE FORMATION AND EVOLUTION OF PLANETARY SYSTEMS

IAU SYMPOSIUM No. 299

COVER ILLUSTRATION: FROM DISKS TO PLANETS

Top left: HST imaging of the protoplanetary disk SSTtau J042021+281349, remarkable for its lateral symmetry and spectacular bipolar jet, highlighted by the F606W filter of HST (Duchêne *et al.*, this work).

Bottom: ALMA image of dust trapped in a ring around the Herbig Ae star Oph IRS 48. The dust and gas cavities of this protoplanetary disk are completely resolved in this $0.2''$ resolution image (van der Marel *et al.*, this work).

Top right: Subaru direct imaging of the planetary companion GJ 504b which orbits a sun-like star 18 pc from the Earth. The projected distance of the planet from the star is 44 AU (Kuzuhara, M *et al.* 2013, ApJ, 774, 11).

IAU SYMPOSIUM PROCEEDINGS SERIES

INTERNATIONAL ASTRONOMICAL UNION
UNION ASTRONOMIQUE INTERNATIONALE

International Astronomical Union

EXPLORING THE FORMATION AND EVOLUTION OF PLANETARY SYSTEMS

PROCEEDINGS OF THE 299th SYMPOSIUM OF THE INTERNATIONAL ASTRONOMICAL UNION HELD IN VICTORIA, CANADA JUNE 2 – 7, 2013

Edited by

MARK BOOTH
Instituto de Astrofísica, Pontificia Universidad Católica de Chile, Vicuña Mackenna 4860, 7820436 Macul, Santiago, Chile
University of Victoria, BC, Canada
National Research Council of Canada Herzberg Astronomy & Astrophysics

BRENDA C. MATTHEWS
National Research Council of Canada, Herzberg Astronomy & Astrophysics
University of Victoria, BC, Canada

and

JAMES R. GRAHAM
University of California, Berkeley, CA, U.S.A.

CAMBRIDGE
UNIVERSITY PRESS

CAMBRIDGE UNIVERSITY PRESS
The Edinburgh Building, Cambridge CB2 2RU, United Kingdom
32 Avenue of the Americas, New York, NY 10013 2473, USA
10 Stamford Road, Oakleigh, Melbourne 3166, Australia

First published 2013

Printed in the United Kingdom by CPI Group (UK) Ltd, Croydon, CR0 4YY

Typeset in System LaTeX 2_ε

A catalogue record for this book is available from the British Library

Library of Congress Cataloguing in Publication data

This journal issue has been printed on FSC-certified paper and cover board. FSC is an independent, non-governmental, not-for-profit organization established to promote the responsible management of the worlds forests. Please see www.fsc.org for information.

ISBN 9781107045200 hardback
ISSN 1743-9213

Table of Contents

Preface

"Exploring the Formation and Evolution of Planetary Systems" is focussed on the evolution of proto-planetary disks and the formation and evolution of the planets themselves. This meeting, which is coincident with the end of observations by the Kepler space observatory, highlights a maturing view of planets detected by the transit and Doppler techniques, results from the first high-resolution imaging with ALMA, and sets the stage for the initial operations for a new generation of high contrast planet imaging instruments.

Circumstellar disks provide a unique window into the process of planet formation around young stars. Characterizing disk structure yields clues to disk evolution by determining the spatial distribution of gas and dust, probing the physical conditions in the disk, and providing snapshots of systems in the process of clearing natal material. The observational characterization of these planet-forming disks by, for example, Herschel, ALMA, and EVLA, is in an era of substantial progress. Together with ever-increasing theoretical attention and computational power, this confluence of developments promises an unusual period of rapid progress in understanding the domain of cool gas and dust in disks.

The timing of this meeting also coincides with advances in high-resolution imaging of planetary systems by interferometers on optical/IR 8-m class telescopes (LBTI and PRIMA) and direct imaging of exoplanets with dedicated high contrast adaptive optics coronagraphs (e.g., P1640, SCExAO, SPHERE, GPI). It is therefore a critical time to bring together the communities working on the earliest phases of planet formation with those studying mature planetary systems and associated debris disks.

While there is strong representation of observational work at this meeting, we solicited review talks that discuss the impact of the observations on models of planet formation and evolution in the broadest terms. No five-day meeting can address the full breadth of each field, but we hope that the legacy of the meeting is the enhanced interaction between those who study the formation of planets and those who study evolved systems.

It is especially fitting that an exoplanet meeting be held in British Columbia. Over thirty years ago, University of British Columbia astronomers Gordon Walker and Bruce Campbell pioneered gas absorption cell spectroscopy with the objective of using precision Doppler measurements to detect exoplanets. Hydrogen fluoride was selected on the advice of the Nobel prize-winning German-Canadian physical chemist Gerhard Herzberg, whose name has long been honored in the name of one of our organizers: the Herzberg Institute of Astrophysics (now NRC Herzberg Astronomy & Astrophysics Programs). By 1988 Campbell and Walker, together with UBC astronomer Stephenson Yang, had achieved a precision of 13 m/s, developed methods to remove differences in the velocity zero point between observing runs, and seen the Doppler signature of γ Cephei b. A quarter of a century later, and with Doppler measurement of orbits for 500 exoplanets in hand, their vision has been realized. The current proceedings reveal a diversity of planetary systems and an array of detection and characterization methods unimaginable twenty five years ago.

James R. Graham, chair SOC
Berkeley, CA, USA, 9 September 2013

THE ORGANIZING COMMITTEE

Scientific

James R. Graham (Chair, USA)
France Allard (France)
Antonio Hales (Chile)
Paul Kalas (USA)
Matthew Kenworthy (The Netherlands)
Anne-Marie Lagrange (France)
Doug Lin (USA)
Bruce Macintosh (USA)
Sarah Maddison (Australia)
Brenda Matthews (Canada)
Dimitri Mawet (Chile)
Amaya Moro-Martín (Spain)
Ruth Murray-Clay (USA)
Don Pollacco (United Kingdom)
Didier Queloz (Switzerland)
Motohide Tamura (Japan)
David Wilner (USA)

Local

Brenda Matthews (Chair, NRC)
Mark Booth (University of Victoria/NRC)
Alice Chow (DI)
James Di Francesco (NRC)
Wesley Fraser (NRC)
Raphaël Galicher (NRC/Grenoble)
James R. Graham (DI/UC Berkeley)
JJ Kavelaars (NRC)
Christian Marois (NRC)
Brenda Parrish (NRC)
Gerald Schieven (NRC)

Acknowledgements

The symposium was sponsored and supported by the IAU Divisions III (Planetary Systems Sciences, VI (Interstellar Matter) and X (Radio Astronomy) and also by the North American ALMA Science Center.

The Local Organizing Committee operated under the auspices of the National Research Council of Canada (NRC) and the Dunlap Institute for Astronomy & Astrophysics (DI), University of Toronto.

Financial support from the
  International Astronomical Union
  Dunlap Institute for Astronomy & Astrophysics, University of Toronto
  National Research Council of Canada
  University of Victoria
  Gemini Observatory
  Royal Astronomical Society of Canada, Victoria chapter
is gratefully acknowledged.

CONFERENCE PHOTOGRAPH

Participants

Brett **Addison**, Australia b.addison@student.unsw.edu.au
Yukako **Aimi**, Japan aimi@iral.ess.sci.osaka-u.ac.jp
Richard **Alexander**, United Kingdom richard.alexander@leicester.ac.uk
France **Allard**, France fallard@ens-lyon.fr
David **Anderson**, United Kingdom d.r.anderson@keele.ac.uk
Sean **Andrews**, USA sandrews@cfa.harvard.edu
Fabien **Anthonioz**, France Fabien.Anthonioz@obs.ujf-grenoble.fr
David **Armstrong**, United Kingdom d.j.armstrong@warwick.ac.uk
Jean-Charles **Augereau**, France augereau@obs.ujf-grenoble.fr
Vanessa **Bailey**, USA vbailey@as.arizona.edu
Kévin **Baillie**, France kevin.baillie@univ-paris-diderot.fr
Gaspar **Bakos**, USA gbakos@astro.princeton.edu
Nick **Ballering**, USA ballerin@email.arizona.edu
Michele **Bannister**, Canada michele.t.bannister@gmail.com
Sara **Barber**, USA sara.d.barber@me.com
Jason W. **Barnes**, USA jwbarnes@uidaho.edu
Mary **Barsony**, USA mbarsony@seti.org
Jean-Loup **Baudino**, France jean-loup.baudino@obspm.fr
Daniel **Bayliss**, Australia daniel@mso.anu.edu.au
Thomas **Beatty**, USA tbeatty@astronomy.ohio-state.edu
Will **Best**, USA wbest@ifa.hawaii.edu
Beth **Biller**, Germany biller@mpia.de
Anthony **Boccaletti**, France anthony.boccaletti@obspm.fr
David **Bohlender**, Canada david.bohlender@nrc-cnrc.gc.ca
Emeline **Bolmont**, France Emeline.Bolmont@obs.u-bordeaux1.fr
Mariangela **Bonavita**, Italy mariangela.bonavita@oapd.inaf.it
Mickaël **Bonnefoy**, Germany bonnefoy@mpia.de
Amy **Bonsor**, France amy.bonsor@gmail.com
Mark **Booth**, Canada mark.booth@nrc.ca
Kim **Bott**, Australia k.bott@unsw.edu.au
Brendan **Bowler**, USA bpbowler@ifa.hawaii.edu
Tim **Brandt**, USA tbrandt@astro.princeton.edu
Hannah **Broekhoven-Fiene**, Canada broekhov@uvic.ca
Joanna **Brown**, USA joannabrown@cfa.harvard.edu
Ludmila **Carone**, Belgium ludmila.carone@wis.kuleuven.be
John **Carpenter**, USA jmc@astro.caltech.edu
Sebastien **Charnoz**, France charnoz@cea.fr
Eugene **Chiang**, USA echiang@astro.berkeley.edu
Elodie **Choquet**, USA choquet@stsci.edu
Duncan **Christie**, USA dac5zm@astro.virginia.edu
Christian **Clanton**, USA clanton@astronomy.ohio-state.edu
Laird **Close**, USA lclose@as.arizona.edu
Christophe **Cossou**, France Christophe.Cossou@u-bordeaux1.fr
Nicolas **Cuello**, France nicolas.cuello@univ-lyon1.fr
Thayne **Currie**, Canada currie@astro.utoronto.ca
Claire **Davies**, United Kingdom cd54@st-andrews.ac.uk
Rebekah **Dawson**, USA rdawson@cfa.harvard.edu
Maria **de Juan Ovelar**, Netherlands mjovelar@strw.leidenuniv.nl
Ramiro **de la Reza**, Brazil delareza@on.br
Robert **De Rosa**, USA rjderosa@asu.edu
Denis **Defrère**, USA ddefrere@email.arizona.edu
James **Di Francesco**, Canada james.difrancesco@nrc-cnrc.gc.ca
Jack **Dobinson**, United Kingdom jack.dobinson@bristol.ac.uk
Sally **Dodson-Robinson**, USA sdr@astro.as.utexas.edu
Ruobing **Dong**, USA rdong@astro.princeton.edu
René **Doyon**, Canada doyon@astro.umontreal.ca
Zachary H **Draper**, Canada zhd@uvic.ca
Gaspard **Duchene**, USA gduchene@berkeley.edu
Rudolf **Dvorak**, Austria dvorak@astro.univie.ac.at
Jo **Barstow**, United Kingdom j.barstow1@physics.ox.ac.uk
Keigo **Enya**, Japan enya@ir.isas.jaxa.jp
Steve **Ertel**, France steve.ertel@obs.ujf-grenoble.fr
Catherine **Espaillat**, USA cespaillat@cfa.harvard.edu
Neal **Evans**, USA nje@astro.as.utexas.edu
Francesca **Faedi**, United Kingdom F.Faedi@warwick.ac.uk
Jackie **Faherty**, Chile jfaherty17@gmail.com
Virginie **Faramaz**, France Virginie.Faramaz@obs.ujf-grenoble.fr
Jay **Farihi**, United Kingdom jfarihi@ast.cam.ac.uk
Julien **Faure**, France julien.y.faure@gmail.com
Stephen **Fendyke**, United Kingdom s.m.fendyke@qmul.ac.uk
Debra **Fischer**, USA debra.fischer@yale.edu
Mike **Fitzgerald**, USA mpfitz@ucla.edu
Kate **Follette**, USA kfollette@as.arizona.edu
Jonathan **Fortney**, USA jfortney@ucolick.org
Wes **Fraser**, Canada wesley.fraser@nrc.ca
Jeffrey **Fung**, Canada fung@astro.utoronto.ca
Boris **Gaensicke**, United Kingdom boris.gaensicke@warwick.ac.uk
Eric **Gaidos**, USA gaidos@hawaii.edu
Raphael **Galicher**, France raphael.galicher@obspm.fr
Antonio **Garufi**, Switzerland antonio.garufi@phys.ethz.ch
Scott **Gaudi**, USA gaudi@astronomy.ohio-state.edu
Yilen **Gómez Maqueo Chew**, United Kingdom y.gomez@warwick.ac.uk
Jean-Franois **Gonzalez**, France Jean-Francois.Gonzalez@ens-lyon.fr
James **Graham**, Canada/USA jrg@berkeley.edu
Phil **Gregory**, Canada gregory@phas.ubc.ca
Pin-Gao **Gu**, Taiwan gu@asiaa.sinica.edu.tw
Janis **Hagelberg**, Switzerland janis.hagelberg@unige.ch
Antonio **Hales**, Chile ahales@alma.cl
Stephan **Hartmann**, Germany hartmann@astro.uni-tuebingen.de

Paul M. **Harvey**, USA pmh@astro.as.utexas.edu
Yasuhiro **Hasegawa**, Taiwan yasu@asiaa.sinica.edu.tw
Korey **Haynes**, USA korey.n.haynes@nasa.gov
Raphaëlle **Haywood**, United Kingdom rdh4@st-andrews.ac.uk
Sasha **Hinkle**, USA shinkley@astro.caltech.edu
Tobias C. **Hinse**, South Korea tchinse@gmail.com
John **Hodgson II**, USA John.Hodgson.71@my.csun.edu
Wayne **Holland**, United Kingdom wayne.holland@stfc.ac.uk
Derek **Homeier**, France derek.homeier@ens-lyon.fr
Jonti **Horner**, Australia j.a.horner@unsw.edu.au
Xu (Chelsea) **Huang**, USA xuhuang@princeton.edu
Elsa **Huby**, France elsa.huby@obspm.fr
Meredith **Hughes**, USA amhughes@wesleyan.edu
Jason **Hwang**, USA jasonhwang2014@u.northwestern.edu
Mike **Ireland**, Australia michael.ireland@mq.edu.au
Alan **Jackson**, United Kingdom ajackson@ast.cam.ac.uk
Marshall **Johnson**, USA mjohnson@astro.as.utexas.edu
Daniel **Jontof-Hutter**, USA daniel.s.jontof-hutter@nasa.gov
Marc **Joos**, France marc.joos@cea.fr
Andres **Jordan**, Chile ajordan@astro.puc.cl
Paul **Kalas**, USA kalas@berkeley.edu
JJ **Kavelaars**, Canada JJ.Kavelaars@nrc-cnrc.gc.ca
Grant **Kennedy**, United Kingdom gkennedy@ast.cam.ac.uk
Matthew **Kenworthy**, Netherlands kenworthy@strw.leidenuniv.nl
Jacques **Kluska**, France jacques.kluska@obs.ujf-grenoble.fr
Mihoko **Konishi**, Japan konishi@iral.ess.sci.osaka-u.ac.jp
Quinn **Konopacky**, Canada konopacky@di.utoronto.ca
Taisiya **Kopytova**, Germany kopytova@mpia.de
Quentin **Kral**, France quentin.kral@obspm.fr
Adam L. **Kraus**, USA akraus@cfa.harvard.edu
Alexander **Krivov**, Germany krivov@astro.uni-jena.de
Sylvestre **Lacour**, France sylvestre.lacour@obspm.fr
Anne-Marie **Lagrange**, France lagrange@obs.ujf-grenoble.fr
Michiel **Lambrechts**, Sweden michiel@astro.lu.se
Samantha **Lawler**, Canada lawler@astro.ubc.ca
Zoë **Leinhardt**, United Kingdom Zoe.Leinhardt@bristol.ac.uk
Min-Kai **Lin**, Canada mklin924@cita.utoronto.ca
Douglas **Lin**, USA lin@ucolick.org
Yoram **Lithwick**, USA y-lithwick@northwestern.edu
Michael **Liu**, USA mliu@ifa.hawaii.edu
Joe **Llama**, United Kingdom joe.llama@st-andrews.ac.uk
Torsten **Löhne**, Germany tloehne@astro.uni-jena.de
Meredith **MacGregor**, USA mmacgreg@cfa.harvard.edu
Bruce **Macintosh**, USA macintosh1@llnl.gov
Sarah **Maddison**, Australia smaddison@swin.edu.au
Thomas I. **Maindl**, Austria thomas.maindl@univie.ac.at
Anne-Lise **Maire**, France Anne-Lise.Maire@obspm.fr
Jerome **Maire**, Canada maire@di.utoronto.ca
Jared **Males** , USA jrmales@email.arizona.edu
Carlo Felice **Manara**, Germany cmanara@eso.org
Rosemary **Mardling**, Australia rosemary.mardling@monash.edu
Naibi **Marias**, USA marinas@astro.ufl.edu
Christian **Marois**, Canada christian.marois@nrc-cnrc.gc.ca
Jonathan P. **Marshall**, Spain jonathan.marshall@uam.es
David **Martin**, Switzerland david.martin@unige.ch
Rebecca **Martin**, USA rebecca.martin@jila.colorado.edu
Brenda **Matthews**, Canada brenda.matthews@nrc-cnrc.gc.ca
Dimitri P. **Mawet**, Chile dmawet@eso.org
Kyle **Mede**, Japan kylemede@astron.s.u-tokyo.ac.jp
Heloise **Meheut**, France heloise.meheut@cea.fr
Jonathan **Menu**, Belgium jonathan.menu@ster.kuleuven.be
Farzana **Meru**, Switzerland farzana@phys.ethz.ch
Dino **Mesa** , Italy dino.mesa@oapd.inaf.it
Tiffany **Meshkat**, Netherlands meshkat@strw.leidenuniv.nl
Stanimir **Metchev**, USA stanimir.metchev@stonybrook.edu
Brian **Metzger**, USA bmetzger@phys.columbia.edu
Cezary **Migaszewski**, Poland migaszewski@astri.umk.pl
Max **Millar-Blanchaer**, Canada maxmb@astro.utoronto.ca
Julien **Milli**, France julien.milli@obs.ujf-grenoble.fr
Caroline **Morley**, USA cmorley@ucolick.org
Katie **Morzinski**, USA ktmorz@arizona.edu
Fatemeh **Motalebi**, Switzerland fatemeh.motalebi@unige.ch
Matthias **Müller**, Germany mmueller@aip.de
Ruth **Murray-Clay**, USA rmurray-clay@cfa.harvard.edu
Erick **Nagel**, Mexico erick@astro.ugto.mx
Valerio **Nascimbeni**, Italy valerio.nascimbeni@unipd.it
Henry **Ngo**, USA hngo@caltech.edu
Eric **Nielsen**, USA enielsen@ifa.hawaii.edu
ke **Nordlund**, Denmark aako@nbi.dk
Jack **O'Malley-James**, United Kingdom jto5@st-andrews.ac.uk
Chris **Ormel**, USA ormel@astro.berkeley.edu
Mayra **Osorio**, Spain osorio@iaa.es
Rahul **Patel**, USA rahul.patel.1@stonybrook.edu
Jenny **Patience**, USA jennifer.patience@asu.edu
Matthew **Penny**, USA penny@astronomy.ohio-state.edu
Marshall **Perrin**, USA mperrin@stsci.edu
Cristobal **Petrovich**, USA cpetrovi@princeton.edu
Neil **Phillips**, Chile nphillip@alma.cl
Rafael **Pinotti**, Brazil rpinotti@astro.ufrj.br
Giampaolo **Piotto**, Italy giampaolo.piotto@unipd.it
Ana-Maria **Piso**, USA apiso@cfa.harvard.edu
Ralph **Pudritz**, Canada pudritz@mcmaster.ca

Chunhua **Qi**, USA cqi@cfa.harvard.edu
Andreas **Quirrenbach**, Germany A.Quirrenbach@lsw.uni-heidelberg.de
Roman **Rafikov**, USA rrr@astro.princeton.edu
Abhi **Rajan**, USA arajan6@asu.edu
Julien **Rameau**, France julien.rameau@obs.ujf-grenoble.fr
Emily **Rauscher**, USA rauscher@astro.princeton.edu
Francoise **Remus**, France francoise.remus@obspm.fr
Elisabetta **Rigliaco**, USA rigliaco@lpl.arizona.edu
Paul B. **Rimmer**, United Kingdom pr33@st-andrews.ac.uk
Adam **Ritchey**, USA aritchey@astro.washington.edu
Timothy J. **Rodigas**, USA rodigas@as.arizona.edu
Leslie **Rogers**, USA larogers@caltech.edu
Katrin **Ros**, Sweden katrin.ros@astro.lu.se
Katherine **Rosenfeld**, USA krosenfeld@cfa.harvard.edu
John **Rostron**, United Kingdom J.W.Rostron@warwick.ac.uk
Raquel **Salmeron**, Austalia raquel@mso.anu.edu.au
Graeme **Salter**, Australia g.salter@unsw.edu.au
Dmitry **Savransky**, USA savransky1@llnl.gov
Gerald H. M. **Schieven**, Canada gerald.schieven@nrc-cnrc.gc.ca
Hilke **Schlichting**, USA hilke@ucla.edu
Cory **Shankman**, Canada cshankm@uvic.ca
Andrew **Shannon**, United Kingdom shannon@ast.cam.ac.uk
Evgenya **Shkolnik**, USA shkolnik@lowell.edu
Mar **Sierra**, Spain Mar.Sierra@sciops.esa.int
Jacob **Simon**, USA jbsimon@jila.colorado.edu
Niki **Sipos**, Switzerland siposn@phys.ethz.ch
Andy **Skemer**, USA askemer@gmail.com
Ian **Skillen**, Spain ian.skillen@gmail.com
Alexis **Smith**, Poland amss@camk.edu.pl
Craig **Stark**, United Kingdom crs21@st-andrews.ac.uk
Christopher **Stark**, USA cstark@dtm.ciw.edu
Elad **Steinberg**, Israel elad.steinberg@mail.huji.ac.il
Kate **Su**, USA tinnin@as.arizona.edu
Motohide **Tamura**, Japan motohide.tamura@nao.ac.jp
Stuart F. **Taylor**, Hong Kong astrostuart@gmail.com
Susan **Terebey**, USA sterebe@calstatela.edu
Johanna **Teske**, USA jteske@as.arizona.edu
Philippe **Thebault**, France philippe.thebault@obspm.fr
Feng **Tian**, China tianfengco@gmail.com
Chris **Tinney** , Australia c.tinney@unsw.edu.au
Amaury **Triaud**, Switzerland amaury.triaud@unige.ch
Trifon **Trifonov**, Germany ttrifono@lsw.uni-heidelberg.de
Anjali **Tripathi**, USA atripath@cfa.harvard.edu
Ana **Uribe**, USA auribe@oddjob.uchicago.edu
Nienke **van der Marel**, Netherlands nmarel@strw.leidenuniv.nl
Matthijs **van der Wiel**, Canada matthijs.vanderwiel@uleth.ca
Arthur **Vigan**, France arthur.vigan@oamp.fr
Simon **Walker**, United Kingdom simon.walker@warwick.ac.uk
Kimberly **Ward-Duong**, USA kwardduo@asu.edu
Alycia **Weinberger**, USA weinberger@dtm.ciw.edu
Peter **Wheatley**, United Kingdom P.J.Wheatley@warwick.ac.uk
Sloane **Wiktorowicz**, USA sloanew@ucolick.org
David **Wilner**, USA dwilner@cfa.harvard.edu
Rob **Wittenmyer**, Australia rob@phys.unsw.edu.au
Schuyler **Wolff**, USA swolff@pha.jhu.edu
Yanqin **Wu**, Canada wu@astro.utoronto.ca
Mark **Wyatt**, United Kingdom wyatt@ast.cam.ac.uk
Ji-Wei **Xie**, Canada jwxie@astro.utoronto.ca
Chao-Chin **Yang**, Sweden ccyang@astro.lu.se
Xiaojia **Zhang**, China xiaojia.f.zhang@gmail.com
Xiaochen **Zheng**, China x.c.zheng1989@gmail.com
Alice **Zurlo**, France alice.zurlo@oamp.fr

Address from the Local & Science Organizing Committee Chairs

Dear colleagues,

When we began to plan this meeting, we anticipated approximately 200 attendees might come to Victoria, British Columbia to enjoy a joint meeting of "planet hunters" with those who study the processes that form planets in disks. We clearly underestimated the appeal such a meeting, of Victoria, or both. We had to raise the cap on the total number of registrants – twice – finally reaching 250, with interest in joining a wait list persisting until just weeks before the meeting.

The result of all our preparations was a spendid meeting with 54 science talks and 168 poster presentations. Though the delegates enjoyed a week of glorious sun in Victoria, the sessions were well attended to the very end.

This meeting would not have been the success it was without the dedicated efforts of the Science and Local Organizing Committees. The Science Organizing Committee selected invited speakers and ranked over 180 abstracts submitted for contributed talks to select 46 for the oral program. In all, 219 scientists presented reviews or their latest results at the meeting. The completed program provided a healthy balance of speakers based on science, gender, geography and professional status; 60% of the contributed talks were given by students or postdoctoral fellows. The Science Organizing Committee, along with the Local Organizing Committee, also read and provided comments to the proceedings' contributors, which helped ensure clarity of content and consistency of format for all the articles.

The Local Organizing Committee was distributed, with major organizational effort provided in Victoria by members at NRC and in Toronto, by members at the Dunlap Institute for Astronomy & Astrophysics. Particular acknowledgement is owed to Brenda Parrish (NRC), Alice Chow (DI) and Mark Booth (U. Vic/NRC), all of whom were able to rise to the call whenever their assistance was needed. Marlene Olsen and Jason Shrivell (NRC) also provided on-site support during the meeting. Acknowledement is also due for the efforts of Chris Sasaki (DI) who designed and maintained the webpage, registration and abstract forms and Gary Berry (NRC) who managed all the computing support during the meeting. Special thanks to Christian Marois, who recorded all the talks and made them available online, James Di Francesco who coordinated press during the meeting, and Gerald Schieven who assembled the program. JJ Kavelaars (NRC) helped organize the public talk and has created a permanent DOI for the meeting presentations and videos (10.11570/13.0003) to which we refer anyone interested in these records from the meeting.

The public talk was given by Prof. Debra Fischer (Yale U.), and presented jointly by the meeting organizers and the Royal Astronomical Society of Canada's Victoria chapter. We gratefully acknowledge the RASC's efforts in providing financial support and advertizing for Debra's excellent talk, entitled "HABITABLE WORLDS: The Search Continues".

The delegates selected their top three student talks and posters at the end of the meeting. The competition was strong, given the high quality of oral and poster presentations. The awardees were Joe Llama (St Andrews) and Julien Faure (CEA/Saclay) for Best Student Talk and Best Student Poster, respectively.

We also wish to thank all the organizations that provided generous financial support to the meeting: the International Astronomical Union, the Dunlap Institute for Astronomy & Astrophysics, the National Research Council of Canada, the University of Victoria, Gemini Observatory and the Royal Astronomical Society of Canada's Victoria chapter.

Brenda C. Matthews (LOC Chair) & James R. Graham (SOC Chair)
Victoria, BC, Canada & Berkeley, CA, USA, 9 September 2013

Exploring the Formation and Evolution of Planetary Systems
Proceedings IAU Symposium No. 299, 2013
M. Booth, B. C. Matthews & J. R. Graham, eds.

doi:10.1017/S1743921313007667

Detecting and Characterizing Exoplanets with Direct Imaging: Past, Present, and Future

Beth Biller

Max-Planck-Institut für Astronomie,
Königstuhl 17, 69117 Heidelberg, Germany
email: biller@mpia.de

Abstract. The last decade has yielded the first images of exoplanets, considerably advancing our understanding of the properties of young giant planets. In this talk I will discuss current results from ongoing direct imaging efforts as well as future prospects for detection and characterization of exoplanets via high contrast imaging. Direct detection, and direct spectroscopy in particular, have great potential for advancing our understanding of extrasolar planets. In combination with other methods of planet detection, direct imaging and spectroscopy will allow us to eventually: 1) study the physical properties of exoplanets (colors, temperatures, etc.) in depth and 2) fully map out the architecture of typical planetary systems. Direct imaging has offered us the first glimpse into the atmospheric properties of young high-mass (3-10 M_{Jup}) exoplanets. Deep direct imaging surveys for exoplanets have also yielded the strongest constraints to date on the statistical properties of wide giant exoplanets. A number of extremely high contrast exoplanet imaging instruments have recently come online or will come online within the next year (including Project 1640, SCExAO, SPHERE, GPI, among others). I will discuss future prospects with these instruments.

Keywords. planetary systems; instrumentation: adaptive optics; techniques: high angular resolution

---

1. Introduction

In the last decade, ~10 companions and candidate companions with estimated masses <13 M_{Jup} have been imaged in orbit around young stars and brown dwarfs (c.f. Chauvin *et al.* 2005, Marois *et al.* 2008, Kalas *et al.* 2008, Lafrenière *et al.* 2008, Lagrange *et al.* 2009, Lagrange *et al.* 2010, Marois *et al.* 2010, Todorov *et al.* 2010, Ireland *et al.* 2011, Luhman *et al.* 2011, Kraus & Ireland 2012, Rameau *et al.* 2013, Quanz *et al.* 2013, and Kuzuhara *et al.* 2013). In total, ~30 companions with estimated masses <25 M_{Jup} have been imaged. (See http://exoplanet.eu for a compilation of these objects.) These discoveries have provided a wealth of new information about young giant planets, as well as some surprises. Prior to these detections, models predicted that young gas giant planets at young ages (10-300 Myr) would likely have cool photospheres with prominent methane absorption features (Baraffe *et al.* 2003, Burrows *et al.* 2003), i.e. that these objects would be spectral analogs to T-type brown dwarfs. However, all known directly-imaged planets at these ages (for instance, 2MASS 1207b and the HR 8799 planets, Chauvin *et al.* 2005, Marois *et al.* 2008, Marois *et al.* 2010) have lacked methane absorption and show extremely red colors, likely due to dust clouds and/or non-equilibrium chemistry in their atmospheres (Bowler *et al.* 2010, Skemer *et al.* 2011, Barman *et al.* 2011a, Barman *et al.* 2011b, Currie *et al.* 2011).

Additionally, all of these companions except for β Pic b (Lagrange *et al.* 2009, Lagrange *et al.* 2010), HR 8799e (Marois *et al.* 2010), and LkCa 15b (Kraus & Ireland 2012) lie at

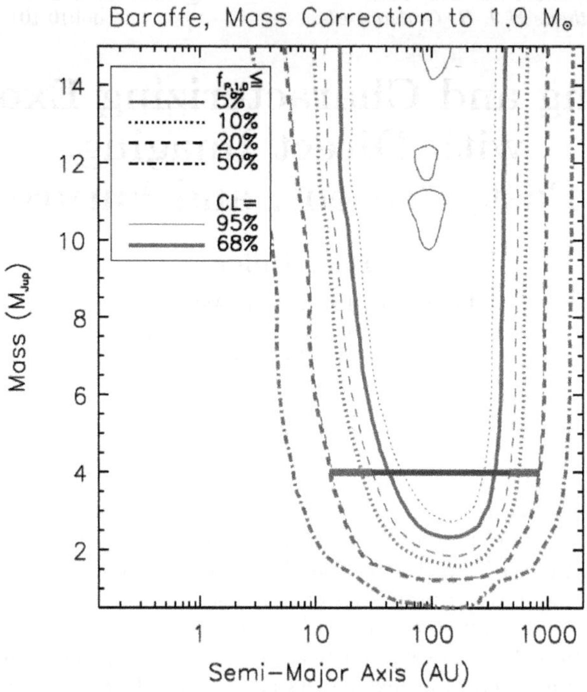

Figure 1. Figure from Nielsen & Close 2010. Upper limit on planet fraction for 1 $M_\odot$ stars from a compilation of 118 stars observed at high contrasts, taking into account the mass scaling from Johnson *et al.* (2007) and using the COND models of Baraffe *et al.* (2003). Less than 20% of sunlike stars host >4 M_{Jup} planets between 30 and 466 AU at the 95% confidence level.

projected separations greater than 20 AU, considerably wider than giant planets in our own solar system. Such widely separated companions are unlikely to be produced by in-situ formation by core-accretion, which likely formed the closer-in (<10 AU) population of planets detected to date via radial velocity studies (Mordasini *et al.* 2009, Janson *et al.* 2012, Dodson-Robinson *et al.* 2009). However, given that only ∼10 such companions have been imaged to date, it is premature to make statements based on such a small sample. Thus, it is a priority to discover additional companions as well as to constrain the distributions of their semimajor axes, eccentricities, masses, etc.

Further detections as well as characterization of known planets will allow us to 1) explore the physical properties of these planets (colors, temperature, and compositions) as well as 2) the architecture of their planetary systems. After discussing some of the technical aspects of direct imaging of planets and setting the stage with a quick discussion of previous and ongoing surveys, I will return to these two key themes. I will conclude with a discussion of future prospects with instruments such as GPI, SPHERE, and SCExAO.

2. Setting the Stage – Technical Aspects and Previous Work

2.1. *Technical Aspects of Direct Imaging*

A mature gas giant planet such as Jupiter will be $>10^8\times$ fainter than its host star. Young (<100 Myr) planets are somewhat brighter, especially in the infrared, and are 10^4 - $10^7\times$ fainter than their host stars (e.g. Baraffe *et al.* 2003, Burrows *et al.* 2003, Fortney *et al.* 2008). These contrasts are in principle achievable in space or from the ground with high-order adaptive optics correction. However, even after AO correction,

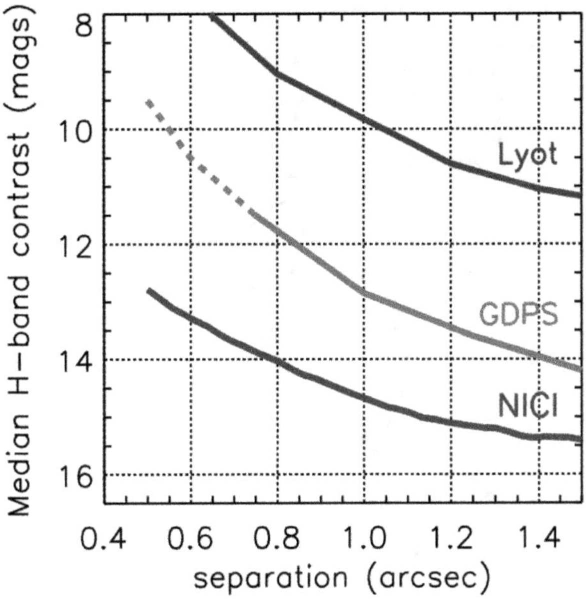

Figure 2. Figure courtesy of M. Liu and the NICI Campaign. Comparison of achieved contrast for a number of completed and ongoing surveys. Relative to the completed Lyot project (Leconte *et al.* 2010) and GDPS surveys (Lafrenière *et al.* 2007b), the NICI campaign reaches >2 mag deeper at separations of 0.8". Other ongoing coronagraphic surveys such as SEEDS, LEECH, and IDPS are reaching comparable contrasts.

speckle noise remains in high-contrast images. These "super-speckles" are generally due to imperfections in the optics themselves and appear in both ground and space data.

A number of techniques may be combined in order to attenuate starlight and suppress quasi-static speckles for direct detection of faint companions to stars. In low-strehl ratio applications, coronagraphs can be used to block out starlight. At high-strehl ratios, use of a coronagraph can correct the Airy pattern and remove a considerable amount of quasi-static pinned speckle as well, thus significantly reducing speckle noise (Sivaramakrishnan & Oppenheimer 2006). In addition, instrumental speckle suppression techniques (e.g. azimuthal differential imaging (ADI) and spectral differential imaging (SDI)) can be used to suppress quasi-static speckles (Racine *et al.* 1999, Marois *et al.* 2006). Both SDI and ADI techniques seek to decorrelate real objects from starlight and speckles. For instance, in ADI observations the rotator is turned off at the Cassegrain focus or set to follow the elevation angle at the Nasmyth focus, allowing the field of view to rotate with parallactic angle on the sky. In a sequence of images taken at different parallactic angles, a real object will track on the sky with the parallactic angle, while speckles will move randomly. From a series of images, a reference PSF can be constructed for and subtracted from each individual image, attenuating quasi-static speckle structure. A number of different algorithms have been developed to optimize this PSF construction and subtraction, including LOCI (Locally Optimized Combination of Images) (Lafrenière *et al.* 2007a) as well as PCA (Primary Component Analysis) methods such as KLIP (Soummer *et al.* 2012), and PYNPOINT (Amara & Quanz 2012).

2.2. *Earlier Generations of Surveys*

In the last decade, a number of deep, adaptive-optics surveys with sample sizes >30 stars have been completed at 8-m telescopes to search for planetary companions. Many

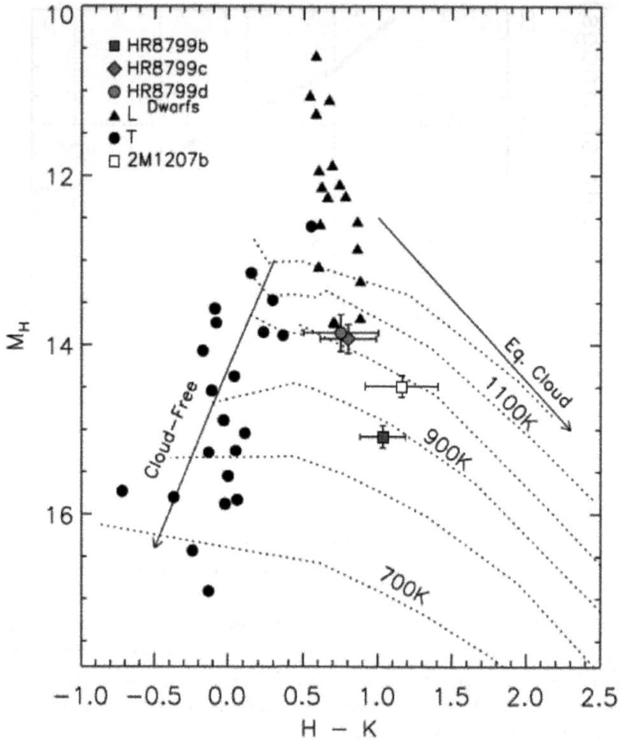

Figure 3. Figure from Barman *et al.* 2011b. M_H vs. $H-K$ for a number of young planets as well as comparison brown dwarfs drawn from Leggett *et al.* 2002 and Knapp *et al.* 2004. Tracks for chemical equilibrium models (radius = 1.4 R_{Jup}, log(g) = 3.5, mean particle size 5 μm, T_{eff} from 7001200K) are plotted as dotted lines. The arrows indicate the cloud-free locus (on the left side of the plot) and the location of extremely thick cloud models (on the right side of the plot). Cloud thickness increases towards the right. The extreme red colors of young planets place them closer to the L-dwarf sequence than the T-dwarf sequence due to the persistence of red dusty clouds at low temperatures and surface gravities.

of these have been conducted in the 1.6 μm *H*-band or 2.2 μm *K*-band (Masciadri *et al.* 2005, Biller *et al.* 2007, Lafrenière *et al.* 2007b, Apai *et al.* 2008, Chauvin *et al.* 2010), while others have focused further into the infrared (3.5-5 μm) in the *L*, L', or *M* bands (Kasper *et al.* 2007, Heinze *et al.* 2010a, Heinze *et al.* 2010b, Rameau *et al.* 2013a.) These surveys generally median attained contrasts of $\gtrsim$10 mag at $\gtrsim$1" (e.g. Biller *et al.* 2007) and are sensitive to >5 M_{Jup} planets at separations of >20 AU. Based on a sample of 118 stars compiled from the surveys of Masciadri *et al.* (2005), Biller *et al.* (2007), and Lafrenière *et al.* (2007b), Nielsen & Close (2010) found that planets more massive than 4 M_{Jup} are found around <20% of FGKM stars in orbits between 22 and 507 AU, at 95% confidence (see Fig. 1). Chauvin *et al.* (2010) find a similar result based on a sample of 88 stars, constraining the fraction of stars with giant planets to <10% at semi-major axes >40 AU for a planet distribution extended from radial velocity power laws.

2.3. *Ongoing or Recently Completed Surveys*

A number of large scale surveys (>100 stars) are ongoing or recently completed, including the NICI Campaign at Gemini-South (Liu *et al.* 2010, Biller *et al.* 2013, Wahhaj *et al.* 2013, Nielsen *et al.* 2013), the NACO Large Program using NACO at the VLT (Buenzli *et al.* 2010, Chauvin *et al.* in prep), SEEDS (Strategic Exploration of Exoplanets and Disks with Subaru) using HiCIAO at Subaru (Thalmann *et al.* 2009, Carson *et al.* 2013,

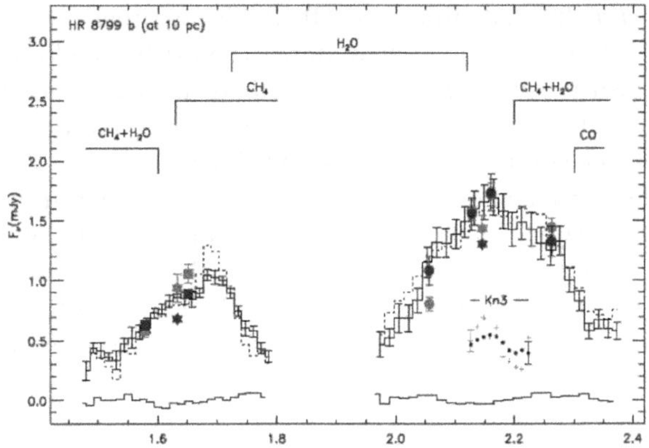

Figure 4. Figure from Barman *et al.* 2011a. Keck OSIRIS H and K-band spectra of HR 8799b. A CO absorption feature is seen in K, but no methane absorption is apparent in either band.

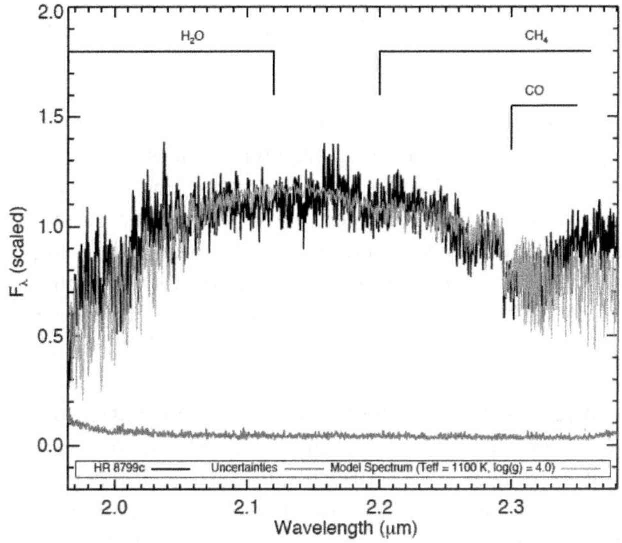

Figure 5. Figure from Konopacky *et al.* 2013. R = 4000 Keck OSIRIS K band spectrum of HR 8799c, showing a clear CO absorption feature. From spectral fitting to this absorption feature, the C/O ratio of HR 8799c appears to be supersolar, implying that this planet likely formed via core accretion.

Janson *et al.* 2013), LEECH at the LBT, and the International Deep Planet Survey (IDPS) using primarily Gemini and Keck (Vigan *et al.* 2012). These surveys combine advanced speckle suppression techniques such as ADI with coronagraphic imaging to achieve contrasts of Δmag>14 at 1" (c.f. Biller *et al.* 2013). A comparison of current and previous surveys is presented in Fig. 2.

While the earlier generation of surveys primarily focused on solar analogues, this generation of surveys has often targeted higher mass A stars. The reasons for this are twofold: 1) a notable number of directly imaged exoplanets have been imaged around A stars (for instance, the HR 8799 planets, β Pic, Fomalhaut) and 2) results from radial velocity surveys suggest that A stars more frequently host Jupiter mass planets than lower mass stars (Johnson *et al.* 2007, 2010). Another under-surveyed sample are lower mass M stars; a number of surveys are seeking to remedy this (Delorme *et al.* 2012, 2013, Bowler *et al.* 2012a, 2012b). Significant efforts are also being made to search for planets around debris disk stars (Wahhaj *et al.* 2013, Janson *et al.* 2013). The existence of a dust disk even after the primordial gas disk has dissipated may imply the existence of a planetary system of some form to replenish the dust. However, recent work suggests that debris disks that still retain some gas can form sharp eccentric rings solely due to dust-gas interactions and without the presence of a planet (Lyra & Kuchner 2013).

Many ongoing surveys focus on moving group stars. Moving groups are associations of young stars (10–300 Myr) that are unconnected to regions of ongoing star-formation. These associations were not discovered until the late 1990's, as moving group members are often dispersed across a wide part of the sky (Zuckerman & Song 2004). Moving group stars provide a unique nearby young sample with well-constrained ages and distances, vital for accurately estimating the mass, luminosity, etc. of any detected companion.

3. Physical Properties

The lack of methane absorption and the red IR colors of the first directly imaged planets came as a surprise (Fig. 3), as brown dwarfs with similar T_{eff} (thus, expected to share similar atmospheric properties and hence colors) show strong methane absorption and blue IR colors. The cause of this divergence is likely the considerable lower surface gravity of young planets compared to older, higher mass brown dwarfs. Indeed, very young, low surface gravity free-floating brown dwarfs have spectra and near-IR colors very comparable to young directly imaged planets (Faherty *et al.* 2013). Low surface gravity translates to very different cloud structure compared to higher surface gravity objects at the same temperatures; low surface gravity allows young planets and brown dwarfs to retain thick dust clouds in their atmospheres at temperatures at which such clouds have dissipated or sunk below the photosphere in higher surface gravity objects (Marley *et al.* 2012).

Direct spectroscopy of exoplanets yields even more information on the atmospheres of these objects. Spectra of HR 8799b in both the H and K bands show a clear lack of methane absorption as well as a CO absorption feature (Barman *et al.* 2011b, Bowler *et al.* 2010, see Fig. 4). The predominance of CO over CH_4 strongly implies the existence of non-equilibrium chemistry. Oppenheimer *et al.* 2013 recently obtained simultaneous low resolution spectra of all 4 planets around HR 8799. These planets all share similar masses and formed in the same disk, yet they display intriguing variations in spectral shape, even though the broadband infrared colors for the planets are relatively similar.

The spectrum of a planet may also reveal the manner in which it formed. Konopacky *et al.* (2013) measure a strong CO absorption feature in their R = 4000 spectrum of HR 8799c (Fig. 5). From careful modeling of the spectrum, they derive C abundance, O abundance, and C/O ratio for HR 8799c. Assuming solar metallicity for the stellar host, HR 8799, and assuming the disk shares the same metallicity as the star, HR 8799c possesses a high C/O ratio relative to solar, with C and O both comparatively depleted relative to solar. This implies strongly that the planet formed via core accretion, as a planet which formed via disk instability should share similar C/O ratio, and C and O abundances as the natal star and disk.

Directly imaged planets also offer the opportunity to measure accurate orbits and improve mass estimates for exoplanets. Most current imaged planets are too widely separated (>20 AU) to yield good orbital determinations on reasonable (<20 year) timescales. However, β Pic b, at a separation similar to Saturn in our own solar system (∼8.5 AU), holds the promise of yielding a full orbit in the relatively near future. Ongoing orbital

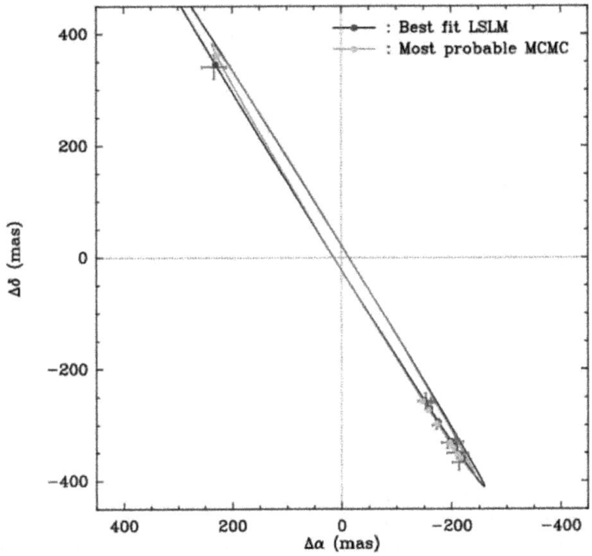

Figure 6. Figure from Chauvin *et al.* 2012. Best orbital fits to current β Pic b astrometry relative to the position of β Pic A and shown on the plane of the sky.

monitoring has already yielded estimates for a number of orbital parameters (See Fig. 6, from Chauvin *et al.* 2012).

4. Architecture of Planetary Systems

Beyond studying the physical properties of individual planets, it is important to understand the distribution of planets as a function of mass and semi-major axis in their planetary systems – i.e. system architecture. Comparatively few planets have been directly imaged considering the large sample sizes and excellent achieved contrasts (Δmag$\sim$14 at 1", Biller *et al.* 2013) of current direct imaging surveys. However, even this lack of detections provides us key information about planetary system architecture. Previous surveys demonstrated that wide giant planets (>4 M_{Jup}, >30 AU from the host star) are rare (e.g. Nielsen & Close 2010). Current surveys deepen and extend this result to smaller separations and lower masses. Statistical analyses have recently been published for a number of large surveys (Vigan *et al.* 2012, Wahhaj *et al.* 2013, Nielsen *et al.* 2013, Biller *et al.* 2013). As one example, I present the results from Biller *et al.* (2013).

Biller *et al.* (2013) present a subset of 80 young, moving group stars which are part of the larger ($\sim$230 star) NICI Campaign (Liu *et al.* 2010). No new planetary companions were found in this sample. To statistically interpret these results, Biller *et al.* (2013) adapt the method established in Allen (2007) for studying stellar binarity in the context of a direct imaging survey of exoplanets, extending the power law distributions in mass and semi-major axis (with indices α and β respectively) derived for RV planets by Cumming *et al.* (2008) to the wider separation regime (>10 AU) probed by direct imaging. The power law distribution in semi-major axis is only extended out to an outer cutoff, since a large reservoir of planets at very wide separations does not exist (Nielsen & Close). The model then has four parameters: power law indices α and β, semi-major axis cutoff a_{max}, and average number of planets per star, F (used to calculate the normalization).

Using Bayesian inference techniques, constraints can be placed on these parameters based on the data from the 73 stars from Biller *et al.* (2013) with completed followup

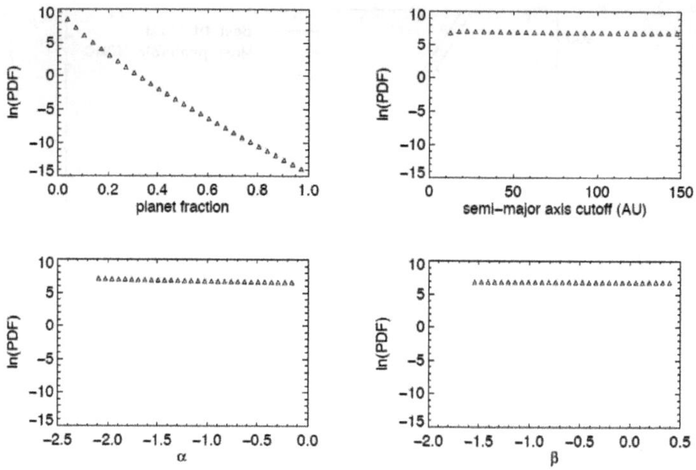

Figure 7. Figure from Biller *et al.* 2013. 1-d marginalized posterior pdfs including 4 free parameters and using the COND models of Baraffe *et al.* (2003). Posterior probability is plotted in logarithmic units. Other than for the planet frequency F, these marginalized posteriors remain unconstrained (i.e. no clear peak or trailing off to 0). Thus, we have only put confidence intervals on the planet fraction for semi-major axes between 10-150 AU. At a 95.4% confidence level, planet fraction must be <6%, marginalized over the ranges α = [-2.1, 0.2], β = [-1.5, 0.4], and cutoff = [10 AU, 150 AU]. Repeating the analysis with the DUSTY models of Baraffe *et al.* (2002) and marginalizing over the same parameter ranges, planet fraction must be <18% at a 95.4% confidence level.

of all candidate companions and no a priori known planetary companions. Results are presented in Fig. 7. Using the COND models of Baraffe *et al.* (2003), F must be <6%, marginalized over the ranges $\alpha = [-2.1, 0.2]$, $\beta = [-1.5, 0.4]$, and cutoff = [10 AU, 150 AU] at a 95.4% confidence level. Repeating the analysis with the DUSTY models of Baraffe *et al.* (2002) and marginalizing over the same parameter ranges, F must be <18% at a 95.4% confidence level. No strong constraints can be placed on the other 3 parameters.

5. Future Prospects

The current generation of surveys place strong constraints on $\gtrsim$1 M_{Jup} planets at separations >10 AU. The next generation of instruments will push down to smaller separations and masses. While only a handful of planets have been directly imaged to date, new instruments with much tighter inner working angles will likely yield significant numbers of discoveries. A number of $\geqslant$1 M_{Jup} planets have been indirectly directed via the radial velocity method at semi-major axes of 5-10 AU (see compilation at exoplanet.eu). While these radial velocity detected planets are too old and faint to be directly imaged, their young counterparts should be accessible with these instruments.

A number of advanced planet-finder instruments will be on the sky within the next year. GPI at Gemini-South (Macintosh *et al.* 2012) will ship to Chile in summer 2013, with SPHERE at VLT (Beuzit *et al.* 2008) following in the autumn. SCExAO is currently undergoing commissioning at Subaru (Guyon *et al.* 2011). Project 1640 at the Palomar 5-m telescope has already published its first science result (Oppenheimer *et al.* 2013). These instruments combine extreme coronagraphy with integral field spectrographs, to enable unprecedented contrasts and novel characterization of exoplanets.

In the longer term, the extremely large telescopes (GMT, TMT, and the E-ELT) and JWST will revolutionize direct imaging and characterization of exoplanets. The

ELTs may in fact yield the first direct image of a terrestrial sized planet. Combining results from Kepler on the frequency of short-period planets around main-sequence stars with contrast predictions for an extreme AO coronagraph on an ELT, Crossfield (2013) predicts that ~10 planets with radii of 1-8 Earth radius and equilibrium temperatures $\lesssim$400 K will be detectable around stars <8 pc from the Sun.

6. Conclusions

The previous and current generation of direct imaging surveys have yielded a number of interesting results regarding both the properties of extrasolar giant planets and the architectures of the planetary systems which they inhabit. Young exoplanets imaged to date have red colors indicative of clouds and/or non-equilibrium chemistry. These red colors are likely a result of the low-surface gravities of these objects, which allow red, dusty clouds to persist even at low temperatures relative to their older, high-gravity brown dwarf counterparts. Statistical analyses of large-scale planet imaging surveys have demonstrated that hot-start gas-giant (>4 M_{Jup}) planets are rare at >10 AU.

References

Allen, P. R. 2007, *ApJ*, 668, 492
Amara, A., & Quanz, S. P. 2012, *MNRAS*, 427, 948
Apai, D., Janson, M., Moro-Martín, A., *et al.* 2008, *ApJ*, 672, 1196
Baraffe, I., Chabrier, G., Allard, F., & Hauschildt, P. H. 2002, *A&A*, 382, 563
Baraffe, I., Chabrier, G., Barman, T. S., Allard, F., & Hauschildt, P. H. 2003, *A&A*, 402, 701
Barman, T. S., Macintosh, B., Konopacky, Q. M., & Marois, C. 2011, *ApJ*, 733, 65
Barman, T. S., Macintosh, B., Konopacky, Q. M., & Marois, C. 2011, *ApJ* (Letters), 735, L39
Beuzit, J.-L., Feldt, M., Dohlen, K., *et al.* 2008, *proc. SPIE*, 7014
Biller, B. A., Close, L. M., Masciadri, E., *et al.* 2007, *ApJS*, 173, 143
Biller, B. A., Liu, M. C., Wahhaj, Z., *et al.* 2013, *ApJ*, 777, 160
Bowler, B. P., Liu, M. C., Dupuy, T. J., & Cushing, M. C. 2010, *ApJ*, 723, 850
Bowler, B. P., Liu, M. C., Shkolnik, E. L., *et al.* 2012, *ApJ*, 753, 142
Bowler, B. P., Liu, M. C., Shkolnik, E. L., & Tamura, M. 2012, *ApJ*, 756, 69
Buenzli, E., Thalmann, C., Vigan, A., *et al.* 2010, *A&A*, 524, L1
Burrows, A., Sudarsky, D., & Lunine, J. I. 2003, *ApJ*, 596, 587
Carson, J., Thalmann, C., Janson, M., *et al.* 2012, arXiv:1211.3744
Chauvin, G., Lagrange, A.-M., Dumas, C., *et al.* 2005, *A&A*, 438, L25
Chauvin, G., Lagrange, A.-M., Bonavita, M., *et al.* 2010, *A&A*, 509, A52
Chauvin, G., Lagrange, A.-M., Beust, H., *et al.* 2012, *A&A*, 542, A41
Crossfield, I. J. M. 2013, *A&A*, 551, A99
Currie, T., Thalmann, C., Matsumura, S., *et al.* 2011, *ApJ* (Letters), 736, L33
Cumming, A., Butler, R. P., Marcy, G. W., *et al.* 2008, *PASP*, 120, 531
Delorme, P., Gagné, J., Malo, L., *et al.* 2012, *A&A*, 548, A26
Delorme, P., Gagné, J., Girard, J. H., *et al.* 2013, *A&A*, 553, L5
Dodson-Robinson, S. E., Veras, D., Ford, E. B., & Beichman, C. A. 2009, *ApJ*, 707, 79
Faherty, J. K., Rice, E. L., Cruz, K. L., Mamajek, E. E., & Núñez, A. 2013, *AJ*, 145, 2
Fortney, J. J., Marley, M. S., Saumon, D., & Lodders, K. 2008, *ApJ*, 683, 1104
Guyon, O., Martinache, F., Clergeon, C., *et al.* 2011, *proc. SPIE*, 8149,
Heinze, A. N., Hinz, P. M., Sivanandam, S., *et al.* 2010, *ApJ*, 714, 1551
Heinze, A. N., Hinz, P. M., Kenworthy, M., *et al.* 2010, *ApJ*, 714, 1570
Ireland, M. J., Kraus, A., Martinache, F., Law, N., & Hillenbrand, L. A. 2011, *ApJ*, 726, 113
Janson, M., Bonavita, M., Klahr, H., *et al.* 2011, *ApJ*, 736, 89
Janson, M., Bonavita, M., Klahr, H., & Lafrenière, D. 2012, *ApJ*, 745, 4

Janson, M., Brandt, T. D., Moro-Martin, A., *et al.* 2013, arXiv:1306.0581
Johnson, J. A., Butler, R. P., Marcy, G. W., *et al.* 2007, *ApJ*, 670, 833
Johnson, J. A., Aller, K. M., Howard, A. W., & Crepp, J. R. 2010, *PASP*, 122, 905
Kalas, P., Graham, J. R., Chiang, E., *et al.* 2008, *Science*, 322, 1345
Kasper, M., Apai, D., Janson, M., & Brandner, W. 2007, *A&A*, 472, 321
Knapp, G. R., Leggett, S. K., Fan, X., *et al.* 2004, *AJ*, 127, 3553
Konopacky, Q. M., Barman, T. S., Macintosh, B. A., & Marois, C. 2013, *Science*, 339, 1398
Kraus, A. L., Ireland, M. J., Martinache, F., & Hillenbrand, L. A. 2011, *ApJ*, 731, 8
Kraus, A. L., & Ireland, M. J. 2012, *ApJ*, 745, 5
Kuzuhara, M., Tamura, M., Kudo, T., *et al.* 2013, *ApJ*, accepted, arXiv:1307.2886
Lafrenière, D., Marois, C., Doyon, R., Nadeau, D., & Artigau, É. 2007, *ApJ*, 660, 770
Lafrenière, D., Doyon, R., Marois, C., *et al.* 2007, *ApJ*, 670, 1367
Lafrenière, D., Jayawardhana, R., & van Kerkwijk, M. H. 2008, *ApJ*l, 689, L153
Lagrange, A.-M., Gratadour, D., Chauvin, G., *et al.* 2009, *A&A*, 493, L21
Lagrange, A.-M., Bonnefoy, M., Chauvin, G., *et al.* 2010, *Science*, 329, 57
Leconte, J., Soummer, R., Hinkley, S., *et al.* 2010, *ApJ*, 716, 1551
Leggett, S. K., Golimowski, D. A., Fan, X., *et al.* 2002, *ApJ*, 564, 452
Liu, M. C., Wahhaj, Z., Biller, B. A., *et al.* 2010, *Proc. SPIE*, 7736,
Lyra, W., & Kuchner, M. 2013, *Nature*, 499, 184
Luhman, K. L., Burgasser, A. J., & Bochanski, J. J. 2011, *ApJ*l, 730, L9
Macintosh, B. A., Anthony, A., Atwood, J., *et al.* 2012, *Proc. SPIE*, 8446,
Marley, M. S., Saumon, D., Cushing, M., *et al.* 2012, *ApJ*, 754, 135
Marois, C., Doyon, R., Nadeau, D., *et al.* 2005, *PASP*, 117, 745
Marois, C., Lafrenière, D., Doyon, R., Macintosh, B., & Nadeau, D. 2006, *ApJ*, 641, 556
Marois, C., Macintosh, B., Barman, T., *et al.* 2008, *Science*, 322, 1348
Marois, C., Zuckerman, B., Konopacky, Q. M., *et al.* 2010, *Nature*, 468, 1080
Masciadri, E., Mundt, R., Henning, T., *et al.* 2005, *ApJ*, 625, 1004
Mordasini, C., Alibert, Y., & Benz, W. 2009, *A&A*, 501, 1139
Nielsen, E. L., & Close, L. M. 2010, *ApJ*, 717, 878
Nielsen, E. L., Liu, M. C., Wahhaj, Z., *et al.* 2013, *ApJ*, 776, 4
Oppenheimer, B. R., Baranec, C., Beichman, C., *et al.* 2013, *ApJ*, 768, 24
Quanz, S. P., Amara, A., Meyer, M. R., *et al.* 2013, *ApJ*(Letters), 766, L1
Racine, R., Walker, G. A. H., Nadeau, D., Doyon, R., & Marois, C. 1999, *PASP*, 111, 587
Rameau, J., Chauvin, G., Lagrange, A.-M., *et al.* 2013, *A&A*, 553, A60
Rameau, J., Chauvin, G., Lagrange, A.-M., *et al.* 2013, arXiv:1305.7428
Sivaramakrishnan, A., & Oppenheimer, B. R. 2006, *ApJ*, 647, 620
Skemer, A. J., Close, L. M., Szűcs, L., *et al.* 2011, *ApJ*, 732, 107
Soummer, R., Pueyo, L., & Larkin, J. 2012, *ApJ* (Letters), 755, L28
Thalmann, C., Carson, J., Janson, M., *et al.* 2009, *ApJ*l, 707, L123
Todorov, K., Luhman, K. L., & McLeod, K. K. 2010, *ApJ*l, 714, L84
Vigan, A., Patience, J., Marois, C., *et al.* 2012, *A&A*, 544, A9
Wahhaj, Z., Liu, M. C., Nielsen, E. L., *et al.* 2013, *ApJ*, 773, 179

Discussion

MARDLING: You didn't seem to take into account the mass of the star with your limit of 10 AU for core accretion models.

BILLER: The limit of 10 AU I gave is specifically for solar mass stars. Higher mass stars with more massive disks may be able to form planets via core accretion at separations greater than 10 AU.

CURRIE: You mentioned the important tools needed to image planets but didn't mention powerful image processing techniques like those Christian Marois helped develop, like

LOCI, that increase contrast. And the new techniques building on this like the SOSIE pipeline or A-LOCI can give giant gains in contrast. So within the NICI Campaign survey, how did your results leverage on exactly how you do image processing?

BILLER: Image processing techniques are certainly vital for getting the most out of ADI data. However, one thing that we found with the NICI Campaign is that different image processing techniques yielded pretty similar results on our combined ADI + SDI data. This may be because the SDI reduction is already removing most of the speckles that, say, LOCI processing of the ADI data would usually remove.

TIAN: Detection of CO in Earth-like planet's atmosphere, especially in the habitable zone: if one can detect O_2 and O_3 in such a planet's atmosphere, is CO detectable? Can we decide its concentration?

BILLER: If one can detect O_2 and O_3 in an exo-Earth atmosphere, CO is likely detectable too. However, such detections are quite far off. While ELTs might be able to directly image a very nearby exo-Earth, spectroscopy of such an object is certainly more than a decade away.

HOMEIER: How may the claimed match to T dwarf templates for a planet like 8799c be reconciled with their still very red (L dwarf-like) SED?

BILLER: Oppenheimer *et al.* (2013) find L5 spectral types for HR 8799b and d. They claim a spectral type of T2 for HR 8799c based on their J band fit, however, there is clearly no methane absorption in the H band for this object and the object does not look very "T"-like at all in this band. As methane absorption is what gives T dwarfs their characteristic blue IR colors, the absence of methane absorption in HR 8799c explains this object's red "L"-like colors. The spectrum for the closest-in planet, HR 8799e, is quite noisy, so I'd imagine any spectral type assigned to it would be quite preliminary.

KENWORTHY: How many stars would you need to look at with an ELT M-dwarf survey to detect 3-5 Earths?

BILLER: Crossfield (2013) predicts the detection of up to 10 1-8 Earth radius planets using an ELT among the $\sim$120 main sequence stars with spectral type earlier than M7 and within 8 pc from the Earth, of which 92 are M stars. Of course, many of these detected planets will be "super-Earths" rather than truly Earth mass planets. In terms of actual Earth analogues, Crossfield (2013) predicts a 40% chance of detecting such a planet in the same 8 pc sample.

Exploring the Formation and Evolution of Planetary Systems
Proceedings IAU Symposium No. 299, 2013
M. Booth, B. C. Matthews & J. R. Graham, eds.

doi:10.1017/S1743921313007679

SEEDS : Strategic Explorations of Exoplanets and Disks with Subaru

Motohide Tamura[1] and the SEEDS team[2]

[1]Dept. of Astronomy, Graduate School of Science, The University of Tokyo,
7-3-1 Hongo, Bunkyo-ku, Tokyo 113-0033, Japan
email: motohide.tamura@nao.ac.jp / motohide.tamura@astron.s.u-tokyo.ac.jp

Abstract. SEEDS is the first Subaru Strategic Program, whose aim is to conduct a direct imaging survey for giant planets as well as protoplanetary/debris disks at a few to a few tens of AU region around 500 nearby solar-type or more massive young stars devoting 120 Subaru nights for 5 years. The targets are composed of five categories spanning the ages of ~1 Myr to ~1 Gyr. Some RV-planet targets with older ages are also observed. The survey employs the new high-contrast instrument HiCIAO, a successor of the previous NIR coronagraph camera CIAO for the Subaru Telescope. We describe the outline of this survey and present its first three years of results. The survey has published ~20 refereed papers by now. The main results are as follows: (1) detection and characterization of the most unequivocal and lowest-mass planet via direct imaging. (2) detection of a super-Jupiter around the most massive star ever imaged, (3) detection of companions around a retrograde exoplanet system, which supports the Kozai mechanism for the origin of retrograde orbit (not in this proceedings, but see Narita *et al.* 2010, 2012). We also report (4) the discovery of unprecedentedly detailed structures of more than a dozen of protoplanetary disks and some debris disks. The detected structures such as wide gaps and spirals arms of a Solar-system scale could be signpost of planet.

Keywords. exoplanet; circumstellar disks; coronagraph, infrared

---

1. Introduction

The first 18 years of indirect observations have provided us more 850 exoplanets and more than 2500 promising candidates from the Kepler mission. This success has largely been due to the accurate radial velocity (RV) method and the transit (TR) method. However, these discoveries are inherently limited in several ways: Since both RV and TR studies are confined to the inner regions of exoplanetary systems (<6 au for a 15-yr survey), we still know very little about the planetary constituents in the outer regions. Both of the above indirect surveys are complicated due to the high level of intrinsic stellar activity for stars; thus, they have traditionally targeted old and quiet stars. Furthermore, they are also not suitable for planet searches around massive stars due to the paucity of stellar absorption lines and the reduced amplitude of the RV motion or the reduced TR depth.

In contrast to the RV or TR technique, direct imaging can be applied to both young and old stars, and can allow measurements of colors, luminosities and spectra, thereby providing temperatures and compositions. As a result, direct imaging is currently the best way to investigate Jupiter-mass or massive planets in the outer (>5 au) regions around "young" stars.

Furthermore, recent 8-meter-class telescopes with adaptive optics (AO) and HST have had revealed the complexity and diversity of protoplanetary disks around very young stars at a scale greater than 100 au (e.g., spirals, banana-split). However, its inner regions (r<100 au) have not been explored due to a limited contrast.

With this background, the Subaru Telescope is promoting the directly imaging project of exoplanets and their formation sites since 2009 for 5 years. This project is named SEEDS (Strategic Explorations of Exoplanets and Disks with Subaru Telescope) and is the first Subaru Strategic Program, a new open use category. With using both new coronagraph camera HiCIAO (Tamura *et al.* 2006; Hodapp *et al.* 2006; Suzuki *et al.* in prep.) commissioned in 2009 and adaptive optics AO188, it will observe exoplanets, companions, and circumstellar structures around about 500 stars in total. HiCIAO utilizes one 2k×2k HgCdTe array with a pixel scale of ∼0.01 arcsec/pixel. The ASIC "Sidecar" array controller allows us very flexible readout. HiCIAO contrast is one order better than CIAO.

SEEDS is a Japanese-led but international project; it consists of some 120 members; 2/3 are domestic researchers, and 1/3 are in US and EU. The project is managed by The University of Tokyo and NAOJ. The completeness and uniformity of this systematic survey will provide important statistical results to be obtained, as well as enabling the study of individual objects of particular interest. The SEEDS data set will be a dominant one in this important field of research for a period of many years.

2. SEEDS Targets

We have selected three main kinds of targets for planet searches (YSOs, open cluster stars, and nearby main-sequence stars) and two for disk/planet searches (protoplanetary and debris disks). About 100 targets are selected for each category. The three categories for planet searches are intended to cover the age ranges of 1–10 Myr, ∼100 Myr (see Yamamoto *et al.* 2013), and 100–1000 Myr, respectively. See Janson *et al.* (2013), Thalmann *et al.* (2011, 2013) for debris disk targets and Brandt *et al.* (2013b) for moving group targets. See Narita *et al.* (2010, 2012) for RV targets. SEEDS observations can be conducted with the full use of new observing modes for the Subaru telescope: polarization differential imaging (PDI), spectral differential imaging (SDI) modes, and angular differential imaging (ADI) mode. The basic concept of such differential imaging (PDI/SDI) is to split up the image into two or more images and then use either different planes of polarization or different spectral filter band-passes to produce a signal that distinguishes faint objects near a bright central object from scattered halo or residual speckles. Great care is being taken to minimize the non-common-path error of the differential optics and its critical components (single and double Wollaston prisms made from high-quality birefringent YLF). However, most observations utilize ADI modes or PDI+ADI mode for a survey because of the sensitivity limit in the SDI mode. See also Brandt *et al.* (2013) for the data reduction.

3. Planet results

SEEDS has reported three (candidate) planets so far. The first one is GJ 758 b (Thalmann *et al.* 2009; Janson *et al.* 2011) and the second one is kappa And b (Carson *et al.* 2013). The former is 10–30 M_J companion orbiting around a Sun-like star and the latter is ∼13 M_J planet around a B star. The latter is the planet imaged around the most massive star. The third one, which is mainly described here, is GJ 504 b (Kuzuhara *et al.* 2013). This is a very faint planet orbiting a Sun-like star. It is at a distance about 18 pc from Earth. The central star itself is bright, visible to naked-eye (V∼5 mag). However, the planet is very dim, 17–20 mag at infrared wavelengths (JHKL′). The age of the central star is estimated from the gyrochronology to be 100–300 Myr, and 100–500 Myr (including other methods) even in the most conservative case. It is most likely that the planet mass is only 3–5.5 Jupiter masses, estimated from its luminosity and age. If so, it can be the lightest-mass planet ever imaged.

From seven epochs of observations, GJ 504b has been confirmed not to be a background stars but in fact orbits around GJ 504A.The apparent distance between the central star and planet is 44 au. Subsequent JHKL′ observations have also reveal that the planet has a unique color. This provides useful information of its atmosphere. It appears that GJ 504b has less cloudy atmosphere, while other imaged planets have cloudy atmosphere.

In direct imaging, planet masses are estimated from its luminosity and age. All the imaged planets so far are younger than 50 million years, which introduce uncertainties in mass estimates. This is because we do not know the planet origin and resultantly the models we use could provide different mass estimates. Therefore, if we simply use the ~14 Jupiter-mass mass boundary between planets and brown dwarfs, their distinction suffers from this uncertainty. For example, if we consider all the imaged planets are formed in a low-energy (more accurately entropy) state, they are all more than 14 Jupiter masses. On the other hand the planet age is old enough, this uncertainty is mitigated. Since the age of GJ 504 is old enough (100–500 Myr), either model predicts a planetary mass rather than a brown dwarf mass. If we simply use the conventional hot-start model and the age estimate based on stellar rotation, the planet mass of GJ 504b is as small as 3 Jupiter masses.

How such a giant planet around Sun-like star formed? The standard planet formation model based on our Solar system is thought to be difficult to explain the formation of these outer planets. It is likely they cannot obtain enough mass in such outer regions. It is also unlikely that a new model of gravitational instability can form outer planets because they require very massive disks around Sun-like stars. Here we note that we have detected many gaps and spiral-arm structures at the same radial distance in the disks where outer planets discovered. These structures can be formed by the presence of planets within the disks. Therefore, the formation of outer planets can be possible (at least at or near its birth), one can guess. Therefore, we consider that now is a time for revisit the planet formation theory to include the results of outer planets and disk structures as SEEDS results demonstrate.

4. Disk results

SEEDS has detected interesting fine-structures of the disks around dozens of young stars so far; they show gaps, spiral arms, rings, and so on with a similar radial distance where those "outer" planets imaged. These structures can be considered as "signpost of planets". Our results on disks also support a requirement of a new planet formation model. Fine structures of many disks are revealed in AB Aur, LkCa 15, HD 169142, SAO 206462, PDS 70, UX Tau, and Upper Sco J1604 (protoplanetary disks), HR 4796A and HIP 79977 (debris disks). For AB Aur (Hashimoto *et al.* 2011), complicated and asymmetrical structures are seen in the inner part (r<140 au) of the disk while confirming the previously reported outer (r>200 au) spiral structure. A double ring structure at ~40 and ~100 au and a ring-like gap between the two are discovered. The observed structures, including a bumpy double ring, a ring-like gap, and a warped disk in the innermost regions, provide essential information for understanding the formation mechanism of wide-orbit (r>20 au) planets. For LkCa 15 (Thalmann *et al.* 2010), sharp elliptical contours delimiting the nebulosity on the inside as well as the outside, consistent with the shape, size, ellipticity, and orientation of starlight reflected from the far-side disk wall, whereas the near-side wall is shielded from view by the disks optically thick bulk. This discovery provides confirmation of the disk geometry that has been proposed to explain the spectral energy distributions of such systems, comprising an optically thick disk with an inner truncation radius of 4–6 au enclosing a largely evacuated gap. An offset of the nebulosity contours along the major axis is also seen. This reinforces the leading theory that dynamical clearing by at least one orbiting body is the cause of the gap. For SAO 169142 (Muto *et al.* 2012), two small-scale spiral structures lying within 0.5″ (70 au) are discovered for the first time. Models for the spiral structures using the spiral density wave theory are used to derive a disk aspect ratio of h∼0.1. This model can potentially give estimates of the temperature and rotation profiles of the disk based on dynamical processes, independently from sub-mm observations. It also predicts the evolution of the spiral structures, which can be observable on timescales of 10–20 years, providing conclusive tests of the model. Assuming a planet, we can make predictions on the locations and, possibly, the masses of the unseen planets. For PDS 70 (Hashimoto *et al.* 2012), a weak-lined T Tauri star of K5 type, 0.82 M_{SUN} mass, <10 Myr age, taking into account the presence of the large and sharp gap, we suggest that the gap could be formed by dynamical interactions of sub-stellar companions or multiple unseen giant planets in the gap. A generic disk model that can simultaneously account for the general features in Spitzer, SMA, and Subaru observations. Particularly, the scattered light images for this model are computed, which agree with the general trend seen in Subaru data. Decoupling between the spatial distributions of the μm-sized dust and mm-sized dust inside the cavity is suggested by the model (Dong *et al.* 2012a,b). See also Takami *et al.* (2013), Follette *et al.* (2013), Grady *et al.* (2013), Tanii *et al.* (2012), Mayama *et al.* (2012), and Kusakabe *et al.* (2012).

References

Brandt, T. D. *et al.* 2013a, *ApJ*, 764, 183
Brandt, T. D. *et al.* 2013b, *ApJ* in press
Carson, J. *et al.* 2013, *ApJ*, 763, L32
Dong, R. *et al.* 2012a, *ApJ*, 750, 161
Dong, R. *et al.* 2012b, *ApJ*, 760, 111
Follette, K. *et al.* 2013, *ApJ*, 767, 10
Grady, C. *et al.* 2013, *ApJ*, 762, 48
Hashimoto, J. *et al.* 2011, *ApJ*, 729, L17
Hashimoto, J. *et al.* 2012, *ApJ*, 758, L19
Hodapp, K. *et al.* 2006, *Proc. SPIE*, 6269, 62693V
Janson, M. *et al.* 2011, *ApJ*, 728, 85
Janson, M. *et al.* 2013, *ApJ*, in press
Kusakabe, N. *et al.* 2012, *ApJ*, 753, 153
Kuzuhara, M. *et al.* 2013, *ApJ*, in press
Mayama, S. *et al.* 2012, *ApJ*, 760, L26
Muto, T. *et al.* 2012, *ApJ*, 748, L22
Narita, N. *et al.* 2010, *PASJ*, 62, 779
Narita, N. *et al.* 2012, *PASJ*, 64, 124
Tamura, M. *et al.* 2006, *Proc. SPIE*, 6269, 62690V
Takami, M. *et al.* 2013, *ApJ*, 772, 145
Tanii, R. *et al.* 2012, *PASJ*, 64, 124
Thalmann, C. *et al.* 2012, *ApJ*, 707, L123
Thalmann, C. *et al.* 2010, *ApJ*, 718, L87
Thalmann, C. *et al.* 2011, *ApJ*, 743, L6
Thalmann, C. *et al.* 2013, *ApJ*, 763, L29
Yamamoto, K. *et al.* 2013, *PASJ*, in press

Discussion

LIN: In addition to HAT-P-7, were distant stellar or planetary companions discovered around the host stars of any other hot Jupiters?

TAMURA: To my knowledge, HAT-P-7 b is so far the only case with a proper motion confirmation.

KAVELAARS: Can you comment on the statistical likelihood of your one detection of a few M_J planet at 40 AU around a sun-like star?

TAMURA: Yes. This is a relatively young star, around 200 Myr, and from our survey we would say such systems are very rare.

Exploring the Formation and Evolution of Planetary Systems
Proceedings IAU Symposium No. 299, 2013
M. Booth, B. C. Matthews & J. R. Graham, eds.
doi:10.1017/S1743921313007680

Results of the NaCo Large Program: probing the occurrence of exoplanets and brown dwarfs at wide orbit

A. Vigan[1], G. Chauvin[2], M. Bonavita[3], S. Desidera[3], M. Bonnefoy[4], D. Mesa[3], J.-L. Beuzit[2], J.-C. Augereau[2], B. Biller[4], A. Boccaletti[5], E. Brugaletta[6], E. Buenzli[7], J. Carson[8], E. Covino[9], P. Delorme[2], A. Eggenberger[2], M. Feldt[4], J. Hagelberg[10], T. Henning[4], A.-M. Lagrange[2], A. Lanzafame[6], F. Ménard[2], S. Messina[6], M. Meyer[7], G. Montagnier[11], C. Mordasini[4], D. Mouillet[2], C. Moutou[1], L. Mugnier[12], S. P. Quanz[7], M. Reggiani[7], D. Ségransan[10], C. Thalmann[13], R. Waters[13] and A. Zurlo[1]

[1]Aix Marseille Université, CNRS, LAM (Laboratoire d'Astrophysique de Marseille) UMR 7326, 13388, Marseille, France. email: `arthur.vigan@oamp.fr`;
[2]UJF-Grenoble 1 / CNRS-INSU, IPAG, UMR 5274, Grenoble, F-38041, France;
[3]INAF - Osservatorio Astronomico di Padova, Vicolo dell' Osservatorio 5, 35122, Padova, Italy;
[4]Max Planck Instiute für Astronomy, Königsthul 17, D-69117 Heidelberg, Germany;
[5]LESIA, Observatoire de Paris, CNRS, University Pierre et Marie Curie Paris 6 and University Denis Diderot Paris 7, 5 place Jules Janssen, 92195 Meudon, France;
[6]INAF - Osservatorio Astrofisico di Catania, via S. Sofia 78, 95123 Catania, Italy;
[7]Institute for Astronomy, ETH Zurich, Wolfgang-Pauli-Strasse 27, 8093 Zurich, Switzerland;
[8]College of Charleston, Department of Physics & Astronomy, USA;
[9]INAF - Osservatorio Astronomico di Napoli, Salita Moiariello 16, 80131 Napoli, Italy;
[10]Geneva Observatory, Univ. of Geneva, Chemin des Mailettes 51, 1290, Versoix, Switzerland;
[11]Observatoire de Haute-Provence, CNRS/OAMP, 04870 Saint-Michel-l'Observatoire, France;
[12]Onera - The French Aerospace Lab, 92322, Chatillon, France;
[13]University of Amsterdam, Postbus 94249, 1090 GE, Amsterdam, The Netherlands

Abstract. Over the past decade, a growing number of deep imaging surveys have started to provide meaningful constraints on the population of extrasolar giant planets at large orbital separation. Primary targets for these surveys have been carefully selected based on their age, distance and spectral type, and often on their membership to young nearby associations where all stars share common kinematics, photometric and spectroscopic properties. The next step is a wider statistical analysis of the frequency and properties of low mass companions as a function of stellar mass and orbital separation. In late 2009, we initiated a coordinated European Large Program using angular differential imaging in the H band (1.66 μm) with NaCo at the VLT. Our aim is to provide a comprehensive and statistically significant study of the occurrence of extrasolar giant planets and brown dwarfs at large (5-500 AU) orbital separation around $\sim$150 young, nearby stars, a large fraction of which have never been observed at very deep contrast. The survey has now been completed and we present the data analysis and detection limits for the observed sample, for which we reach the planetary-mass domain at separations of $\gtrsim$50 AU on average. We also present the results of the statistical analysis that has been performed over the 75 targets newly observed at high-contrast. We discuss the details of the statistical analysis and the physical constraints that our survey provides for the frequency and formation scenario of planetary mass companions at large separation.

Keywords. instrumentation: adaptive optics, instrumentation: high angular resolution, stars: imaging, methods: statistical

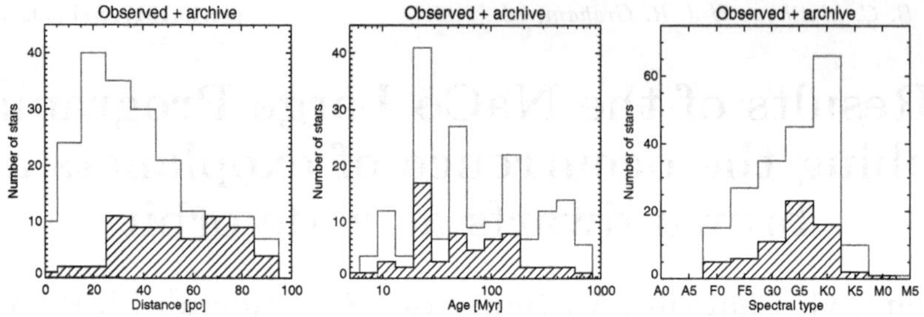

Figure 1. Distribution of NaCo-LP target sample in distance (left), age (center) and spectral type (right). The observed sample is plotted as the hatched black histogram, and the full sample for the statistical analysis as the blue histogram.

1. Introduction

The NaCo Large Program (NaCo-LP) project was initiated in 2008 as a large-scale European effort to measure the occurrence of giant exoplanets and brown dwarf at wide orbits (50–500 AU). This survey was also designed to prepare the near-infrared survey (NIRSUR) to be conducted with the SPHERE instrument (Beuzit *et al.* 2008). The NaCo-LP would provide useful information such as first epoch observation for background contaminants and false alarm probability estimation. A total of 18 nights was allocated on NaCo at the VLT†. We are now concluding the survey.

2. Sample selection

The target selection for the NaCo-LP was initially done within the SPHERE NIRSUR database. The following criteria were used: $\delta \leqslant 25^{\text{o}}$, d $\leqslant$ 100 pc, age $\leqslant$ 200 Myr, R $\leqslant$ 9.5 mag, 0.4 mag $\leqslant B - V \leqslant$ 1.2 mag, no visual binaries within 6", no spectroscopic binaries. The R magnitude threshold was designed to ensure that the targets observed in the NaCo-LP will also be observable in the future with the SPHERE instrument. The $B - V$ cut in color was designed to select only solar-type stars. This selection resulted in an initial sample of ~200 stars, which was then cross-correlated with previous and on-going surveys at the time (e.g. Masciadri *et al.* (2005); Biller *et al.* (2007); Lafrenière *et al.* (2007a); Chauvin *et al.* (2010), ...) in order to select only stars that had never been observed at high-contrast. The second selection step resulted in a sample of 110 stars. Both samples are illustrated in Fig. 1.

3. Observations and data analysis

The strategy was to observe in H-band with the S13 objective (13 mas/pixel, FoV 14"×14"), pupil-tracking mode for proper implementation of angular differential imaging (Marois *et al.* 2006), and the 0.7" Lyot coronagraph. Each observations consisted of a reference PSF (without coronagraph) obtained with a neutral density filter, followed by 45 minutes of deep coronagraphic images. Unfortunately, after this first observing run in December 2009, we discovered a problem with the pupil-tracking mode that induced a slow drift of the star behind the coronagraph that could reach several pixels per minutes. For the subsequent observing runs and follow-ups, the observing strategy was modified to saturated imaging with shorter integration times and a shift-and-addprocedure.

† ESO program 184.C-0567, PI J.-L. Beuzit, "Probing the Occurrence of Exoplanets and Brown Dwarfs at Wide Orbits".

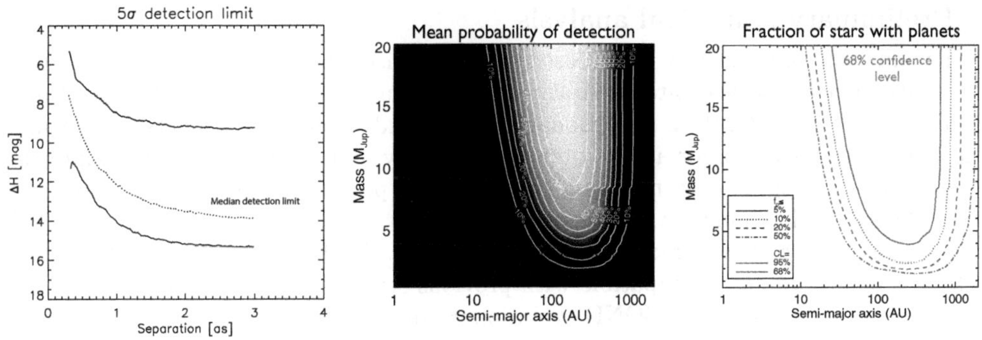

Figure 2. *Left:* Best (top), worst (bottom) and median (middle) 5σ detection limits of the survey within 3" given as the contrast with respect to the star in H-band. *Center:* mean probability of detection of the observed sample as a function of mass and semi-major axis. *Right:* Upper limit to the fraction of stars that can harbor a substellar companion, as a function of mass and semi-major axis, at a 68% confidence level, assuming a null-result for the survey.

The data analysis was performed using four independent pipelines, and two of them were used to reduce and analyze the complete survey (see descriptions in Vigan *et al.* 2012 and Rameau *et al.* 2013). The general scheme of the analysis consisted in five main steps: cleaning of the data, frame centering, bad frames rejection, ADI analysis, and astrometry and photometry of the candidates. The results presented here are based on an analysis using LOCI (Lafrenière *et al.* 2007b) within 3" from the star and spatial filtering in a 5×5 λ/D box outside of this radius.

The detection limits where estimated as the azimuthal RMS in λ/D-width annuli normalized to the reference PSF peak. ADI flux losses where estimated by introducing fake planets into the data cubes at various orientations and separations. The best, worst and median detection limits of the survey are plotted in Fig. 2 (left).

From the observations, 17 previously unknown binaries were identified and rejected during the acquisition. Several hundreds of companion candidates (cc) were identified in the final images and detection maps. Follow-up observations allowed obtaining second epochs for the vast majority of them, confirming no new substellar companions.

4. First results of the survey

The survey produced nonetheless two interesting results so far. First, the debris disk around the young G8V star HD 61005 was resolved for the first time as a narrow ring from our NaCo observations (Buenzli *et al.* 2010). We determined that the ring center is offset by at least 2.75 ±0.85 AU from star. The offset, together with a relatively steep inner rim, could indicate a planetary companion that perturbs the remnant planetesimal belt. From our imaging data we set upper mass limits for companions that exclude any object above the deuterium-burning limit for separations down to 0.3 arcsec. HD 61005 will be a favored target for the forthcoming high-contrast imagers.

The second interesting result is the white dwarf (WD) companion detected around the star HD 8049 (Zurlo *et al.* 2013). This star was originally estimated as young (< 400 Myr), leading to a false estimate of the mass of the companion. Further analysis of the stellar parameters revealed discrepancies between age indicators based on stellar activity and other indicators (lithium abundance, galactic motion), leading to a reevaluation of the age. Moreover, radial velocity trend over almost 30 years was only compatible with a more massive object, leading to the assumption that the companion was in fact a WD, which was later confirmed with VLT/SINFONI JHK spectroscopy.

5. Preliminary statistical analysis

The null result of the survey can be used to infer statistical properties on the population of massive planet at wide orbital separation. For this purpose, we performed Monte-Carlo simulations using the QMESS tool (Bonavita *et al.* 2013) to generate planets on a regular grid of mass and semi-major axis to obtain probability of detection maps for each of the observed target (Fig. 2, center). Using the formalism previously described in Lafrenière *et al.* (2007a), and assuming a null-result of the survey, we determine that less than 20% of stars can harbor planets more massive than 4 M_{Jup} in the range 55–800 AU (Fig. 2, right). This estimation is compatible with previous studies (e.g. Nielsen & Close 2010). This analysis is based on the COND evolutionary models (Baraffe *et al.* 2003).

References

Baraffe, I., Chabrier, G., Barman, T. S., Allard, F., & Hauschildt, P. H. 2003, A&A, 402, 701
Beuzit, J.-L., Feldt, M., Dohlen, K., *et al.* 2008, in SPIE Conference Series, Vol. 7014
Biller, B. A., Close, L. M., Masciadri, E., *et al.* 2007, ApJ, 173, 143
Bonavita, M., de Mooij, E. J. W., & Jayawardhana, R. 2013, PASP, 125, 849
Buenzli, E., Thalmann, C., Vigan, A., *et al.* 2010, A&A, 524, L1
Chauvin, G., Lagrange, A.-M., Bonavita, M., *et al.* 2010, A&A, 509, A52+
Lafrenière, D., Doyon, R., Marois, C., *et al.* 2007a, ApJ, 670, 1367
Lafrenière, D., Marois, C., Doyon, R., Nadeau, D., & Artigau, É. 2007b, ApJ, 660, 770
Marois, C., Lafrenière, D., Doyon, R., Macintosh, B., & Nadeau, D. 2006, ApJ, 641, 556
Masciadri, E., Mundt, R., Henning, T., Alvarez, C., & Barrado y Navascués, D. 2005, ApJ, 625, 1004
Nielsen, E. L. & Close, L. M. 2010, ApJ, 717, 878
Rameau, J., Chauvin, G., Lagrange, A.-M., *et al.* 2013, A&A, 553, A60
Vigan, A., Patience, J., Marois, C., *et al.* 2012, A&A, 544, A9
Zurlo, A., Vigan, A., Hagelberg, J., *et al.* 2013, A&A, 554, A21

Discussion

L. CLOSE: Can you speak about the fact that HD 8049 looked young in its rotation index, but was in fact old. Is it spin up?

A. VIGAN: Yes, our assumption is that it is spin up because of material accreted from the WD progenitor

J. FARIHI: Interpretation of rejuvenation by AGB wind accretion at $a \geqslant 50$ AU should be re-examined. Low mass MS companions to WD, even those in $a < 1$ AU orbits, do not exhibit any signs of mass accretion. There are now ∼1000 such WD+MS systems known with no evidence for AGB wind accretion.

S. METCHEV: Your effort to exclude stars previously observed at high-contrast is very laudable, but may have introduced unknown biases in sample selection. In particular, no BD discoveries among your 110 stars seems very significant compared to several surveys showing BD frequency of a few percent.

A. VIGAN: Our parent sample was 200 stars, from which we excluded 90 previously observed ones, but the statistical analysis will be done on the full sample. In the parent sample, some stars may have already known BD companions (e.g. AB Pic).

Exploring the Formation and Evolution of Planetary Systems
Proceedings IAU Symposium No. 299, 2013
M. Booth, B. C. Matthews & J. R. Graham, eds.

doi:10.1017/S1743921313007692

Search for cool extrasolar giant planets combining coronagraphy, spectral and angular differential imaging

Anne-Lise Maire[1], Anthony Boccaletti[2], Julien Rameau[3], Gaël Chauvin[3], Anne-Marie Lagrange[3], Mickaël Bonnefoy[4], Silvano Desidera[5], Mélody Sylvestre[2], Pierre Baudoz[2], Raphaël Galicher[2], and David Mouillet[3]

[1]LUTH, Obs. Paris/CNRS/Univ. Paris 7, 92195 Meudon, France
email: anne-lise.maire@obspm.fr; [2]LESIA, Obs. Paris/CNRS/Univ. Paris 6/Univ. Paris 7, 92195 Meudon, France; [3]Institut de Planétologie et d'Astrophysique de Grenoble, University Joseph Fourier/CNRS, 38041 Grenoble, France; [4]Max Planck Institute for Astronomy, D-69117 Heidelberg, Germany; [5]INAF - Osservatorio Astronomico di Padova, I-35122 Padova, Italy

Abstract. Spectral differential imaging (SDI) is part of the observing strategy of current and on-going high-contrast imaging instruments on ground-based telescopes. Although it improves the star light rejection, SDI attenuates the signature of off-axis companions to the star, just like angular differential imaging (ADI). However, the attenuation due to SDI has the peculiarity of being dependent on the spectral properties of the companions. To date, no study has investigated these effects. Our team is addressing this problem based on data from a direct imaging survey of 16 stars combining the phase-mask coronagraph, the SDI and the ADI modes of VLT/NaCo. The objective of the survey is to search for cool (Teff<1000-1300 K) giant planets at separations of 5-10 AU orbiting young, nearby stars (<200 Myr, <25 pc). The data analysis did not yield any detections. As for the estimation of the sensivity limits of SDI-processed images, we show that it requires a different analysis than that used in ADI-based surveys. Based on a method using the flux predictions of evolutionary models and avoiding the estimation of contrast, we determine directly the mass sensivity limits of the survey for the ADI processing alone and with the combination of SDI and ADI. We show that SDI does not systematically improve the sensitivity due to the spectral properties and self-subtraction of point sources.

Keywords. Planetary systems, techniques: high angular resolution, techniques: image processing

---

1. Introduction

Most of the direct imaging searches for young giant exoplanets conducted so far use adaptive optics (AO) combined with broad-band imaging in the near-infrared (for instance, Lafrenière *et al.* 2007, Chauvin *et al.* 2010, Vigan *et al.* 2012). Biller *et al.* (2007) instead performed a survey using narrow-band adaptive optics imaging and spectral differential imaging (SDI, Racine *et al.* 1999). More recently, the NICI team on Gemini (Liu *et al.* 2010) and our team on VLT/NaCo (Maire *et al.*, in prep.) completed surveys to search for cool (<1 300 K) and close-in (>5–10 AU) young giant exoplanets combining several state-of-the-art high-contrat imaging techniques: coronagraphy, SDI, and angular differential imaging (ADI, Marois *et al.* 2006). This observing strategy is similar to the strategy which will be employed for SPHERE (Beuzit *et al.* 2008) and GPI (Macintosh *et al.* 2008). Biller *et al.* (2007) and Nielsen *et al.* (2008) determined SDI sensitivity limits by converting the contrast levels measured in the reduced images into masses using evolutionary models. In this proceeding, we show that this method is optimistic for

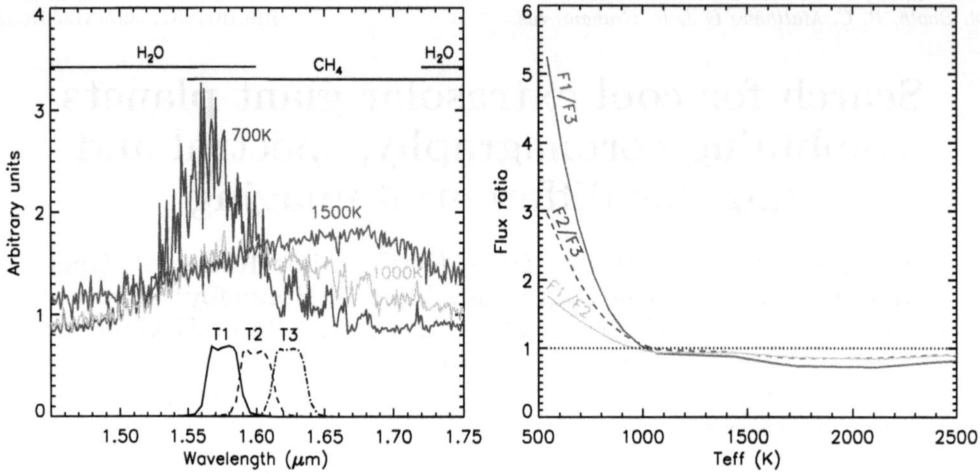

Figure 1. Left: Spectra of model atmospheres of giant planets for different effective temperatures (colored solid lines, BT-Settl models from Allard *et al.* 2011). Each spectrum is normalized to its value at 1.6 μm and is vertically shifted by a constant. The transmission of the three SDI filters of NaCo are shown (black curves). Theoretical absorption bands of water and methane are also indicated. Right: Flux ratio as a function of the effective temperature derived for the NaCo SDI filters. F_1, F_2, and F_3 refer to the fluxes in the filters at 1.575, 1.6, and 1.625 μm.

the analysis of SDI-processed images and instead we propose a straightforward method based on the injection of synthetic planets in the raw data, combined with the use of evolutionary models. This framework should serve as a basis for the data reduction and analysis of SPHERE and GPI. We present the observing strategy, the procedure used to analyze and interprete the data, and the detection limits of the survey.

2. Observing strategy

We use an observing strategy combining the four-quadrant phase mask, the ADI and SDI modes of VLT/NaCo. SDI exploits the natural wavelength dependence of a star image (Racine *et al.* 1999). Two simultaneous images taken at different wavelengths are rescaled spatially to correct for the PSF chromaticity and rescaled in intensity to correct for the filter transmission before being subtracted. SDI is intended to take advantage of the presence of a methane absorption band at $\sim$1.6 μm in the spectra of cool (<1300 K) giant planets (Fig. 1, left), that is not present in the star's spectrum. Based on a complete compilation of young and nearby stars recently identified in young co-moving groups and from systematic spectroscopic surveys, we selected a sub-sample of 16 stars, mostly AFGK spectral types, according to their age ($\lesssim$200 Myr), distance ($\lesssim$25 pc) and R-band brightness ($\lesssim$ 9.5). The age cut-off ensures that the cool companions detected will have planetary masses. The distance cut-off guarantees that these targets are the most favorable for the detection of cool companions. The targets are brighter than $R = 9.5$ in the visible to allow good AO efficiency.

3. Analysis of spectral differential imaging data

In practice, SDI produces a significant attenuation of the planet itself that has to be quantified to derive accurate photometry. First, this attenuation depends on the planet's spectral properties, which are determined in first approximation by the effective temperature, as illustrated in the right panel of Fig. 1. We represent the flux ratios between the

NaCo SDI filters as a function of their effective temperature. For temperatures higher than the methane condensation temperature (1 300 K), the flux ratios are close to 1, so the flux differences are minimum and the self-subtraction of point sources is high. Interestingly, the flux ratios remain close to 1 below 1 300 K down to 1 000 K. Thus, the SDI domain is constrained to temperatures lower than those what we expect from the methane condensation temperature. Below 1 000 K, the flux ratios increase and reach the highest values for the F_1/F_3 ratio. Consequently, we consider for the data analysis only the image subtraction $I_1 - I_3$, with I_1 (I_3) is the image at 1.575 (1.625) μm. The second factor to account for when deriving the SDI attenuation is the geometrical overlap between the two planet images in I_1 and I_3 due to the spatial rescaling. This attenuation is maximum near the image center and decreases with angular separation.

In the general context of high-contrast imaging with a single spectral filter, the detection limits are measured on contrast maps. In most cases, we use the azimuthal standard deviation to derive a 1-dimensional contrast plot, which after correction from various attenuations (ADI and/or coronagraph) is then converted into mass limits at 5 σ according to a given evolutionary model. The problem with SDI is different as we measure a differential intensity. It can be expressed as follows (no coronagraph and no ADI):

$$F_{\rm rs} = F_{\rm r} - F_{\rm s} \times \alpha \times \phi(\vec{r}) \tag{3.1}$$

where $F_{\rm r}$ and $F_{\rm s}$ are the object intensities at respectively $\lambda_{\rm r}$ and $\lambda_{\rm s}$, α the intensity rescaling factor, $\phi(\vec{r})$ the attenuation due to the spatial rescaling ($\phi \sim 1$ close to the image center and $\phi \sim 0$ at large separations) and $\vec{r} = (r, \theta)$. Since the residual flux of a point source can exhibit positive or negative values, like the residual noise measured in the processed image, it becomes difficult to disentangle the noise from a planet signal. The method used for single-band surveys (ADI, coronagraphy) is no longer valid. From Eq. (3.1), we see that, providing $\phi(\vec{r})$ is calibrated (by using the point-spread functions measured in the different filters), we have to test the individual intensities $F_{\rm r}$ and $F_{\rm s}$ for all planet masses (taken from an evolutionary model) which reproduce the measured $F_{\rm rs}$. We expect that several values of $F_{\rm r}$ and $F_{\rm s}$ can match the observations, resulting in degeneracies in planet mass. The number of degeneracies and the values of the mass solutions depend on $\phi(\vec{r})$, i.e. the position in the image field. To break these degeneracies, we need to compare the SDI sensitivity limits to the limits derived in broad-band imaging. For the analysis of our survey, we consider and compare two processings: classical ADI (noted ADI, Marois *et al.* 2006) and an algorithm combining SDI as a first step and classical ADI as a second step (refered as ASDI).

4. Sensitivity limits of the survey

The data analysis did not yield any detections. We focus instead on the interpretation of the detection limits. For this, we applied a straightforward and robust method based on the injection of fake planets in the data set at the cost of a longer computing time and sparsity in the detection map. The use of fake planets assumes that the PSF is the same in the coronagraphic (or saturated) and the out-of mask (or unsaturated) images. This hypothesis strongly depends on the AO-loop and photometric stabilities, and we should expect variations from one data set to another. The fake planets are injected simultaneously at several separations and position angles in the raw datacubes with the flux predicted by the BT-Settl models (Allard *et al.* 2011) and processed with ADI and ASDI. For each separation, we measured the planet fluxes using aperture photometry and averaged them. The noise was measured on the data processed without the fake planets

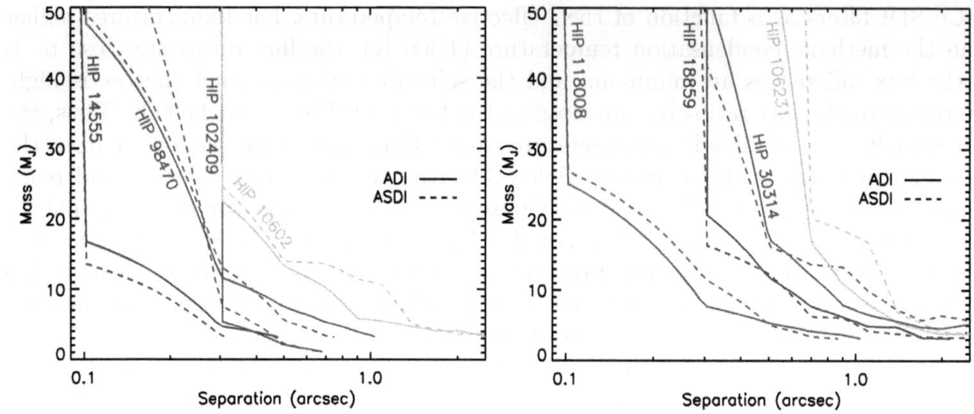

Figure 2. Sensitivity limits at 5 σ of the survey in Jupiter masses of detectable companions (BT-Settl models, Allard *et al.* 2011) for the youngest stars of the survey ($\leqslant$70 Myr). The curves are derived for two pipelines: classical ADI (solid curves) and ASDI (dashed curves).

in rings of 1 FWHM and scaled to the same aperture size. The process was repeated for each model mass. Finally, we interpolate for each separation the signal-to-noise ratio vs. mass relation to derive the corresponding mass achieved at 5 σ.

Figure 2 presents the mass sensivity limits at 5 σ of the survey as a function of the angular separation for the youngest stars of the survey ($\leqslant$70 Myr). The detection limits are cut when the minimum effective temperature available in the BT-Settl grids ($\sim$500 K) is reached. The main result is that SDI can either improve or degrade the sensitivity (ASDI curve reaching either lower or larger masses than the corresponding ADI curve). There are no particular trends with the star age and the angular separation. We are analyzing possible trends with parameters representative of the quality of the observations.

5. Conclusion

We show that the interpretation of SDI-processed images requires a different analysis than that used for broad-band direct imaging surveys. In particular, we need to compare the sensivity limits to those obtained in broad-band imaging in order to break the degeneracies in planet mass. This analysis demonstrates that SDI does not improve systematically the sensitivity because of the spectral properties and self-subtraction of point sources. SDI on NaCo gives the best performance for separations of 0.5–1″. We expect improved performances with SPHERE and GPI thanks to their extreme AO systems.

References

Allard, F., Homeier, D., & Freytag, B. 2011, *Proc. ASP*, Vol. 448, 91
Beuzit, J.-L., Feldt, M., Dohlen, K., *et al.* 2008, *Proc. SPIE*, Vol. 7014, 701418
Biller, B. A., Close, L. M., Masciadri, E., *et al.* 2007, *ApJS*, 173, 143
Chauvin, G., Lagrange, A.-M., Bonavita, M., *et al.* 2010, *A&A*, 509, A52
Lafrenière, D., Doyon, R., Marois, C., *et al.* 2007, *ApJ*, 670, 1367
Liu, M. C., Wahhaj, Z., Biller, B. A., *et al.* 2010, *Proc. SPIE*, Vol. 7736, 77361K
Macintosh, B., Graham, J., Palmer, D., *et al.* 2008, *Proc. SPIE*, Vol. 7015, 701518
Marois, C., Lafrenière, D., Doyon, R., Macintosh, B., & Nadeau, D. 2006, *ApJ*, 6, 556
Nielsen, E. L., Close, L. M., Biller, B. A., Masciadri, E., & Lenzen, R., *ApJ*, 674, 466
Racine, R., Walker, G. A. H., Nadeau, D., Doyon, R., & Marois, C., 1999, *PASP*, 111, 587
Vigan, A., Patience, J., Marois, C., *et al.* 2012, *A&A*, 544, A9

Discussion

GRAHAM: How does SDI improve with increased spectral resolution?

MAIRE: When you increase the spectral resolution, the planet signals in the spatially rescaled images will be less separated (the spatial rescaling factor is proportional to the wavelength ratio). As a result, the self-subtraction of the planet will be higher and the SDI performance will be worse.

Exploring the Formation and Evolution of Planetary Systems
Proceedings IAU Symposium No. 299, 2013
M. Booth, B. C. Matthews & J. R. Graham, eds.

doi:10.1017/S1743921313007709

The Large Binocular Telescope Interferometer & Adaptive Optics System: On-sky Performance and Results

Vanessa Bailey[1], Philip Hinz[1], Vidhya Vaitheeswaran[1], Andrew Skemer[1], Denis Defrère[1], Timothy Rodigas[1], Simone Esposito[2], Enrico Pinna[2] and Alfio Puglisi[2]

[1]Steward Observatory, University of Arizona, 933 N Cherry Ave., Tucson AZ, 85719

[2]Istituto Nazionale di Astrofisica, Osservatorio Astrofisico di Arcetri, Largo Enrico Fermi 5, I - 50125, Florence Italy
email: vbailey@as.arizona.edu

Abstract. Increasing spatial resolution and contrast capabilities will make possible new direct detections of exoplanets, exozodis, and circumstellar disks. The Large Binocular Telescope Interferometer (LBTI) has been engineered to sit at the combined focus of the Large Binocular Telescope's two 8.4m apertures. Both apertures are equipped with 672-actuator deformable secondary mirrors, the first of the next generation of "extreme" adaptive optics (AO) systems. We present an overview of the LBTI AO instrument suite and detail current on-sky performance.

Keywords. instrumentation: adaptive optics, instrumentation: interferometers, instrumentation: high angular resolution

---

Direct imaging of exoplanets and circumstellar disks delivers exciting scientific results, but requires high-quality, stable Point Spread Functions (PSFs). Direct imaging allows for detailed study of the atmospheric properties of planets akin to gas giants in our own outer solar system (e.g.: Konopacky *et al.* 2013). In order to lay the framework for future imaging searches for exo-Earths, we will first need to understand the properties of exozodis, dust belts akin to our own zodical dust (e.g.: Millan-Gabet *et al.* 2011). In each case we are limited by the planet-to-star contrast ratio and angular resolution that can be delivered by the current generation of adaptive optics and imaging systems.

In light of these requirements, the Large Binocular Telescope Interferometer (LBTI) has been designed to sit at the combined focus of the Large Binocular Telescope's (LBT) two 8.4m, adaptive optics-fed apertures. Each aperture independently delivers stable, diffraction-limited images, ideal for exoplanet and cicumstellar disk searches at $3-4\mu$m. When the beams are combined interferometrically, the 23m baseline between the apertures allows LBTI to resolve faint exozodi emission around nearby stars at 10μm. Both of these operational modes are made possible by LBTI's dual, high-contrast adaptive optics (AO) systems.

The LBTI natural guide star AO system is one of only two AO systems in the world which use adaptive secondary mirrors (ASMs) and pyramid wavefront sensors (PWFSs; Ragazzoni 1996), the other being the Magellan Clay telescope (Close *et al.* 2012). LBTI's PWFSs are clones of the design used in the LBT FLAO system (Esposito *et al.* 2011). The two LBT ASMs themselves are 0.9m in diameter, each with 672 actuators. The high actuator density allows for correction of up to 400 aberration modes. By replacing a fixed secondary with a deformable one, no additional elements are added to the optical path. This design is particularly beneficial at wavelengths $> 3\mu$m, where each warm optic

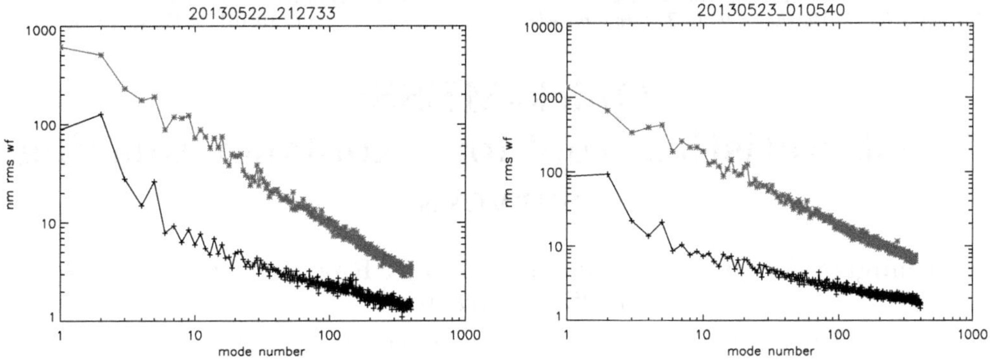

Figure 1. LBTI AO performance (RMS wavefront error) as a function of mode number. Low mode numbers correspond to low spatial frequencies (e.g.: tip, tilt, focus, astigmatism...), while high mode numbers correspond to high spatial frequencies. The upper line (-*-) is the atmospheric power spectrum, while the lower line (-+-) is the power spectrum of the corrected wavefront. *Left:* Natural seeing $\sim 0.8''$. *Right:* Natural seeing $\sim 1.3''$.

contributes to the overall background flux, decreasing the science camera sensitivity (Lloyd-Hart *et al.* 2000). Another benefit to using an ASM on a Gregorian telescope is the ability to calibrate the AO system after it has been installed on the telescope. By calibrating in-situ, the the mirror actuators and wavefront sensor may be precision-aligned and their on-sky configuration duplicated.

The PWFS was chosen because it provides improved performance and flexibility as compared to traditional Shack-Hartmann sensors. One especially advantageous aspect of PWFS, compared to Shack-Hartmann sensors, is the ability to bin the WFS CCD for improved sensitivity on faint stars. The LBTI AO system can guide on stars as faint as $R = 15$, with an ultimate goal of $R = 16$ or fainter, for improved sky coverage.

Excellent on-sky performance has been demonstrated in the last year, with up to 400 spatial aberration modes corrected. Figure 1 shows the improvement in wavefront error for bright targets ($R = 5 - 6$) as measured by the AO system. We achieve a factor of 5 improvement in cumulative wavefront error in good seeing (0.8"), and a factor of 9 in moderate seeing (1.3"). The greatest net gains are made at low modes (like tip/tilt, astigmatism, etc.), though even at the highest spatial frequencies, the system still reduces the RMS error by a factor of $3 - 7$.

The quality and stability of the resulting PSF have enabled exceptional high-contrast imaging results and will be the basis for the next generation nulling interferometery survey. LBTI has already achieved an L' contrast of 13 magnitudes at 0.75" and its nulling interferometry channel will ultimately reach sensitivities of 10 *zodis* (see companion papers by D. Defrere *et al.*, T. Rodigas *et al.*, and A. Skemer *et al.*).

References

Close, L., *et al.* 2012, *SPIE*, 8447, 84470X
Esposito, S., *et al.* 2011, *SPIE*, 8149, 814902
Lloyd-Hart, M. 2000, *PASP*, 768, 264
Konopacky, Q., *et al.* 2013, *Science*, 339, 1398
Millan-Gabet, R., *et al.* 2011, *ApJ*, 734, 67
Ragazzoni, R. 1996, *JModOpt*, 43, 289
Lloyd-Hart, M. 2000, *PASP*, 112, 264

Exploring the Formation and Evolution of Planetary Systems
Proceedings IAU Symposium No. 299, 2013
M. Booth, B. C. Matthews & J. R. Graham, eds.

doi:10.1017/S1743921313007710

Quick-MESS: A fast statistical tool for Exoplanet Imaging Surveys

Mariangela Bonavita[1,2], Ernst De Mooij[2], Ray Jayawardhana[2] and Raffaele Gratton[1]

[1]Osservatorio Astronomico di Padova - INAF Vicolo dell'Osservatorio, 5 35141 Padova (Italy)
email: mariangela.bonavita@oapd.inaf.it

[2]Dept. of Astronomy & Astrophysics, University of Toronto, 50 St. George St. M5S 3H4 Toronto ON (Canada)

Abstract. Several tools have been developed for the analysis of the results of direct imaging exoplanet surveys, mostly using a combination of Monte-Carlo simulations or a Bayesian approach. Here we present a novel approach to the statistical analysis of Direct Imaging surveys, called Quick-MESS, which allows for a much faster and flexible analysis.

Keywords. stars: low-mass, brown dwarfs,(stars:) planetary systems, methods: statistical

1. Quick MESS

Upcoming direct imaging (DI) surveys will enable the statistics of planets at large ($a > 5$ AU) separations to be determined, complementing those from transit and RV surveys . Tools to investigate these surveys have been developed and typically use Monte-Carlo simulations or a Bayesian approach (see e.g. Chauvin *et al.* 2010, Lafreniere *et al.* 2008, Nielsen *et al.* 2010, Bonavita *et al.* 2012). Here we present Quick-MESS, a novel tool that uses a grid-based approach to analyzing DI surveys.

The main steps of the code, explained in detail in Bonavita *et al.* (2013) can be summarized as follows:

(*a*) Evaluate the probability of detection as a function of the eccentricity and the normalized separation. By using the normalized separation rather than the projected separation (ρ) or the physical separation (a), the resulting projection probability map is scale-free and only needs to be calculated once

(*b*) Convert the contrast curve of the instrument into a minimum detectable planet mass as a function of projected separation using planetary evolutionary models (e.g. Baraffe *et al.* (2003), Burrows *et al.* 2003).

(*c*) For each star in sample, calculate the expected probability of finding a planet as a function of semi-major axis (a) and planetary mass (M_p), by converting the normalized separation of the projection probability map in to a projected separation for a given a using the distance to the star and the contrast curve.

(*d*) Choose a probability function for planetary mass and semi-major axis and fold those into the planet-probability as found in step (c). This step can be repeated with a different choice of planet parameter distributions, and using the same planet-probability, this being the key for the speed and flexibility of QMESS.

This approach leads to a substantial reduction in the required computational time with respect to other tools based on Monte-Carlo sampling of the planet distribution. QMESS

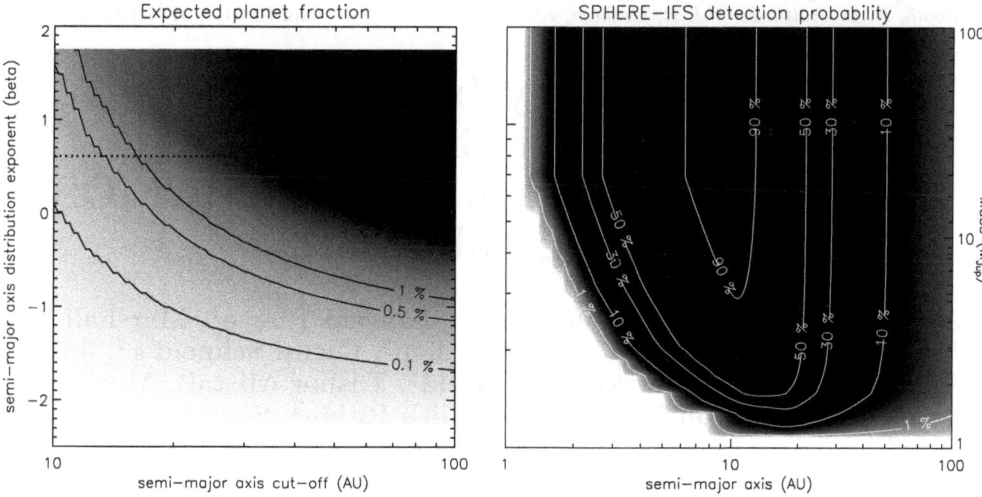

Figure 1. Left: Expected planet fraction for the GDPS survey (Lafreniere *et al.* 2008) assuming different semi-major axis distributions (power law with index beta) extrapolated to up to several maximum values (cut-off). The dotted line highlights the beta value found from the RV results ($\beta = -0.61$, see Cumming *et al.* 2008). **Right:** Predicted detection probability for a 10 Myr old A0V star at 20 pc, if observed with SPHERE-IFS (Beuzit *et al.* 2008).

is also an extremely flexible tool, enabling the study of a large range of parameter space for the mass and semi-major axes distributions (see left panel of Fig. 1) without the need of re-simulating the planet distribution. In addition to the analysis of a survey, QMESS can also be used to assess the performances of, and to select the most suitable targets for new surveys, instruments and/or different observing strategies (see right panel of Fig. 1)

References

Baraffe, I., Chabrier, G., Barman, T. S., Allard, F., & Hauschildt, P. H. 2003, *A&A*, 402, 701
Beuzit, J.-L., Feldt, M., Dohlen, K., *et al.* 2008, *SPIE*, 7014
Bonavita, M., Chauvin, G., Desidera, S., *et al.* 2012, *A&A*, 537, A67
Bonavita, M., de Mooij, E. J. W., & Jayawardhana, R. 2013, *PASP*, 125, 849
Burrows, A., Sudarsky, D., & Lunine, J. I. 2003, ApJ, 596, 587
Chauvin, G., Lagrange, A.-M., Bonavita, M., *et al.* 2010, *A&A*, 509, A52
Cumming, A., Butler, R. P., Marcy, G. W., *et al.* 2008, *PASP*, 120, 531
Lafrenière, D., Jayawardhana, R., & van Kerkwijk, M. H. 2008, *ApJL*, 689, L153
Nielsen, E. L. & Close, L. M. 2010, *ApJ*, 717, 878
Vigan, A., Patience, J., Marois, C., *et al.* 2012, *A&A*, 544, A9

Exploring the Formation and Evolution of Planetary Systems
Proceedings IAU Symposium No. 299, 2013
M. Booth, B. C. Matthews & J. R. Graham, eds.

doi:10.1017/S1743921313007722

Archival Legacy Investigation of Circumstellar Environments using KLIP algorithm on HST NICMOS coronagraphic data

Elodie Choquet[1], J. Brendan Hagan[1], Laurent Pueyo[1], Marshall D. Perrin[1], Dean C. Hines[1], Christine Chen[1], Glenn Schneider[2], John Debes[1], David Golimowski[1], Neill Reid[1], Tushar Mittal[1], Margaret Moerchen[1], Mamadou N'Diaye[1], Abhijith Rajan[3], Sean Lonsdale and Remi Soummer[1]

[1]Space Telescope Science Institute,
3700 San Martin Drive, Baltimore MD 21218, USA
[2] University of Arizona [3] Arizona State University

Abstract. The Archival Legacy Investigation of Circumstellar Environments (ALICE) project (AR-12652) is currently conducting a comprehensive and consistent reprocessing of HST-NICMOS coronagraphic survey data to search for point sources and disks using advanced PSF subtraction. The KLIP algorithm (Karhunen-Loève Image Projection) was developed for this project, and has proven very effective at processing the hundreds of selected archival images. This project has already been very successful with numerous detections of previously unseen point sources and several resolved debris disks that we are currently following up by multiple avenues. We give an overview of the project including preliminary scientific results with companion candidates and improved images of known disks

Keywords. techniques: image processing, planetary systems

1. Karhunen-Loève Image Projection (KLIP) algorithm

The KLIP algorithm was developed for the ALICE project. It uses a Karhunen-Loève (KL) decomposition of the library of reference PSFs, and generates a model star PSF from the projection of the target on the KL vectors (Soummer *et al.* 2012). KLIP is a Principal Component Analysis (PCA) method and is very similar to LOCI (Lafreniere *et al.* 2007). The main advantages of KLIP over LOCI is the possibility of direct forward modeling and significant speed increase.

2. KLIP versus Classical subtractions on NICMOS disks

KLIP also provides dramatically superior sensitivity to circumstellar disks for archival NIC2 data compared to classical PSF subtraction, as shown in Fig. 1 with the debris disk HD 181327 (top) and the young transitional Herbig Ae system HD141569 (bottom). The right panels show the KLIP reprocessing of the same data. Note that for HD141569 the KLIP processing only included the central portion of the field of view. For both targets, KLIP substantially reduces PSF subtraction residuals at small separations, enabling the inner clearings to be seen much better.

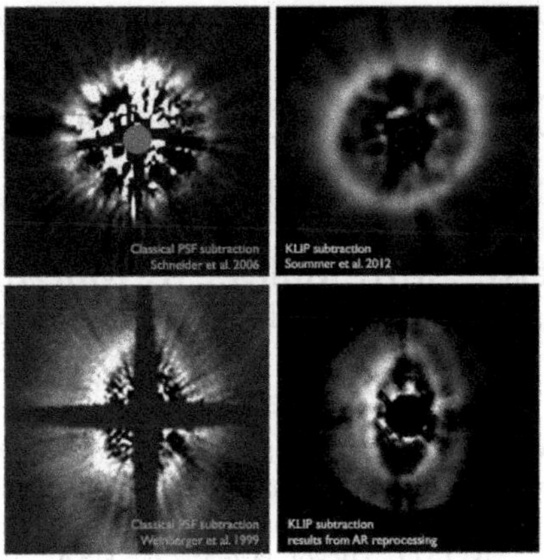

Figure 1. HD181327 (top) and HD141569 (bottom) processed with classical PSF subtraction (left panels) and with the KLIP algorithm (right panels).

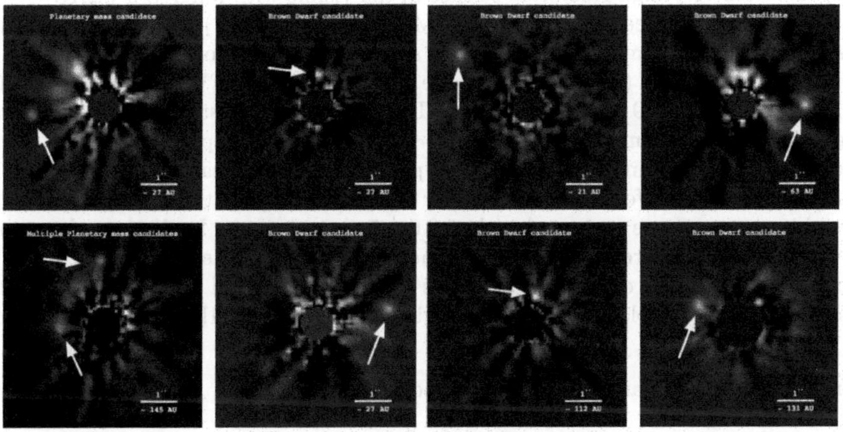

Figure 2. Companion candidates discovered in the NIC2 archives using the KLIP algorithm.

3. Candidate point sources in the NICMOS coronagraphic archive

Previously unseen point-source candidates are now revealed by our reprocessing of NIC2 coronagraphic data with the KLIP algorithm (see Fig. 2). These sources have absolute J magnitudes from 10-17 at separations of 1-3 arcsec with age range 10-100 MYr. This makes these candidates consistent with planet to brown dwarf regimes. Statistical analyses of background source counts indicate that it is highly unlikely for all of these to be background stars.

Acknowledgements

Support for Program No. HST-AR-12652.01-A was provided by NASA through a grant from the Space Telescope Science Institute, which is operated by the Association of Universities for Research in Astronomy, Incorporated, under NASA contract NAS5-26555.

References

Lafreniere, D., Marois, C., Doyon, R., Nadeau, D., & Artigau, E. 2007, *ApJ*, 660, 770
Schneider, G., Silverstone, M. D., Hines, D. C., *et al.* 2006, *ApJ*, 650, 414
Soummer, R., Pueyo, L., & Larkin, J. 2012, *ApJ*, 755L, 28
Weinberger, A. J., Becklin, E. E., Schneider, G., *et al.* 1999, *ApJ*, 525, L53

Exploring the Formation and Evolution of Planetary Systems
Proceedings IAU Symposium No. 299, 2013
M. Booth, B. C. Matthews & J. R. Graham, eds.

doi:10.1017/S1743921313007734

Visible AO Observations at Halpha for Accreting Young Planets

L.M. Close[1], K. Follette[1], J.R. Males[1], K. Morzinski[1], T.J. Rodigas[1], P. Hinz[1], Y-L. Wu[1], D. Apai[1], J. Najita[4], A. Puglisi[2], S. Esposito[2], A. Riccardi[2], V. Bailey[1], M. Xompero[2], R. Briguglio[2] and A. Weinberger[3]

[1]Steward Observatory, University of Arizona, Tucson, AZ 85721, USA
[2]INAF - Osservatorio Astrofisico di Arcetri, I-50125, Firenze, Italy
[3]Carnegie DTM, Washingtion DC. USA
[4]NOAO, Tucson AZ 85721, USA

Abstract. We utilized the new high-order (250-378 mode) Magellan Adaptive Optics system (MagAO) to obtain very high-resolution science in the visible with MagAO's VisAO CCD camera. In the good-median seeing conditions of Magellan (0.5 − 0.7") we find MagAO delivers individual short exposure images as good as 19 mas optical resolution. Due to telescope vibrations, long exposure (60s) r' (0.63μm) images are slightly coarser at FWHM = 23-29 mas (Strehl $\sim$ 28%) with bright ($R < 9$ mag) guide stars. These are the highest resolution filled-aperture images published to date. Images of the young ($\sim$ 1 Myr) Orion Trapezium θ^1 Ori A, B, and C cluster members were obtained with VisAO. In particular, the 32 mas binary θ^1 Ori C_1C_2 was easily resolved in non-interferometric images for the first time. Relative positions of the bright trapezium binary stars were measured with $\sim 0.6 - 5$ mas accuracy. In the second commissioning run we were able to correct 378 modes and achieved good contrasts (Strehl> 20% on young transition disks at $H\alpha$). We discuss the contrasts achieved at $H\alpha$ and the possibility of detecting low mass ($\sim 1 - 5$ Mjup) planets (past 5AU) with our new SAPPHIRES survey with MagAO at $H\alpha$.

Keywords. Extra-solar Planets, Adaptive Optics

---

1. Introduction to Visible AO and the Magellan AO System

To reach the maximum resolution (and use important recombination lines) with modern telescopes one must work blueward of the NIR in the visible. So called "visible AO" can theoretically reach 19 mas resolutions on an 6.5m telescope at 0.656 μm ($H\alpha$). We note that AO with "lucky" imaging in the visible has been successfully used at the somewhat smaller 5m Palomar (Law *et al.* 2009) and has reached resolutions of 35 mas. In the near future some polarization work will be done in the visible with the 8m VLT with the SPHERE AO system and ZIMPOL (Bazzon *et al.* 2012). Yet, MagAO is the first large ($D \geqslant 6.5$ m) telescope AO system designed to work in the visible –complete with a facility CCD AO science camera (VisAO). The MagAO commissioning results presented here inform us on the utility of large telescope visible AO performance.

We have developed an AO system (inspired, in large part, by LBT's FLAO system; Esposito *et al.* (2012)) that can reach 20 mas resolutions with just 250-378 modes at 1 kHz sampling speeds.

MagAO, with its VisAO camera†, is the first large telescope ($\geqslant$ 6.5m) facility AO system deployed that is targeting science in the visible (0.6-1.1 μm). As is shown in

† see *http://visao.as.arizona.edu/* for web resources, SPIE publications, and guides for observers

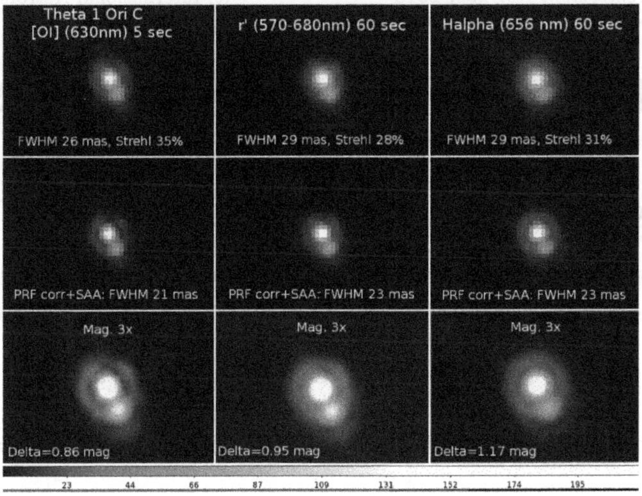

Figure 1. *Top row:* the central ionizing binary of the Trapezium: θ^1 Ori C as imaged with MagAO's VisAO CCD camera in different filters. Note the excellent resolution in the raw 60 second image. We note that no post-detection shift and add (SAA) was applied, nor was there any frame selection used to produce these top row images. Typically we achieved resolutions of $0.026 - 0.029$" and Strehls of 28-35% in $0.5 - 0.7$" V-band seeing. *Middle row:* the same data as the top row, except the images have been post-detection aligned (SAA) and the pixel response function (PRF) has been removed. This improved image resolution by $\sim 5-6$ mas. *Bottom row:* the row above is magnified by $3\times$ to better display the data of the middle row. These are the highest resolution, deep, images ever obtained to our knowledge. Figure reproduced from Close *et al.* (2013a).

figure 1, MagAO (at first light) produced long exposure (60s) diffraction-limited (110 nm WFE; 28% Strehl) $0.63\mu m$ images. For more technical details about MagAO itself please see Close *et al.* (2012a).

2. Contrasts at $H\alpha$ and the SAPPHIRES $H\alpha$ Survey

We estimate the contrasts achieved ($\Delta H\alpha = 9.5$ mag at 70 mas) at $H\alpha$ from almost 2 hours of open shutter observations of a young Tauri star. The data were reduced with an ADI reduction. See Close *et al.* (2013b) for more details of these contrasts.

We have started a **S**urvey for **A**ccreting **P**roto **P**lanets at **H**alpha and in **IR E**mission in transitional disk**S**: SAPPHIRES. This will be a survey in both $H\alpha$ and simultaneously in the L' (3.8 microns) band. In deep 2 hour on-source images we expect to detect ~ 7 accreting planets in the gaps of 24 nearby transitional disks. Again please see Close *et al.* (2013b) for more details about SAPPHIRES.

References

Bazzon, Andreas *et al.* (2012) Proceedings of the SPIE, Volume 8446, id. 844693-844693 14

Close, L. M. *et al.* 2012a Proc. SPIE, Volume 8447, id. 84470X-84470X-16

Close, L. M. *et al.* 2012b ApJ, 749, article id. 180.

Close, L. M. *et al.* 2013 ApJ, 774, 94

Close, L. M. *et al.* 2013 ApJ Lett, submitted

Esposito, S. *et al.* 2012 proc. SPIE Volume 8447, article id. 84470U

Law, N. M.; Mackay, C. D.; Dekany, R. G.; Ireland, M.; Lloyd, J. P.; Moore, A. M.; Robertson, J. G.; Tuthill, P.; Woodruff, H. C. 2009, ApJ 692, 924.

Exploring the Formation and Evolution of Planetary Systems
Proceedings IAU Symposium No. 299, 2013
M. Booth, B. C. Matthews & J. R. Graham, eds.

doi:10.1017/S1743921313007746

SCExAO: First Results and On-Sky Performance

Thayne Currie[1], Olivier Guyon[2], Frantz Martinache[2], Christophe Clergeon[2], Michael McElwain[3], Christian Thalmann[4], Nemanja Jovanovic[2], Garima Singh[2], and Tomoyuki Kudo[2]

[1]Department of Astronomy and Astrophysics, University of Toronto
email: **currie@astro.utoronto.ca**

[2]National Astronomical Observatory of Japan

[3]NASA-Goddard Space Flight Center

[4]Astronomical Institute "Anton Pannekoek," University of Amsterdam

Abstract. We present new on-sky results for the Subaru Coronagraphic Extreme Adaptive Optics imager (SCExAO) verifying and quantifying the contrast gain enabled by key components: the closed-loop coronagraphic low-order wavefront sensor (CLOWFS) and focal plane wavefront control ("speckle nulling"). SCExAO will soon be coupled with a high-order, Pyramid wavefront sensor which will yield > 90% Strehl ratio and enable 10^6–10^7 contrast at small angular separations allowing us to image gas giant planets at solar system scales. Upcoming instruments like VAMPIRES, FIRST, and CHARIS will expand SCExAO's science capabilities.

Keywords. instrumentation: adaptive optics, detectors, stars: planetary systems

1. Introduction

SCExAO is a next-generation extreme-AO platform designed to image and characterize the spectra of self-luminous jovian planets at solar system-like scales around nearby, young stars (Martinache and Guyon 2009). Components comprising SCExAO are being commissioned in several phases. Already in Phase 1 we have verified the performance of the *Phase Induced Amplitude Apodization* (PIAA) coronagraph, which improves our ability to image planets at $\approx 2\ \lambda/D$ and with full efficiency by better suppressing diffraction at the telescope pupil edge (Martinache *et al.* 2012a). Here, we describe on-sky verification of other key "Phase 1" components: the CLOWFS (Guyon *et al.* 2011), which corrects for tip-tilt errors to yield $\approx 10^{-3}\ \lambda/D$ pointing accuracy, and "speckle nulling", which uses a deformable mirror (DM) to cancel out static speckles limiting planet detections at small separations. Finally, we describe the timeline for implementing SCExAO's high-order wavefront control and new instrumentation.

2. Phase 1 Engineering Observations

On November 14, 2012, we observed Pollux with SCExAO combined with the facility adaptive optics system (AO-188) and HiCIAO to test and verify the on-sky performance of the CLOWFS and speckle nulling components of SCExAO. After a close-loop sequence of speckle nulling, using images acquired by the SCExAO internal science camera (XS), we obtained a sequence of images acquired with HiCIAO alternating between the DM flat volt-map and the DM speckle nulling volt-map (e.g. Martinache *et al.* 2012b).

Figure 1 displays the combined image obtained without (left) and with (right) speckle nulling. Despite the XS's high read noise, the speckle nulling algorithm successfully (blindly) calibrated the brightest diffraction features of the long exposure image. For individual images, contrast gain is moderate ($\sim 2\times$ for the brightest part of the first

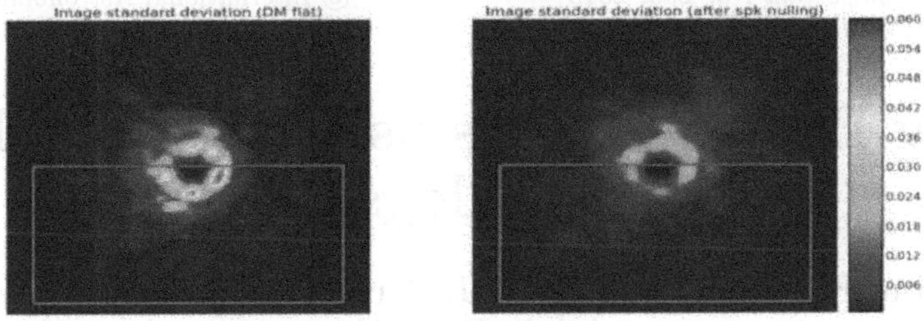

Figure 1. *SCExAO noise map of the combined image without (left) and with speckle nulling.*

Figure 2. *SCExAO as configured in Phase 1 (left) and Phase 2 (and beyond) (right).*

ring), but real and persistent. Speckle nulling reduces the combined image's standard deviation over the entire frame by ~ one-half. Inside the control region (boxed region), speckle nulling lowers the image standard deviation by about a factor of ~ 7. Thus, at the smallest angular separations (2-4 λ/D), speckle nulling alone improves our detection limits by ≈ 1-2 magnitudes (Martinache *et al.* 2013, in prep.).

3. Current Timeline and Future Instrumentation

SCExAO has now passed its Phase 1 requirements. Phase 2 of SCExAO commissioning in Summer/Fall 2013 will incorporate the 2000-actuator DM controlled by a high-order Pyramid wavefront sensor, which will yield > 90% Strehl at 1.6 μm (Figure 2). Combined with *PIAA*; *CLOWFS*; "speckle nulling" calibrated from a faster, lower noise internal camera; and *angular differential imaging* plus aggressive post-processing (Marois *et al.* 2006, Currie *et al.* 2012); SCExAO should then achieve 10^6–10^7 contrasts at ≈ 0.1–0.5" and thus image numerous young jovian planets at solar system-like scales. Interferometric instruments VAMPIRES and FIRST capable of studying planet formation at even smaller scales will be commissioned at the same time. Finally, CHARIS will eventually replace HiCIAO as the primary SCExAO high-contrast instrument and yield near-IR spectra of young planets (McElwain *et al.* 2013) to probe the planets' atmospheres.

References

Currie, T., *et al.,* 2012, *Ap. Lett.*, 760, L32
Guyon, O., *et al.,* 2011, in Proc. *SPIE*, Vol. 8149
Marois, C., *et al.,* 2006, *Ap. J*, 641, 556
Martinache, F., *et al.,* 2009, in Proc. *SPIE*, Vol. 7440
Martinache, F., *et al.,* 2012a, in Proc. *SPIE*, Vol. 8447
Martinache, F., *et al.,* 2012b, *PASP*, 124, 1288
McElwain, M., *et al.,* 2012, in Proc. *SPIE*, Vol. 8446

Exploring the Formation and Evolution of Planetary Systems
Proceedings IAU Symposium No. 299, 2013
M. Booth, B. C. Matthews & J. R. Graham, eds.

doi:10.1017/S1743921313007758

Young Brown Dwarfs as Giant Exoplanet Analogs

Jacqueline K. Faherty[1,2], Kelle L. Cruz[2,4], Emily L. Rice[2,3] and Adric Riedel[2,4]

[1]Department of Astronomy, Universidad de Chile Cerro Calan, Las Condes

[2]Department of Astrophysics, American Museum of Natural History, Central Park West at 79th Street, New York, NY 10023

[3] Department of Engineering Science & Physics, College of Staten Island, 2800 Victory Blvd., Staten Island, NY 10301 USA

[4] Department of Physics & Astronomy, Hunter College, 695 Park Avenue, New York, NY 10065, USA

Abstract. Young brown dwarfs and directly-imaged exoplanets have enticingly similar photometric and spectroscopic characteristics, indicating that their cool, low gravity atmospheres should be studied in concert. Similarities between the peculiar shaped H band, near and mid-IR photometry as well as location on color magnitude diagrams provide important clues about how to extract physical properties of planets from current brown dwarf observations. In this proceeding we discuss systems newly assigned to 10-150 Myr nearby moving groups, highlight the diversity of this uniform age-calibrated brown dwarf sample, and reflect on their implication for understanding current and future planetary data.

Keywords. Astrometry– stars: low-mass– brown dwarfs

1. Introduction

Despite different formation mechanisms, brown dwarfs and giant exoplanets share many physical properties, including overlapping temperature regimes and condensate clouds in their atmospheres. Recent studies have revealed a striking resemblance between observations of directly imaged giant exoplanets and young low-temperature brown dwarfs (e.g. Faherty *et al.* 2013). Both populations deviate significantly from older, equivalent temperature objects, and it has been proposed that thick clouds present in the young objects but not in the old ones could explain anomalous observables (Barman *et al.* 2011, Madhusudhan *et al.* 2011). While only a handful of planetary systems can be directly studied with current technology, young brown dwarfs are relatively numerous, bright, and isolated in the field. As such, they are readily accessible to ground-based facilities and lend themselves to detailed, extensive studies not currently possible for exoplanets (Cruz *et al.* 2009, Rice *et al.* 2010ab, Faherty *et al.* 2012, 2013, Allers *et al.* 2013).

2. Assigning Ages to Low-Gravity Brown Dwarfs

We have identified 65 low-surface gravity brown dwarfs with spectroscopic, photometric, luminosity, and/or kinematic properties that make them excellent candidates for an age calibrated sample. Many of these sources are coincident with the locations of individual members of nearby moving groups (e.g. Argus–30-50 Myr, β Pictoris–12-22 Myr, Tucana Horologium–10-40 Myr, AB Doradus–50-120 Myr; Malo *et al.* 2013). To conclusively assign membership, we require precise astrometric measurements. Consequently, as part of the Brown Dwarf Kinematics Project (BDKP–Faherty *et al.* 2009, 2010, 2011, 2012), we have prioritized collecting parallax, proper motion and radial velocity data on the low-surface gravity brown dwarfs. We have enough information on our sources to compute space velocity, positions, and preliminary membership in nearby moving groups.

3. Diversity of Young Brown Dwarfs

Using estimated UVW velocities and XYZ positions in combination with a convergent point and Bayesian analysis (Rodriguez *et al.* 2013) we deduce membership for 30 low-surface gravity brown dwarfs. We assign two sources to Argus, four to AB Doradus, 12 to β Pictoris and 12 to Tucana Horologium. We see a diversity among the spectral gravity classifications within each group. For instance, β Pictoris has eight γ (very low gravity) and four β (intermediate gravity) sources. The diversity that is emerging among equal age sources is not surprising. Faherty *et al.* 2013 discuss the optical γ sources and find that this subpopulation shows a large range in the extent of red photometric color (in both the NIR and the MIR). For those sources with parallax measurements we find that the M dwarfs are overluminous in the NIR whereas the L dwarfs are normal to underluminous in the NIR regardless of the age calibration. We postulate that this diversity among the age calibrated sample is due to atmospheric chemistry which can differ between sources.

4. Discussion: Connection to Exoplanets

The estimated T_{eff} of the directly imaged planetary-mass companions 2M1207b (TW Hydrae ~ 8-20 Myr; Chauvin *et al.* 2004) and HR8799b (Columba ~ 10-40 Myr; Marois *et al.* 2010) are ~1100K and 1600K, respectively, corresponding to mid L and early T spectral types (e.g. Barman *et al.* 2011). Detailed studies of their near-IR spectra and photometric data reveal that both planets (1) are 1-2 mag underluminous on color-magnitude diagrams, (2) have unusually red near-IR colors, and (3) display sharply peaked H-band spectra. Our population of 30 brown dwarfs, now kinematically associated with 10-150 Myr groups share in these deviations from field brown dwarfs. Emerging as the most probable explanation for these features are enhanced clouds induced by gravity effects. Understanding in detail the variations in observable features of the plethora of data available for young brown dwarfs that can be attributed to gravity and/or atmosphere features is critical for interpreting future exoplanet data.

References

Allers, K. N. & Liu, M. C. 2013, *ApJ*, 772, 79A
Barman, T. S., Macintosh, B., Konopacky, Q. M., & Marois, C. 2011, *ApJ*, 733, 65
Chauvin, G., Lagrange, A.-M., Dumas, C., *et al.*, P. 2004, *A&A*, 425, L29
Cruz, K. L., Kirkpatrick, J. D., & Burgasser, A. J. 2009, *AJ*, 137, 3345
Faherty, J. K., Burgasser, A. J., Bochanski, J. J., *et al.* 2011, *AJ*, 141, 71
Faherty, J. K., Burgasser, A. J., Cruz, K. L., *et al.* 2009, *AJ*, 137, 1
Faherty, J. K., Burgasser, A. J., Walter, F. M., *et al.* 2012,*ApJ*, 752, 56
Faherty, J. K., Burgasser, A. J., West, A. A., *et al.* 2010, *AJ*, 139, 176
Faherty, J. K., Rice, E. L., Cruz, K. L., Mamajek, E. E., & Núñez, A. 2013, *AJ*, 145, 2
Malo, L., Doyon, R., Lafrenière, D., *et al.* 2013, *ApJ*, 762, 88
Madhusudhan, N., Burrows, A., & Currie, T. 2011,*ApJ*, 737, 34
Marois, C., Macintosh, B., Barman, T., *et al.*, 2008, Science, 322, 1348
Rice, E. L., Barman, T., Mclean, I. S., Prato, L., & Kirkpatrick, J. D. 2010a,*ApJs*, 186, 63
Rice, E. L., Faherty, J. K., & Cruz, K. L. 2010b,*ApJl*, 715, L165
Rodriguez, D. R.,Zuckerman, B., Kastner, J. H., *et al.*, 2013, *ApJ*, 774, 101

Exploring the Formation and Evolution of Planetary Systems
Proceedings IAU Symposium No. 299, 2013
M. Booth, B. C. Matthews & J. R. Graham, eds.

doi:10.1017/S174392131300776X

GRAPHIC: The Geneva Reduction and Analysis Pipeline for High-contrast Imaging of planetary Companions

Janis Hagelberg, Damien Ségransan, Stéphane Udry, and François Wildi

Observatoire de Genève, University of Geneva,
51 Chemin des Maillettes, 1290, Versoix, Switzerland
email: janis.hagelberg@unige.ch

Abstract. We present a new analysis and reduction pipeline for the detection of planetary companions using Angular Differential Imaging. The pipeline uses Fourier transforms for image shifting and rotation in order to achieve very low signal loss. Furthermore it is parallelised in order to run on computer clusters of up to 1024 cores. The pipeline was developed in Geneva for the ongoing direct imaging campaign for stars with radial velocity drifts in the HARPS and CORALIE radial-velocity planet-search surveys. In addition to that, a disk mode has been implemented in the context of observations of the protoplanetary disk around HD142527.

Keywords. instrumentation: adaptive optics, techniques: high angular resolution, techniques: image processing, methods: data analysis, planetary systems, stars: low-mass, brown dwarfs

1. Introduction

Direct imaging techniques based on Angular Differential Imaging (ADI) (Marois *et al.* 2006) which use field rotation in order to subtract the stellar point-spread-function, are currently the techniques achieving the highest contrast detections at small separations. ADI relies on shifts of well aligned frames and image derotation to correct for the field rotation. We developed a pipeline called GRAPHIC (Hagelberg *et al.* in prep) based on Fourier transforms to achieve high fidelity image transformations in order to push detections limits of ADI even further.

2. ADI using Fourier transforms

The main advantage of Fourier-based image processing over the conventional direct interpolation based methods is that interpolation is done on spatial frequencies. As a result the original signal isn't changed and only some additional very high frequency noise gets injected: conversely spline interpolation algorithms act as low-pass filters spreading out the noise and putative companions signal.

The two main hurdles to overcome in order to use Fourier-based image processing is on one hand the additional computation time linked to the intensive use of FFTs and on the other hand the Gibbs oscillations caused by sharp image features such as bad pixels or image borders. The processing time can drastically be reduced thanks to massive parallelisation, however removing sharp image features whitout loosing signal needs extra care.

Using the translation property of Fourier transforms, we shift each frame by applying a two-dimensional phasor. This way, to recenter a frame by a sub-pixel step a in x and b in y, we apply a phasor of the form $e^{-i2\pi(\nu_x \cdot a+\nu_y \cdot b)}$ to the Fourier transformed image.

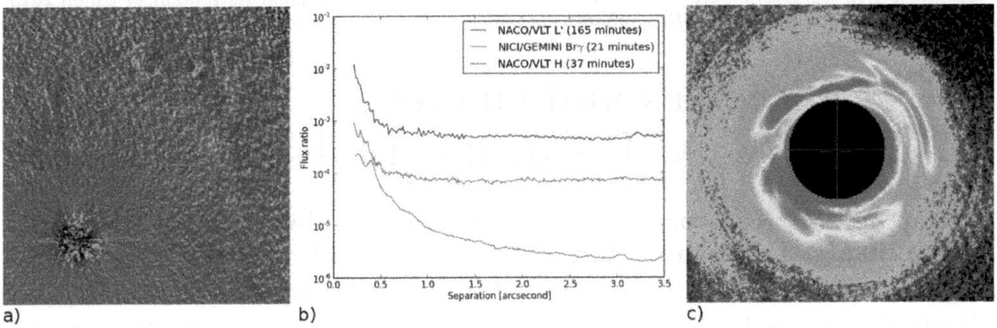

Figure 1. Applications of GRAPHIC: a) Direct detection of companion candidates revealed by radial velocity trends in the HARPS and CORALIE data. b) 5σ detection limits for various instruments, filters, and total integration time combinations. c) NICI/GEMINI Ks band imaging of the protoplanetary disk around HD142527 reduced using the prototype GRAPHIC pipeline Casassus *et al.* (2013).

The Fourier-based rotation is implemented using the three shears algorithm described by Larkin *et al.* (1997) and Unser *et al.* (1995). It is based on the property that a rotation matrix can be decomposed into a product of three shear matrices.

3. Applications of GRAPHIC

The pipeline was primarily developed to reduce data from the ongoing L' band ADI campaign on NACO/VLT for stars with radial velocity drifts in the HARPS and CORALIE radial-velocity planet-search surveys.

GRAPHIC was also applied to some data from the coordinated European Large Program using H band ADI with NACO which aims to provide a comprehensive and statistically significant study of the occurrence of extrasolar giant planets and brown dwarfs at large (5-500 AU) orbital separation around 150 young, nearby stars (Zurlo *et al.* 2013).

The previous version of the pipeline (PADIP) was used to reduce NICI/GEMINI data of the protoplanetary disk around HD142527, and to search for companions within the protoplanetary gap (Casassus *et al.* 2013). In this framework a disk mode has been developed to minimise the self-subtraction artefacts which tend to appear when using ADI for disk imaging (Milli *et al.* 2012). Performance comparison with other pipeline and further details on the disk mode are given in Casassus *et al.* 2013.

References

Casassus, S., Plas, G. van der, M, S. P., Dent, W. R.F., Fomalont, E., Hagelberg, J., Hales, A., Jordán, A., Mawet, D., Ménard, F., Wootten, A., Wilner, D., Hughes, A. M., Schreiber, M. R., Girard, J. H., Ercolano, B., Canovas, H., Román, P. E., & Salinas, V. 2013, *Nature*, 493, 191

Hagelberg J., Ségransan D., Udry S., Wildi F., *in prep.*

Larkin, K. G., Oldfield, M. A., & Klemm, H. 1997, *Opt. Commun.*, 139, 99

Marois, C., Lafrenière, D., Doyon, R., Macintosh, B., & Nadeau, D., 2006 *ApJ*, 641, 556

Milli, J., Mouillet, D., Lagrange, A.-M., Boccaletti, A., Mawet, D., Chauvin, G., Bonnefoy, M. 2012, *A&A*, 545, 111

Unser, M., Thevenaz, P., Yaroslavsky, L. , 1995 *IEEE IP*, 4, 1371

Zurlo, A., Vigan, A., Hagelberg, J., Desidera, S., Chauvin, G., Almenara, J. M., Biazzo, K., Bonnefoy, M., Carson, J. C., Covino, E., Delorme, P., D'Orazi, V., Gratton, R., Mesa, D., Messina, S., Moutou, C., Ségransan, D., Turatto, M., Udry, S., & Wildi, F., 2013 *A&A*, 554, 21

Exploring the Formation and Evolution of Planetary Systems
Proceedings IAU Symposium No. 299, 2013
M. Booth, B. C. Matthews & J. R. Graham, eds.

doi:10.1017/S1743921313007771

Successes and challenges of the APP Coronagraph

Matthew A. Kenworthy[1], Sascha Quanz[2], Gilles Otten[1], Tiffany Meshkat[1], Johanan Codona[3], Frans Snik[1], Michael E. Meyer[2], Markus Kasper[4] and Julien Girard[5]

[1]Leiden Observatory, Leiden University, P.O. Box 9513, 2300 RA Leiden, The Netherlands
email: kenworthy@strw.leidenuniv.nl

[2]Institute for Astronomy, ETH Zurich, CH-8093 Zurich, Switzerland

[3]Steward Observatory, 933 N. Cherry Avenue, Tucson, AZ 85721

[4]European Southern Observatory, Garching bei Munchen, Germany

[5]European Southern Observatory, Vitacura, Cassilla 19001, Santiago, Chile

Abstract.
The Apodizing Phase Plate (APP) coronagraph has been used to image the exoplanet β Pictoris b and the protoplanet candidate around HD 100546, and is currently in use in surveys with NaCo at the VLT. Its success is due to its tolerance to tip-tilt pointing errors in current AO systems, which degrade the performance of nearly all other coronagraphs. Currently the sensitivity of the APP is limited by non-common path errors in the science camera systems and by its chromatic behaviour. We present the achromatized Vector APP coronagraph and address how we will measure and minimise non-common path errors with Focal Plane Wavefront Sensing algorithms.

Keywords. instrumentation: high angular resolution, stars: planetary systems, stars: individual (β Pic, HD 100546)

1. Introduction

Coronagraphs are used to suppress the diffraction halo around bright stars, revealing the presence of faint companions (brown dwarfs and planets) next to their primary star. There are many coronagraph designs that work at small inner working angles (Guyon *et al.* 2006) and continuing technological developments are ongoing (for a recent review see Mawet *et al.* 2012).

The Apodizing Phase Plate (APP) Coronagraph is a pupil-plane optic that is a realization of Phase Apodization Coronagraphy (Codona & Angel 2004; Codona *et al.* 2006). In common with other pupil plane reformatters (Carlotti *et al.* 2012; Carlotti 2013), it is therefore relatively insensitive to residual tip-tilt motions in the focal plane of the science camera from residual atmospheric correction and from vibrations in the telescope superstructure. A first light demonstrator on the MMTO 6.5m telescope at $\lambda = 4.85\mu m$ (Kenworthy *et al.* 2007), was followed by an L' band APP on the Very Large Telescope (Kenworthy *et al.* 2010a,c; Girard *et al.* 2010) and most recently at L' band on the Large Binocular Telescope in LMIRCam (Wilson *et al.* 2008; Kenworthy *et al.* 2010b; Leisenring *et al.* 2012). Characterizations of planets include β Pictoris b (Quanz *et al.* 2010) and the discovery of a low mass companion to HD 100546 (Quanz *et al.* 2013), along with its inclusion in several VLT surveys for extrasolar planets around nearby stars with debris disks (Meshkat *et al.* 2013) and A/F stars (Quanz *et al.* 2011).

2. Current Performance and Future Directions

APP observations have placed strong constraints on low mass objects at Solar System scales around Fomalhaut (Kenworthy *et al.* 2013), typically with contrasts of 12 mag at 0.4 arcseconds. The refinement of Principal Component Analysis (PCA) techniques (Amara & Quanz 2012; Soummer *et al.* 2012) lead to an additional ~ 0.5 mag of sensitivity (see Meshkat *et al.* these proceedings; Meshkat *et al.* (2013) submitted).

The performance of the APP coronagraph is limited by (i) the non-common path aberrations (NCPA) between the AO camera and the wavefront sensor path in the science instrument, and (ii) its chromatic design. With the APP we will use Phase Sorting Interferometry (Codona & Kenworthy 2013) which uses phase diversity from the AO-corrected atmosphere to sense NCPA. Another development is the Vector APP (VAPP), an implementation of the APP that uses multiple stacked layers of polarization phase plates to provide a broadband wavelength solution (Snik *et al.* 2012) that also provides two simultaneous and complementary PSFs with the dark D region on opposing sides around the star. New versions of this optic are currently being characterized at Leiden Observatory, and the APP is being considered as the baseline coronagraph for ERIS, the new thermal infrared imager for the VLT, and for METIS, the thermal infrared imager and spectrograph for the European ELT.

References

Amara A. & Quanz S. P., 2012, MNRAS, 427, 948

Carlotti A., 2013, A&A, 551, A10

Carlotti A., Kasdin N. J., Vanderbei R. J., & Delorme J.-R., 2012, SPIE 8442.

Codona J. L. & Angel R., 2004, ApJ, 604, L117

Codona J. L. & Kenworthy M., 2013, ApJ, 767, 100

Codona J. L., Kenworthy M. A., Hinz P. M., Angel J. R. P., & Woolf N. J., 2006, SPIE 6269, pp. 62691N

Girard J. H. V., *et al.,* 2010, SPIE 7736

Guyon O., Pluzhnik E. A., Kuchner M. J., Collins B., & Ridgway S. T., 2006, ApJS, 167, 81

Kenworthy M., Quanz S., Meyer M., Kasper M., Girard J., Lenzen R., Codona J., & Hinz P., 2010a, The Messenger, 141, 2

Kenworthy M. A., Codona J. L., Hinz P. M., Angel J. R. P., Heinze A., & Sivanandam S., 2007, ApJ, 660, 762

Kenworthy M. A., Hinz P. M., Codona J. L., Wilson J. C., Skrutskie M. F., & Solheid E., 2010b, SPIE 7734

Kenworthy M. A., Meshkat T., Quanz S. P., Girard J. H., Meyer M. R., & Kasper M., 2013, ApJ, 764, 7

Kenworthy M. A., Quanz S. P., Meyer M. R., Kasper M. E., Lenzen R., Codona J. L., Girard J. H., & Hinz P. M., 2010c, SPIE 7735

Leisenring J. M., *et al.,* 2012, SPIE 8446

Mawet D., *et al.,* 2012, SPIE 8442

Meshkat T., Kenworthy M. A., Su K. Y., & Bailey V. P., 2013, AAS 221, 149.28

Quanz S. P., Amara A., Meyer M. R., Kenworthy M. A., Kasper M., & Girard J. H., 2013, ApJ, 766, L1

Quanz S. P., Kenworthy M. A., Meyer M. R., Girard J. H. V., & Kasper M., 2011, ApJ, 736, L32

Quanz S. P., Meyer M. R., Kenworthy M. A., Girard J. H. V., Kasper M., Lagrange A.-M., Apai D., Boccaletti A., Bonnefoy M., Chauvin G., Hinz P. M., & Lenzen R., 2010, ApJ, 722, L49

Snik F., Otten G., Kenworthy M., Miskiewicz M., Escuti M., Packham C., Codona J., 2012, SPIE 8450

Soummer R., Pueyo L., & Larkin J., 2012, ApJ, 755, L28

Wilson J. C., Hinz P. M., Skrutskie M. F., Jones T., Solheid E., Leisenring J., Garnavich P., Kenworthy M., Nelson M. J., & Woodward C. E. 2008, SPIE 7013

Exploring the Formation and Evolution of Planetary Systems
Proceedings IAU Symposium No. 299, 2013
M. Booth, B. C. Matthews & J. R. Graham, eds.

doi:10.1017/S1743921313007783

Estimation of Contaminants for Direct Imaging of Exoplanets: Constraint on the Stellar Distribution Model with both NIR and Deep Imaging Data

Mihoko Konishi[1], Hiroshi Shibai[1], Taro Matsuo[2], Kodai Yamamoto[1], Jun Sudo[1], Matthias S. Samland[1], Misato Fukagawa[1], Takahiro Sumi[1] and SEEDS team

[1]Department of Earth and Space Science, Graduate School of Science, Osaka University, 1-1 Machikaneyama-cho, Toyonaka-shi, Osaka, 560-0043, Japan
email: `konishi@iral.ess.sci.osaka-u.ac.jp`

[2]Department of Astronomy, Faculty of Science, Kyoto University, Kitashirakawa-Oiwake-cho, Sakyo-ku, Kyoto, 606-8502, Japan

Abstract. There are faint contaminants near primary stars in the direct imaging of exoplanets. Our goal is to estimate statistically the ratio of exoplanets in the detected batch of point sources by calculating the fraction of contamination. In this study, we compared the detected number of stars with the number of contaminants predicted by our model. We found that the observed number of faint stars were fewer than the predicted results towards the Pleiades and GOODS-South field when the parameters of the conventional stellar distribution models were employed. We thus estimated new model parameters in correspondence to the results of the observations.

Keywords. infrared: stars - Galaxy: stellar content - Galaxy: fundamental parameters - stars: low-mass, brown dwarfs - stars: luminosity function, mass function

Motivation

The goal of our study is to estimate statistically the ratio of exoplanets out of the detected number of point sources. In other words, we want to know strictly how many stars are contaminating the images. The number of contaminants can be found by using stellar distribution models of the Galaxy. Present stellar distribution models, however, are not constructed using both NIR and deep imaging data. So, we compared the observed number of stars to the number of stars predicted by the stellar distribution models of the Galaxy due to check whether they are consistent or not.

Models

For contaminants, we focused on late-type main-sequence field stars. We employed the stellar distribution models composed of a thin disk, thick disk, and halo to predict the number of contaminants for every sub-spectral type from K2 to T7. According to previous works (Gilmore & Reid 1983; Gould *et al.* 1998; Cabalello *et al.* 2008; Chang *et al.* 2010; Chang *et al.* 2012),

$$Thin\,\&\,Thick\ Disk:\ \mathcal{D}_{i,j}(R,Z) = n_{\odot,i} \times f_{d,j} \times \exp\left[-\frac{R-R_\odot}{h_{R,j}} - \frac{|Z-Z_\odot|}{h_{Z,j}}\right]$$

$$Halo:\ \mathcal{H}_i(R,Z) = n_{\odot,i} \times f_h \times \left[\frac{R^2+(Z/\kappa)^2}{R_\odot^2+(Z_\odot/\kappa)^2}\right]^{\frac{p}{2}},$$

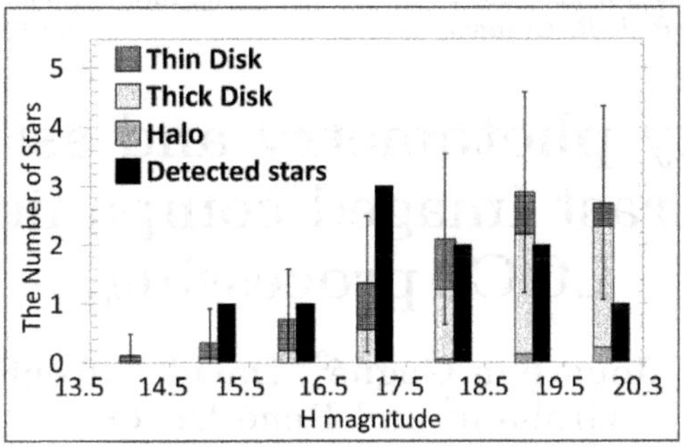

Figure 1. Observed and predicted numbers of contaminants towards the Pleiades open cluster

where j indicates the thin or thick disk, $n_{\odot}$ is the local stellar density, $f_{d,j}$ and f_h indicate the ratio of each component, $h_{R,j}$ and $h_{Z,j}$ indicate the scale height of the disk, κ is the flattening parameter of the halo, p is the power low index of the halo, and $(R_{\odot}, Z_{\odot})$ is the Sun's position. For local stellar densities, we adopted values derived by the previous works (Cabalello *et al.* 2008; Reid *et al.* 2004; Reid *et al.* 2008).

Data

We used observational results of two areas; the Pleiades open cluster (Yamamoto *et al.* 2013) and the GOODS-South field (Windhorst *et al.* 2011). The Pleiades data were obtained with Subaru+HiCIAO/AO188. The FoV was $19''.5 \times 19''.5$. Twenty Pleiades member stars were observed. The median detection limit was 20.3 mag in H-band. The GOODS-South field data were obtained with HST+WFC3. The FoV was $9'.75 \times 4'.5$. The detection limit with F160W filter was 25 mag.

Results and Discussion

Figure 1 shows the observed and predicted numbers of stars with increasing magnitude towards the Pleiades. We found that the detected number of faint stars was different from the number predicted from existing Galactic models. The predicted numbers of faint stars towards the Pleiades and the GOODS-South field were larger than observed numbers by two and about a few thousand, respectively. We thus adjusted parameters of the Galactic models so as to fit them to our observational results. Based on the results of the Pleiades observations, the scale height of the thick disk may be smaller than that previously thought. Also, from the observations of the GOODS-South field, the number of M-type stars in the halo was remarkably smaller than predicted from existing Galactic models.

The stellar distribution of M-type stars in the Galaxy is made clearer by both NIR and deep imaging.

References

Caballero, J. A., Burgasser, A. J., & Klenment, R. 2008, *A&A*, 488, 188
Chang, C.-K., Ko, C.-M., & Peng, T.-H. 2010, *ApJ*, 724, 182
Chang, C.-K., Lai, S.-Y., Ko, C.-M., & Peng, T.-H 2012, *ApJ*, 759, 94
Gilmore, G. & Reid, N. 1983, *MNRAS*, 202, 1025
Gould, A., Flynn, C., & Bahcall, J. N. 1998, *ApJ*, 503, 798
Reid, I. N., Cruz, K. L. *et al.* 2004, *AJ*, 128, 463
Reid, I. N., Cruz, K. L. *et al.* 2008, *AJ*, 136, 1290
Windhorst, R. A., Cohen, S. H. *et al.* 2011, *ApJS*, 193, 27
Yamamoto, K., Matsuo, T. *et al.* 2013, *PASJ*, 65, 90

Exploring the Formation and Evolution of Planetary Systems
Proceedings IAU Symposium No. 299, 2013
M. Booth, B. C. Matthews & J. R. Graham, eds.

doi:10.1017/S1743921313007795

High-fidelity photometry and astrometry of high-contrast imaged companions using LOCI processing.

Jérôme Maire[1], Jonathan Gagné[2], David Lafrenière[2], James R. Graham[3] and René Doyon[2]

[1]Dunlap Institute for Astronomy and Astrophysics, University of Toronto, 50 St. George Street, Toronto, Ontario. M5S 3H4, Canada
email: maire@dunlap.utoronto.ca

[2] Université de Montréal, Département de physique C.P.6128, Succ. Centre-Ville H3C 3J7, Montréal, Canada

[3] Astronomy Department, University of California, Berkeley, CA 94720

Résumé. Direct imaging and spectroscopy of exoplanets is a key element for understanding planet formation and migration. Such direct detections and characterizations remains technologically challenging, since a very high contrast ratio and small angular separation are involved, and futhermore speckle noise limits the high-contrast imaging performance. We further discuss a speckle subtraction and suppression technique that fully takes advantage of spectral and time-domain information on quasi-static speckles to measure the highest-fidelity photometry as well as accurate astrometry of detected companions.

Keywords. techniques : image processing, techniques : high angular resolution, instrumentation : spectrographs

1. Introduction

We propose an extension of the LOCI speckle-suppression algorithm, that simultaneously reduces the contribution of speckles in the reduced images and delivers accurate photometry and astrometry of the companions. This method preserves the companion signal by including its presence in images during the speckle noise suppression. First, it preserves companion brightness during the speckle suppression and delivers accurate photometry and astrometry without any aperture-dependent photometry on residual images. The proposed method is also very fast in terms of computing time as it needs only a single optimization in a limited region. Furthermore, this method is generalized to spectral datacube as delivered by an IFS, taking advantage of both ADI and SDI processing, giving powerful speckle suppression.

Spectral Differential Imaging (SDI, Marois *et al.* 2000) or Angular Differential Imaging (ADI) algorithm like Locally Optimized Combination of Images (LOCI, Lafrenière *et al.* 2007) are very efficient for suppressing the speckle noise in high-contrast images, improving the effective contrast ratio and companion detectability. However, the resultant speckle suppression is not ideal for high-fidelity companion photometry, as the signal of the companion itself is affected by these methods which focus on the speckle noise suppression.

2. Data processing

The LOCI-PSF method (Maire *et al.* 2012) have been tested by using simulated data in which we injected fake planets of known brightness. We applied the LOCI-PSF algorithm

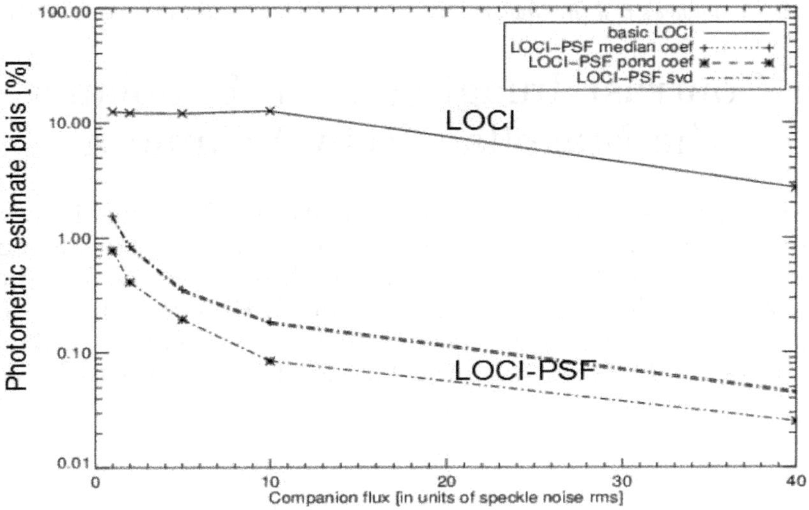

Figure 1. Relative difference between injected and measured companion brightness as a function of companion brightness (temporal sequence of 40 ADI images, with photon noise and varying speckle pattern).

to measure the companion brightness which is compared to the injected one. We placed all our simulations in the case of the Gemini Planet Imager (GPI) at Gemini South Observatory.

GPI coronagraphic AO-corrected H-band PSFs are generated through simulations of the AO control loop, woofer-tweeter correction, and Fresnel wavefront propagation through a high-fidelity optical model of GPI to properly simulate out-of-pupil-plane and finite size optics and model long-lived semi-static speckles produced by internal wavefront aberrations (Poyneer *et al.* 2006, Poyneer *et al.* 2007, Véran *et al.* 1997, Jolissaint *et al.* 2004).

Results from these simulations show that the LOCI-PSF method produces smaller biais in photometric and astrometric measurement than for standard post-LOCI photometry (Fig. 1). As for standard LOCI algorithms, the photometry given by the LOCI-PSF method is still affected by the variability of the speckle pattern, but at a lower level. Confidence interval can be used to indicate the reliability of the LOCI-PSF companion brightness estimate.

A powerful combination the ADI and LOCI methods into a single one, the Spectral LOCI-PSF method (Maire *et al.* 2012) can be used to take advantage of simultaneous measurement of the speckle pattern over several spectral channels as done by an IFS in addition to standard ADI speckle suppression.

References

Jolissaint, L., Véran, J.-P., *et al.* 2004, *SPIE*, 5490, 151–163

Lafrenière, D., Marois, C., Doyon, R., & Nadeau, D. 2007, *ApJ*, 660, 770–780

Maire, J., Gagné, J., Lafrenière, D., Doyon, R., Graham, J. R., Véran, J.-P., & Poyneer, L. A. 2012, *SPIE*, Volume 8447, article id. 844760, 14 pp. (2012)

Marois, C., Doyon, R., Racine, R., & Nadeau, D. 2000, *PASP*, v112, 91-96

Poyneer, L. A., Véran, J.-P., *et al.* 2006, *SPIE*, 6272

Poyneer, L. A. Macintosh, B., *et al.* 2007, *JOSAA*, 24, 2645

Véran, J.-P., Rigaut, F., *et al.* 1997, *JOSAA*, 14, 11, 3057-3069

Exploring the Formation and Evolution of Planetary Systems
Proceedings IAU Symposium No. 299, 2013
M. Booth, B. C. Matthews & J. R. Graham, eds.

doi:10.1017/S1743921313007801

High Contrast Imaging of an Exoplanet with the Magellan VisAO Camera

Jared R. Males[1†], Laird M. Close[1], Katie M. Morzinski[1†], Derek Kopon[2], Alfio Puglisi[3], Victor Gasho[1], Kate Follette[1], Simone Esposito[3], Armando Riccardi[3], Enrico Pinna[3], Marco Xompero[3], Runa Briguglio[3], Carmelo Arcidiacono[4], Phil M. Hinz[1], Alan Uomoto[5], Tyson Hare[5], Fernando Quiros-Pacheco[3], Javier Argomedo[3], Lorenzo Busoni[3], T.J. Rodigas[1] and Ya-Lin Wu[1]

[†]NASA Sagan Fellow

[1]Steward Observatory, University of Arizona, 933 N. Cherry Ave., Tucson, AZ 85721, USA
email: jrmales@as.arizona.edu

[2]Max-Planck-Institut für Astronomie, Königstuhl 17, D-69117 Heidelberg, Germany

[3]INAF - Osservatorio Astrofisico di Arcetri, Largo E Fermi 5, 50125 Firenze, Italy

[4]INAF - Osservatorio Astronomico di Bologna, Via Ranzani 1, 40127 Bologna, Italy

[5]Carnegie Observatories, OCIW, 813 Santa Babara St. Pasadena, CA 91101, USA

Abstract. The Magellan Adaptive Optics (MagAO) system saw first light in November 2012 at Las Campanas Observatory (LCO) on the 6.5m Clay telescope. Here we present an introduction to MagAO's visible wavelength diffraction limited imager, VisAO. VisAO delivers Strehl ratios greater than 30% from 0.62 microns (r') through 1 micron, where Strehl is even higher, and achieved resolutions as small as 20 milli-arcseconds. We took advantage of the excellent performance of MagAO/VisAO to conduct high contrast observations of an exoplanet in the optical. With VisAO, we are, for the first time, able to begin characterizing exoplanet atmospheres in the optical from the ground.

Keywords. Planets and satellites: individual (Beta Pictoris b); Stars: individual (Beta Pictoris); Stars: planetary systems; Instrumentation: adaptive optics

1. Introduction

During first light commissioning of MagAO (Close *et al.* 2012) and its VisAO science camera (Males *et al.* 2012; Kopon *et al.* 2012) we observed the known exoplanet host star β Pictoris (Lagrange *et al.* 2009). MagAO consists of a 585 actuator adaptive secondary mirror and 28x28 pyramid wavefront sensor (PWFS). VisAO is a CCD camera co-mounted with the PWFS, with a specialized compliment of filters, a wollaston prism, and an occulting mask coronagraph. We obtained over 2 hours of open-shutter at 0.986μm on β Pic, using the VisAO CCD and the occulting mask to prevent bleeding.

2. Results

We show the point spread function (PSF) measured during this observation in Figure 1. MagAO/VisAO delivered a Strehl ratio of 40% to the VisAO CCD in typical seeing conditions at LCO. One can also observe the effect of the coronagraphic occulting mask in Figure 1. With such good Strehl ratios, MagAO/VisAO is capable of high contrasts. We reduced the β Pic observations with the KLIP principle component analysis (PCA) algorithm (Soummer *et al.* 2012). The 5σ detection threshold at $0.5''$, shown in Figure 2, is $\sim 3\times 10^{-5}$. The details of our detection of the $\sim 10 M_J$ exoplanet β Pictoris b are provided

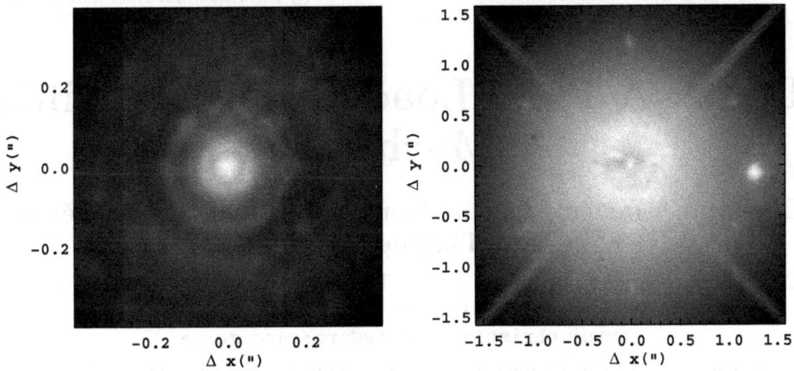

Figure 1. MagAO/VisAO 1μm PSFs. At left we show the off-coronagraph PSF, which has a Strehl ratio of 40%. At right we show a 2 hour exposure, behind the occulting mask.

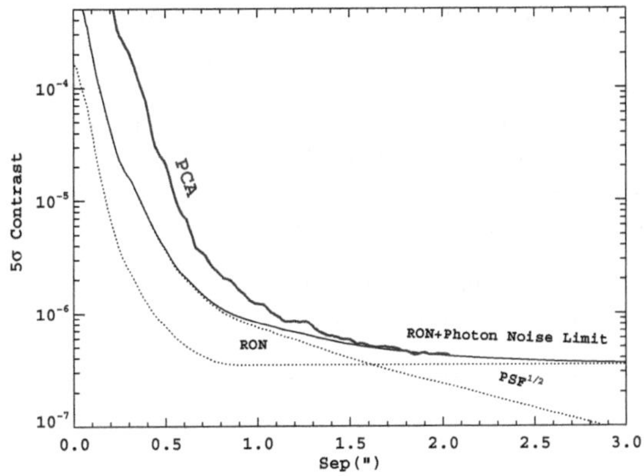

Figure 2. MagAO/VisAO 1μm 5σ Contrast Limits. In this plot we show the contribution of readout noise (RON) and PSF halo photon noise to the detection limit of this observation. The detection limit due to RON is not flat due to the occulting mask. We also show the 5σ profile achieved with principle component analysis (PCA) using the KLIP algorithm.

in Males *et al.* (2013). We also observed this exoplanet with MagAO's IR camera, Clio — see Morzinski *et al.* (2013) in these proceedings.

Acknowledgements MagAO development was supported by the NSF MRI program, the TSIP program, and the Magellan partners. JRM is grateful for the generous support of the Phoenix ARCS Foundation. LMCs research was supported by NSF AAG and NASA Origins of Solar Systems grants. KMM was supported under contract with the California Institute of Technology (Caltech) funded by NASA through the Sagan Fellowship Program.

References

Close, L. M., *et al.* 2012, *SPIE*, 8447, 84470X
Kopon, D., *et al.* 2012, *SPIE*, 8447, 84473D
Lagrange, A. M., *et al.* 2009, *A&A*, 491, L21
Males, J. R., *et al.* 2012, *SPIE*, 8447, 844742
Males, J. R., *et al.* 2013, *In Prep*
Morzinski, K. M., *et al.* 2013, *IAUS 299*
Soummer, R., *et al.* 2012, *ApJ*, 755, L28

Exploring the Formation and Evolution of Planetary Systems
Proceedings IAU Symposium No. 299, 2013
M. Booth, B. C. Matthews & J. R. Graham, eds.

doi:10.1017/S1743921313007813

TLOCI: A Fully Loaded Speckle Killing Machine

Christian Marois[1], Carlos Correia[1,2], Jean-Pierre Véran[1] and Thayne Currie[3]

[1]National Research Council of Canada, Dominion Astrophysical Observatory
5071 West Saanich Rd, Victoria, BC, V9E 2E7, Canada
email: christian.marois@nrc-cnrc.gc.ca

[2]Centre for Astrophysics of the University of Porto, Rua das Estrelas 4150-762 Porto, Portugal
[3]Dept. of Physics and Astronomy, University of Toronto

Abstract. A new high-contrast imaging subtraction algorithm (TLOCI) is presented to maximize a planet signal-to-noise ratio. The technique uses an input spectrum and template PSFs to optimize the reference image coefficient determination to minimize the flux contamination via self-subtraction (thus maximizing its throughput wavelength per wavelength) of any planet that have a similar spectrum to the template spectrum in the image, while trying, at the same time, to maximize the speckle noise subtraction. The optimization is performed by a correlation matrix conditioning. Using laboratory Gemini Planet Imager data, the new algorithm is shown to be superior to the simple/double difference, polynomial fit and original LOCI algorithm.

Keywords. stars: imaging, instrumentation: adaptive optics, methods: data analysis, techniques: image processing, (stars:) planetary systems

1. Introduction

Direct exoplanet imaging is a challenging endeavour, as planets are located very close to their host stars and they are orders of magnitude fainter. The most successful technique to suppress the star light has been the angular differential imaging (ADI, where the intrinsic field-of-view rotation is used to decouple the light of quasi-static aberrations from real objects) in conjunction with the locally optimized combination of images (LOCI) algorithm (Marois *et al.* 2006, Lafrenière *et al.* 2007). With the development of extreme adaptive optics system integral field spectrographs, another technique that uses the intrinsic spectral differences between stars and cool planets by differencing polychromatic images (spectral differential imaging SDI, Marois *et al.* 2000, Sparks & Ford 2002) can now be used. One challenge with SDI is the possible large planet flux variation as a function of wavelength. This important variation is inducing a large amount of planet self-subtraction when trying to use the original LOCI algorithm with SDI, complicating the planet detection and spectral extraction. In this proceeding, we are presenting a new algorithm (TLOCI) to perform an optimal image subtraction using any type of data acquisition scheme (such as ADI and SDI) that maximizes a planet SNR.

The new subtraction algorithm (called TLOCI for "template LOCI") differs from the original LOCI (minimizing the noise) or SDI damped-LOCI (minimizing the noise while trying to maximize the flux in an aperture of a known planet, Pueyo *et al.* 2012), by trying to maximize the signal-to-noise ratio (SNR) of any point sources having a specific spectrum in an image. For each LOCI section, an hypothetical planet is simulated in a noiseless image having a spectrum of interest (methane, dusty, etc). A signal correlation matrix ($A_{\rm PSF}$) is then obtained from these simulated noiseless planet images and this matrix is added, with a proper normalizing Λ parameter, to the standard LOCI

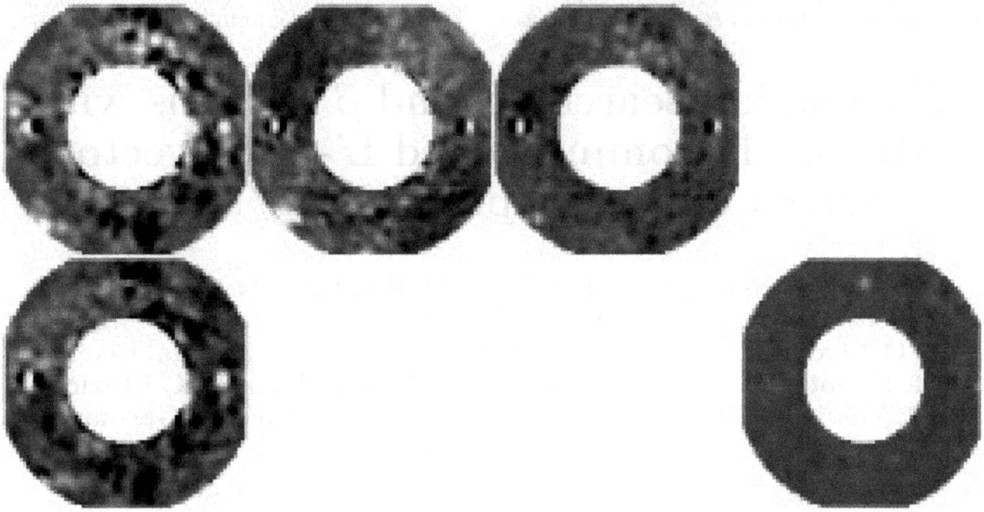

Figure 1. *A ΔH 17.3 methane spectrum planet discovery comparison (the planet is located North of the star; ΔH 14.8 dusty (West) and flat spectrum (East) planets are also added). The first column shows a single difference (top) and double difference (bottom). A 2nd order polynomial fit with a $0.5\lambda/D$ gap (spectral deconvolution technique) is presented in the second column. The third columns show a LOCI analysis using a $0.5\lambda/D$ gap. The TLOCI methane-optimized algorithm is shown in the last column. TLOCI is achieving 5x the LOCI SNR on the methane planet (the dusty and flat spectrum planets have low SNRs in this subtraction due to the TLOCI algorithm not being optimized for these spectra).*

correlation matrix (A). The regularized correlation matrix ($[1 - \Lambda]A + \Lambda A_{\rm PSF}$) is then inverted to obtain the LOCI coefficients that are used to compute a linear combination of images for optimized subtraction.

2. Observations, Data Analysis & conclusion

The algorithm is tested using Gemini Planet Imager (GPI, Macintosh *et al.* 2008) laboratory data (a single 60s × 5 coadd exposure, no turbulence). After initial reduction using the GPI pipeline (Maire *et al.* 2012), images are spatially magnified to align speckles and registered at the image center. Three planets are then implemented at 0.6 arcsec with three different spectra. One artificial planet has a flat spectrum (East), one has a methane-like spectrum (North) and the last one has a dusty-like spectrum (West). The images are then processed using various algorithms and the template-based LOCI algorithm (see Fig. 1).

The TLOCI algorithm was presented to optimally combine a set of images. This more optimal planet finding algorithm can be used for SDI, ADI or archive-based image subtraction (all these techniques performed as a single speckle subtraction step). TLOCI spectral extraction accuracy will be discussed in Marois *et al.,* in prep.

References

Lafrenière, D., *et al.,* 2007, *ApJ*, 660, 770
Macintosh, B., *et al.,* 2008, *Proceedings of the SPIE*, 7015, 701518
Maire, J., *et al.,* 2012, *Proceedings of the SPIE*, 7735, 773531
Marois, C., *et al.,* 2000, *PASP*, 112, 91
Marois, C., *et al.,* 2006, *ApJ*, 641, 556
Pueyo, L., *et al.,* 2012, *ApJS*, 199, 6
Sparks, W. & Ford, H., 2002, *ApJ*, 578, 543

Exploring the Formation and Evolution of Planetary Systems
Proceedings IAU Symposium No. 299, 2013
M. Booth, B. C. Matthews & J. R. Graham, eds.

doi:10.1017/S1743921313007825

Companion search around β Pictoris with the newly commissioned L'-band vector vortex coronagraph on VLT/NACO

D. Mawet[1], O. Absil[2], J. Milli[1,3], P. Baudoz[4], A. Boccaletti[4], G. Chauvin[3], C. Delacroix[2], J. H. Girard[1], A. M. Lagrange[3], J. O'Neal[1], P. Bourget[1], P. Forsberg[5], F. Gonte[1], S. Habraken[2], C. Hanot[2], M. Karlsson[5], M. Kasper[6], J.-L. Lizon[6], K. Muzic[1], R. Olivier[7], E. Peña[1], N. Slusarenko[1], L. E. Tacconi-Garman[3] and J. Surdej[2]

[1]ESO, Alonso de Cordova 3107, Casilla 19001, Vitacura, Santiago 19, Chile
email: dmawet@eso.org

[2]AGO, Université de Liège, 17 Allée du Six Août, B-4000 Liège, Belgium

[3]IPAG, University Joseph Fourier, CNRS, BP 53, 38041, Grenoble, France

[4]LESIA, Observatoire de Paris, 5 pl. J. Janssen, F-92195 Meudon, France

[5]Ångström Laboratory, Uppsala University, Box 534, SE-751 21 Uppsala, Sweden

[6]ESO, Karl-Schwarzschild-Str. 2, 85748 Garching bei München, Germany

[7]GDTech s.a., LIEGE Science Park, rue des Chasseurs Ardennais, B-4031 Liège, Belgium

Abstract. Here we present the installation and successful commissioning of an L'-band Annular Groove Phase Mask (AGPM) coronagraph on VLT/NACO. The AGPM is a vector vortex coronagraph made from diamond subwavelength gratings tuned to the L' band. The vector vortex coronagraph enables high contrast imaging at very small inner working angle (here $0''.09$, the diffraction limit of the VLT at L'), potentially being the key to a new parameter space. During technical and science verification runs, we discovered a late-type companion at two beamwidths from an F0V star (Mawet *et al.* 2013), and imaged the inner regions of β Pictoris down to the previously unexplored projected radius of 1.75 AU. The circumstellar disk was also resolved from $\simeq 1''$ to $5''$ (see J. Milli *et al.*, these proceedings). These results showcase the potential of the NACO L-band AGPM over a wide range of spatial scales.

Keywords. High contrast imaging, vortex coronagraphy, β Pictoris b

---

1. Introduction

High contrast imaging has thoroughly combed through the limited parameter space accessible with first-generation ground-based adaptive optics instruments and the HST. Only a few objects were discovered, and many non-detections reported and statistically interpreted. The field is now in need of a technological breakthrough (Absil & Mawet 2013). We aim at opening a new parameter space with first-generation systems such as NACO at the Very Large Telescope, by providing ground-breaking inner working angle (IWA) capabilities in the L' band. This mid-infrared wavelength range is a sweet spot for high contrast coronagraphy since the planets-to-star brightness ratio is favorable, while Strehl ratio is naturally higher.

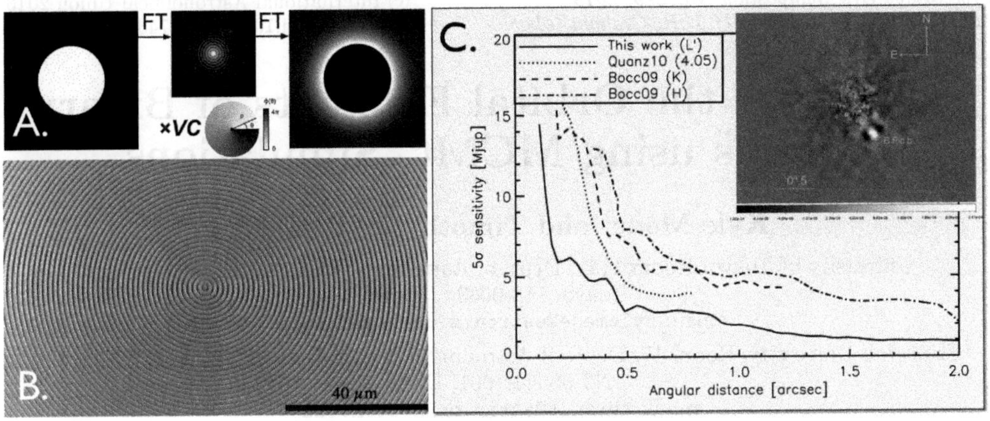

Figure 1. A. Illustration of the diffraction effect of the vortex phase mask on a filled aperture (left). All of the on-axis coherent light appears outside of the downstream pupil (right). A circular aperture (Lyot stop) then blocks it all. FT stands for Fourier transform. B. Scanning electron microscope (SEM) images of the NACO AGPM made from diamond sub-wavelength gratings. C. Detection limits in terms of M_{Jup} with the L' NACO AGPM after removing β Picb from the data. We used the BT-Settl model assuming 12 Myr. These detection limits are the best ever presented, despite our conservative "small sample statistic" corrections to contrast at small angles. We show recent work at shorter wavelengths and with the L' band Apodized Phase Plate (APP) for comparison.

2. Methods and results

An annular groove phase mask (AGPM) vector vortex coronagraph (Fig. 1A) optimized for the L' band, made from diamond subwavelength gratings (Fig. 1B) has been manufactured and qualified in the lab (Delacroix *et al.* 2013). The AGPM enables high contrast imaging at very small IWA (here $0''.09$), potentially being the key to a new parameter space. We present the results of the installation and successful commissioning of an L'-band AGPM on VLT/NACO (Mawet *et al.* 2013). For the science verification run (31 January 2013), we imaged the inner regions of β Pictoris down to the previously unexplored projected radius of 1.75 AU with unprecedented point source sensitivity. The planet was detected at very high SNR (Fig. 1C, inset), enabling precise astrometry and photometry. With the fake negative companion technique used in a multi-parametric optimization (Amoeba), we measured the following parameters: $r = 0.448 \pm 0.01$ (first hint that the planet might be past quadrature), PA $= 211° \pm 1°$, and $\Delta L' = 8 \pm 0.3$ mag. The disk was also clearly resolved (see J. Milli *et al.*, these proceedings).

3. Conclusions

No new object is detected with sufficient confidence level. There are a few hot spots close in and within the disk, to be followed up. Despite the average to mediocre conditions of the run, the NACO L' AGPM mode provides the deepest detection limits at the smallest IWA ever explored (down to 1.75 AU, see Fig. 1C).

References

Absil, O. & Mawet, D. 2010, *A&ARv*, 18, 317
Delacroix, C., Absil, O., Forsberg, P., *et al.* 2013, *A&A*, 553, 98
Mawet, D., Absil, O., Delacroix, *et al.* 2013, *A&AL*, 552, 13

Exploring the Formation and Evolution of Planetary Systems
Proceedings IAU Symposium No. 299, 2013
M. Booth, B. C. Matthews & J. R. Graham, eds.

doi:10.1017/S1743921313007837

Solving for the Orbital Elements of Binary Systems using MCMC Simulations.

Kyle Mede[1] and Timothy D. Brandt[2]

[1]University of Tokyo, Room 1121, Dept. of Astronomy, Hongo 7-3-1, Bunkyo-ku, Tokyo, 113-0033, Japan
email: kylemede@astron.s.u-tokyo.ac.jp

[2]Princeton University, Room 17, Dept. of Astronomy, Peyton Hall, 4 Ivy Lane, Princeton, NJ 08544-1001, USA
email: tbrandt@astro.princeton.edu

Abstract. Recent simulation and observational data have been used to investigate the ability of Kozai oscillations to explain the formation of "hot Jupiter" planetary systems. One of the first exoplanets discovered, τ Boo Ab, orbits a star with a binary companion, making it an excellent testbed for this scenario. We have written a three-dimensional Markov Chain Monte Carlo (MCMC) simulator to constrain the orbit of the distant stellar companion τ Boo B, and are currently deriving orbital parameters and confidence intervals. These orbital parameters will confirm or reject Kozai oscillations as a plausible formation mechanism for τ Boo Ab.

Keywords. techniques: radial velocities, high angular resolution, astrometry, objects: τ Boo

---

Over 800 extrasolar planets (or exoplanets) have been confirmed so far, with nearly 2500 more candidates awaiting ground-based follow-up. Thanks to the large number of known planets, we are beginning to understand how planetary systems form and how abundant solar-like systems are. The first formation models assumed that planets formed on approximately coplanar and circular orbits, as we observe in our solar system today. However, as more planets were observed, it became clear that many (~17%) are in close-in orbits, while others are eccentric or misaligned (Nagasawa & Ida 2011). For these planets, new mechanisms were proposed to explain their existence. One leading mechanism is Kozai oscillations, in which a stellar mass companion exchanges energy and angular momentum with the planet, producing oscillations in the planet's eccentricity and inclination.

Kozai oscillations may explain the origin of the short-period exoplanet τ Boo Ab. In this scenario, the distant (~225 AU) ~$0.4M_\odot$ companion star τ Boo B would have excited τ Boo Ab to high eccentricity (and close passages to its host star, τ Boo A), where tidal dissipation could damp the planet onto its observed short-period (~3.3 day), nearly circular orbit (Brogi *et al.* 2012). With strongly constrained orbital parameters the equations of Takeda *et al.* (2005) can be applied to compare the period of the Kozai oscillations to the ~2Gyr age of τ Boo A, allowing for investigation if they could have played a role in the planet's formation.

Three previous investigations of τ Boo B's orbit have estimated its period to be anywhere from 390 to 2000 years, and its eccentricity to be from 0.42 to 0.91 (Hale 1994, Popović & Pavlović 1996, Roberts *et al.* 2011), see figure (1). Even with 163 years of observations, τ Boo B has traversed less than half of the shortest proposed orbit. Investigations of such long-period systems require careful analyses, preferably with explicit priors and well-characterized posterior probability distributions. We therefore re-analyze the τ Boo system using Markov Chain Monte Carlo (MCMC).

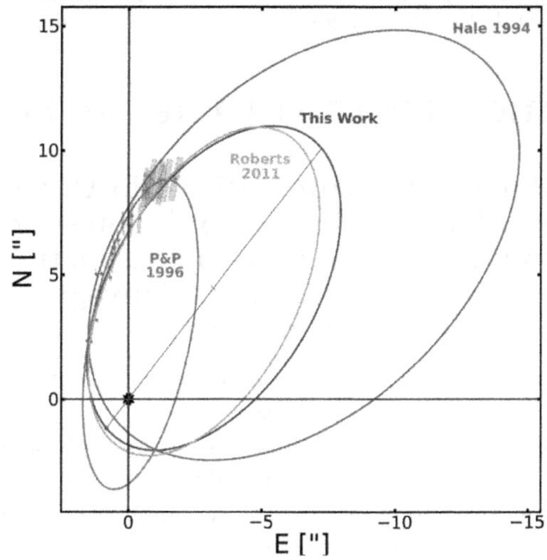

Figure 1. Three previously proposed orbits for τ Boo B (Hale 1994, P&P 1996, and Roberts 2011), together with our best-fit. The locations of τ Boo B are shown as the small stars, with their associated errors, and τ Boo A at the origin. The proposed orbits have not converged to a consensus solution, which is needed for a complete assessment of the Kozai formation scenario.

One of the benefits of MCMC is that, in the limit of a long chain, its samples are drawn directly from the posterior probability distribution. A standard Monte Carlo simulation, by contrast, will draw many points in regions of parameter space with exceptionally low likelihood. MCMC can therefore be significantly more efficient at exploring a high-dimensional parameter space in spite of the fact that its samples are not independent. Because MCMC directly probes the posterior probability distribution, calculating 68% and 95% confidence intervals for each parameter also becomes trivial.

Previous investigations of τ Boo B's orbit used only the positions measured from direct imaging data. However, radial velocity measurements of τ Boo A show a trend from τ Boo B superimposed on τ Boo Ab's 3.3 day orbit. We are taking advantage of these data, along with more recent astrometry, in a full three-dimensional simulator, which promises tighter orbital constraints and well-defined posterior probability distributions of the orbital parameters. Our initial results show a more eccentric orbit for τ Boo B than the most recent investigation, Roberts *et al.* (2011), with $e = 0.79$ +/–0.01.

Our new constraints on the orbit of τ Boo B will allow dynamical calculations of plausible orbital histories of the system over its 2 Gyr life, and will confirm or reject the possibility of a Kozai origin.

References

Brogi, M., Snellen, I. A. G., de Kok, R. J., Albrecht, S., Birkby, J., & de Mooij,E. J. W. 2012, *AAP*, 486, L25

Donati, J.-F., Moutou, C., Farès, R., Bohlender, D., Catala, C., Deleuil, M., Shkolnik, E., Collier Cameron, A., Jardine, M. M., & Walker, G. A. H. 2008, *MNRAS*, 385, 1179

Hale, A. 1994, *AJ*, 107, 306

Popović, G. M. & Pavlović, R. 1996, *BAB*, 153, 57

Nagasawa, M. & Ida, S. 2011, *ApJ*, 742, 72

Roberts, Jr., L. C., Turner, N. H., ten Brummelaar, T. A., Mason, B. D., & Hartkopf, W. I. 2011, *AJ*, 142, 175

Takeda, G. & Rasio, F. A. 2005, *ApJ*, 627, 1001T

Exploring the Formation and Evolution of Planetary Systems
Proceedings IAU Symposium No. 299, 2013
M. Booth, B. C. Matthews & J. R. Graham, eds.

doi:10.1017/S1743921313007849

Performance tests on the SPHERE-IFS

Dino Mesa[1], Raffaele Gratton[1], Riccardo U. Claudi[1], Silvano Desidera[1], Enrico Giro[1], Alice Zurlo[1,2], Anne Costille[2], Arthur Vigan[2], Claire Moutou[2], Jean-Luc Beuzit[3], Kjetil Dohlen[2], Markus Feldt[4], David Mouillet[3], Jean-Francois Sauvage[3], Markus Kasper[5] and Jacopo Antichi[6]

[1]INAF - Osservatorio Astronomico di Padova - Vicolo dell'Osservatorio 5, I-35122, Padova, Italy

[2]Aix Marseille Universite, CNRS, LAM (Laboratoire d'Astrophysique de Marseille) UMR 7326, 13388, Marseille, France

[3]Institut de Planetologie et d'Astrophysique de Grenoble, UJF, CNRS, 414 rue de la piscine, 38400, Saint Martin d'Heres, France

[4]Max-Planck-Institut fur Astronomie, Konigstuhl 17, D-69117 Heidelberg, Germany

[5]European Southern Observatory, Karl Schwarzschild Strasse, 2, D-85748 Garching bei Munchen, Germany

[6]INAF - Osservatorio Astrofisico di Arcetri - L.go E. Fermi 5, I-50125 Firenze, Italy

Abstract. Until now, just a few extrasolar planets (30 out of 860) have been found through the direct imaging method. This number should greatly improve when the next generation of High Contrast Instruments like Gemini Planet Imager (GPI) at Gemini South Telescope or SPHERE at VLT will became operative at the end of this year. In particular, the Integral Field Spectrograph (IFS), one of the SPHERE subsystems, should allow a first characterization of the spectral type of the found extrasolar planets. Here we present the results of the last performance tests that we have done on the IFS instrument at the Institut de Planetologie et d'Astrophysique de Grenoble (IPAG) in condition as similar as possible to the ones that we will find at the telescope. We have found that we should be able to reach contrast down to 5×10^{-7} and make astrometry at sub-mas level with the instrument in the actual conditions. A number of critical issues have been identified. The resolution of these problems could allow to further improve the performance of the instrument.

Keywords. instrumentation, Direct imaging

1. Introduction

Imaging of extrasolar planets is a very challenging goal because of the very large luminosity contrast (of the order 10^{-6} for young giant planets and of the order of 10^{-8}-10^{-10} for old giant and rocky planets) and the small angular separation (few tenths of arcsec for a planet at ~10 AU around a star at some tens of pc) between the host star and the companion objects. However, a number of different project are either now running (e.g. Project 1640 at the 5 m Palomar Telescope - see Crepp *et al.* 2011) or are going to begin like the Gemini Planet Imager (GPI) at the Gemini South Telescope (Macintosh *et al.* 2006) or SPHERE at the ESO Very Large Telescope (Beuzit *et al.* 2006). This last instrument, in particular, includes three scientific channels that are a differential imager and dual band polarimeter called IRDIS operating in the near infrared between the Y and the Ks band (Dohlen *et al.* 2008), a polarimeter called ZIMPOL that will look for old planets at visible wavelengths (Thalmann *et al.* 2008) and an Integral Field Spectrograph

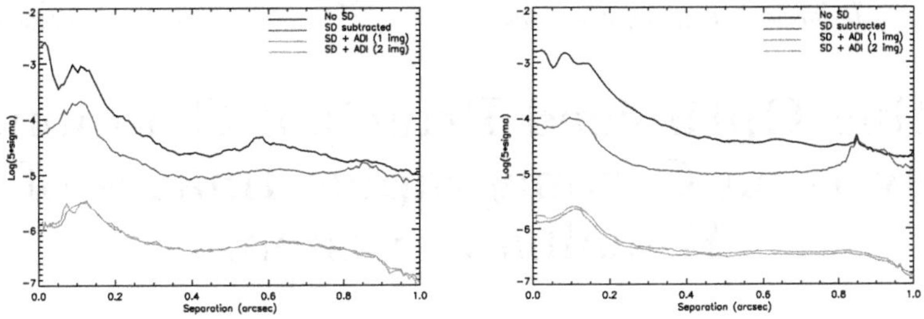

Figure 1. Contrast plot for IFS operating in the YJ-mode (left panel) and in the YH-mode (right panel).

(IFS) operating in the near infrared between the Y and the H band (Claudi *et al.* 2008). In the next paragraphs we will present the results of the laboratory tests on the IFS.

2. Test description

Tests on the IFS instrument were held in January and February 2013 at the *Institut de Planetologie et d'Astrophysique de Grenoble* (IPAG) facility with the aim to validate functionality of the science and calibration templates and to preliminary estimate the performances of the instrument. The tests were performed both in the YJ (0.95 - 1.35 micron) and in YH (0.95 - 1.65 micron) mode using the appropriate combination of Lyot coronagraph and apodized mask.

Data were then reduced exploiting the Data Reduction and Handling (DRH) software that allows to perform all the required calibrations and the speckle subtraction procedure through the spectral deconvolution (SD) method (Sparks & Ford 2002). A further speckle suppression can be obtained applying angular differential imaging (ADI) (Marois *et al.* 2006). Given that we do cannot perform any rotation of the field of view during our tests, we can just perform a simulation of the method so that our results have to be regarded as just an estimation of the contrast that we will be able to get.

3. Results

In Figure 1 we display the 5σ contrast plot that we can get for the IFS operating both in the YJ-mode (left panel) and in the YH-mode (right panel). A contrast better than 10^{-6} can be obtained for both the modes appropriately combining SD and ADI. To further confirm this results we add a number of simulated planets to the raw data at different separations and with luminosity contrast of 10^{-5} and 10^{-6} and reduced these data following the same procedure. All simulated planets are visible with a S/N greater than 5.

References

Beuzit, J.-L., Mouillet, D., Moutou, C., *et al.* 2006, *tafp.conf*, 353
Crepp, J. R., Pueyo, L., Brenner, D., *et al.* 2011, *ApJ*, 729, 132
Claudi, R. U., Turatto, M., Gratton, R. G., *et al.* 2008, *SPIE*, 7014, 111
Dohlen, K., Langlois, M., Saisse, M., *et al.* 2008, *SPIE*, 7014, 118
Macintosh, B., Graham, J., Palmer, D., *et al.* 2006, *SPIE*, 6272, 18
Marois, C., Lafreniere, D., Doyon, R., *et al.* 2006, *ApJ*, 641, 556
Sparks, W. B. & Ford, H. C., 2002, *ApJ*, 578, 543
Thalmann, C., Schmid, H. M., Boccaletti, A., *et al.* 2008, *SPIE*, 7014, 112

Exploring the Formation and Evolution of Planetary Systems
Proceedings IAU Symposium No. 299, 2013
M. Booth, B. C. Matthews & J. R. Graham, eds.

doi:10.1017/S1743921313007850

Testing Optimized Principal Component Analysis on Coronagraphic Images of the Fomalhaut System

Tiffany Meshkat[1], Matthew Kenworthy[1], Sascha P. Quanz[2] and Adam Amara[2]

[1]Sterrewacht Leiden,
P.O. Box 9513, Niels Bohrweg 2, 2300 RA Leiden, The Netherlands
email: meshkat@strw.leidenuniv.nl

[2]Institute for Astronomy, ETH Zurich,
Wolfgang-Pauli-Strasse 27, 8093 Zurich, Switzerland

Abstract. We present the results of a study to optimize the principal component analysis (PCA) algorithm for planet detection, a new algorithm complementing ADI and LOCI for increasing the contrast achievable next to a bright star. We apply PCA to our Fomalhaut VLT NACO Apodizing Phase Plate NB4.05 data.

Keywords. planets and satellites: detection – data analysis techniques: image processing, PCA, LOCI – stars: individual (Fomalhaut)

1. Data

Data were obtained at the VLT/UT4 with NaCo (Lenzen *et al.* 2003, Rousset *et al.* 2003) in July and August 2011 (087.C-0701(B)). The Fomalhaut data were first analyzed and published in Kenworthy *et al.* 2013. Fomalhaut, a young nearby A4V star, is surrounded by an inclined dust debris belt with a planet that has been imaged in the optical, 119 AU away from the star (Kalas *et al.* 2008). Kenworthy *et al.* 2013 searched for planets close in around Fomalhaut, from 4-10 AU, using the LOCI algorithm.

Fomalhaut was used as the natural guide star with the visible band wavefront sensor. The L27 camera on NaCo was used with the NB4.05 filter ($\lambda = 4.051\mu m$ and $\Delta\lambda = 0.02\mu m$) and the Apodizing Phase Plate coronagraph (APP, Kenworthy *et al.* 2010, Quanz *et al.* 2010) to provide additional diffraction suppression. We used pupil tracking mode to perform ADI (Marois *et al.* 2006). The PSF core is intentionally saturated to increase the signal from any potential companions.

The unsaturated data is used to calibrate the photometry in the saturated images. Fake planets are subsequently used to determine the limiting magnitude after image processing with LOCI, PCA, and variations on PCA in this work.

2. Results

Optimized PCA yields a more sensitive contrast curve (up to 1 magnitude gain) than the LOCI algorithm at small inner working angles for our Fomalhaut dataset (Figure 1). See Meshkat *et al.* (2013, submitted), for more details on this work.

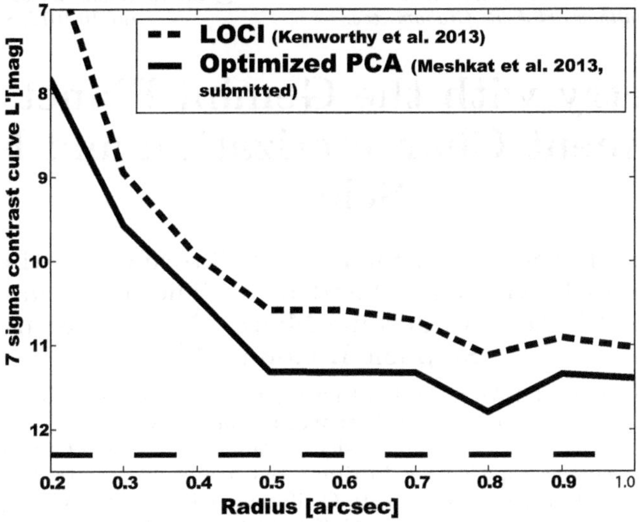

Figure 1. Contrast curves for a 7σ detection of a point source in our Fomalhaut APP data processed with LOCI and PCA. The LOCI dashed curve is adapted from Kenworthy *et al.* 2013 to a 7σ detection. The PCA solid contrast curve is the mean value for artificial planets inserted at two PAs on opposite sides of the star (PA = 45° and 225°). The wide dashed straight line is the background limit.

References

Kalas, P., Graham, J. R., Chiang, E., Fitzgerald, M. P., Clampin, M., Kite, E. S., Stapelfeldt, K., Marois, C., & Krist, J. 2013, *Sci*, 322, 1345K

Kenworthy, M. A., Meshkat, T., Quanz, S. P., Girard, J. H., Meyer, M. R., & Kasper, M. 2013, *ApJ*, 764, 7K

Kenworthy, M. A., Quanz, S. P., Meyer, M. R., Kasper, M. E., Lenzen, R., Codona, J. L., Girard, J. H., & Hinz, P. M. 2010, *SPIE*, 7735E, 103K

Lenzen, R., Hartung, M., Brandner, W., Finger, G., Hubin, N. N., Lacombe, F., Lagrange, A.-M., Lehnert, M. D., Moorwood, A. F. M., & Mouillet, D. 2003, *SPIE*, 4841, 944L

Marois, C., Lafrenière, D., Doyon, R., Macintosh, B., & Nadeau, D. 2006, *ApJ*, 641, 556M

Rousset, G., Lacombe, F., Puget, P., Hubin, N. N., Gendron, E., Fusco, T., Arsenault, R., Charton, J., Feautrier, P., Gigan, P., Kern, P. Y., Lagrange, A.-M., Madec, P.-Y., Mouillet, D., Rabaud, D., Rabou, P., Stadler, E., & Zins, G. 2003, *SPIE*, 4839, 140R

Quanz, S. P., Meyer, M. R., Kenworthy, M. A., Girard, J. H. V., Kasper, M., Lagrange, A.-M., Apai, D., Boccaletti, A., Bonnefoy, M., Chauvin, G., Hinz, P. M., & Lenzen, R. 2010, *ApJ*, 722L, 49Q

Exploring the Formation and Evolution of Planetary Systems
Proceedings IAU Symposium No. 299, 2013
M. Booth, B. C. Matthews & J. R. Graham, eds.

doi:10.1017/S1743921313007862

Polarimetry with the Gemini Planet Imager: Instrument Characterization and Future Science

Max Millar-Blanchaer [1], Sloane J. Wiktorowicz[2], Marshall D. Perrin[3], James R. Graham[4], Sandrine J. Thomas[5], Daren Dillon[6], Michael P. Fitzgerald[7], Jérome Maire[8], Bruce Macintosh[9] and Stephen J. Goodsell[10]

[1]Dept of Astronomy and Astrophysics, University of Toronto, ON, M5S 3H4, Canada
email: maxmb@astro.utoronto.ca
[2]Astronomy and Astrophysics Dept., University of California at Santa Cruz, CA, 95064, USA
[3]Space Telescope Science Institute, Baltimore, MD, 21218, USA
[4]Astronomy Department, University California at Berkeley, CA, 94720, USA
[5]NASA Ames Research Center, Moffett Field, CA 94035, United States
[6]UC Observatories, University of California, Santa Cruz, CA, 95064, USA
[7]Department of Physics & Astronomy, University of California, Los Angeles, CA, 90095, USA
[8]Dunlap Institute for Astronomy & Astrophysics, University of Toronto, ON, M5S 3H4, Canada
[9]Lawrence Livermore National Laboratory, CA, 94551, USA
[10]Gemini Observatory, Hilo, Hawaii, 96720, USA

Abstract. The Gemini Planet Imager (GPI) is a high contrast coronagraph designed to directly image exoplanets and circumstellar disks. GPI includes a polarimetry mode designed to characterize dust grains and enhance the contrast of scattered, polarized light by a factor of 100. Reflections and birefringence of optics within the optical train induce a polarization signature that needs to be measured a priori and calibrated out during data reduction. Here we report on the results of an extensive laboratory characterization campaign of the polarimetry mode. The linear instrumental polarization has been measured in 4 GPI passbands and found to be between 3.5 ± 0.3 % at 1.0 micron and 1.1 ± 0.3 % at 2.0 microns. Modulation efficiency has been measured to be 94% at 1.0 micron increasing to 97% at 2.0 microns. Stability has been shown to better than 0.6% over timescales of $\sim$ 3 months and over cool down cycles. The tests show that GPI passes all polarimetry design requirements and should be able to measure circumstellar disk linear polarization to 1% accuracy.

1. Disk Science with GPI

The study of debris disks has been shown to be an invaluable tool in the characterization of planetary systems. Disk morphology has been used to predict the presence of unseen companions - e.g. Kalas *et al.* 2005, Lagrange *et al.* 2009 and Marois *et al.* 2008 - and provide constraints on planetary masses and orbital elements. In addition, the polarization signature of disks can provide information on the size distribution of dust grains (e.g. Graham *et al.* 2007). The GPI Exoplanet Survey (GPIES) campaign has been granted a total of 890 hours of observing time, 35 of which has been earmarked for debris disk polarimetry: 10 hours for snapshots of stars predicted to have detectable debris disks and 25 hours for high SNR images of disks with apparent planet-induced structure. GPIs polarimetry mode is implemented via a Wollaston prism, a half wave plate, and software for reconstructing Stokes parameter datacubes (Perrin *et al.* 2010 and Maire *et al.* 2010).

Table 1. The instrumental polarization is stable to 0.6%. over 3 month timescales

GPI Passband	Instrumental Linear Polarization	Instrumental Phase Retardance
Y	$3.54 \pm 0.35\%$	-10.6 ± 1.1^o
J	$3.00 \pm 0.50\%$	-5.2 ± 0.50^o
H	$2.37 \pm 0.28\%$	-3.58 ± 0.95^o
K1	$1.14 \pm 0.33\%$	-2.34 ± 0.60^o

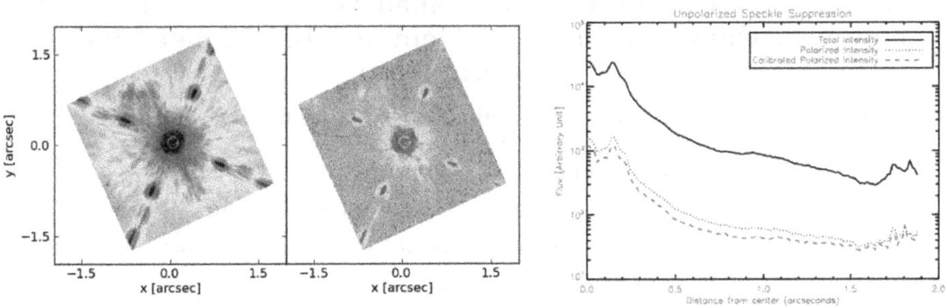

Figure 1. Polarimetry mode total intensity (left) and linearly polarized intensity (center). Intensity as a function of position from the central spot (right).

2. Instrumental Polarization

Calibration of the subsystem was carried out at UC Santa Cruz. The instrumental polarization has been measured and we have reached our performance goals in all cases except the Y band phase retardance, see Table 1. The measurements indicate that we should be able to achieve on-sky accuracy of $\sim 1\%$ (Perrin *et al.* 2010).

3. Speckle Suppression

In a set of coronagraphic images the intensity within $1''$ of the central spot was reduced by a factor of 20 when using linearly polarized intensity opposed to total intensity, see Figure 1. Performance here is limited by the polarization of the laboratory light source ($\sim 10\%$) and high read noise, so these values should be taken as a lower limit on expected performance.

4. Conclusions

The GPI Polarimetry subsystem has met all laboratory performance requirements and is expected to achieve an on-sky accuracy of 1% when measuring linear polarization. These results have been replicated over timescales of 3 months. The use of linear polarized light has been shown to provide an effective means of suppressing residual speckles.

References

Graham, J. R., Kalas, P. G., & Matthews, B. C. 2007, *ApJ*, 654, 595
Kalas, P., Graham, J. R., & Clampin, M. 2005, *Nature*, 435, 1067
Lagrange, A.-M., Gratadour, D., Chauvin, G., *et al.* 2009, *A&A*, 493, L21
Macintosh, B. A., Graham, J. R., Palmer, D. W., *et al.* 2008, *SPIE*, 7014, 31
Maire, J., Perrin, M. D., Doyon, R., *et al.* 2010, *SPIE*, 7735, 102
Marois, C., Macintosh, B., Barman, T., *et al.* 2008, *Science*, 322, 1348
Perrin, M. D., Graham, J. R., Larkin, J. E., *et al.* 2010, *SPIE*, 7736, 192

Exploring the Formation and Evolution of Planetary Systems
Proceedings IAU Symposium No. 299, 2013
M. Booth, B. C. Matthews & J. R. Graham, eds.
doi:10.1017/S1743921313007874

The Gemini NICI Planet-Finding Campaign: The Frequency of Giant Planets around Young B and A Stars

Eric L. Nielsen[1], Michael C. Liu[1], Zahed Wahhaj[2], Beth A. Biller[3], Thomas L. Hayward[4] and The Gemini NICI Planet-Finding Campaign Team

[1]Institute for Astronomy, University of Hawaii, 2680 Woodlawn Drive, Honolulu HI 96822, USA
email: enielsen@ifa.hawaii.edu

[2]European Southern Observatory, Alonso de Córdova 3107, Vitacura, Santiago, Chile

[3]Max-Planck-Institut für Astronomie, Königstuhl 17, 69117 Heidelberg, Germany

[4]Gemini Observatory, Southern Operations Center, c/o AURA, Casilla 603, La Serena, Chile

Abstract. We have carried out high contrast imaging of 70 young, nearby B and A stars to search for brown dwarf and planetary companions as part of the Gemini NICI Planet-Finding Campaign. Our survey represents the largest, deepest survey for planets around high-mass stars ($\approx$1.5–2.5 $M_{\odot}$) conducted to date and includes the planet hosts β Pic and Fomalhaut. Despite detecting two new brown dwarfs, our observations did not detect new planets around our target stars, and we present upper limits on the fraction of high-mass stars that can host giant planets that are consistent with our null result.

Keywords. stars: brown dwarfs, instrumentation: adaptive optics, planetary systems, stars: individual (HIP 79797)

The Gemini NICI Planet-Finding Campaign is a 4-year direct imaging survey designed to determine the frequency of giant planets at large and medium separations, and to determine how that frequency depends on stellar mass (Liu *et al.* 2010). As part of the Campaign, we surveyed 70 young B and A stars, and detected a 33^{+12}_{-9} M_{Jup} (M_H = 9.57 $\pm$ 0.13 mags) brown dwarf around the star HD 1160 (Nielsen *et al.* 2012), confirmed the known brown dwarf around HR 7329 (Lowrance *et al.* 2000), and found that the previously-known brown dwarf HIP 79797 B (Huélamo *et al.* 2010) is itself a tight (3 AU) brown dwarf binary, with masses 58^{+21}_{-20} and 55^{+20}_{-19} M_{Jup} (H = 10.16 $\pm$ 0.16 and 10.28 $\pm$ 0.16 mags). Follow-up of all other new candidate companions within 400 AU projected separation for stars not in crowded fields found all these candidates to be background objects. We also demonstrate a new Bayesian technique to determine ages for field B and A stars that more accurately determines ages and uncertainties from color-magnitude diagram position. We find that the typical method for determining the ages of field A stars by comparison to the Pleiades color-magnitude diagram systematically underestimates the ages of these stars, by up to several hundred Myrs. Our results are described in depth in Nielsen *et al.* (2013).

As we do not detect planets around our 70 high-mass stars we are able to place strong upper limits on the fraction of such stars that can host a wide-separation planet, following the Monte Carlo technique used previously in Nielsen *et al.* (2006), Nielsen *et al.* (2008), and Nielsen & Close (2010). Figure 1 shows the 95% confidence upper limits we can place on giant planets orbiting 2 $M_{\odot}$ stars as a function of planet mass and orbital semi-major

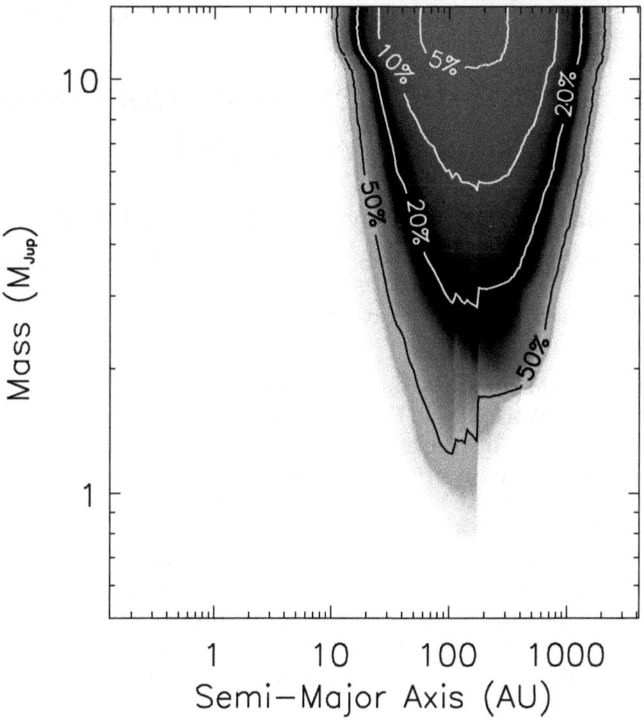

Figure 1. Upper limit on the fraction of 2 $M_{\odot}$ stars that can have planets as a function of mass and orbital semi-major axis given the null result of the NICI Campaign for planets, at 95% confidence. We use the evolutionary models of Baraffe *et al.* (2003) to convert between NIR magnitude and mass as a function of age.

axis. We find that giant planets around such stars are not common, with fewer than 20% of these stars having 4 M_{Jup} planets between 59 and 460 AU. Planets like HR 8799 b (7 M_{Jup}, 68 AU) appear to be rare around high-mass stars: fewer than 10% of B and A stars can have such a planet at 95% confidence. Given the limited number of nearby, young B and A stars improved statistics will likely not come from larger sample sizes but rather from improved instrumentation, as extreme planet-finding instruments like GPI and SPHERE search for planets around these (or similar) high-mass stars at much smaller projected separations between star and planet.

References

Baraffe, I., Chabrier, G., Barman, T. S., Allard, F., & Hauschildt, P. H. 2003, A&A, 402, 701

Huélamo, N., Nürnberger, D. E. A., Ivanov, V. D., *et al.* 2010, A&A, 521, L54

Liu, M. C., Wahhaj, Z., Biller, B. A., *et al.* 2010, Proceedings of SPIE, 7736

Lowrance, P. J., Schneider, G., Kirkpatrick, J. D., *et al.* 2000, ApJ, 541, 390

Nielsen, E. L., Close, L. M., & Biller, B. A. 2006, IAU Colloq. 200: Direct Imaging of Exoplanets: Science & Techniques, 111

Nielsen, E. L., Close, L. M., Biller, B. A., Masciadri, E., & Lenzen, R. 2008, ApJ, 674, 466

Nielsen, E. L. & Close, L. M. 2010, ApJ, 717, 878

Nielsen, E. L., Liu, M. C., Wahhaj, Z., *et al.* 2012, ApJ, 750, 53

Nielsen, E. L., Liu, M. C., Wahhaj, Z., *et al.* 2013, 776, 4

Exploring the Formation and Evolution of Planetary Systems
Proceedings IAU Symposium No. 299, 2013
M. Booth, B. C. Matthews & J. R. Graham, eds.

doi:10.1017/S1743921313007886

Exploring exoplanetary systems beyond 1AU with WFIRST

Matthew T. Penny[1] **and B. Scott Gaudi**[1]

[1]Department of Astronomy, The Ohio State University, 140 W. 18th Ave., Columbus, OH, USA 43210

email: penny@astronomy.ohio-state.edu, gaudi@astronomy.ohio-state.edu

Abstract. The Wide Field InfraRed Survey Telescope (WFIRST) was the top ranked large space mission of the New Worlds, New Horizons Decadal Survey, and is currently under active study by NASA. Its primary instrument will be a large-format high-resolution near-infrared imager and slitless spectrometer. A primary goal of WFIRST will be to perform a high-cadence microlensing survey of the Galactic bulge to search for low-mass exoplanets beyond the ice line. We highlight some of the expected results of the WFIRST exoplanet survey. For example, the survey will probe the abundance of Earth-mass planets from less than 1 AU outwards, including free-floating planets. In its peak sensitivity range of ∼2–5 AU, WFIRST will be sensitive to planets with masses lower than Mercury, and even down to the mass of Ganymede. Overall, WFIRST is expected to detect several thousand bound planets, in addition to several thousand free-floating planets. WFIRST will complete the exoplanet census begun by Kepler, enabling an unprecedented understanding of planetary systems and their formation.

Keywords. planets and satellites: detection, planetary systems, gravitational lensing: micro, space vehicles, telescopes

---

WFIRST will be a wide-field (∼0.3 deg^2), diffraction-limited (∼0.16–0.29 arcsec), near-infrared survey spacecraft using either a 1.3-m telescope (Green *et al.* 2012) or a donated 2.4-m telescope (Spergel *et al.* 2013). A primary mission goal, the microlensing survey will run for 432 days over five years to detect hundreds of free-floating planets and thousands of cold, bound planets down to the mass of solar system moons. Deep, high-resolution imaging over a long time baseline will allow planet hosts to be imaged after they have separated from the source stars they lens, enabling the measurement of host and planet masses for most hosts.

Using the MaBμLS simulator (pronounced *may*-buls, Penny *et al.* 2013) we have simulated the WFIRST microlensing survey. Figure 1 shows two representations of the expected yield of WFIRST in relation to current exoplanet discoveries. Plotted as a function of planet mass and orbital semimajor axis, the blue shaded region shows the region in which the 2.4-m version of WFIRST can be expected to find more than 3 planets if there is one such planet per star, with the dashed line showing the equivalent mass sensitivity for free-floating planets. The red shaded region shows the nominal sensitivity of an extended Kepler mission to transiting planets. Blue points show the expected distribution of WFIRST detections if cold exoplanets follow the mass function measured by Cassan *et al.* (2012) and conservatively extrapolated to masses below 5 $M_\oplus$ (see Penny *et al.* 2013 for details). Black points show currently known exoplanets while red points show Kepler planet candidates. Solar system planets, and three of their largest satellites – Ganymede, Titan and the Moon – are denoted by colourful symbols. The figure also highlights the capabilities of the WFIRST microlensing survey and shows an example of low-mass bound and free-floating planet detections.

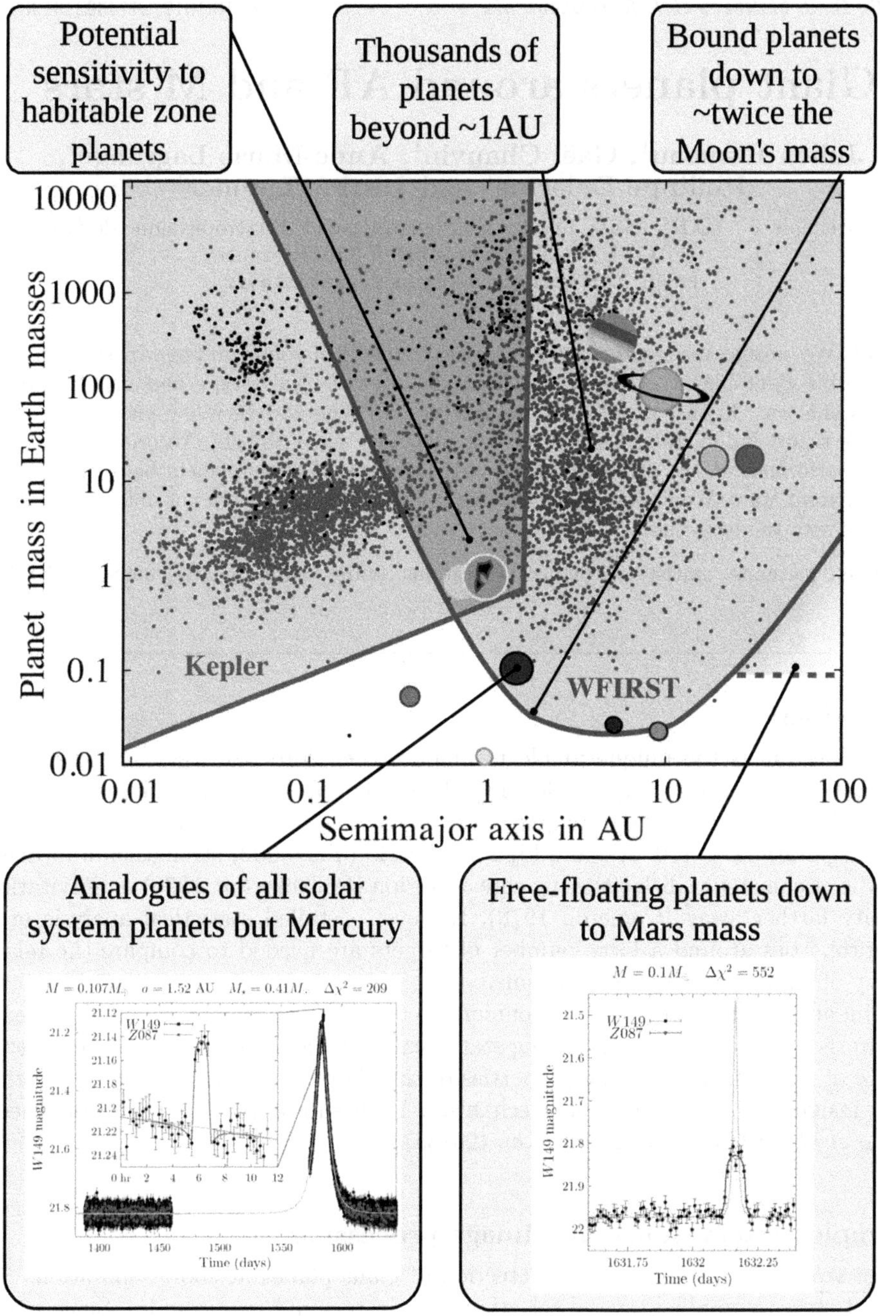

Figure 1. Predictions for the exoplanet microlensing survey of WFIRST and examples of simulated detections of low-mass bound and free-floating planets.

References

Green, J., *et al.* 2012, *arXiv* 1208.4012
Spergel, D., *et al.* 2013, *arXiv* 1305.5422
Penny, M. T., *et al.* 2013, *MNRAS*, 434, 2
Cassan, A., *et al.* 2012, *Nature* 481, 167

Exploring the Formation and Evolution of Planetary Systems
Proceedings IAU Symposium No. 299, 2013
M. Booth, B. C. Matthews & J. R. Graham, eds.

doi:10.1017/S1743921313007898

Giant planets around AF and M stars

Julien Rameau[1], Gaël Chauvin[1], Anne-Marie Lagrange[1], Philippe Delorme[1] and Justine Lannier[1]

[1]UJF-Grenoble 1 / CNRS-INSU, Institut de Planétologie et d'Astrophysique de Grenoble (IPAG) UMR 5274, Grenoble, F-38041, France
email: `julien.rameau@obs.ujf-grenoble.fr`

Abstract. We present the results of two three-year surveys of young and nearby stars to search for wide orbit giant planets. On the one hand, we focus on early-type and massive, namely β Pictoris analogs. On the other hand, we observe late type and very low mass stars, i.e., M dwarfs. We report individual detections of new planetary mass objects. According to our deep detection performances, we derive the observed frequency of giant planets between these two classes of parent stars. We find frequency between 6 to 12 % but we are not able to assess a/no correlation with the host-mass.

Keywords. methods: statistical, planetary systems, stars: individual (HD95086, J01033563-5515561)

1. Introduction

Direct imaging is the unique viable technique to explore planetary systems beyond 10 AU but it is, however, very challenging. The discoveries of giant planets around the young and dusty early-type stars (e.g., Marois *et al.* 2008, Lagrange *et al.* 2009) at physical separations $\leqslant 110$ AU and high-mass ratio ($q \simeq 0.002$) strengthen a formation within the circumstellar disk either by core accretion (Pollack *et al.* 1996) or gravitational instability farther away (Cameron 1978). However, detailed statistical analysis of the planet properties around a large number of targets are needed to compare the relative efficiency of these formation mechanisms.

We aim at detecting substellar companions to two distinct population of stars, i.e. low mass dwarfs and early-type stars. Using statistical formalism, our goal is also to infer the influence of the host mass to the properties of the giant planet population and perhaps give an insight on their formation mechanism. Each subsample has been published in Delorme *et al.* (2012) and Rameau *et al.* (2013a).

2. Sample, observations, and image processing

Target selection. To optimize the detection of giant planet at short separations from their host-star, we applied several selection criteria to build two samples. Since planets are intrinsically brighter when they are young, we selected the stars to be younger than 120 Myr. We limited the volume of the sample to 50 pc for the M dwarfs and 90 pc for the AF stars to be sensitive to few AUs from the star. Our subsample thus contains more than 55 M dwarfs among them 45 have been observed so far and 42 for the AF stars (see Fig. 1, left)

Observing strategies. The observations were done with VLT/NaCo at L′ ($\lambda = 3.8\mu m$) with the pupil tracking mode to enable angular differential imaging (ADI, Marois *et al.* 2006). The data were reduced with the IPAG-ADI pipeline (e.g. Chauvin *et al.* 2012).

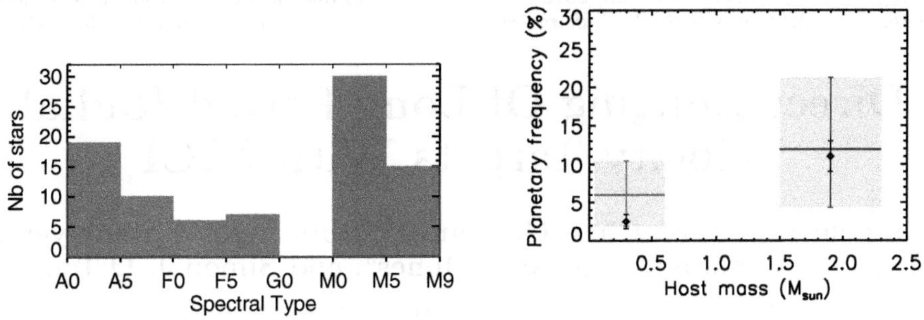

Figure 1. Left: histogram of the spectral types for our two samples. Right: Occurrence of giant planets as a function of the stellar mass in a range [1.,13] M_{Jup} and [1.,300] AU. Black diamonds show the results for close-in giant planets from radial velocity observations (Johnson *et al.* 2007).

3. Occurrence of giant planets as a function of the stellar mass

Discoveries of new substellar companions. Delorme *et al.* (2013) reported the discovery of a $12-14$ M_{Jup} orbiting a pair of young (30 Myr) late M-dwarfs (2M01033563-5515561) at a projected separation of 84 AU (system at 42 pc). Later on, Rameau *et al.* (2013b) detected a probable $4-5$ M_{Jup} companion to the young $(10-17$ Myr) A star HD 95086 at a projected separation of 56 AU. The last discovery lacks of good observing conditions for assessing the true detection but strong arguments tend to favor a (bound) substellar nature for the object.

Frequency of giant planets. On the one hand, for the AF sample, we typically achieved a contrast down to $10-12$ mag at 0.5", i.e., $1-3$ M_{Jup} beyond $5-30$ AU according to the COND models (Baraffe *et al.* 2003). On the other hand, we reached $8.5-11$ mag at 0.5" for our M targets, i.e., below the jovian regime to 2 M_{Jup} beyond $4-20$ AU. We then performed Monte-Carlo simulations to derive the occurrence of giant planets in a uniform grid between [1.,13] M_{Jup} and [1.,300] AU (Rameau *et al.* 2013a). We found that, in our volume-limited samples, $11.9^{+9.3}_{-7.6}$ % of AF stars harbor at least one giant planet while this rate is $6.0^{+4.4}_{-4.1}$ % for the M dwarfs. The large uncertainties are due to the lack of good, i.e., below few Jupiter mass, sensitivity below $5-10$ AU. In Fig.1 (right) the occurrence of giant planets is shown as a function of the stellar masses, one mass-bin corresponding to each subsample. Due to the large uncertainties, we do not assess or exclude any trend. For a comparison, we over plotted the rates of close-in giant planets for the same mass bin from Johnson *et al.* (2007), with the trend clearly assessed.

References

Baraffe, I., Chabrier, G., Allard, F., & Hauschildt, P. 2003, *Brown Dwarfs*, 211, 41
Cameron, A. G. W. 1978, *Moon and Planets*, 18, 5
Chauvin, G., Lagrange, A.-M., Beust, H., *et al.* 2012, *A&A*, 542, A41
Delorme, P., Lagrange, A. M., Chauvin, G., *et al.* 2012, *A&A*, 539, A72
Delorme, P., Gagné, J., Girard, J. H., *et al.* 2013, *A&A*, 553, L5
Johnson, J. A., Butler, R. P., Marcy, G. W., *et al.* 2007, *APJ*, 670, 833
Lagrange, A.-M., Gratadour, D., Chauvin, G., *et al.* 2009, *A&A*, 493, L21
Marois, C., Lafrenière, D., Doyon, R., Macintosh, B., & Nadeau, D. 2006, *APJ*, 641, 556
Marois, C., Macintosh, B., Barman, T., *et al.* 2008, *Science*, 322, 1348
Pollack, J. B., Hubickyj, O., Bodenheimer, P., *et al.* 1996, *Icarus*, 124, 62
Rameau, J., Chauvin, G., Lagrange, A.-M., *et al.* 2013, *A&A*, 553, A60
Rameau, J., Chauvin, G., Lagrange, A.-M., *et al.* 2013, arXiv:1305.7428

Exploring the Formation and Evolution of Planetary Systems
Proceedings IAU Symposium No. 299, 2013
M. Booth, B. C. Matthews & J. R. Graham, eds.

doi:10.1017/S1743921313007904

Direct Imaging Of Long Period Radial Velocity Targets With NICI

Graeme S. Salter[1], Chris G. Tinney[1], Robert A. Wittenmyer[1], James S. Jenkins[2], Hugh R.A. Jones[3], and Simon J. O'Toole[4]

[1]Department of Astrophysics, University of New South Wales, Kensington Campus, 2034, NSW, Australia
email: g.salter@unsw.edu.au

[2]Universidad de Chile, Camino el Observatorio 1515, Las Condes, Santiago, Chile, Casilla 36-D

[3]University of Hertfordshire, Centre for Astrophysics Research, Science and Technology Research Institute, College Lane, AL10 9AB, Hatfield, UK

[4]Australian Astronomical Observatory, PO Box 915, North Ryde 1670, NSW, Australia

Abstract. We are finally entering an era where radial velocity and direct imaging parameter spaces are starting to overlap. Radial velocity measurements provide us with a minimum mass for an orbiting companion (the mass as a function of the inclination of the system). By following up these long period radial velocity detections with direct imaging we can determine whether a trend seen is due to an orbiting planet at low inclination or an orbiting brown dwarf at high inclination. In the event of a non-detection we are still able to put a limit on the maximum mass of the orbiting body. The Anglo-Australian Planet Search is one of the longest baseline radial velocity planet searches in existence, amongst its targets are many that show long period trends in the data. Here we present our direct imaging survey of these objects with our results to date. ADI Observations have been made using NICI (Near Infrared Coronagraphic Imager) on Gemini South and analysed using an in house, LOCI-like, post processing.

Keywords. planetary systems, instrumentation: adaptive optics, high angular resolution, techniques: image processing, stars: imaging, low-mass, brown dwarfs

1. Introduction

The Anglo-Australian Planet Search (AAPS) has been one of the major Doppler search programs, operating on the AAT since 1998. It currently targets some 250 stars for regular observation and has detected over 40 exoplanets to date. Some 13% of our AAPS targets show evidence for long-period companions at periods greater than the baseline of our data. With their orbital periods only constrained to be longer than the AAPS survey time, we are unable to use the observed velocities over that period to constrain the masses of these objects: they could have periods of $\sim$ 20 years and orbital separations of $\sim$ 6AU, and hence planetary masses, or they could have periods of hundreds of years, separations of >10AU and hence brown-dwarf or M-dwarf masses.

Direct imaging is the only means by which we can determine the nature of the long-period candidates in our AAPS sample. We select all AAPS stars within 60pc, giving us 30 targets. Most of the AAPS targets are 4–6 Gy old G dwarfs, inactive and slowly rotating, providing good radial velocity stability. Having such an old target sample means we are unlikely to be able to detect planetary mass bodies, however, a NICI non-detection provides an upper mass limit, pushing targets towards the planetary mass regime.

The detection of brown dwarf companions to these stars is also of great interest. The AAPS targets are mostly G dwarfs, for which metallicity, gravity and age can be better determined than any other class of star. Therefore, any brown dwarf companions discovered would break the age-mass degeneracy typically faced with un-agable field brown dwarfs and spectra/photometry could be used as benchmarks in this field.

2. Observing Strategy and Data Reduction

To perform our direct imaging search we use the Near Infrared Coronagraphic Imager (NICI) on Gemini South. In order to reach the the highest possible contrast we perform Angular Differential Imaging (ADI) (Marois *et al.* (2006)). To do this the telescopes de-rotator is switched off allowing the field to rotate on the detector but keeping the speckle noise stationary. In this way, the speckle noise can be fitted and removed from the observed data.

Our data reduction is described briefly below, for more details see Salter *et al.* in prep.

- Normal calibrations such as flat fielding, dark subtraction, bad pixel removal and de-warping are performed.
- NICI images show significant large scale structure which is fit and removed from each image.
- We perform our own implementation of a Locally Optimised Combination of Images (LOCI) analysis based largely the work of Lafreniere *et al.* (2007).
- Each image is de-rotated and combined to give the final result.

Short exposure images of some targets, prior to the main ADI observations, showed bright companions that have now been confirmed to be co-moving from further short integrations. Long ADI observations were not required for these targets. Three of our targets were not suitable for ADI observations as their location on the sky wouldn't give enough field rotation over a reasonable time frame. For these targets we used regular dithered imaging.

3. Results

To date, in our sample of 30, 11 targets have been found to have co-moving companions. All appear to be M dwarf companions at projected separations varying from $\sim 10 - 500$AU (Salter *et al.* in prep). Twelve targets have been shown to not have any companions observable with NICI. Through analysis of the achieved contrast and inserting fake objects into the data, these observations provide an upper mass limit to the system. A lower mass limit and a minimum orbital period is also provided by the radial velocity measurements (Salter *et al.* in prep).

A further 7 targets have objects that have been detected but second epoch imaging is still required to confirm that they are co-moving.

References

Marois, C., Lafreniere, D., Doyon, R., Macintosh, B., & Nadeau, D. 2006, *ApJ*, vol 641, 556-564

Lafreniere, D., Marois, C., Doyon, R., Nadeau, D., & Artigau, E. 2006, *ApJ*, vol 660, 770-780

Exploring the Formation and Evolution of Planetary Systems
Proceedings IAU Symposium No. 299, 2013
M. Booth, B. C. Matthews & J. R. Graham, eds.

doi:10.1017/S1743921313007916

Campaign Scheduling and Analysis for the Gemini Planet Imager

Dmitry Savransky[1], Bruce A. Macintosh[1], James Graham[2], Quinn M. Konopacky[3] and the GPI science team

[1]Lawrence Livermore National Lab, Livermore, CA USA,
email: savransky1@llnl.gov

[2]University of California Berkeley, Berkeley, CA USA

[3]Dunlap Institute, University of Toronto, Toronto, Canada

Abstract. The Gemini Planet Image (GPI) is a new, high-contrast, exoplanet-imaging, facility instrument for the Gemini South observatory, scheduled to begin science observations in 2014. The GPI Exoplanet Survey (GPIES) has been awarded 890 hours to image and spectrally and polarimetrically characterize young, giant planets within 100 parsecs of the solar system. In preparation for the survey, we have developed a framework for simulating GPI observations and generating end-to-end survey simulations. We present new extensions to this modeling effort and our latest results. We discuss systematic methods for scheduling the survey to ensure that the population of discovered planets is useful in constraining formation models and possibly distinguishing between gravitational collapse and core accretion as the primary formation mechanism.

Keywords. GPI, Exoplanets, Surveys, Survey modeling, Survey analysis

1. GPI and GPIES Overview

GPI is an advanced, precision AO system combined with a coronagraph and integral-field, near-infrared spectrograph designed to image self-luminous, young, giant planets (Macintosh *et al.* 2012). GPI currently achieves raw (5σ) contrasts of 2×10^{-6} in 60 seconds in a single slice of an extracted H band data cube. Simulated, hour-long observing sequences, based on lab data with atmospheric phase plates, predict contrasts of 3×10^{-7} with post-processing (Marois *et al.* 2010).

GPIES aims to produce the first robust census of giant planet populations in the 5–50 AU range. We will observe 600 stars (see Fig. 1(a)) spanning spectral types A-M selected from published young association catalogs and a proprietary list—the result of thousands of hours of telescope observations in preparation for GPI—that adds several hundred new young (<100 Myr, <75 pc) and adolescent (<300 Myr, <35 pc) stars, which will allow our survey to probe in to the 5 AU ice line. The range of separations studied by GPI is inaccessible to Doppler and transit techniques and will provide a new window into planet formation.

2. Survey Simulation

Surveys are automatically scheduled by maximizing the cost function:

$$J = k_1 p_{\text{det}} - k_2 \frac{t_{\text{tot}}}{t_{\text{rem}}} \tag{2.1}$$

where p_{det} is the probability of detection (McBride *et al.* 2011), t_{tot} is the time the target will be observable for the remainder of the survey, t_{rem} is the time remaining,

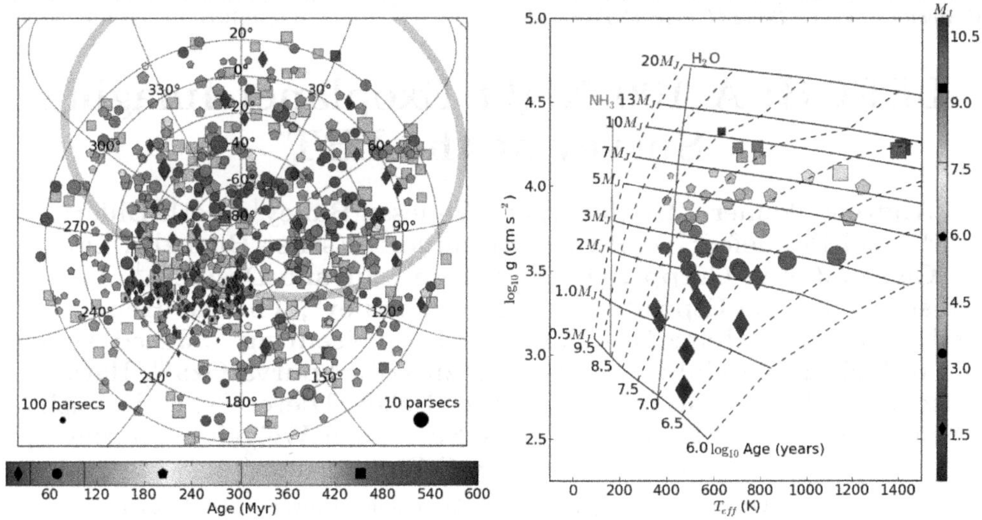

Figure 1. (a) GPIES targets, shaded by age and sized (inversely) by distance. Symbols represent ages in the bins marked on the colorbar. The grey band represents the galactic equator with the galactic center marked by the '+'. (b) Gravity vs. effective temperature with evolutionary tracks (solid black lines) and isochrones (dashed lines) from Burrows *et al.* (2003). Points are simulated planets with size proportional to their radii and shaded by mass. Symbols represent masses in the bins marked on the colorbar.

and k_1 and k_2 are normalized weights. Followups are selected to maximize diversity in mass and semi-major axis space, using contrast as a proxy for mass and the projected separation as an estimate for the semi-major axis (Savransky *et al.* 2011). We produce thousands of sample surveys and analyze the ensembles to predict and optimize survey performance.

3. Simulation Results

Simulations show a yield of 24.8 ± 4.8 unique planet detections, on average, when all planets are modeled as cold start, and 39.6 ± 6 when all are modeled as hot start. We also find that the planets detected in the simulations span a sufficient space in surface gravity and effective temperature to make the sample useful in model selection (Fig. 1(b)) and that we are able to account for survey selection biases in fitting the distributions of masses and semi-major axes with average sample sizes. We also find that simulated astrometric data points for multiple observations of a subset of planets allow us to discriminate between different input distributions of eccentricities, so that the survey will be able to tell whether discovered orbits are predominantly circular or eccentric.

This work performed under the auspices of the U.S. Department of Energy by Lawrence Livermore National Laboratory under Contract DE-AC52-07NA27344.

References

Burrows, A., Sudarsky, D., & Lunine, J. I. 2003, *ApJ*, 596, 587

Macintosh, B. A., *et al.* 2012, in: Proc. SPIE, Vol. 8446, 84461U

Marois, C., Macintosh, B., & Véran, J. P. 2010, in: Proc. SPIE, Vol. 7736, 77361J

McBride, J., Graham, J., Macintosh, B., Beckwith, S., Marois, C., Poyneer, L., & Wiktorowicz, S. 2011, *PASP*, 123, 692

Savransky, D., Cady, E., & Kasdin, N. J. 2011, *ApJ*, 728, 66

Exploring the Formation and Evolution of Planetary Systems
Proceedings IAU Symposium No. 299, 2013
M. Booth, B. C. Matthews & J. R. Graham, eds.

doi:10.1017/S1743921313007928

LEECH: A 100 Night Exoplanet Imaging Survey at the LBT

Andrew Skemer[1], Daniel Apai[1], Vanessa Bailey[1], Beth Biller[2], Mickael Bonnefoy[2], Wolfgang Brandner[2], Esther Buenzli[2], Laird Close[1], Justin Crepp[3], Denis Defrere[1], Silvano Desidera[4], Josh Eisner[1], Simone Esposito[5], Jonathan Fortney[6], Thomas Henning[2], Phil Hinz[1], Karl-Heinz Hofmann[7], Jarron Leisenring[1,8], Jared Males[1], Rafael Millan-Gabet[9], Katie Morzinski[1], Apurva Oza[10], Ilaria Pascucci[11], Jenny Patience[12], George Rieke[1], Dieter Schertl[7], Joshua Schlieder[2], Mike Skrutskie[10], Kate Su[1], Gerd Weigelt[7], Charles E. Woodward[13] and Neil Zimmerman[2]

[1]Steward Observatory, Department of Astronomy, University of Arizona
[2]Max Planck Institute for Astronomy
[3]Department of Physics, Notre Dame University
[4]Osservatorio Astronomico di Padova, Istituto Nazionale di Astrofisica
[5]Osservatorio Astrofisico di Arcetri, Istituto Nazionale di Astrofisica
[6]Department of Astronomy and Astrophysics, University of California, Santa Cruz
[7]Max Planck Institute for Radio Astronomy
[8]Institute for Astronomy, Eldgenossische Technische Hochschule Zurich
[9]NASA Exoplanet Science Institute, California Institute of Technology
[10]Department of Astronomy, University of Virginia
[11]Department of Planetary Sciences, Lunar and Planetary Laboratory, University of Arizona
[12]School of Earth and Space Exploration, Arizona State University
[13]Minnesota Institute for Astrophysics, University of Minnesota

Abstract. In February 2013, the LEECH (LBTI Exozodi Exoplanet Common Hunt) survey began its 100-night campaign from the Large Binocular Telescope atop Mount Graham in Arizona. LEECH neatly complements other high-contrast planet imaging efforts by observing stars in L' band (3.8 microns) as opposed to the shorter wavelength near-infrared bands (1–2.3 microns). This part of the spectrum offers deeper mass sensitivity for intermediate age (several hundred Myr-old) systems, since their Jovian-mass planets radiate predominantly in the mid-infrared. In this proceedings, we present the science goals for LEECH and a preliminary contrast curve from some early data.

Keywords. instrumentation: adaptive optics, (stars:) planetary systems, surveys

---

Discovering Adolescent Exoplanets: Most exoplanet imaging surveys operate in the near-infrared (H-band; 1.65 μm), which provides the best contrasts at small separations. However, self-luminous exoplanets emit most of their light at longer wavelengths, and this becomes more important as exoplanets age and lose the residual energy from their formation. By operating in the mid-infrared (L; 3.8 μm) LBTI is sensitive to planets around older stars (see Figure 1). This will allow LEECH to discover adolescent exoplanets around nearby stars, complementing the young samples discovered by other surveys (see Figure 2), such as GPI, SPHERE, SEEDS, and Project 1640 (Macintosh *et al.* 2008, Beuzit *et al.* 2008, Tamura 2009, Hinkley *et al.* 2011)

Connecting Planets with Disks: Using the two LBTI science cameras in parallel, LEECH and HOSTS (see paper by Defrere *et al.* in these proceedings) will search for giant planets and inner debris disks simultaneously. The LEECH and HOSTS samples include very nearby stars that have been targeted by other relevant surveys (Doppler-RV

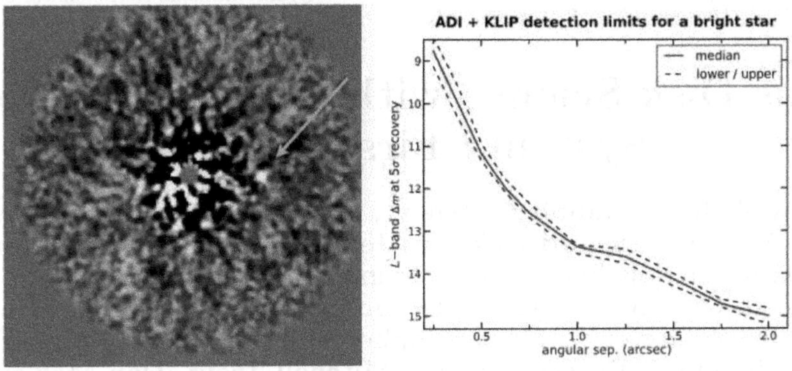

Figure 1. LEFT: A first light image from LEECH with the star removed and an artificial planet inserted. The artificial planet is 13 magnitudes fainter than the star at a separation of 0.75″, equivalent to a 3 Mjup planet 7.5 AU from a 0.5 Gyr solar type star at 10 pc. LEECH's ability to image older exoplanets than other surveys will extend our knowledge of exoplanet evolution. RIGHT: ADI+PCA (KLIP) detection limits for a bright star with 2 hours integration time on 1 telescope (preliminary).

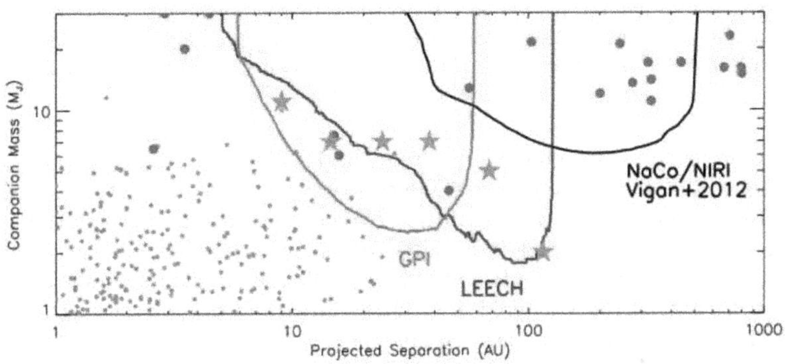

Figure 2. Planet mass versus semi-major axis for the known exoplanets and planetary mass companions listed in www.exoplanet.eu shown as small circles for radial velocity detections, large circles for imaging, and stars for imaged planets around HR 8799, β Pic, and Fomalhaut. The median planet mass sensitivities are shown for the GPI A- and F-star sample, the LEECH sample, and an A-star search (Vigan *et al.* 2012) with NaCo/NIRI. Both GPI and LEECH will explore the critical missing link between radial velocity searches and current AO imaging surveys.

planet searches and outer debris disk studies). Combining this data, will provide the first comprehensive view of exoplanetary systems.

Characterizing Exoplanet Atmospheres: Early observations suggest that directly imaged planets have different colors and spectra than their field brown-dwarf analogs. By studying planets with LBTI's broad wavelength coverage (1-13 μm) and over a wide age range, LEECH will characterize the bulk atmospheric properties and evolution of extrasolar planets.

References

Beuzit, J.-L., Feldt, M., Dohlen, K., *et al.* 2008, *SPIE*, 7014

Hinkley, S., Oppenheimer, B. R., Zimmerman, N., *et al.* 2011, *PASP*, 123, 74

Macintosh, B. A., Graham, J. R., Palmer, D. W., *et al.* 2008, *SPIE*, 7015

Tamura, M. 2009, *AIPC*, 1158, 11

Vigan, A., Patience, J., Marois, C., *et al.* 2012, *A&A*, 544, A9

Exploring the Formation and Evolution of Planetary Systems
Proceedings IAU Symposium No. 299, 2013
M. Booth, B. C. Matthews & J. R. Graham, eds.

doi:10.1017/S174392131300793X

Debris Disk Science with the Palomar ExAO System : First Results

Matthew Wahl[1], Stanimir Metchev[1], Rahul Patel[1], Eugene Serabyn[2], Dimitri Mawet[6], Richard Dekany[3], Jennifer Roberts[2], Rick Burruss[2], Antonin Bouchez[3,4], Tuan Truong[2], Christoph Baranec[3], Stephen Guiwits[2], David Hale[3], John Angione[2], Thang Trinh[2], Jeffry Zolkower[3,5], J. Christopher Shelton[2], Dean Palmer[2], John Henning[3], Ernest Croner[3], Mitchell Troy[2], Dan McKenna[3] and Jonathan Tesch[2]

[1]Dept. of Physics and Astronomy, Stony Brook University, Stony Brook, NY 11794
[2]Jet Propulsion Laboratory, California Institute of Technology, 4800 Oak Grove Blvd., Pasadena, CA 91109, USA
[3]Caltech Optical Observatories, California Institute of Technology, 1200 E. California Blvd., MC 11-17, Pasadena, CA 91125, USA; rgd@astro.caltech.edu
[4]Now at Giant Magellan Telescope Observatory Corp., Pasadena, CA 91106, USA
[5]Now at Department of Astronomy, Cornell University, Ithaca, NY 14853, USA
[6]European Southern Observatory (ESO), Alonso de Cordova 3107 Vitacura, 763 0355, Santiago, Chile

Abstract. We present first imaging results from the PALM-3000 adaptive optics system and PHARO camera on the Hale 5 m telescope. Observations using a vector vortex coronagraph have given us direct detections of the two-ring dusty debris system around the star HD 141569. Our observations reveal the inner clearing in the disk to unprecedentedly small angular separations, and are the most sensitive yet at the H and K bands. We are for the first time able to measure and compare the colors of the scattered light in the inner and outer dust rings, and find that the outer ring is significantly bluer than the inner ring.

---

1. Overview, Data and Analysis

Images of the dusty debris disk around HD 141569 were obtained with PALM-3000 adaptive optics system and PHARO camera on the Palomar 5 m telescope in June 2012. HD 141569 is an A0 type Herbig Ae star located at 99 ± 8 pc (Van Leeuwen 2007) with an age of roughly 5 Myr. The PHARO camera contains a hard edged Lyot coronagraph for high-contrast imaging, and as of 2009, a vector vortex coronagraph (Mawet *et al.* 2009). The vector vortex coronagraph works by inducing a phase change to incident on axis light causing it to interfere with itself destructively inside the pupil, thus nulling it.

Observations of a reference star, HD 142864, were carried out so that accurate point spread function (PSF) subtraction can be achieved. We integrated for 1.0 hr on target, for each band, and 0.5 hr on the calibrator star. The data were analyzed using Karhunen-Loeve Image Processing (KLIP; Soummer *et al.* 2012) which performs principle component analysis. Our images resolve the known two-ring structure of the disk, and are the most sensitive to date in the H and K bands.

2. Results

We observe an inner clearing from $\sim 0\farcs5$ (50 AU) out to the edge of the inner disk at $1\farcs77$ (177 AU). The inner disk is $0\farcs64$ (64 AU) wide and extends out to $2\farcs41$ (241 AU).

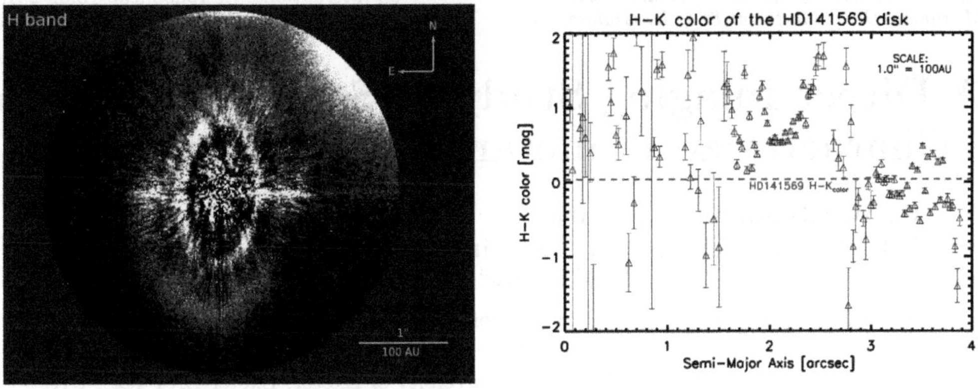

Figure 1. Left : Image of HD 141569 taken in the H band. The two known rings are visible, and there is an extra feature to the east between them. It is characterized as emission at $1''.5$ extending from a PA of 30° to 98°. **Right :** H-K Color plot of the two rings showing that the inner ring is noticeably redder than the outer ring. Our data reduction with KLIP, while optimized for revealing the extended emission from the disk, incurs a dark hole of non-astrophysical negative residuals around the star. As a first-order attempt at removing this effect, we fit a line to the minimum residual emission as a function of radial separation.

The outer disk has emission from $3''.06$ to $3''.87$ (306 AU to 387 AU) with a width of $0''.81$ (81 AU). Both are offset from the central star by measurable amounts, but do not share a common center.

There is also an obvious gap between the two disks, which may be cleared by a forming exoplanet. The gap has alternatively been interpreted as part of a spiral structure formed from perturbations by HD141569 B and C (Clampin *et al.* 2003, Quillen *et al.* 2005, Ardila *et al.* 2005, Wyatt 2005) whose PSF wings are visible in the north-west corner of Fig. 1.

The H-K disk color in Fig. 1 was inferred from the the H and K_s band surface brightness profiles (SBPs) of the HD 141569 debris disk. There are two peaks in the SBPs, one at $2''.0$, and the other, less prominent between $3''.3$-$3''.6$ correspond to the mid-ring semi-major axes of the two circumstellar rings. Our data lose sensitivity inwards of $0''.8$, although for the first time we can rule out $> 3\times$ brighter rings inwards to $0''.5$. Another new result is the determination of the near-infrared color of the circumstellar disk. In particular, the outer ring is 0.5–1.0 mag bluer than the inner ring.

References

Van Leeuwen, F., 2007, *A&A*, 474, 653–664
Mawet, D., *et al.* 2009, Optics Express, 17, 2902
Soummer, R., Pueyo, L., & Larkin, J., 2012, *ApJ*, 755, L28
Clampin, M., Krist, J. E., & Ardila, D. R., 2003, *ApJ*, 126, 385–392
Quillen, A. C., Varnière, P., Minchev, I., & Frank, A. 2005 *ApJ*, 129, 2481–2495
Ardila, D. R., Lubow, S. H., & Golimowski, 2005, *ApJ*, 627, 986–1000
Wyatt, M. C., 2005, *ApJ*, 440, 937–948

Exploring the Formation and Evolution of Planetary Systems
Proceedings IAU Symposium No. 299, 2013
M. Booth, B. C. Matthews & J. R. Graham, eds.

doi:10.1017/S1743921313007941

A Direct Imaging Study to Search for and to Characterize Planetary Mass Companions

K. Ward-Duong[1*], J. Patience[1,2], R. J. De Rosa[1], A. Rajan[1], P. Hinz[3], A. Skemer[3], K. Morzinski[3], J. Males[3], L. M. Close[3], D. W. McCarthy[3] and C. Kulesa[3]

[1]School of Earth and Space Exploration, Arizona State University, Tempe, AZ 85287, USA
*email: kwardduo@asu.edu

[2]Astrophysics Group, School of Physics, University of Exeter, Exeter EX4 4QL, UK

[3]Steward Observatory, The University of Arizona, Tucson, AZ 85721, USA

Abstract. We present preliminary results from two parallel programs to search for new substellar companions to nearby, young M-stars and to characterize the atmospheres of known planetary mass and temperature substellar companions. For the M-star survey, we are analyzing high angular resolution archival data on systems within 15pc, complementing a subset with well-determined young ages based on measurements of several age indicators. The results include stellar and substellar companion candidates, which we are currently pursuing with follow-up second epoch images. The characterization component of the project involves using LBT LMIRCam and MMT ARIES direct imaging and spectroscopy data to investigate the atmospheres of known young substellar companions with masses overlapping the planetary regime. These atmospheric studies will represent an analogous comparison to the atmospheres of young imaged planets, and provide a means to fundamentally test evolutionary models, enhancing our understanding of the overall substellar population.

Keywords. stars: low-mass, stars: binaries, planetary systems, stars: brown dwarfs

1. M-dwarf Companion Search

Despite the ubiquity of M-dwarf stars, their population characteristics have not yet been refined in comparison to recent studies of more massive stars (e.g., Raghavan *et al.* 2010). We present preliminary results from an M-dwarf companion search comprising a volume-limited survey of nearby M-stars within 15 pc, selected using *Hipparcos* parallaxes and B-V and V-K color criteria. For 188 stars within our sample, we reduced and analyzed available near-infrared, high-resolution data from the *Hubble Space Telescope*, Canada-France-Hawaii Telescope, and Very Large Telescope archives, searching projected separations of $\sim$ 1 to 100 AU from the primary stellar components. The combined data reach median sensitivities to companions with a K-band difference of 9 magnitudes from their primaries at separations of $2''$. A total of 19 substellar candidate systems and 30 stellar candidate systems were detected from this survey (two example candidates are shown in the top panel of Figure 1). Follow-up observations, and a comparison with previously published literature measurements, will be carried out to determine the physical association of the detected candidates. From this study, we are deriving the key population statistics of separation distribution, mass ratio distribution, and multiplicity frequency for M-dwarf stars.

2. Substellar Companion Characterization

Measurements of the atmospheric compositions of giant extrasolar planets represent an important pathway to understanding the formation and evolution history of exoplanets.

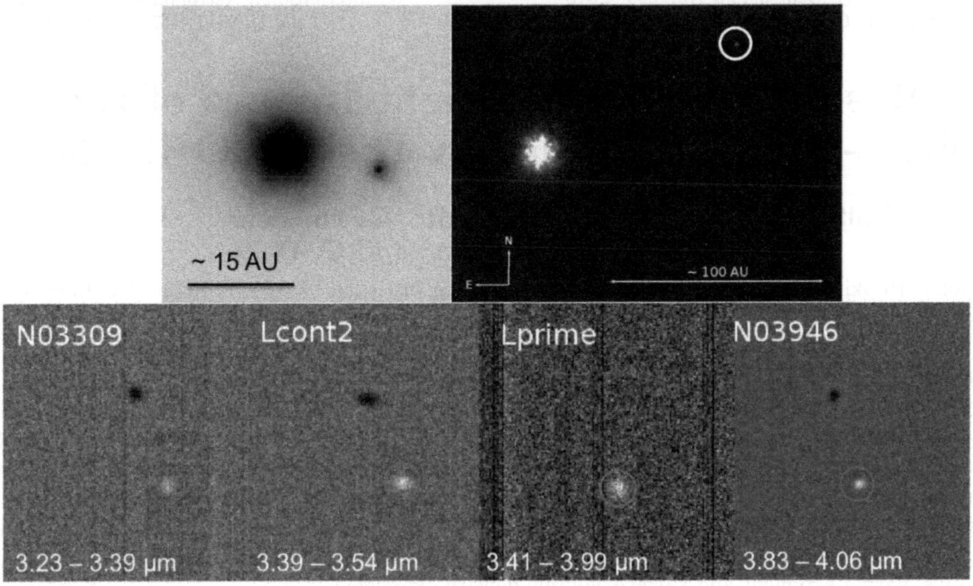

Figure 1. (**Top**) Examples of two new stellar (left, ΔK-mag ∼ 4) and substellar (right, ΔK-mag ∼ 7) candidates identified in this survey. (**Bottom**) LBT narrow and broad L-band difference imaging of G196-3 B, a known substellar companion to an M-dwarf with a projected separation of ∼ 300 AU.

Establishing an empirical sequence of object atmospheres spanning the transition between the brown dwarf and planetary regimes provides a critical means of testing these theories and investigating the formation mechanisms of substellar objects. With infrared spectra obtained with the ARIES instrument at the MMT Observatory, L-band imaging from LBT LMIRCam, and visible photometry, we are comparing the spectral energy distributions of substellar objects with theoretical atmospheric models to constrain effective temperatures and bolometric luminosities. These companions exhibit unusually red infrared colors in common with the imaged planetary companions to HR 8799. The young ages of these systems ($\lesssim$ 400 Myr) make them important analogs to directly imaged planets, and allow us to explore the atmospheric properties of low mass, low surface gravity objects. The LBT measurements with different mid-IR filters are particularly sensitive to the effects of clouds, chemistry, and metallicity, which will enable detailed tests of atmospheric models. By combining MMT infrared spectra with the L-band LBT data, we will have comprehensive wavelength coverage of the bulk of the emergent flux from these young substellar objects, allowing us to define an empirical sequence. We are also measuring the C/O ratios of young brown dwarf companions to test formation hypotheses by searching for similar or different abundances to the host objects, which may depend on formation via core accretion versus gravitational instability (Öberg *et al.* 2011). Example LBT thermal IR data on a known substellar companion to an M-dwarf with a projected separation of ∼ 300 AU (G196-3 B) are shown in the bottom panel of Figure 1. Similar studies will be performed on confirmed detections from our M-dwarf survey, and combining our ongoing observations of known companions will allow for atmospheric characterization over a wide range of substellar objects.

References

Öberg, K. I., Murray-Clay, R., & Bergin, E. A. 2011, *ApJ*, 743, 16
Raghavan, D., *et al.* 2010, *ApJS*, 190, 1

Exploring the Formation and Evolution of Planetary Systems
Proceedings IAU Symposium No. 299, 2013
M. Booth, B. C. Matthews & J. R. Graham, eds.

doi:10.1017/S1743921313007953

Resolved Scattered Light Images of the Edge-On Protoplanetary Disk ESO Hα 569

S. Wolff[1], M. Perrin[2], K. Stapelfeldt[3], G. Duchêne[4,5], J. Krist[6], F. Ménard[5,7], D. Padgett[3] and C. Pinte[5,7]

[1]Johns Hopkins University, [2]Space Telescope Science Institute, [3]NASA GSFC, [4] UC Berkeley, [5]IPAG, [6]JPL, [7]U. de Chile

Abstract. We present detailed models of the edge-on protoplanetary disk ESO Hα 569 (SSTgbs J111110.7-764157) from resolved scattered light images from HST and a complete spectral energy distribution. Data was obtained as part of an HST campaign to catalogue edge-on disks around young stars in nearby star forming regions (HST program 12514, PI: Karl Stapelfeldt). We confirm that this object is an optically thick edge-on disk around a young star with an outer radius of 125 AU. Using full radiative transfer models, we probe the distribution of dust grains and overall shape of the disk (inclination, scale height, dust mass, maximum particle size, inner radius, flaring exponent and surface/volume density exponent).

Keywords. Protoplanetary Disk, Young Stars, Planet Formation

1. Introduction and Observations

The unique geometry of edge-on circumstellar disks provides the best opportunity to study the detailed structure as the bright central star is occulted, providing contrast. The spectral energy distribution can place constraints on the mass and distribution of grain sizes within a disk, while the scattered light images place constraints on the geometry of the disk. ESO-Hα 569 is a member of the Chameleon I star forming region which is known to have a distance of 160 $\pm$ 15 pc and an age of 1-2 Myrs (Whittet *et al.* 1997). The central star has a spectral type of M2.5 (Luhman *et al.* 2007) and is known to be under-luminous (Comerón *et al.* 2004). For more information on this edge-on disks campaign see the proceedings by Stapelfeldt *et al.* and Duchêne *et al.* in this volume.

Scattered light images were obtained using HST ACS/WFC in F814W and F606W (Figure 1; left panel). The disk has a radius of $\sim 0.8'' = 125$ AU. The top nebula's peak surface brightness is $\sim$ 20x the bottom, suggesting an inclination of $\sim 80°$. It is close to left/right symmetric, though the southern side is very slightly brighter. There is a clear jet present in F606W. The SED compiled from various catalogues is shown in Figure 1 (Right Panel). It shows the double-peaked shape typical of edge-on disks, with a minimum near 11 μm and thermal emission peaking at 70 – 100 μm.

2. Radiative Transfer Modeling and Results

Using the MCFOST radiative transfer code (Pinte *et al.* 2006), we construct spectral energy distributions and 0.8 μm scattered light images for a grid of over 200,000 models (Table 1; Column 2). Baysian analysis was performed to arrive at the best fit parameter values (Table 1; Column 3-6). For similar work see Pinte *et al.* 2008 and Duchêne *et al.* 2010. The best fit image and SED are shown in Figure 1.

We found that a sharp edged disk gave the overall shape of the upper disk but failed to get the flux ratio between the top and bottom and didn't yield enough diffuse material

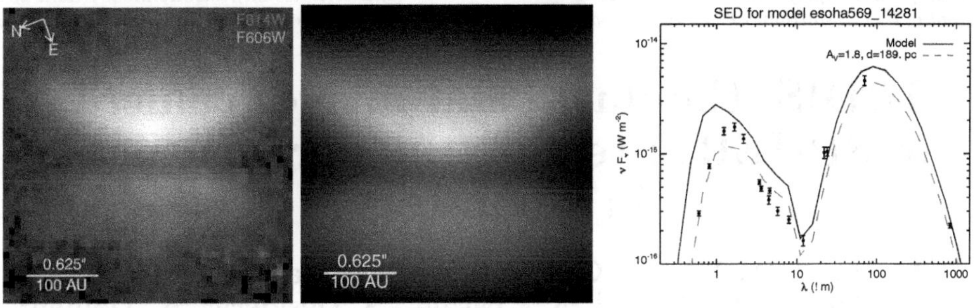

Figure 1. Left: False color composite HST images of the protoplanetary disk ESO-Hα 569 with a log stretch. (F606W and F814W). Both filters show the dust lane and asymmetries between the top and bottom of the disk, while only F606W establishes the presence of an outflow jet. Center: Best fit $0.8\,\mu m$ model image from the combined SED and image χ^2 analysis. Right: Observed (data points) and best fit spectral energy distribution (dashed line).

Table 1. Modeled Disk Parameters and Best fit values from the χ^2 distributions.

Parameters	Range of Values	SED	Image	Weighted	Total
Inclination (degrees)	60 to 90	77.63	75.52	75.52	75.52
Scale Height (AU at r = 100 AU)	10, 15, 20, 25, 30	15	25	20-25	20
Dust Mass ($M_\odot$)	10^{-4}, 3×10^{-4}, 10^{-3}	10^{-3}	3×10^{-4}	3×10^{-4}	3×10^{-4}
Flaring Exponent (Beta)	1.0,1.1, 1.2, 1.3, 1.4,1.5	1.5	1.4	1.4	1.3
Surface Density (Alpha)	-2, -1.5, -1.0, -0.5, 0.0	-2.0	-0.5	-1.5	-0.5
Max Particle size (μm)	10, 100, 1000, 3000	3000	3000	1000-3000	3000
Grain Porosity	0.0, 0.5	0	0.5	0.0	0.0
Structure	Disk, Tapered-edge Disk	sharp	tapered	tapered	tapered

above the disk. This was solved with a tapered edge model where the density Σ, falls off exponentially with some critical radius R_C of material outside of the disk (Hughes *et al.* 2008).

$$\Sigma = \Sigma_c \left(\frac{R}{R_c}\right)^{-\gamma} exp\left[\left(-\frac{R}{R_c}\right)\right]^{2-\gamma}$$

Using the tapered edge disk model, we can simultaneously fit the spectral energy distribution and images well. We are continuing to refine the parameter grid for the fit. The large scale height (relative to typical protoplanetary disks) is consistent with the low mass of the central star.

References

Comerón, F., Reipurth, B., Henry, A., & Fernández, M. 2004, *A&A*, 417, 583
Duchêne, G., McCabe, C., Pinte, C., *et al.* 2010, *ApJ*, 712, 112
Hughes *et al.* 2008, *ApJ*, 678, 1119H
Luhman, K. L. 2007, *ApJS*, 173, 104
Pinte, C., Menard, F., Duchêne, G., & Bastien, P. 2006, *A&A*, 459, 797
Pinte, C., Padgett, D. L., Menard, F., *et al.* 2008, *A&A*, 489, 633
Whittet, D. C. B., Prusti, T., Franco, G. A. P., *et al.* 1997, *A&A*, 327, 1194

Exploring the Formation and Evolution of Planetary Systems
Proceedings IAU Symposium No. 299, 2013
M. Booth, B. C. Matthews & J. R. Graham, eds.

doi:10.1017/S1743921313007965

IRDIS, the dual-band imager camera of SPHERE: testing the performances in laboratory

A. Zurlo[1], A. Vigan[1], C. Moutou[1], D. Mesa[2], R. Gratton[2], M. Langlois[3], J.-L. Beuzit[4], A. Costille[1], S. Desidera[2], K. Dolhen[1], C. Gry[1], F. Madec[1], D. Le Mignant[1], D. Mouillet[4] and J.-F. Sauvage[4]

[1]Aix Marseille Université, CNRS, LAM (Laboratoire d'Astrophysique de Marseille) UMR 7326, 13388, Marseille, France

[2]INAF - Osservatorio Astronomico di Padova - Vicolo dell'Osservatorio 5, I-35122, Padova, Italy

[3]CRAL, UMR 5574, CNRS, Université Lyon 1, 9 avenue Charles André, 69561 Saint Genis Laval Cedex, France

[4]Institut de Planetologie et d'Astrophysique de Grenoble, UJF, CNRS, 414 rue de la piscine, 38400, Saint Martin d'Heres, France

Abstract. Next year the second generation instrument SPHERE will begin science operations at the Very Large Telecope (ESO). This instrument will be dedicated to the search for exoplanets through the direct imaging techniques, with the new generation extreme adaptive optics. In this poster, we present the performances of one of the focal instruments, the Infra-Red Dual-beam Imaging and Spectroscopy (IRDIS). All the results have been obtained with tests in laboratory, simulating the observing conditions in Paranal. We tested several configurations using the sub-system Integral Field Spectrograph (IFS) in parallel and simulating long coronographic exposures on a star, calibrating instrumental ghosts, checking the performance of the adaptive optics system and reducing data with the consortium pipeline. The contrast one can reach with IRDIS is of the order of 10^{-6} at 0.5 arcsec separation from the central star.

Keywords. Direct imaging, instrumentation

1. Test description

IRDIS sub-system is one of the three focal sub-systems of SPHERE, working in different configurations in the range of 0.9–2.3 μm: dual-band imaging, long slit spectroscopy and polarimetric imaging. IRDIS can observe in its dual-band imaging mode in parallel with the other focal sub-instrument IFS. A series of laboratory tests has been performed to find the final capabilities of the instrument and check possible issues. Contrast plots of the NIRSUR (IRDIS in H2H3 and IFS in YJ) configuration is shown in Figure 1. The NIRSUR mode is the one optimized for planets search. It uses the spectral differential imaging (SDI) procedure (Racine *et al.* 1999), a technique that permits the detection of faint companions in two narrow-band filters in which the contrast between primary and planet is very different (e.g. wavelength close to the methane band head).

2. Conclusions

The performance we achieved in laboratory sub-system IRDIS confirmed the previsions of the simulations (Vigan *et al.* 2010). The dual-band imager will allow the detection of planets with a luminosity contrast of 10^{-6} at 0.5 arcsec of separation from the central

Table 1. Observing conditions and data reduction.

Observing conditions & Data reduction		
Turbulence	0.″6 and 0.″8 seeing	Backgroung subtraction + flat-filtering
Filter	Dual-band H2H3	Rescaling
Coronograph	Apodized Lyot	Median recombination of the data-cube frames
Dithering	4×4	Spectral subtraction (SDI)
Exposure time	3584 s	NO ADI
IFS	yes	Injection of synthetic planets of different position angles and separation

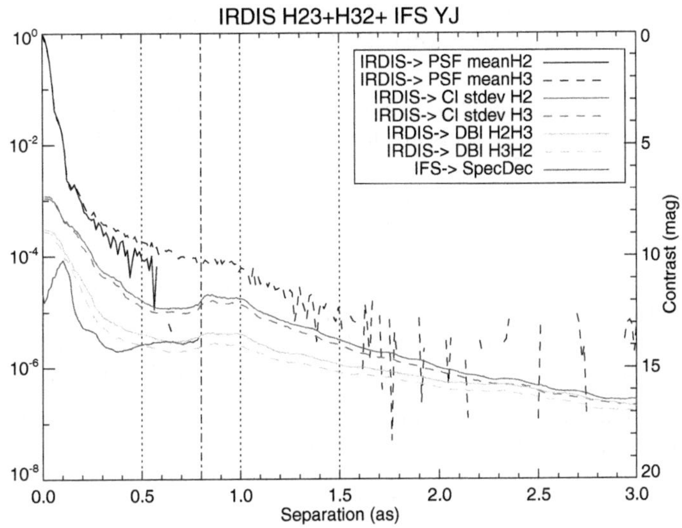

Figure 1. Contrast plot showing the performances of IRDIS and IFS. The dual band filter H2-H3 has been used for IRDIS while IFS worked in YJ mode. The turbulence of 0.″85 seeing. The plot shows the contrast with respect to the host star (black line) vs the separation in arcsec. Apodized Lyot coronographic profile (at 1 σ) is showed by the red lines, while in green is represented the result of the dual band imaging (DBI). The blue curve represents the IFS result after spectral deconvolution. Vertical lines indicates separation from the host star of 0.5, 0.8 (limit of IFS), 1.0 and 1.5 arcsec.

star. Synthetic fake planets have been injected in the raw data to better estimate the detection limits. We found that for planets with a contrast of 1×10^{-4} the signal to noise ratio (SNR) goes from 6.4 for internal to 33.3 for external planets. During the observations at the VLT the exposures will be acquired in pupil-stabilized mode, where the field will rotate, to take advantage of the ADI (Marois *et al.* 2005) technique in addition to the SDI. This configuration has not been tested realistically in the laboratory. Simulations predict a gain of a factor 10 in contrast when using ADI in addition to SDI. The photometric and astrometric characterization of faint companions, down to 14 mag of contrast with respect to their host stars, will be possible using the ADI and SDI techniques.

References

Marois, C., Lafrenière, D., *et al.* 2005, *ApJ*, 641, 556–564

Racine R., Walker G. A.H., Nadeau D., Doyon R., & Marois C. 1993, *PASP*, 111, 587–593

Vigan A., Moutou C., Langlois M., *et al.* 2003, *Monthly Notices of the Royal Astronomical Society*, 407, 71

Exploring the Formation and Evolution of Planetary Systems
Proceedings IAU Symposium No. 299, 2013
M. Booth, B. C. Matthews & J. R. Graham, eds.
doi:10.1017/S1743921313007977

Radio Interferometry Observations of the Hallmarks of Planet Formation

Sean M. Andrews[1]

[1]Harvard-Smithsonian Center for Astrophysics
60 Garden Street, Cambridge, Massachusetts 02138, USA
email: **sandrews@cfa.harvard.edu**

Abstract. Some of the fundamental processes involved in the evolution of circumstellar disks and the assembly of planetary systems are just now becoming accessible to astronomical observations. The new promise of observational work in the field of planet formation makes for a very dynamic research scenario, which is certain to be amplified in the coming years as the revolutionary Atacama Large Millimeter/submillimeter Array (ALMA) facility ramps up to full operations. To highlight the new directions being explored in these fields, this brief review will describe how high angular resolution measurements at millimeter/radio wavelengths are being used to study several crucial aspects of the formation and early evolution of planetary systems, including: the gas and dust structures of protoplanetary disks, the growth and migration of disk solids, and the interactions between a young planetary system and its natal, gas-rich disk.

Keywords. circumstellar matter, planetary systems: formation, planetary systems: protoplanetary disks, radio continuum: stars, radio lines: stars, submillimeter

1. Introduction

Planet formation occurs in the dusty, gas-rich disks that orbit young stars. But, although astronomical observations have been probing these disks for nearly thirty years, it is only relatively recently that the quality and diversity of the available data have become sufficient to facilitate quantitative insights on the planet formation process. In that sense, this is a particularly exciting time for the field: the *direct* study of the formation and early evolution of planetary systems is poised to move from a primarily theoretical topic to one that is richly informed and empirically constrained by a deluge of new observational input. Fundamental to these high expectations is the revolutionary enhancement in sensitivity and resolution to circumstellar disk material that will be provided by the new ALMA interferometer, operating at wavelengths from ~0.35–3.5 mm.

The excitement surrounding the ALMA project in the disk community is due to the long-awaited high spatial and spectral resolution access to some of the most fundamental tracers of disk material. The wavelength range from 350 μm to ~1 cm (herefafter termed "radio" for convenience) is incredibly rich in physical diagnostics of disk material, primarily because the vast majority of the disk volume is literally quite cool (~10-100 K). Longward of ~500 μm, the thermal continuum emission radiated by dust grains is relatively bright (>1000× the stellar photosphere; so no contrast issues) and optically thin (Beckwith *et al.* 1990). Therefore, the continuum surface brightness profile is a good proxy for the radial structure of the dust disk: $I_\nu \propto \kappa_\nu \Sigma_d T_d$, where κ_ν are the dust mass opacities, Σ_d the surface densities, and T_d the (midplane) temperatures (Andrews *et al.* 2009; Isella *et al.* 2009). Moreover, the radio spectrum is richly populated with the pure rotational transitions of simple molecules, providing a wealth of spectral line data that inform our understanding of gas disk structures, chemistry, and kinematics.

Despite the remarkable physical diagnostic capabilities of disk tracers in the radio, the volume of data on the subject has so far been small due to the limited capabilities of current interferometers. As of this writing, the continuum emission from ~50 disks has been imaged at sub-arcsecond resolution; a sparse subsample of only ~10 of those disks have been probed in molecular line emission on similar scales (and those usually only in the most abundant tracer, CO, and its primary isotopologues). To give some perspective on future expectations, I would estimate that with the enhanced capabilities of ALMA we will eventually see these numbers grow into the thousands. In essence, the fundamental promise of ALMA in the field of planet formation is really its ability to deliver on the key demographic properties of circumstellar disks. But what is it we want to learn?

In this brief review, I will quickly highlight three issues that are at the forefront of the field (and driving intense interest with ALMA in its early science mode), based on the current state-of-the-art radio interferometry observations of protoplanetary disks: (1) the signposts of young planetary systems and disk-planet interactions; (2) the growth and migration of disk solids; and (3) prospects for quantifying the gas reservoirs in disks.

2. Transition Disks: Evidence for Young Planets?

Theoretical models of the planet formation process tend to impose disk properties (sometimes informed by observational constraints) as initial conditions, and then run a complex set of physical and/or statistical simulations in an effort to reproduce some of the key demographic features of the exoplanet population (e.g., Ida & Lin 2004, 2005; Alibert *et al.* 2005; Mordasini *et al.* 2009). The implicit assumption in these models is that the physics involved in the formation and early evolution of a planetary system is understood well enough to be effectively parametric and deterministic. In a statistical sense, that assumption might be valid: but the behavior of a young planetary system in its early stages, while still embedded in its natal disk reservoir, can be quite complex (see the review by Kley & Nelson 2012). Once a planet has formed, strong feedback between it and the surrounding disk material will modify the planet mass and orbit on a relatively short timescale (e.g., Papaloizou *et al.* 2007; Baruteau & Masset 2012). Indeed, these early evolutionary effects imply that the demographic features in the exoplanet population do not truly reflect the system properties at the formation epoch: therefore, in a sense it is these planet-disk interactions that are the "initial conditions" worth worrying about. Ideally, we want to place more concrete, empirical constraints on these interactions to facilitate improvements in the planet formation models. A direct search for young planets is difficult (although see the contribution by Ireland & Kraus in these proceedings), but an indirect approach aimed at interpreting the signatures of planet-disk interactions manifested in perturbations to the remnant disk structure is more promising.

A planet that forms in a disk will open a narrow gap around its orbital radius, a_p, and slightly enhance the local gas pressure outside the gap (e.g., Bryden *et al.* 1999; Crida *et al.* 2006). Dust grains that drift inwards with respect to the gas (see §3) can be trapped in this gas pressure maximum, creating a dense, concentrated ring of mm/cm-sized particles centered at ~1.2–2.0 a_p (depending on the companion mass and the disk viscosity; Pinilla *et al.* 2012; Paardekooper & Mellema 2006). The properties of the disk interior to the gap depend strongly on the inward flow of material past the planet, which is partially tied to the planetary accretion efficiency (Lubow *et al.* 1999; Crida & Morbidelli 2007; Zhu *et al.* 2011). While some gas and small dust grains may reach the inner disk (Lubow & D'Angelo 2006; Dodson-Robinson & Salyk 2010), larger particles will be filtered from this flow by the gas pressure maximum (Rice *et al.* 2006; Zhu *et al.* 2012; Pinilla *et al.* 2012). This theoretical scenario makes some key observational

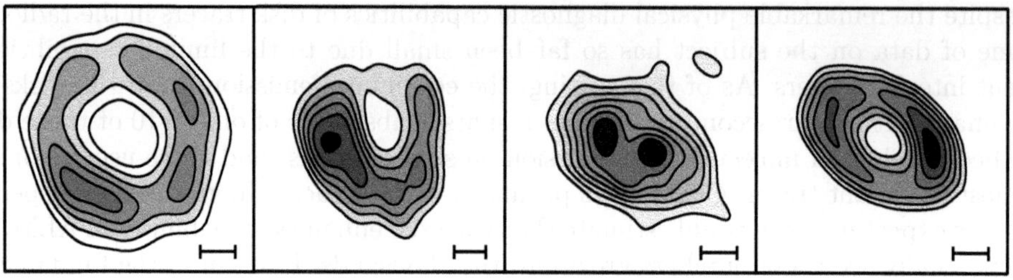

Figure 1. A selection of striking Submillimeter Array (SMA) 880 μm dust continuum images of transition disks: (*from left*) RX J1604.3−2130 (Mathews *et al.* 2012), SAO 206462 (Brown *et al.* 2009), GM Aur (Hughes *et al.* 2009), and LkCa 15 (Andrews *et al.* 2011b). A scale bar in the bottom right corner of each panel marks a projected physical distance of 40 AU for each disk.

predictions: disks interacting with a young planet (or planetary system) should exhibit a depletion of material at their centers and reduced accretion rates onto their host stars (depending on the planet accretion efficiency), with a narrow ring of large particles that trace the pressure maximum outside the gap (e.g., Wolf & D'Angelo 2005).

A category of young disks identified ~25 years ago (Strom *et al.* 1989; Skrutskie *et al.* 1990) offers compelling empirical evidence for such planet-disk interactions. These so-called "transition" disks usually exhibit a relative deficit in their infrared spectrum, consistent with a diminished reservoir of warm dust within (at least) a few AU of the host star. However, it is difficult to robustly quantify these cavities with only unresolved spectrophotometric measurements, especially at optically thick infrared wavelengths. Interferometric observations in the radio were employed to first resolve the transition disks (Piétu *et al.* 2006; Hughes *et al.* 2007, 2009; Brown *et al.* 2008, 2009), which were found to exhibit limb-brightened "ring"-like morphologies in the dust continuum. Figure 1 illustrates these features for some particularly striking examples. Current limitations in sensitivity and resolution produce substantial selection biases on these radio observations, in favor of the most massive (i.e., brightest) disks with the largest cavities ($r \gtrsim 15$ AU for the nearest star-forming regions). Still, these transition disk morphologies are found to be surprisingly common: at least 1 in 4 disks in the bright half of the radio continuum luminosity distribution has a large, dust-depleted cavity (Andrews *et al.* 2011a).

Although depleted, these cavities are decidedly *not* empty. The currently available radio continuum data constrain the reduction of mm-sized grains to a factor $\gtrsim$100, but the presence of a deficit in the infrared spectrum is often a more stringent constraint on the dust content in the inner few AU (e.g., Andrews *et al.* 2011a; Espaillat *et al.* 2007; 2010). Most of the well-studied transition disk hosts are actively accreting gas (e.g., Fang *et al.* 2009; Espaillat *et al.* 2012), and direct tracers of small (~μm-sized) dust grains (Dong *et al.* 2012; Follette *et al.* 2013) and molecular gas (e.g., Salyk *et al.* 2009; Rosenfeld *et al.* 2012) are being found well inside the radio-wave dust continuum rings.

Even at these early stages in the study of transition disks, their observational characteristics provide firm constraints on the physical mechanisms responsible for their origins. In most cases, the data are difficult to explain with either alternative disk evolution processes (see, e.g., Owen *et al.* 2011 and Birnstiel *et al.* 2012b regarding the issues with photoevaporation or particle growth, respectively) or more massive stellar/brown dwarf companions (Pott *et al.* 2010; Kraus *et al.* 2011). That said, a verified association of the observations with the planet formation theory through direct detections of faint, long-period companions located interior to the radio continuum rings would be beneficial (e.g., Huélamo *et al.* 2011; Kraus & Ireland 2012). Ultimately, I expect that ALMA observations will provide two key demographic constraints on planet formation through transition disks. First, the incidence of ring-like continuum emission geometries will offer a relatively unbiased estimate (compared to infrared colors) of the frequency and orbital distribution of giant planets at their formation epoch. And second, detailed probes of the gas content inside these dust rings, when coupled with accretion rates onto the host (and potentially direct imaging of the companions) can help estimate the mass distribution of these planets and will give new insights on their gaseous envelope accretion rates.

3. The Evolution of Disk Solids

The collisional evolution and migration of solids embedded in a gas-rich disk are fundamental processes in the core accretion model for planet formation. But building a quantitative, predictive theoretical framework for these processes is challenging, due to the complex interplay of the relevant physical mechanisms (coagulation, aerodynamics, turbulence, etc.) and enormous range of scales involved: solids need to grow $\sim$14 orders of magnitude in size (>40 orders of magnitude in mass) – from sub-μm grains to 1000 km embryos – within a few Myr. To further develop such a framework, we need to better understand two obstacles to the early stages of the planetary growth process. The first is related to the efficiency of growth by collisions: once dust grains grow to a certain size, their relative velocities can be sufficiently high that collisions are either non-productive (the particles bounce off one another; e.g., Zsom *et al.* 2010) or outright destructive (the particles fragment; e.g., Brauer *et al.* 2008, Birnstiel *et al.* 2010). The second is related to the rapid, inward migration of solids due to aerodynamic drag interactions with the gaseous reservoir in which they are embedded (radial drift; Weidenschilling 1977).

Numerical simulations of the evolution of solids in protoplanetary disks (informed by laboratory collision experiments) that account for these growth "barriers" make two robust, qualitative predictions that in principle can be tested with astronomical observations (e.g., Birnstiel *et al.* 2010a,b, 2011, 2012a): (1) the spatial distribution of solids should be radially size-sorted, such that larger particles are preferentially concentrated closer to the central stellar host; and (2) the abundance of $\sim$mm/cm-sized solids should be strongly depleted relative to the gas at large disk radii, due to the effects of radial drift (e.g., see also Takeuchi & Lin 2002, 2005; Brauer *et al.* 2007). Although the remote sensing of disk solids is limited to particles smaller than a few cm and stellocentric distances larger than $\sim$10 AU (presently), such observational data is readily scalable as an effective (albeit indirect) probe of the evolution of larger bodies in the inner disk.

The shape of the radio continuum spectrum has long been exploited to learn about the size distribution of disk solids (Beckwith & Sargent 1991). For an optically thin medium in the Rayleigh-Jeans limit, the radio spectrum scales like $I_\nu \propto \nu^{2+\beta}$, where β is the spectral index of the dust opacities ($\kappa_\nu \propto \nu^\beta$). For reference, $\beta \approx$ 1.7-2.0 in the ISM (Hildebrand 1983; Finkbeiner *et al.* 1999), but is only $\sim$0.5-1.0 in the integrated spectra of protoplanetary disks (e.g., Ricci *et al.* 2010a, 2010b). Draine (2006) demonstrated that, for a wide range of mineralogical compositions, a lower β is produced by a more top-heavy grain size distribution (see also Miyake & Nakagawa 1993; D'Alessio *et al.* 2001). However, the global, integrated β measurements derived from unresolved photometry offer only relatively weak empirical constraints on detailed models of particle evolution (Birnstiel *et al.* 2010b). The potential for quantitative development of these models instead relies on spatially resolved constraints on the radio colors – i.e., measurements of the $\beta(r)$ profile – to track how the particle size distribution varies as a function of the local physical conditions in the disk (e.g., Isella *et al.* 2010; Guilloteau *et al.* 2011).

Such measurements are notoriously challenging, since they require quality interferometric datasets with high sensitivity, resolution, and fidelity over a wide range of observing

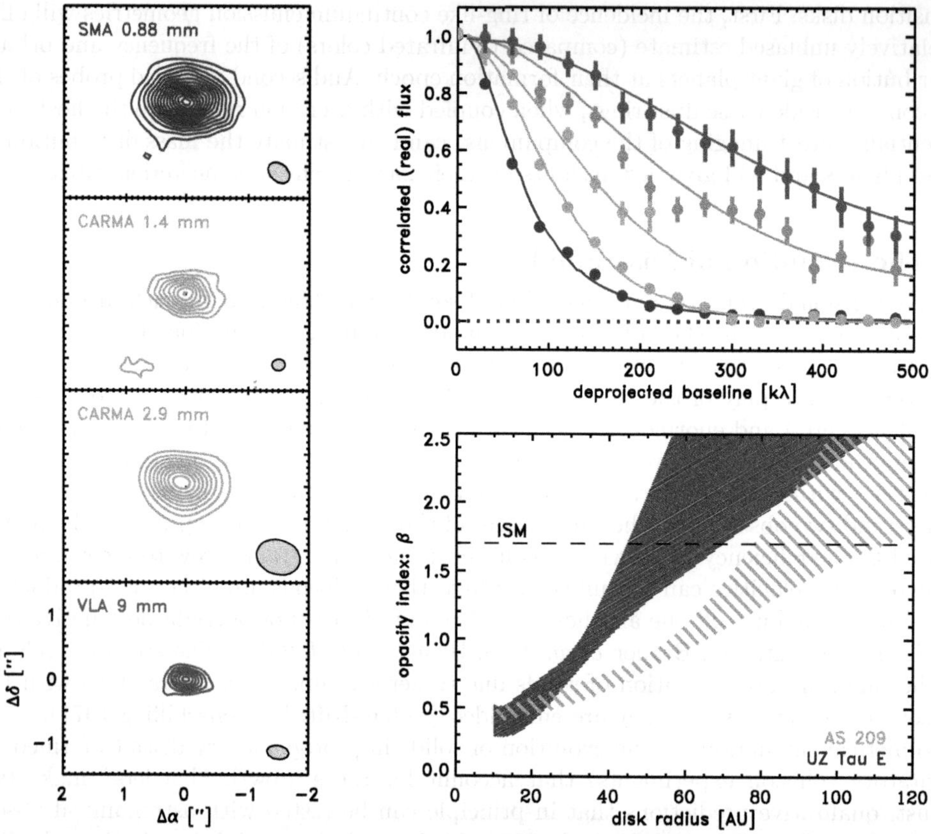

Figure 2. (*left*) A sequence of dust continuum images of the UZ Tau E disk at different radio wavelengths (adapted from Harris *et al.* 2013). (*top right*) The normalized, deprojected visibility profiles for the UZ Tau E disk continuum emission, where colors correspond to different observing wavelengths: (*from left to right*) 0.88, 1.4, 2.9, 8.0, and 9.8 mm. This panel demonstrates unequivocally that the size of the dust emission is anti-correlated with the observing wavelength. (*bottom right*) The derived radial profile of the dust opacity index, $\beta(r)$, for the UZ Tau E (*solid*) and AS 209 (*hatched*) disks (the regions correspond to 95% confidence intervals), from Harris *et al.* (2013) and Pérez *et al.* (2012), respectively. In both cases, the increasing $\beta(r)$ is consistent with the size-sorting of disk solids expected from models of particle growth and radial drift.

frequencies. However, an ongoing key project at the upgraded VLA ("Disks@EVLA", PI Claire Chandler) is facilitating a vast improvement in this work by providing excellent high-resolution observations of the dust continuum emission from disks at cm wavelengths. An illustrative example from this project – the disk around UZ Tau E – is highlighted in Figure 2. In a recent analysis by Harris *et al.* (2013), continuum emission from this disk was resolved at five distinct wavelengths from 0.88 to 9.8 mm, and was found to exhibit a clear anti-correlation between the spatial extent of the continuum emission (i.e., the apparent size) and the observing wavelength. Since thermal emission peaks at a wavelength comparable to the representative particle size, these data hint that larger solids are preferentially concentrated near the central host star. That behavior was quantified by modeling the radial optical depth profile at each wavelength, and then fitting a power-law spectrum at each radius to derive $\beta(r)$ (see Isella *et al.* 2010 for details). This $\beta(r)$ profile (Fig. 2; bottom right) shows a pronounced increase, confirming the presence of a radial variation in the particle size distribution. A similar constraint on $\beta(r)$ for the AS 209 disk is also shown for reference (see Pérez *et al.* 2012). A comparison to the theoretical models of Birnstiel *et al.* suggest that these data are consistent with particle growth by more than an order of magnitude, and can be readily explained by a growth process that is limited by radial drift (as opposed to turbulent fragmentation). Moreover, the larger scale Disks@EVLA project is finding that this firm evidence for the radial size-sorting of disk solids is a *generic feature* of protoplanetary disks. While that unfortunately implies that one cannot measure the dust surface densities, Σ_d, using data at only a single wavelength, it does suggest that we have an exciting opportunity to empirically characterize the formation of planetesimals in these disks.

According to simulations of particle evolution, the observational signature described above should be accompanied by a radial variation in the dust-to-gas mass ratio, ζ, such that the outer disk is selectively depopulated of mm/cm-sized particles. In essence, $\zeta(r)$ should be a decreasing function. Unfortunately, we do not know how to robustly measure disk gas densities (see §4), so a *direct* observational constraint on $\zeta(r)$ is not possible. Nevertheless, we can employ an analogous technique to the one described above to indirectly search for a non-constant $\zeta(r)$ by comparing the spatial distributions of gas and dust tracers. When the apparent sizes of a disk are measured in a CO rotational transition and the neighboring dust continuum, observers typically find a notable discrepancy: the molecular gas appears much more spatially extended than the dust (e.g., Piétu *et al.* 2007; Isella *et al.* 2007). However, an interpretation of this discrepancy requires caution: bright gas tracers (like CO emission lines) are optically thick, so they are sensitive to very small amounts of material in the outer disk, whereas the dust continuum emission is optically thin, and therefore is subject to sensitivity or model interpretation issues at large disk radii (see Hughes *et al.* 2008 for a discussion of these issues). With a sufficiently sensitive dataset, a modeling procedure based on a logical negation can be used to indirectly infer a decreasing $\zeta(r)$: (1) construct a model structure that reproduces the observations of dust; (2) assume a constant ζ at all radii; (3) simulate the corresponding model of the molecular line emission; (4) compare to observations; (5) invariably find that the model line emission is much too compact, and therefore infer that assumption (2) is incorrect – so, $\zeta(r)$ must be a decreasing function. This is a difficult task, requiring extensive radiative transfer modeling, but there is mounting evidence that indeed confirms the expected gas–dust size discrepancy (by a factor of $\sim$2-5 in radius) in some disks, providing strong, qualitative support for the predictions of radial drift models (Panić *et al.* 2009; Andrews *et al.* 2012; Rosenfeld *et al.* 2013b). In the near future, I expect that the combination of this approach to characterize $\zeta(r)$ with the constraints on $\beta(r)$ described above will offer new insights on the general migration of disk solids.

4. Toward New Constraints on Gas Disk Structures

Ultimately, in the context of constraining the initial conditions available for giant planet formation, we want to observationally infer gas densities in protoplanetary disks. However, this is a remarkably challenging measurement. The absence of a permanent dipole moment for the H_2 molecule means that the primary constituent of the gas reservoir is effectively "dark." To date, most studies have depended on dust as an overall density tracer: but given the discussion in §3 regarding the mounting evidence for spatial variations in the particle size distribution and dust-to-gas ratio, this is clearly not a reliable approach. We could instead rely on trace molecular species as a proxy for H_2, but in general a given molecular abundance (relative to H_2) is not known *a priori*. Even with guidance from a chemical network model, the optically thin (usually isotopologue) emission lines for a given species are quite weak, and will require a considerable time investment even with ALMA. Moreover, a self-consistent model of spatially resolved

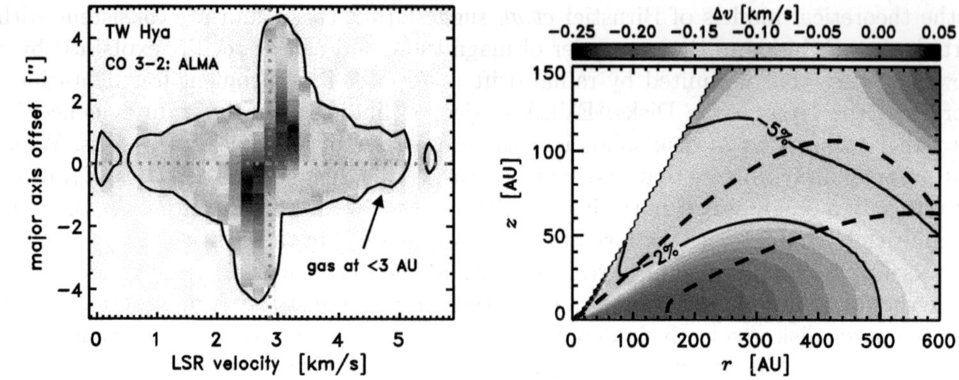

Figure 3. (*left*) A position-velocity diagram for the CO $J = 3{-}2$ line in the TW Hya disk, made from ALMA science verification data. Since this disk is observed with a nearly pole-on orientation, the presence of emission wings extending a few km s^{-1} from the systemic velocity is a clear indication of gas residing inside the inferred dust cavity even though the angular resolution of the data is much larger than the cavity size (see Rosenfeld *et al.* 2012). (*right*) A two-dimensional map of the deviations from the Keplerian velocity for a nominal structure model of the HD 163296 disk, based on an analysis of ALMA science verification observations of a range of CO emission lines (Rosenfeld *et al.* 2013a). Dashed contours mark the CO emission region, indicating that non-Keplerian motions caused by geometric and pressure effects (and perhaps self-gravity) in the disk atmosphere are on the scale of the observational spectral resolution. In the future, such kinematic effects can be used to help better constrain the gas structure.

line emission needs to consider the complex inter-dependence of the multi-dimensional temperature and density structures (e.g., see Rosenfeld *et al.* 2013a).

Admittedly, this paints a somewhat bleak picture for the prospects of measuring gas densities. At least for large samples, that might be valid. However, there is an incremental path toward achieving this goal for individual disks. The first step is to try and reconstruct the temperature distribution of the gas, as a function of both radius and height in the disk. In principal, this can be achieved in two (complementary) ways: (1) spectral mapping of a sequence of line ratios, using transitions with a range of optical depths to probe the temperatures at a range of different layers in the disk atmosphere (e.g., Dartois *et al.* 2003); and (2) locating molecular depletion boundaries directly (e.g., Qi *et al.* 2011; see also the contribution from Qi in this volume). Such observations would then need to be coupled to a suite of optically thin lines (e.g., a $C^{18}O$ ladder), and the structure iterated to ensure a physically consistent coupling between the gas densities and temperatures. To my knowledge, this has not yet been convincingly accomplished. But, in principal, the molecular excitation, radiative transfer, and chemical network tools required for the job are all available or under active development: it is all just a matter of accumulating the data with sufficient quality and resolution before putting them to use.

In most analyses of spatially and spectrally resolved line data, the co-dependence of the disk velocity field and gas structure is a complication that is often left untreated (not for invalid reasons: the current data does not typically warrant addressing it). But the close relationship between disk structure and kinematics will hopefully provide some mutually beneficial opportunities in the ALMA era (see Figure 3). One such example is the exploitation of the regular differential (Keplerian) rotation of the disk to probe gas content at spatial scales much smaller than the angular resolution affords: in effect, a kinematic super-resolution based on the wings of an emission line. So far, this approach has been used to assess whether gas is present inside the dust cavities of transition disks (see §2; Dutrey *et al.* 2008; Rosenfeld *et al.* 2012), although it could certainly be used in more general cases to probe physical conditions in the inner regions of a disk. Another example would be to take advantage of the small deviations from Keplerian rotation expected for emission lines that form high above the disk midplane, due to both geometric and pressure effects (and potentially self-gravity; see Rosenfeld *et al.* 2013a), to better understand the vertical distribution of gas densities and temperatures; in effect, to use kinematics as an additional constraint when constructing a gas structure model.

5. Summary

The anticipation of mature ALMA operations comes at a dynamic time in circumstellar disk research. This review highlighted a few key subjects being pursued in this field, including: (1) the use of disk structures as signposts of early giant planet formation; (2) the mounting evidence for the observable signatures of the growth and migration of disk solids; and (3) some prospects for better characterizing the gas content of these disks, in the context of initial conditions for planet formation. The results reviewed above should be considered the first steps in an era of new precision – and probably surprises – in the development of an observational component to the study of planet formation.

I would like to thank the organizers for inviting me to give this review, and to acknowledge the useful input and hard work of my colleagues that participated in much of the analysis and results that are presented here, especially: Katherine Rosenfeld, Robert Harris, David Wilner, Meredith Hughes, Til Birnstiel, Catherine Espaillat, Joanna Brown, Paola Pinilla, Geoff Mathews, and the Disks@EVLA team.

References

Alibert, Y., Mordasini, C., Benz, W., & Winisdoerffer, C. 2005, *A&A*, 434, 343

Andrews, S. M., Wilner, D. J., Hughes, A. M., Qi, C., & Dullemond, C. P. 2009, *ApJ*, 700, 1502

Andrews, S. M., *et al.* 2011, *ApJ*, 732, 42 (2011a)

Andrews, S. M., Rosenfeld, K. A., Wilner, D. J., & Bremer, M. 2011, *ApJ*, 742, L5 (2011b)

Andrews, S. M., *et al.* 2012, *ApJ*, 744, 162

Baruteau, C., & Masset, F. 2013, *Tides in Astronomy and Astrophysics, Lecture Notes in Physics*, eds. Souchay, Mathis, & Tokieda (Springer-Verlag: Berlin), 861, 201

Beckwith, S. V.W., Sargent, A. I., Chini, R. S., & Güsten, R. 1990, *AJ*, 99, 924

Beckwith, S. V.W. & Sargent, A. I. 1991, *ApJ* 381, 250

Birnstiel, T., Dullemond, C. P., & Brauer, F. 2010, *A&A*, 513, 79 (2010a)

Birnstiel, T., *et al.* 2010, *A&A*, 516, L14 (2010b)

Birnstiel, T., Ormel, C. W., & Dullemond, C. P. 2011, *A&A*, 525, 11

Birnstiel, T., Klahr, H., & Ercolano, B. 2012, *A&A*, 539, 148 (2012a)

Birnstiel, T., Andrews, S. M., & Ercolano, B. 2012, *A&A*, 544, 79 (2012b)

Brauer, F., *et al.* 2007, *A&A*, 469, 1169

Brauer, F., Dullemond, C. P., & Henning, T. 2008, *A&A*, 480, 859

Brown, J. M., Blake, G. A., Qi, C., Dullemond, C. P., & Wilner, D. J. 2008, *ApJ*, 675, L109

Brown, J. M., *et al.* 2009, *ApJ*, 704, 496

Bryden, G., Chen, X., Lin, D. N.C., Nelson, R. P., & Papaloizou, J. C.B. 1999, *ApJ*, 514, 344

Crida, A., Morbidelli, A., & Masset, F. 2006, *Icarus*, 181, 587

Crida, A. & Morbidelli, A. 2007, *MNRAS*, 377, 1324

D'Alessio, P., Calvet, N., & Hartmann, L. 2001, *ApJ*, 553, 321

Dartois, E., Dutrey, A., & Guilloteau, S. 2003, *A&A*, 399, 773

Dodson-Robinson, S. E. & Salyk, C. 2011, *ApJ*, 738, 131

Dong, R., *et al.* 2012, *ApJ*, 750, 161

Draine, B. T. 2006, *ApJ*, 636, 1114

Dutrey, A., *et al.* 2008, *A&A*, 490, L15

Espaillat, C., *et al.* 2007, *ApJ*, 670, L135
Espaillat, C., *et al.* 2010, *ApJ*, 717, 441
Espaillat, C., *et al.* 2012, *ApJ*, 747, 103
Fang, M., *et al.* 2009, *A&A*, 504, 461
Finkbeiner, D. P., Davis, M., & Schlegel, D. J. 1999, *ApJ*, 524, 867
Follette, K. B., *et al.* 2013, *ApJ*, 767, 10
Guilloteau, S., Dutrey, A., Piétu, V., & Boehler, Y. 2011, *A&A*, 529, 105
Harris, R. J., *et al.* 2013, *ApJ*, submitted
Hildebrand, R. H. 1983, *QJRAS*, 24, 267
Huélamo, N., Lacour, S., Tuthill, P., Ireland, M., Kraus, A., & Chauvin, G. 2011, *A&A*, 528, L7
Hughes, A. M., *et al.* 2007, *ApJ*, 664, 536
Hughes, A. M., Wilner, D. J., Qi, C., & Hogerheijde, M. R. 2008, *ApJ*, 678, 1119
Hughes, A. M., *et al.* 2009, *ApJ*, 698, 131
Ida, S., & Lin, D. N.C. 2004, *ApJ*, 604, 388
Ida, S., & Lin, D. N.C. 2005, *ApJ*, 626, 1045
Isella, A., Testi, L., Natta, A., Neri, R., Wilner, D., & Qi, C. 2007, *A&A*, 469, 213
Isella, A., Carpenter, J. M., & Sargent, A. I. 2009, *ApJ*, 701, 260
Isella, A., Carpenter, J. M., & Sargent, A. I. 2010, *ApJ*, 714, 1746
Kraus, A. L., Ireland, M. J., Martinache, F., & Hillenbrand, L. A. 2011, *ApJ*, 731, 8
Kraus, A. L. & Ireland, M. J. 2012, *ApJ*, 745, 5
Lubow, S. H., Seibert, M., & Artymowicz, P. 1999, *ApJ*, 526, 1001
Lubow, S. H. & D'Angelo 2006, *ApJ*, 641, 526
Kley, W. & Nelson, R. P. 2012, *ARAA*, 50, 211
Mathews, G. S., Williams, J. P., & Ménard, F. 2012, *ApJ*, 753, 59
Mordasini, C., Alibert, Y., Benz, W. 2009, *A&A*, 501, 1139
Owen, J. E., Ercolano, B., & Clarke, C. J. 2011, *MNRAS*, 412, 13
Paardekooper, S.-J., & Mellema, G. 2006, *A&A*, 453, 1129
Panić, O., Hogerheijde, M. R., Wilner, D., & Qi, C. 2009, *A&A*, 501, 269
Papaloizou, J. C.B., Newlson, R. P., Kley, W., Masset, F. S., & Artymowicz, P. 2007, in *Protostars & Planets V*, eds. B. Reipurth, D. Jewitt, & K. Keil (Univ. Arizona Press: Tucson), 655
Pérez, L. M., *et al.* 2012, *ApJ*, 760, L17
Piétu, V., Dutrey, A., Guilloteau, S., Chapillon, E., & Pety, J. 2006, *A&A*, 460, L43
Piétu, V., Dutrey, A., & Guilloteau, S. 2007, *A&A*, 467, 163
Pinilla, P., Benisty, M., & Birnstiel, T. 2012, *A&A*, 545, 81
Pott, J.-U., *et al.* 2010, *ApJ*, 710, 265
Qi, C., *et al.* 2011, *ApJ*, 740, 84
Rice, W. K.M., Armitage, P. J., Wood, K., & Lodato, G. 2006, *MNRAS*, 373, 1619
Ricci, L., *et al.* 2010, *A&A*, 512, 15 (2010a)
Ricci, L., Testi, L., Natta, A., & Brooks, K. J. 2010, *A&A*, 521, 66 (2010b)
Rosenfeld, K. A., *et al.* 2012, *ApJ*, 757, 129
Rosenfeld, K. A., *et al.* 2013, *ApJ*, 774, 16 (2013a)
Rosenfeld, K. A., *et al.* 2013, *ApJ*, 775, 136 (2013b)
Salyk, C., Blake, G. A., Boogert, A. C. A., & Brown, J. M. 2009, *ApJ*, 699, 330
Skrutskie, M. F., *et al.* 1990, *AJ*, 99, 1187
Strom, K. M., Strom, S. E., Edwards, S., Cabrit, S., & Skrutskie, M. F. 1989, *AJ*, 97, 1451
Takeuchi, T. & Lin, D. N.C. 2002, *ApJ*, 581, 1344
Takeuchi, T. & Lin, D. N.C. 2005, *ApJ*, 623, 482
Weidenschilling 1977, *MNRAS*, 180, 57
Wolf, S. & D'Angelo, G. 2005, *ApJ*, 619, 1114
Zhu, Z., Nelson, R. P., Hartmann, L., Espaillat, C., & Calvet, N. 2011, *ApJ*, 729, 47
Zhu, Z., Nelson, R. P., Dong, R., Espaillat, C., & Hartmann, L. 2012, *ApJ*, 755, 6
Zsom, A., Ormel, C. W., Güttler, C., Blum, J., & Dullemond, C. P. 2010, *A&A*, 513, 57

Discussion

GAIDOS: Are any of the lines that ALMA can access so intrinsically narrow that pressures, and hence gas mass surface densities, could be estimated from pressure broadening?

ANDREWS: No. So long as the disk is less massive than the stellar host, the gas densities are always sufficiently low that thermal broadening will always dominate pressure broadening (by orders of magnitude) for any reasonable molecular cross section.

RAFIKOV: You find high values of the opacity index β, at the level of $\beta \gtrsim 3$, in the outer parts of the disk. What material properties of grains can give rise to such behavior?

ANDREWS: This is essentially an artifact of the modeling process, and not necessarily produced by a real dust population. In practice, any estimate of β increases rapidly at large radii where we do not detect any cm-wave emission with the VLA.

LIN: Is there any inconsistency between the rarity of companions around young stars from the NICI survey and the detection of transitional disks around 1/3 of the systems?

ANDREWS: Not yet, but I agree that the statistics are uncomfortably close to a conflict. If we consider the total population of a young cluster (rather than the biased samples imaged so far) – including the apparently disk-less members – the transition disk fraction seems consistent with the limits placed on gas giant planets at large radii from direct imaging surveys (i.e., $\lesssim$5–10%). However, if future surveys with ALMA end up finding a lot more transition disks, then we would need to re-consider the planet-disk interaction hypothesis as a primary cause for the observed disk structures. One mechanism worth considering is the potential for migration to alleviate the tension: perhaps the planets that open the cavities will later migrate to smaller stellocentric distances.

DODSON-ROBINSON: Have there been any attempts to model the evolution of the gas/solids ratio in disks based on grain growth and even planetesimal formation models?

ANDREWS: This is really an ongoing effort by the community, but I can refer you to the preliminary models by Birnstiel *et al.* (2012) as a good place to start. I expect a great deal more work in this direction in the coming few years.

PUDRITZ: Do we have any observational constraints on the growth timescales for dust in disks?

ANDREWS: The short answer is "not really." It is clear that large particles (mm/cm sizes) have already formed in ~Myr-old disks, but pinpointing timescales at finer time resolution is a challenge. Theoretical models expect such growth to be very fast indeed, so the data are not quite yet offering robust constraints on this question.

HALES: You showed an example of a disk that showed different morphologies in the 1.3 mm SMA map compared to your recent 450 μm ALMA data (a central peak versus a ring at 1.3 mm). Could you please comment on this differentce?

ANDREWS: I believe this is an optical depth effect, where the inner disk is optically thick in the ALMA Band 9 data, but not in the SMA 1.3 mm data: but this has yet to be examined quantitatively. (Note: these data are not discussed or displayed in this review. The question and response are included for the sake of completeness.)

Exploring the Formation and Evolution of Planetary Systems
Proceedings IAU Symposium No. 299, 2013
M. Booth, B. C. Matthews & J. R. Graham, eds.

doi:10.1017/S1743921313007989

Planet formation in action: resolved gas and dust images of a transitional disk and its cavity

Nienke van der Marel[1], Ewine F. van Dishoeck[1,2], Simon Bruderer[2], Til Birnstiel[3], Paola Pinilla[4], Cornelis P. Dullemond[4], Tim A. van Kempen[1,5], Markus Schmalzl[1], Joanna M. Brown[3], Gregory J. Herczeg[6], Geoffrey S. Mathews[1] and Vincent Geers[7]

[1] Leiden Observatory, Leiden University, P.O. Box 9513, 2300 RA Leiden, The Netherlands,

[2] Max-Planck-Institut für Extraterrestrische Physik Giessenbachstrasse 1, 85748 Garching, Germany,

[3] Harvard-Smithsonian Center for Astrophysics, 60 Garden Street, Cambridge, MA 02138, USA

[4] Heidelberg University, Center for Astronomy, Institute for Theoretical Astrophysics, Albert Ueberle Str. 2, 69120, Heidelberg, Germany,

[5] Joint ALMA Offices, Av. Alonso de Cordova, Santiago, Chile

[6] Kavli Institute for Astronomy and Astrophysics, Peking University, Yi He Yuan Lu 5, Hai Dian Qu, 100871 Beijing, PR China

[7] Dublin Institute for Advanced Studies, 31 Fitzwilliam Place, Dublin 2, Ireland

Abstract. Planet formation and clearing of protoplanetary disks is one of the long standing problems in disk evolution theory. The best test of clearing scenarios is observing systems that are most likely to be actively forming planets: the transitional disks with large inner dust cavities. We present the first results of our ALMA (Atacama Large Millimeter/submillimeter Array) Cycle 0 program using Band 9, imaging the Herbig Ae star Oph IRS 48 in CO 6–5 and the submillimeter continuum in the extended configuration. The resulting $\sim 0.2''$ spatial resolution completely resolves the cavity of this disk in the gas and the dust. The gas cavity of IRS 48 is half as large as the dust cavity, ruling out grain growth and photoevaporation as the primary cause of the truncation. On the other hand, the continuum emission reveals an unexpected large azimuthal asymmetry and steep edges in the dust distribution along the ring, suggestive of dust trapping. We will discuss the implications of the combined gas and dust distribution for planet formation at a very early stage. This is one of the first transition disks with spatially resolved gas inside the cavity, demonstrating the superb capabilities of the Band 9 receivers.

Keywords. planetary systems: formation,planetary systems: protoplanetary disks

---

1. Introduction

Transitional disks are protoplanetary disks around young stars with a hole or gap in their small dust grain distribution. These disks possibly form the evolutionary link between optically thick protoplanetary disks around T Tauri stars and optically thin debris disks around older systems (Williams & Cieza 2011). There are three different mechanisms in this phase of disk evolution that could cause the appearance of such a dust hole (e.g. Najita *et al.* 2007): (i) increased grain growth in the central region of the disk that lowers the emitting surface area below the detectable limit; (ii) photoevaporation by the radiation from the star; (iii) the clearing by either a stellar or (proto)planetary

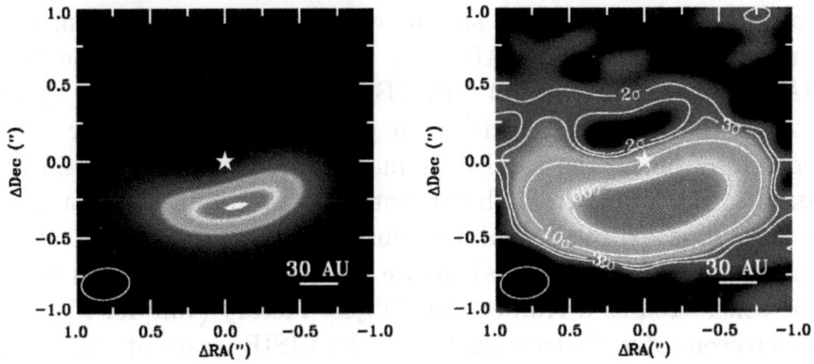

Figure 1. The inclined disk around IRS 48 in dust as observed with ALMA Band 9 observations. The ALMA beam during the observations is indicated with a ellipse. The figure shows the 685 GHz continuum emission in linear (left) and logarithmic color scale (right), with white contours at 2, 3, 10 and 100 σ ($\sigma = 0.82$ mJy/beam).

companion. It is therefore not possible to assign the origin of the hole by only looking at the dust distribution. However, the gas distribution will depend on the mechanism: in the case of grain growth, the gas density is not affected at all and there will still be gas present inside the dust hole; in the case of photoevaporation, the gas and dust are cleared simultaneously from the inside out; in the case of a companion the dust hole is expected to be larger than the gas hole and an inner (dust) disk may still be present. The key to understanding the origin of the hole in a transitional disk is therefore resolving both the gas and dust distribution at scales of <50 AU, which ALMA finally allows us to do.

We present ALMA Cycle 0 results of one of these transitional disks, Oph IRS 48 (van der Marel *et al.* 2013). This Herbig Ae star, located in the star forming region Ophiuchus, was already known as a peculiar disk: VLT-VISIR resolved mid infrared imaging shows a large dust ring at 18.7 μm with a radius of ~55 AU (Geers *et al.* 2007), whereas VLT-CRIRES near infrared observations of the rovibrational CO v = 1–0 line reveal a hot gas ring at 30 AU radius (Brown *et al.* 2012a). However, both observations only address the hot dust and gas in the inner disk that is directly heated by the star: submillimeter observations are required for the bulk cold disk material. At the distance of Ophiuchus (120 pc), the spatial resolution of ALMA ($\sim 0.2''$) provides a beam radius of ~14 AU, sufficient to distinguish between these two reported hole sizes and resolve the disk in both the gas and the dust. The ALMA observations were taken in July 2012, using the Band 9 instrument (~ 690 GHz) in the Extended Configuration, tracing both the ^{12}CO 6–5 line and the 685 GHz dust continuum.

2. Results

The ^{12}CO data show a full Keplerian gas disk at 50° inclination, except for a strong absorption around $v_{\rm LSR} = 3.5$ km s^{-1} due to foreground clouds. Otherwise, the spectral profile is symmetric around the source velocity (4.5 km s^{-1}) and the highest velocity channels are consistent with a gas hole with a radius of 20 AU (van der Marel *et al.* 2013, Bruderer *et al.* 2013). On the other hand, the dust continuum emission that was expected to be ring-shaped, similar to the mid infrared emission, reveals a highly asymmetric structure on the southern side of the disk (see Figure 1). This crescent feature has a peak intensity of 390σ, with $\sigma = 0.8$ mJy/beam, with no significant detectable emission in the northern part of the disk, indicating a contrast of a factor >130.

The crescent could be fitted with an arc with an azimuthal extent of ~100° and a radial extent between 45 and 80 AU from the star. Combining the total 690 GHz flux with SMA fluxes at 345 and 230 GHz (Brown *et al.* 2012b), the spectral slope α ($F_\nu \sim \nu^\alpha$) is 2.7 ± 0.25, consistent with dust growth up to millimeter sizes. Assuming a dust temperature of 60 K, the total dust mass of the millimeter grains is derived as ~9 Earth masses. Considering the high concentration of millimeter dust particles in one region of the disk, it is expected that the dust grains in this region will grow up to a kilometer in size (planetesimal sizes) before being scattered around the ring, so the crescent can be considered as a Kuiper Belt Object factory (van der Marel *et al.* 2013). Comparing the crescent of millimeter dust with the VISIR image at 18.7 μm (tracing the hot micron-sized dust particles), there is not only a separation between the gas and the millimeter dust, but also between the small and large dust particles (Figure 1).

3. Implications: dust trapping

The size of the gas hole in combination with the radial extent of the dust continuum implies that a companion is responsible for the hole in this transitional disk. This can be understood in the framework of *dust trapping*. In a 'normal' protoplanetary disk without holes or gaps, dust evolution is governed by a balance between dust coagulation and fragmentation (by collisions) and radial inward drift (Weidenschilling 1977).

Radial drift is caused by the interaction with the gas: the gas in a Keplerian disk moves at sub-Keplerian speeds, since it feels the force of its own pressure gradient which is usually pointing inwards. Small dust particles are coupled to the gas and move along the same orbits, while larger dust grains decouple from the gas and approach Keplerian velocities. Therefore the larger dust particles feel a constant head wind, which causes the particle to lose angular momentum and drift inwards. As a consequence, large dust particles will always move towards regions of higher pressure. One of the main problems in dust evolution modeling is that dust particles in the outer disk can not grow beyond millimeter sizes because of their rapid inward drift ('radial drift barrier').

One of the proposed solutions for the radial drift barrier is dust trapping in local pressure maxima in the disk. Pinilla *et al.* (2012) used a combination of FARGO hydrodynamic models in combination with dust evolution modeling (Birnstiel *et al.* 2010) to show that when a planet is placed in orbit in a gas disk, it will quickly clear its path and create an overdensity (radial pressure bump) at the edge of the gap, where the large dust particles accumulate within 10^5 years and remain trapped, allowing further dust growth. The created dust hole will always be larger than the gas hole, with their sizes depending on the viscosity and the planet mass. The radial separation between gas and dust in IRS 48 indicates that the companion has to be massive, >10 $M_{\rm Jup}$ (van der Marel *et al.* 2013). It was demonstrated that for such a massive companion within a disk of moderate viscosity the pressure bump becomes Rossby unstable and quickly forms a long-lived vortex in the gas. Such a vortex (a small overdensity in the gas) can trap the dust in both the radial and azimuthal direction, where the trapping efficiency depends on the size of the dust particle and the viscosity (Birnstiel *et al.* 2013). An overdensity of a factor 2–3 in the gas can lead to more than a factor of 100 overdensity in the millimeter dust. The high asymmetry in the dust crescent in IRS 48 could thus be reproduced using this scenario (van der Marel *et al.* 2013).

We propose the dust trapping scenario with a 10 $M_{\rm Jup}$ companion at 20 AU as an explanation for the separation between the millimeter-sized dust, gas and micron-sized dust in IRS 48. A dust trap located at 60 AU distance from the star has an orbital period of 330 years, so the orbital shift will be measurable with ALMA within 20 years. This dust trap, acting as Kuiper Belt Object factory, is analogous to a planet factory at smaller radii around other stars. The observational evidence for the dust trapping mechanism may provide a new view on planet formation and disk evolution models. The question remains whether IRS 48 is a special case, or if dust trapping is a common phenomenon in transition disks. ALMA Cycle 0 observations of the transitional disk HD142527 also show a high dust asymmetry without large azimuthal variations in the gas (Casassus *et al.* 2013). ALMA will be able to provide answers for other transition disks, providing even higher spatial resolution in the coming years, allowing both the gas and dust to be resolved.

References

Birnstiel, T., Dullemond, C. P., & Brauer, F. 2010, *A&A*, 513, A79

Birnstiel, T., Dullemond, C. P., & Pinilla, P. 2013, *A&A*, 550, L8

Brown, J. M., Herczeg, G. J., Pontoppidan, K. M., & van Dishoeck, E. F. 2012, *ApJ*, 744, 116

Brown, J. M., Rosenfeld, K. A., Andrews, S. M., Wilner, D. J., & van Dishoeck, E. F. 2012, *ApJ*, 758, 30

Bruderer, S., van der Marel, N., van Dischoeck, E. F., & van Kempen, T. A., 2013, *A&A*, submitted

Casassus, S., van der Plas, G., M, S. P., Dent, W. R. F., Fomalont, E., Hagelberg, J., *et al.* 2013, *Nature*, 493, 191

Geers, V. C., Pontoppidan, K. M., van Dishoeck, E. F., Dullemond, C. P., Augereau, J.-C., *et al.* 2007, *A&A*, 469, L35

Najita, J. R., Strom, S. E., & Muzerolle, J., 2007, *MNRAS*, 378, 369

Pinilla, P., Benisty, M., & Birnstiel, T., 2012, *A&A*, 545, A81

van der Marel, N., van Dishoeck, E. F., Bruderer, S., Birnstiel, T., Pinilla, P., *et al.* 2013, *Science*, 340, 1199

Williams, J. & Cieza, L. 2011, *ARAA*, 49, 67

Weidenschilling, S. J. 1977, *MNRAS*, 180, 57

Discussion

BARSONY: I had e-mailed Nienke an image of IRS 48 at 12.5 micron that is very similar to the 18.7 micron image. I just wanted to note that systems like this may be common or not, but in terms of detecting the small particles by mid infrared continuum, that is going to be incredibly difficult. We observed a 100 sources in ρ Oph with Keck with a mid infrared instrument and this was the only source that showed this kind of structure. Since it is an A0 star it has enough UV to excite the small PAHs and the very small grains so you can see it reradiating in the mid infrared, but for lower mass systems it would be impossible to find where the small dust particles are.

HALES GEBRIM: You show that the Rossby instability can reproduce the submillimeter data well. Have you tried plugging the same model in radiative transfer code to see if it can also reproduce the scattered light or mid infrared images?

VAN DER MAREL: No, we haven't done that yet, but it's something we would like to try.

DONG: Is it possible to confirm the presence of the planet?

VAN DER MAREL: It would be difficult, because this disk is in a region of very high optical extinction, so optical direct imaging will be impossible. We hope there will be some result from a NACO program with Sparse Aperture Masking, looking for a companion in the near infrared. But this is still work in progress.

Exploring the Formation and Evolution of Planetary Systems
Proceedings IAU Symposium No. 299, 2013
M. Booth, B. C. Matthews & J. R. Graham, eds.

doi:10.1017/S1743921313007990

The VLTi/PIONIER survey of southern TTauri disks†

Fabien Anthonioz[1] F. Ménard[2,1], C. Pinte[1], W-F. Thi[1], J.-B. Lebouquin[1], J.-P. Berger[3], M. Benisty[1], O. Absil[4], G. Duchène[5,1], B. Lazareff[1], F. Malbet[1], R. Millan-Gabet[6], W. Traub[7], and G. Zins[1]

[1]Institut de Planétologie et d'Astrophysique de Grenoble, 38400 St Martin d'Hères, France.
[2]UMI-FCA, CNRS/INSU, France (UMI 3386), and Universidad de Chile, Santiago, Chile.
[3]European Southern Observatory, 85748 Garching, Germany.
[4] Institut d'Astrophysique et de Géophysique, Université de Liège, Sart Tilman, Belgium.
[5]Astronomy Department, University of California, Berkeley, CA 94720-3411 USA.
[6]California Institute of Technology, Pasadena, CA 91125, USA.
[7]Jet Propulsion Laboratory (NASA/JPL), Pasadena, CA, 91109, USA.
email: `Fabien.Anthonioz@obs.ujf-grenoble.fr`

Abstract. Studying the inner regions of protoplanetary disks (1-10 AU) is of importance to understand the formation of planets and the accretion process feeding the forming central star. Herbig AeBe stars are bright enough to be routinely observed by Near IR interferometers. The data for the fainter T Tauri stars is much more sparse. In this contribution we present the results of our ongoing survey at the VLTI. We used the PIONIER combiner that allows the simultaneous use of 4 telescopes, yielding 6 baselines and 3 independent closure phases at once. PIONIER's integrated optics technology makes it a sensitive instrument. We have observed 22 T Tauri stars so far, the largest survey for T Tauri stars to this date.

Our results demonstrate the very significant contribution of an extended component to the interferometric signal. The extended component is different from source to source and the data, with several baselines, offer a way to improve our knowledge of the disk geometry and/or composition. These results validate an earlier study by Pinte *et al.* 2008 and show that the dust inner radii of T Tauri disks now appear to be in better agreement with the expected position of the dust sublimation radius, contrary to previous claims.

Keywords. techniques: interferometric, stars: pre–main-sequence, interferometric survey

1. Introduction

The inner parts of the disks are intensively studied for the formation of terrestrial planets, the accretion process and more generally the evolution of the disk shape and composition. While near-infrared (NIR) interferometry is able to resolve these regions, the first data sets published were limited to a few baselines (generally 1 or 2) at low sensitivity. Simple geometrical models were used to interpret the data, e.g., the (gaussian) ring model. That model is based on the assumption that all of the NIR emission comes from the heated disk's inner rim thus creating a luminous ring around the star. Inner radii inferred by such model are coherent with the sublimation radius for the more luminous Herbig stars but suggest significant deviations for the less luminous T Tauri stars. Various physical mechanisms have been proposed to explain this unexpected discrepancy (Eisner *et al.* 2007). However, Pinte *et al.* (2008) resolved this discrepancy

† Data obtained at the ESO VLTI as part of programmes 086. C-0433, 087. C-0703, 088. C-0670, 089. C-0769 and 091.C-0570.

by taking into account scattered light in details in the models allowed to resolve the apparent discrepancy between the inferred inner radius and the sublimation radius for low luminosity stars. Scattered light has a further interesting observable consequence: it is expected to produce a rapid decrease of the visibilities at short baselines, which is due to the fact that the scattered "halo" is very extended. Detecting such "drop-off" requires a good u-v coverage of faint targets. We present the results of our survey below.

2. PIONIER observations

The interferometric data were obtained with the four relocatable Auxiliary Telescopes (AT) of the Very Large Telescope Interferometer (Haguenauer *et al.* 2010) and the 4-telescope combiner PIONIER (Lebouquin *et al.* 2011). We observed 22 southern T Tauri stars brighther than 8.5 in H band during 25 nights distributed over 5 semesters. The weather condition were average for the majority of the nights with a total of 13/25 nights lost due to bad weather. The stars were selected based on available evidence for the presence of a circumstellar disk, usually NIR excess in the SED or evidence from radio interferometry.

The use of 4 telescopes simultaneously allows for measurements of 6 different baselines, resulting also in 3 independent closure phase measurements. PIONIER is also more sensitive than its predecessors, allowing a significant sample of the relatively fainter T Tauri stars to be observed. During the first 4 runs, the ATs were stationed in an extended array configuration, providing separations on the ground between telescopes ranging from 47 to 129 meters The last run has been carried out in compact configuration (separations between 11 and 36 meters).

The observation strategy was designed to intertwine the science target between different calibrators. A typical observation sequence was then CAL1 - SCI - CAL2 - SCI - CAL1 and repeat as needed, with each observation block (either science or calibrator) typically composed of 5 exposures each of which composed of 100 scans. Data reduction and calibration was performed by running the dedicated PNDRS package† (Lebouquin *et al.* 2011).

3. Description of the data sets

For the data collected in the extended array configuration, only the intermediate baselines of the visibility curve is sampled, i.e., neither the drop-off at short baselines nor the "fully resolved range" at very long baselines where the visibilities plateau are measured. In that middle range, the visibilities vary roughly as a straight line for simple star+disk sources. If an extended structure is present, causing a quick drop-off at short baselines, then extrapolation of the observed mid-range visibility curve to zero spacings should yield values below unity. We performed a linear fit on our sample (Figure 1). A large fraction of the visibility curves has an intercept at 0m baseline below V^2=1, as expected from the presence on an extended source of incoherent light, e.g., extended scattered light over the disk surface.

To explain that "drop-off", consider an object composed of a unresolved star + its disk, which visibility can be splitted (to first order) into the thermal emission coming from the inner rim and the scattered emission coming from all over the surface. With

† http://apps.jmmc.fr/~swmgr/pndrs

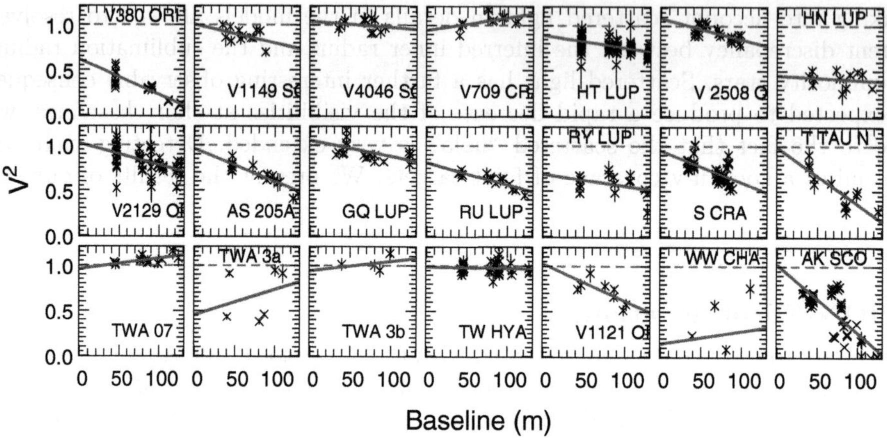

Figure 1. Visibility data (black points) of our sample and their associated linear regression (solid line).The dashed horizontal line indicate $V^2 = 1$. For the majority of the resolved non binary stars with low inclination (that is to say all the stars which visibility profile differ from one, minus TWA 3a and WW Cha), the intercept of the linear fit is below 1, indicating an extended component, varying from star to star.

these assumptions, the total visibility can be written :

$$V_{tot} = \frac{V_{star}F_{star} + V_{therm}F_{therm} + V_{ext}F_{ext}}{F_{star} + F_{therm} + F_{ext}}$$

with V the visibility and F the flux coming from the star (X_{star}), the disk's thermal emission (X_{therm}) and the disk's scattered light (X_{ext}).Because scattered light is spatially extended over a large area, its visibility curves drops rapidly with increasing baseline. The short baseline data now available provide constraints on the size of the scattering region as the drop-off is detected up to ≈10m baselines, therefore the angular scale of the scattering region must be larger than 33mas. For the 2 objects which have short baseline data (HT and RU Lupi which are at 150 parsecs), it corresponds to structures larger than 5 AU .

4. Detailed modelling

While the simple linear fitting of our data provides some information about the large scattering emission, it is a *qualitative* tool limited to simple systems with low inclinations.Moreover, that kind of analysis provides little information about the disk's structure and composition. To derive more quantitative information, we are running disk models and radiative transfer, including more data as well, for the individual sources.

As an exemple we present the model created to adjust both SED and the interferometric data of the TTauri star HT Lupi. To do so, we used the radiative transfert code *MCFOST* (Pinte *et al.* 2006), including full scattered light treatment. The modelled disk is composed of an inner dusty disk, a gap (to fit the 70μm MIPS photometric point and to take into account the clearing that might be operated by the close companion discovered by Ghez *et al.* 1997) and a massive outer disk. This model also matches the Spitzer IRS spectra when the inner disk is filled with small astrosilicate grains (as defined by Draine & Lee 1984). To constrain the inclination and the position angle, we calculated the values of the visibility and closure phase of our model (at the same baselines and for the same closed triangles) by rotating our image model in order to find the best position angle value, and doing this for each inclination. As we can see in the Figure 2 our model

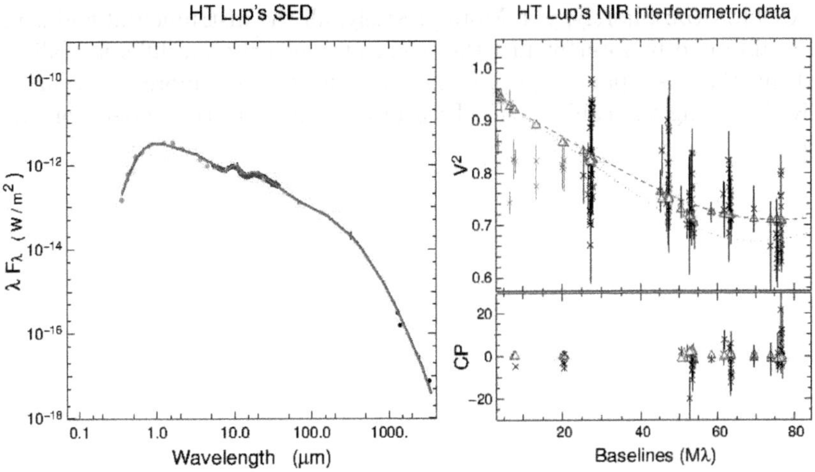

Figure 2. *Left*: fit of HT Lupi's SED. Our model is displayed by the bold red line, the dark blue line is the Spitzer spectrum and the other points represent the visible, NIR, IRAS and mm photometry. *Right*: modelling or the visibility profile (upper panel) and closure phase (lower panel). The data and error bars are displayed in black (or in blue for the recently acquired short baseline data). The dashed and dotted line are the visibility profiles of our model along the minor and major axis. the triangles are the visibility profile of our model taking into account the baselines orientation.

successfully reproduces the SED and long baseline interferometric data, but has to be refined further to fit our recently acquired short baseline data.

Acknowlegdment

We acknowledge funding from the EU FP7-2011 under Grant Agreement nr. 284405 (PERG06-GA-2009-256513) by the intermediate of the Disk Analysis (DIANA) program. This research has made use of the Simbad database operated at CDS, Strasbourg, France. Most of the data presented herein were obtained at the Very Large Telescope observatory, made possible by the generous financial support of the European Space Observatory.

References

Pinte, C., Ménard, F., Berger, J. P. & Benisty, M. 2008, *APJ*, 663, L63

Pinte, C., Ménard, F., Duchêne, G. & Bastien, P. 2006, *AAP*, 459, 797

Lebouquin, J.-B. *et al.* 2011, *AAP*, 535, 67

Draine, B.-T. & Hyung Mok Lee 1984, *AAP*, 285, 89

Eisner, J.-A., Hillenbrand, L.-A., White, R.-J., Bloom, J.-S., Akeson, R.-L., & Blake, C.-H. 1984, *AAP*, 669, 1072

Haguenauer, P. 2003, in: Alonso, J. *et al.* (eds.), *The very large telescope Interferometer: 2010 edition*, Proc. Society of Photo-Optical Instrumentation Engineers (SPIE), vol 7734

Ghez, A.-M., White, R.-J. and Simon, M. 1997, *AAP*, 490, 353

Discussion

Alycia Weinberger: Could you describe whether the uncertainties on the visibilities include all calibration uncertainties that is, is the absolute level of the visibilities reliable at the level of the error bars shown?

Fabien Anthonioz: Shortly, yes. More presisely, all the instrumental and atmospheric biases are calibrated by intertwining the science target between different calibrators (as explained in the last paragraph of the section 2). For more information about PIONIER (including the calibration), I warmly recommend the paper from Lebouquin *et al.* 2011.

Exploring the Formation and Evolution of Planetary Systems
Proceedings IAU Symposium No. 299, 2013
M. Booth, B. C. Matthews & J. R. Graham, eds.

doi:10.1017/S1743921313008004

HST Imaging of New Edge-on Circumstellar Disks in Nearby Star-forming Regions

K.R. Stapelfeldt[1], G. Duchêne[2,3], M. Perrin[4], S. Wolff[5], J.E. Krist[6], D.L. Padgett[1], F. Ménard[7,3], and C. Pinte[7,3]

[1]NASA GSFC, [2]UC Berkeley, [3]IPAG/Grenoble, [4]STScI, [5]JHU, [6]JPL/Caltech, [7]UMI 3386 CNRS / U. de Chile

Abstract. Edge-on, optically thick circumstellar disks have been previously imaged at subarcsecond resolution around about a dozen nearby young stellar objects. In these systems the central star is occulted from direct view, bright star image artifacts are absent, and the disk reflected light is clearly seen. Comparison of *Hubble Space Telescope (HST)* edge-on disk images with scattered light models has allowed key disk structural parameters and dust grain properties to be determined. Edge-on disks have been systematically undercounted to date: while 10% of young stars should statistically be occulted by their disk, the observed frequency is much less. Thus there is a significant potential for discovering and imaging new examples. *Spitzer Space Telescope* legacy science programs have provided the first good spectral energy distribution (SED) measurements for the previously known edge-on disks. These can be used as templates to identify new candidates in far-infrared survey datasets.

We report on the results of our *HST* program to image twenty-one edge-on disk candidates mostly selected from their SEDs. Eleven are well-resolved with radii ranging from 30-400 AU, nine for the first time and six showing highly collimated jets. Outstanding individual sources include a large and symmetric new template object, a highly flattened disk not accreting onto its central star, and an asymmetric disk with a misaligned jet which likely traces tidal perturbations in a binary system. Follow-up work to obtain ancillary data and perform scattered light modeling of the most symmetric disks is now being pursued. The results of this program will guide a new round of searches for these rare but important snapshots of protoplanetary disk evolution.

Keywords. Protoplanetary disks, Stars: pre-main sequence, ISM: jets and outflows

1. Background and Previous Work

The nearest large samples of protoplanetary disks are located in star-forming regions at distances of ~150 pc. Their internal structure provides important clues to the planet formation processes, but spatial resolutions below 0.2″ are required to resolve the relevant size scales. Scattered light imaging at 0.05″ resolution with *HST* at optical wavelengths, and with ground-based adaptive optics in the near-IR, has produced the most detailed images to date of these disks (Watson *et al.* 2007). The reflected light of ~35 young circumstellar disks has been reported; see http://circumstellardisks.org for an updated compilation. About 15 have been detected against the bright glare of directly visible starlight (e.g., TW Hya, Krist *et al.* 2000; IM Lupi, Pinte *et al.* 2008; J1604-2130, Mayama *et al.* 2012). In the other cases, nearly edge-on disks have been resolved as absorption lanes bisecting an extended reflection nebulosity. In addition to these local young stars, more than thirty backlit absorption disks have been imaged at these wavelengths toward the Orion Nebula at three times greater distance (Ricci *et al.* 2008).

Edge-on disks offer a unique opportunity for scattered light imaging, as the bright central star is occulted from view and thus does not pose a contrast problem. Furthermore, in an edge-on system the disk's thickness can be directly seen, tracing the local

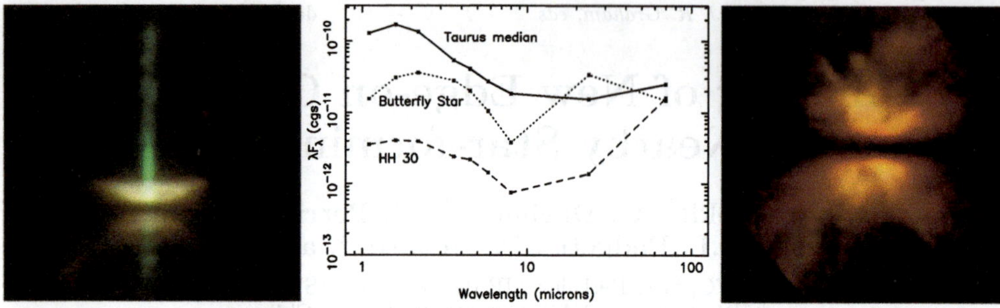

Figure 1. Left: *HST/WFPC2* VRI image of the HH 30 edge-on disk (Watson & Stapelfeldt 2004), shown in log stretch. The disk is 450 AU in diameter. Center: *Spitzer* SEDs of HH 30 (dashed line) and the Butterfly Star (dotted line) compared with the median *Spitzer* SED of young stars in the Taurus star-forming region (solid line; Rebull *et al.* 2010). Right: *HST/NICMOS* JHK image of the Butterfly Star (IRAS 04302+2247; Padgett *et al.* 1999), shown in log stretch. The dark lane extends 840 AU in diameter.

disk temperature and the degree of dust settling toward the disk mid-plane. Detailed comparisons of the images with scattered light models, starting with HH 30 (Burrows *et al.* 1996; Wood *et al.* 1998; Cotera *et al.* 2001; Watson & Stapelfeldt 2004; see Fig. 1), have allowed measurements of disk radii, vertical scale heights, radial flaring, and the properties of the scattering dust. Interesting variations between disks are apparent: small scale heights indicative of dust settling are seen in HK Tau B (McCabe *et al.* 2011), ISM-like dust grains are found in Class I sources such as the Butterfly Star (Wolf *et al.* 2003) and CB 26 (Sauter *et al.* 2009) but larger grains are required in the case of HH 30. In multiple systems disk outer radii appear to be systematically smaller (Stapelfeldt *et al.* 1998, 2003; Perrin *et al.* 2006; Neuhäuser *et al.* 2009). Scattered light dominates the SED of edge-on disks out to mid-infrared wavelengths (McCabe *et al.* 2003, Duchêne *et al.* 2010), indicating both high column densities and the presence of large grains. To fully reveal the diversity of young disk properties, it would be highly valuable to image additional examples.

2. A New Target Sample

The edge-on disks known up to now have largely been imaged by accident during companion searches or studies of Herbig-Haro jet sources. A new path to identifying these objects has been opened by the availability of sensitive far-IR surveys of nearby star-forming regions conducted by the *Spitzer*, *Herschel*, and *WISE* missions. The *Spitzer* "c2d" (Evans *et al.* 2009), Taurus (Rebull *et al.* 2010), and Gould Belt Legacy Surveys (Allen *et al.* 2007) have captured the SEDs of the known edge on disks in nearby molecular clouds. In comparison to the median SED for young stars, the edge-on disk SEDs (Fig. 1) are underluminous by factors of 10-100 in the optical and near-IR (but not highly extincted), have otherwise normal far-IR and submm luminosities, and typically show distinct peaks of scattered light and thermal emission. Using these as templates, we searched through ~2000 *Spitzer* young star SEDs in the aforementioned surveys and identified dozens of them with similar characteristics. This list was narrowed to the 21 best-characterized and optically visible candidates. Each was observed with the Advanced Camera for Surveys (ACS) for a single orbit in our *HST* Cycle 19 Program 12514. Images were made in the F606W and F814W filters in order to distinguish disk reflected light (present in both bands) from jet line emission (which is nearly absent in F814W).

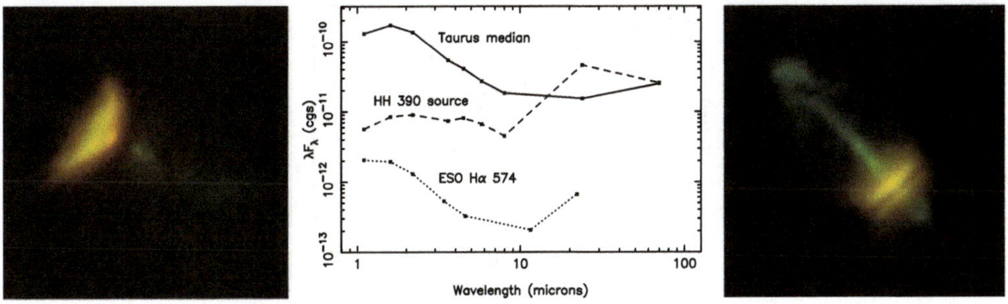

Figure 2. Left: *HST/ACS* RI image of the HH 390 source, shown in log stretch. The one-sided nebula is 220 AU in diameter. Center: *Spitzer* SED of HH 390 source (dashed line) and *WISE* SED of Hα574 (dotted line) compared with the median Spitzer SED of young stars in Taurus (solid line; Rebull *et al.* 2010). Right: *HST/ACS* RI image of ESO Hα574, shown in log stretch. The disk is 150 AU in diameter and the outermost jet knots extend to 7.5″ .

3. *HST* Survey Results

Eleven objects are resolved as edge-on disks on our survey, nine of these for the first time. Highly collimated jets are seen in six of these systems. Six other objects proved to be compact nebulae with odd or cometary morphology, and four targets appeared as point sources with no nebulosity. The latter may have suffered from confused *Spitzer* SEDs. Due to space limitations only a subset of the results are discussed here.

The HH 390 star is located in Taurus. Its flat SED from 1-10 μm is consistent with a scattered light disk inclined away from edge-on, and this is confirmed by the appearance of a one-sided nebula in the *HST* image (Fig. 2). The flared nebulosity extends perpendicular to the outflow axis with a diameter of 220 AU.

ESO Hα574 is a very compact disk in Chamaeleon with a collimated outflow first studied by Comerón & Reipurth (2006). The low source luminosity and almost edge-on orientation make this one of the faintest edge-on disks yet recognized. The disk is very flat and shows an unusual diffuse halo of reflected light along the base of the jet (Fig. 2).

Oph 163131.2 is a new object in Ophiuchus discovered by Evans *et al.* (2009). The disk is very close to edge-on with a classic double-peaked SED. It is 320 AU in diameter and is highly flattened (Fig. 3), suggesting the possibility of vertical dust settling. There is no extended jet emission, and the star surprisingly lacks measurable Hα emission.

The HH 48 source in Chamaeleon is revealed as a 2.3″ binary (Fig. 3). The primary component is the source of the known outflow and appears pointlike, but its faintness and the absence of strong reddening suggests it is seen in scattered light. The secondary has an edge-on disk with its own bipolar jets along PA –10° and a unique internal structure. This disk is 200 AU in diameter; the dark lane appears asymmetrically thicker on its western side; the intersection of the jet axis and the disk plane is offset ~20 AU from the center of the disk; and its jet axis is tilted 6° from an axis perpendicular to the disk plane. These features, along with the outer radius $< 1/3$ the projected binary separation, suggest tidal disturbances from the primary acting on the outer disk of the secondary.

Tau 042021.4 is a large and symmetric edge-on disk that may become a new prototype for the class. Our results for this object can be found in Duchêne *et al.* (these proceedings).

ESO Hα569 is a compact flared disk with an unusually large scale height consistent with a low stellar mass. For our results on this object, see Wolff *et al.* (these proceedings).

The $> 50\%$ success rate of our program demonstrates that SED selection is an effective way to identify new edge-on disks for study. Our results increase the number of edge-on disks imaged with *HST* in nearby molecular clouds by more than 50%. Several

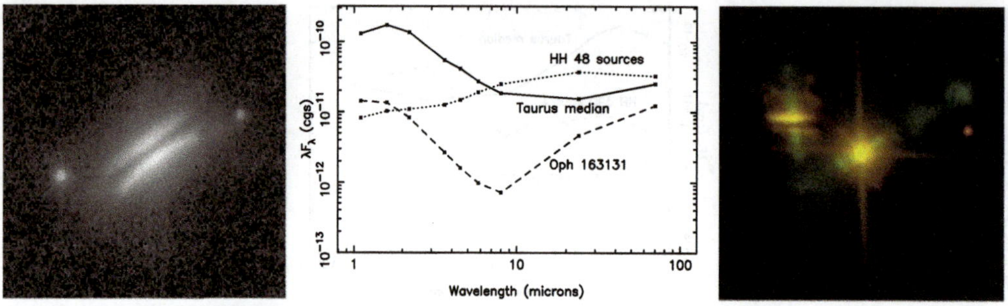

Figure 3. Left: *HST/ACS* R image of the edge-on disk of SSTc2dJ163131.2-242628 shown in log stretch. The disk is 320 AU in diameter. Center: *Spitzer* SEDs of SSTc2dJ163131.2 (dashed line) and the HH 48 system (dotted line) compared with the median *Spitzer* SED of young stars in Taurus (solid line; Rebull *et al.* 2010). Right: *HST/ACS* RI image of the HH 48 outflow sources, shown in log stretch. The secondary star has an edge-on disk 200 AU in diameter.

outstanding objects were found with unique characteristics, and there are many more candidates to pursue in the *Spitzer*, *Herschel*, and *WISE* datasets.

Combined scattered light and SED modeling is being pursued for the most symmetric sources, with the goal of placing specific constraints on their density distributions and dust grain properties. Goals for additional datasets include expanding the wavelength range of scattered light imaging (to measure the dust lane chromaticity), filling out mm/submm photometry, and imaging with ALMA.

Acknowledgements

This work was supported in part by *HST* GO grant 12514 to NASA/Goddard, UC Berkeley, and JPL/Caltech. Funding was provided by NASA, but was restricted such that the NASA and JPL co-Is were forbidden to attend this Symposium. This work was presented in oral session by G. Duchêne.

References

Allen, L., & *Spitzer* Gould Belt Legacy Team 2007, BAAS, 39, 881
Burrows, C. J., Stapelfeldt, K. R., Watson, A. M., *et al.* 1996, ApJ, 473, 437
Comerón, F. & Reipurth, B. 2006, A&A, 458, L21
Cotera, A. S., Whitney, B. A., Young, E., *et al.* 2001, ApJ, 556, 958
Duchêne, G., McCabe, C., Pinte, C., *et al.* 2010, ApJ, 712, 112
Evans, N. J., *et al.* 2009, ApJS, 181, 321
Krist, J. E., Stapelfeldt, K. R., Ménard, F., *et al.* 2000, ApJ, 538, 793
Mayama, S., Hashimoto, J., Muto, T., *et al.* 2012, ApJ, 760, L26
McCabe, C., Duchêne, G., & Ghez, A. M. 2003, ApJ, 588, L113
McCabe, C., Duchêne, G., Pinte, C., *et al.* 2011, ApJ, 727, 90
Neuhäuser, R., Krämer, S., Mugrauer, M., *et al.* 2009, A&A, 496, 777
Padgett, D. L., Brandner, W., Stapelfeldt, K. R., *et al.* 1999, AJ, 117, 1490
Perrin, M. D., Duchêne, G., Kalas, P., & Graham, J. R. 2006, ApJ, 645, 1272
Pinte, C., Padgett, D. L., Ménard, F., *et al.* 2008, A&A, 489, 633
Rebull, L. M., Padgett, D. L., McCabe, C.-E., *et al.* 2010, ApJS, 186, 259
Ricci, L., Robberto, M., & Soderblom, D. R. 2008, AJ, 136, 2136
Sauter, J., Wolf, S., Launhardt, R., *et al.* 2009, A&A, 505, 1167
Stapelfeldt, K. R., Krist, J. E., Menard, F., *et al.* 1998, ApJ, 502, L65
Stapelfeldt, K. R., Ménard, F., Watson, A. M., *et al.* 2003, ApJ, 589, 410
Watson, A. M. & Stapelfeldt, K. R. 2004, ApJ, 602, 860

Watson, A. M., Stapelfeldt, K. R., Wood, K., & Ménard, F. 2007, Protostars & Planets V, 523
Wolf, S., Padgett, D. L., & Stapelfeldt, K. R. 2003, ApJ, 588, 373
Wood, K., Kenyon, S. J., Whitney, B., & Turnbull, M. 1998, ApJ, 497, 404

Discussion

TRIAUD: Is the misaligned jet you detected the first such occurrence?

DUCHÊNE: As far as I know, Yes, but it's also one of the best defined: the disk and the jet orientations can be estimated precisely. This is not the case for all objects. Furthermore, Right now it's not possible to know if there is a warp and if a putative inner disc would be aligned with the jet.

LIN: Are the binaries with misaligned jets physically associated?

DUCHÊNE: Most likely. They both are members of the Cha cloud, whose population is relatively sparse, so it is not likely that they are un-related. However, no definitive evidence is available as proper motions is slow at these distances and orbital motion too slow to monitor.

Exploring the Formation and Evolution of Planetary Systems
Proceedings IAU Symposium No. 299, 2013
M. Booth, B. C. Matthews & J. R. Graham, eds.

doi:10.1017/S1743921313008016

TW Hydrae: multi-wavelength interferometry of a transition disk

J. Menu[1,2]†, R. van Boekel[2], T. Henning[2], M. Benisty[3], C. J. Chandler[4], H. Linz[2], C. Waelkens[1], S. M. Andrews[5], N. Calvet[6], J. M. Carpenter[7], S. A. Corder[8], A. T. Deller[9], C. P. Dullemond[10], J. S. Greaves[11], R. J. Harris[5], A. Isella[7], W. Kwon[12], J. Lazio[13], L. G. Mundy[14], L. M. Perez[7], L. Ricci[7], A. I. Sargent[7], S. Storm[14], L. Testi[15,16] & D. J. Wilner[5]

[1]Instituut voor Sterrenkunde, KU Leuven, Belgium
[2]Max Planck Institut für Astronomie, Heidelberg, Germany
[3]Institut de Planétologie et d'Astrophysique de Grenoble, France
[4]National Radio Astronomy Observatory, Socorro, USA
[5]Harvard-Smithsonian Center for Astrophysics, Cambridge, USA
[6]Department of Astronomy, University of Michigan, Ann Arbor, USA
[7]Astronomy Department, California Institute of Technology, Pasadena, USA
[8]Joint ALMA Observatory, Santiago, Chile
[9]ASTRON, The Netherlands Institute for Radio Astronomy, Dwingeloo, The Netherlands
[10]Institute for Theoretical Astrophysics, University of Heidelberg, Germany
[11]School of Physics and Astronomy, University of St. Andrews, UK
[12]Department of Astronomy, University of Illinois, Urbana, USA
[13]Jet Propulsion Laboratory, California Institute of Technology, Pasadena, USA
[14]Department of Astronomy, University of Maryland, College Park, USA
[15]European Southern Observatory, Garching, Germany
[16]INAF-Osservatorio Astrofisico di Arcetri, Firenze, Italy

Abstract. For over a decade, the structure of the inner "hole" in the transition disk around TW Hydrae has been a subject of debate. To probe the innermost regions of the protoplanetary disk, observations at the highest possible spatial resolution are required. We present new interferometric data of TW Hya from near-infrared to millimeter wavelengths. We confront existing models of the disk structure with the complete data set and develop a new, detailed radiative-transfer model. This model is characterized by: 1) a spatial separation of the largest grains from the small disk grains; and 2) a smooth inner rim structure, rather than a sharp disk edge.

Keywords. stars: individual (TW Hya), planetary systems: protoplanetary disks, techniques: interferometric

1. Introduction

The potential discovery of (proto)planetary bodies inside gaps of transition disks (e.g. Quanz *et al.* 2013) provides the ultimate justification for calling circumstellar disks around young stars "protoplanetary". Conversely, it also suggests that a transition-disk phase might be fundamental in the formation of planetary systems.

A well-known example of a protoplanetary disk with an inner hole is TW Hydrae (TW Hya). Being the closest accreting T Tauri star, it has been the target of many studies on the geometry of the circumstellar dust distribution. Calvet *et al.* (2002) were the first to conclude that the low near-infrared emission from the disk indicates a lack of dust close to the star. Their proposed model had a gap radius of 4 AU, with only a very low amount

† PhD fellow of the Research Foundation – Flanders; email: `jonathanm@ster.kuleuven.be`

of dust inside the central hole. Whereas millimeter observations seemed to confirm this geometry (Hughes *et al.* 2007), a model with a much smaller central hole ($\sim 0.7\,$AU in radius) was used to explain mid-infrared interferometry (Ratzka *et al.* 2007). Later models proposed hybrid structures with an additional dust ring in the hole (Akeson *et al.* 2011) or even a self-luminous companion (Arnold *et al.* 2012).

Clearly, the disk of TW Hya turns out to be more complex than initial models suggested, and a final consensus on what currently is the "best" model is far from being reached. Understanding the disk structure is essential for answering at least 2 questions:

- *What are the dominant processes that determine the disk dispersal?* The inner-disk structure might bear the signposts of planet formation.
- *What is the gas structure of the disk?* The dust distribution can be interpreted as a "boundary condition" for studying the gas dynamics/chemistry.

The goal of this contribution is not to answer the above questions, but to see how well we currently understand the structure of the TW Hya disk. We collected high angular resolution data from near-infrared to millimeter wavelengths (Sect. 2), and give the results of comparing the data with published models in Sect. 3. Finally, in Sect. 4 we describe the new model we propose for the geometry of the TW Hya disk.

2. Interferometric data of TW Hya

Below, we introduce the new and archival interferometric data collected for this study. We also collected the spectral energy distribution (SED) from UV to mm-wavelengths.

Near-infrared interferometry (VLTI/PIONIER). Data was obtained with the PIONIER instrument on the Very Large Telescope Interferometer (VLTI) in 2011. The H-band observations ($\lambda = 1.65\,\mu$m) for baselines up to 130 m have an average visibility of $V^2 = 0.968 \pm 0.005$, indicating that TW Hya is largely unresolved. The 1.6 % of missing correlated flux to reach $V^2 = 1$ can be interpreted as evidence for an overresolved scattering component from the outer disk.

Mid-infrared interferometry (VLTI/MIDI). N-band ($\lambda = 8$-$13\,\mu$m) observations of different epochs indicate a characteristic scale of $R = 0.7\,$AU for the mid-infrared emission. The data indicates no significant inclination for the inner disk region. We combine the observations into a 25-m baseline and a 50-m baseline correlated flux curve.

Sub-mm interferometry (SMA). Our analysis includes the sub-mm interferometry presented in Andrews *et al.* (2012), obtained using the Submillimeter Array (SMA; $\lambda = 870\,\mu$m). The data is deprojected as in the original paper.

Millimeter interferometry (VLA). The final piece of information is a new 9-mm data set obtained with the Karl G. Jansky Very Large Array (VLA), as part of the Disks@EVLA key program (`https://safe.nrao.edu/evla/disks/`; PI C. Chandler). The data are a combination CnB, BnA, and A-array data. As in the case of the SMA data, the visibilities were deprojected. The observations have a much higher signal-to-noise ratio than previous VLA observations (Hughes *et al.* 2007). In particular, we do not confirm the visibility null at $\sim 1000\,$kλ, which was interpreted as essential to confirm the disk structure with a 4-AU inner gap. A compact contribution from free-free emission cannot yet be excluded.

3. Key models revisited

As a first test, we selected 3 published models for the TW Hya disk and made a reimplemented version with a radiative-transfer code (see Sect. 4.1). These models are from Calvet *et al.* (2002), Ratzka *et al.* (2007), and Andrews *et al.* (2012). A full description of the model reimplementation is beyond the scope of this short paper; we refer to an

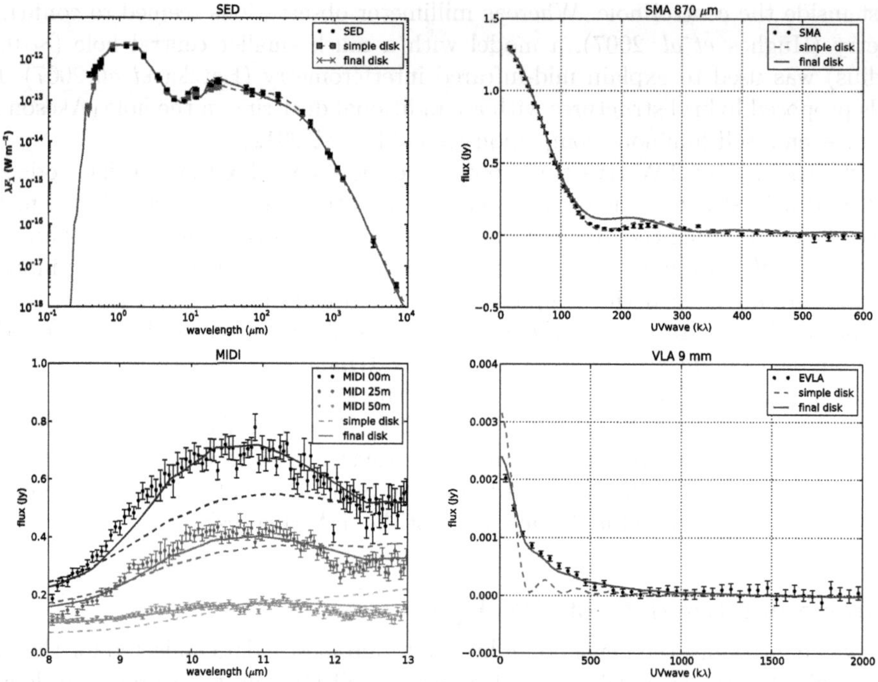

Figure 1. New disk model for the TW Hya disk (*blue, full line*), compared with a simple homogeneous disk model with a vertical inner rim (*red, dashed line*)

oncoming paper for this (Menu *et al.*, in prep.). We implemented the complete model structure, including the dust composition and the inner-rim shape.

Apart from the fact that the models cannot simultaneously reproduce all data sets, four conclusions can be drawn:

(*a*) grains much larger than micron sizes are need to reproduce the SED;
(*b*) a full radiative transfer code is essential for modeling shadowing by the inner disk;
(*c*) the model SEDs all reproduce the data to some extent: degenerate SED-modeling;
(*d*) special care is needed in the millimeter: the new VLA data provides much stronger constraints than the old VLA data.

4. A new model for the TW Hya disk

4.1. *Model structure*

Since none of the three reconsidered models fully reproduces the combined data sets, we set up a new model. We use the Monte Carlo radiative-transfer code MCMax (Min *et al.* 2009), which allows the simulation of an azimuthally symmetric protoplanetary disk around a central star. The code fully solves the vertical hydrostatic equilibrium, and has a built-in implementation of the vertical settling of dust grains based on the α-disk turbulence prescription (Mulders & Dominik 2012).

For the model fitting, we use a genetic algorithm (Cantó *et al.* 2009). Generations of models are calculated, keeping only the best models for offspring generation. This offspring is chosen randomly in the parameter space, close to the parent parameters.

Even using intelligent fitting algorithms, there is a large interest in minimizing the number of free parameters. We kept the stellar properties fixed to literature values

Table 1. Fitted parameters of the full-disk model shown in Fig. 1.

inner radius $R_{\rm in}$	0.42 AU	dust mass$_{<100\,\mu{\rm m}}$	$5\times10^{-5}\,{\rm M}_\odot$	surf. density$_{<100\,\mu{\rm m}}$ p	0.7
outer radius $R_{\rm out}$	64 AU	dust mass$_{>100\,\mu{\rm m}}$	$6\times10^{-5}\,{\rm M}_\odot$	surf. density$_{>100\,\mu{\rm m}}$ $p_{\rm large}$	1.4
trans. radius $R_{\rm exp}$	3.6 AU	width w	0.45	turbulence parameter α	10^{-5}

($M_\star = 0.6\,{\rm M}_\odot$, $T_{\rm eff} = 4000\,{\rm K}$, $L = 0.25\,{\rm L}_\odot$, $d = 51\,{\rm pc}$; see e.g. Ratzka *et al.* 2007). Also the dust composition was kept fixed to a mixture of amorphous olivines and carbonaceous grains (70 % vs. 30 %, respectively; grain size $a = 0.01$-$1000\,\mu$m, size density $n(a) \propto a^{-3.5}$).

4.2. *Results and discussion*

Modeling the disk with a simple disk model with a vertical inner rim (+ an empty central hole) and a homogeneous composition from small to large grains does not lead to a successful reproduction of the data, as shown in Fig. 1. We modify this simple model in two ways. First, we introduce a rounded inner rim by taking a surface density

$$\Sigma(R) = \begin{cases} \Sigma_{\rm exp}\,(R/R_{\rm exp})^{-p}\,\exp\left(-(1-R/R_{\rm exp})^3/w^3\right) & \text{for } R_{\rm in} \leqslant R < R_{\rm exp}, \\ \Sigma_{\rm exp}\,(R/R_{\rm exp})^{-p} & \text{for } R_{\rm exp} \leqslant R \leqslant R_{\rm out} \end{cases} \quad (4.1)$$

(Mulders *et al.* 2013). Here, $R_{\rm in}$, $R_{\rm out}$ and p are the inner radius, outer radius, and surface density power; $\Sigma_{\rm exp}$ is the surface density at a transition radius $R_{\rm exp}$, and w is a dimensionless rim width. Secondly, we decouple the largest grains ($a > 100\,\mu$m) from the smaller grains ($a < 100\,\mu$m), and give the large grains a separate surface density power $p_{\rm large}$ (also applying Eq. 4.1 for the surface density).

The best model obtained for this model structure is shown in Fig. 1. Although the fit to the SED and the SMA data is slightly degraded in quality, a much more accurate fit to the MIDI and VLA data is reached.

The rim shape in Eq. 4.1 follows from modeling mid-infrared interferometry of the (pre-)transitional disk object HD 100546, and is shown to be in agreement with hydrodynamical simulations of a brown dwarf opening the disk gap (Mulders *et al.* 2013). Our radiative transfer model shows that the MIDI data agrees with a similar inner-rim structure. Full hydrodynamical simulations of the inner-disk region would be needed to quantify the mass and orbital radius of a potential companion.

A strong concentration of larger grains towards the inner disk regions is also inferred from millimeter observations of other protoplanetary disks (e.g. Pérez *et al.* 2012). Rather than modeling the data at different wavelengths individually, and inferring inhomogeneous dust growth a posteriori, we assembled a model that immediately incorporates an inhomogeneous dust composition.

References

Akeson, R. L., Millan-Gabct, R., Ciardi, D. R., *et al.* 2011, *ApJ*, 728, 96
Andrews, S. M., Wilner, D. J., Hughes, A. M., *et al.* 2012, *ApJ*, 744, 162
Arnold, T. J., Eisner, J. A., Monnier, J. D., & Tuthill, P. 2012, *ApJ*, 750, 119
Calvet, N., D'Alessio, P., Hartmann, L., *et al.* 2002, *ApJ*, 568, 1008
Cantó, J., Curiel, S., & Martínez-Gómez, E. 2009, *A&A*, 501, 1259
Hughes, A. M., Wilner, D. J., Calvet, N., *et al.* 2007, *ApJ*, 664, 536
Min, M., Dullemond, C. P., Dominik, C., de Koter, A., & Hovenier, J. W. 2009, *A&A*, 497, 155
Mulders, G. D. & Dominik, C. 2012, *A&A*, 539, A9
Mulders, G. D., Paardekooper, S.-J., Panić, O., *et al.* 2013, *A&A*, 557, 68

Pérez, L. M., Carpenter, J. M., Chandler, C. J., *et al.* 2012, *ApJ* (Letters), 760, L17
Quanz, S. P., Amara, A., Meyer, M. R., *et al.* 2013, *ApJ* (Letters), 766, L1
Ratzka, T., Leinert, C., Henning, T., *et al.* 2007, *A&A*, 471, 173

Discussion

VAN DER WIEL: Are all input parameters determined by the model fitting, or are some fixed a priori? Specifically the turbulence parameter could be linked not just to how fast big grains settle, but also to spectral line shapes of molecular gas.

MENU: Even using a genetic algorithm, there is a strong interest in minimizing the number of free parameters for minimizing the computation time. Therefore, we fix parameters that *can* be fixed (to some extent), e.g. the stellar properties, and only let a selection of essential disk parameters vary during the fit. Concerning the turbulence parameter, it remains to be shown that the constraints we get from modeling the dust agree with observations of molecular lines. However, one needs to be careful: the assumption of constant α-value or constant dust-to-gas ratio throughout the disk might not be entirely valid, and the disk regions dominating the α-estimates might differ for the gas and the dust.

WEINBERGER: What do you assume about the (not well constrained) mass and temperature of the star and how does changing these affect the fit parameters?

MENU: It is indeed correct that not only the disk structure is under discussion, but also the stellar parameters. We have done an SED-fit for determining the (single-star) model that agrees best with the optical and near-infrared photometry, and take a literature value for the stellar mass ($0.6\,M_{\odot}$). A few recent works have achieved better fits of the stellar spectrum with mixed spectra with different weights, which also gives clues on the stellar mass, but the uncertainties remain significant. Changing the stellar properties will alter the amount of incident radiation (for a change in luminosity) and the vertical hydrostatic equilibrium (for a change in mass), which in turn changes the temperature structure and amount of flaring. This can be expected to happen in a more or less continuous way.
However, is vertical hydrostatic equilibrium a valid approximation throughout the disk? And is the stellar spectrum intercepted by the disk the same as the one we observe, given the very particular orientation of the system (pole-on) w.r.t. the observer (e.g., possible spots might be distributed very anisotropically)? Taking into account these questions might even make changes to the stellar properties irrelevant.

Exploring the Formation and Evolution of Planetary Systems
Proceedings IAU Symposium No. 299, 2013
M. Booth, B. C. Matthews & J. R. Graham, eds.

doi:10.1017/S1743921313008028

Probing planet nurseries with rare isotopologues of CO

Sarah E. Dodson-Robinson[1], Mo Yu[1] and Karen Willacy[2]

[1]Astronomy Department, University of Texas,
2515 Speedway Dr C1400, Austin, TX 78712, USA

[2]Jet Propulsion Laboratory, California Institute of Technology,
4800 Oak Grove Drive, Pasadena, CA 91109, USA

Abstract. The properties of planet-forming midplanes of protostellar disks remain largely unprobed by observations due to the high optical depths of common molecular lines. However, rotational emission lines from rare isotopologues may have optical depth near unity in the vertical direction, so that the lines are strong enough to be detected, yet remain transparent enough to trace the disk midplane. We have computed chemical models of protostellar disks including different C and O isotopes and detailed photochemical reactions. The CO condensation front is in the giant planet-forming region, within 20 AU of the star. We show that the optical depths of low-order rotational lines of $C^{17}O$ are around unity, which suggests that it may be possible to see into the disk midplane using $C^{17}O$. In lower-mass disks, the slightly more abundant $C^{18}O$ is a possible midplane probe. ALMA observations would provide estimates of the disk midplane temperature if CO ice line were spatially resolved. With our computed $C^{17}O/H_2$ abundance ratio, we would also be able to measure disk surface densities from the fluxes of low-order $C^{17}O$ transitions.

1. Introduction and Methods

The temperature, density and composition of proto-planetary disks determine materials available for planet formation. However fundamental properties of protostellar disks such as midplane temperature, turbulent visosity, and disk mass are currently poorly constrained by observations. Here we seek a molecule that can be used to probe planet-forming protostellar disk midplanes, as opposed to disk surfaces. We focus here on CO isotopologues—molecules with a rare isotope in place of one of the atoms—because their transitions are readily observable with ALMA and they are extremely volatile, which places their condensation fronts as far from the star as possible to facilitate angular resolution of the CO distribution in nearby disks

We have computed chemical models of evolving disks that include C and O isotopes. The evolving disk models from Landry *et al.* (2013) provide the temperature and density used as input to the chemical model. The computational method is similar to that of Dodson-Robinson *et al.* (2009). In the Landry *et al.* (2013) models, the heating is provided by stellar radiation and the magnetorotational instability, which gives a hot disk surface and a cold midplane. After we have computed the abundances of each CO isotopologue as a function of time, we calculate the optical depth of each isotopologue as a function of distance from the star by integrating the absorption efficiency over the vertical disk height.

2. Results and Summary

Figure 1 shows time snapshots of the optical depths in the $J = 3 \rightarrow 2$ transition of the CO isotopologues as a function of distance from the star. We find that the rare isotopologues of CO are good probes of the planet-forming midplane of the disk. A given tracer works best when the optical depth is near unity, as observations can peer into the disk midplane while maximizing the observed signal. $C^{17}O$ is most effective as a probe of the inner 30 AU of the disk while $C^{18}O$ probes the outer regions of the disk.

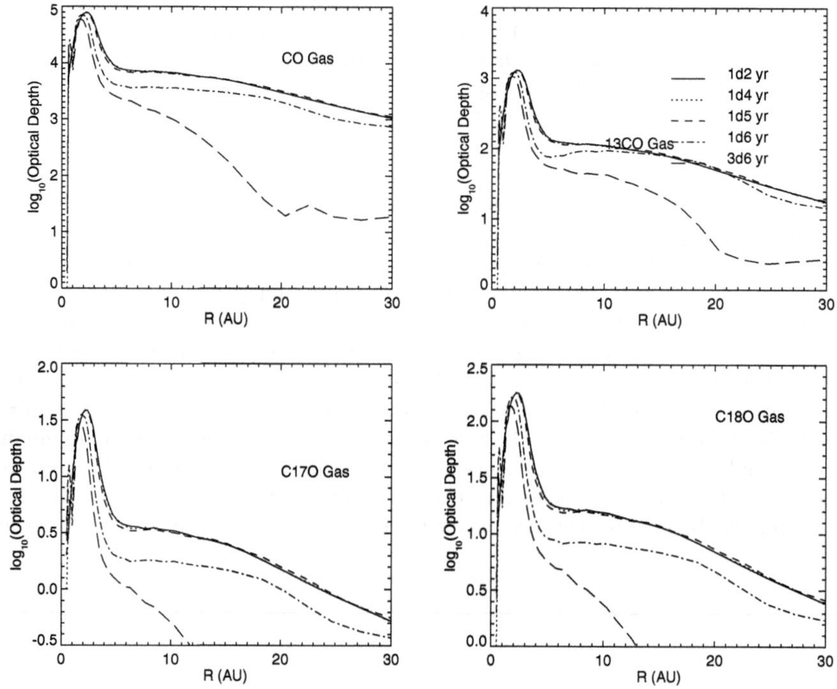

Figure 1. Vertical optical depth at $J = 3 \rightarrow 2$ of CO, ^{13}CO, $C^{17}O$, and $C^{18}O$ at different time snapshots in the disk evolution. The depletion of all isotopologues of CO at late times is due to the formation of hydrocarbons. $C^{17}O$ and $C^{18}O$ probe the disk midplane in the giant planet-forming region of the disk. The model disk has mass $0.03 M_\odot$ spread over 70 AU.

Here we have verified that rare CO isotopologues are good probes of planet-forming protostellar disk midplanes. Such observations are good complements to the chemical imaging techniques of, for example, Qi *et al.* (2013), who search for CO freezeout by locating the inner edge of emission from molecules that are destroyed by CO gas, such as H_2CO and N_2H^+.

This work was funded by NASA grant NNX10AH28G to SDR and KW. KW's work was conducted at the Jet Propulsion Laboratory, California Institute of Technology, under contract with the NASA Aeronautics and Space Administration.

References

Dodson-Robinson, S. E., Willacy, K., Bodenheimer, P., Turner, N. J., & Beichman, C. A. 2009, *Icarus*, 200, 672

Landry, R., Dodson-Robinson, S. E., Turner, N. J., & Abram, G. 2013, *ApJ*, 771, 80

Qi, C., Öberg, K. I., & Wilner, D. J. 2013, *ApJ*, 765, 34

Exploring the Formation and Evolution of Planetary Systems
Proceedings IAU Symposium No. 299, 2013
M. Booth, B. C. Matthews & J. R. Graham, eds.

doi:10.1017/S174392131300803X

Panchromatic imaging and modeling of SSTtau J042021+281349: A new prototypical edge-on protoplanetary disk

G. Duchêne[1,2], K. Stapelfeldt[3], A. Isella[4], M. Perrin[5], F. Ménard[2,6], D. Padgett[3], C. Pinte[2,6], S. Wolff[7], J. Krist[8], A. Ghez[9] & Q. Konopacky[10]

[1]UC Berkeley, [2]IPAG, [3]NASA GSFC, [4]Caltech, [5]STScI, [6]U. de Chile, [7]Johns Hopkins University, [8]JPL/Catltech, [9]UCLA, [10]Dunlap Institute

Abstract. We present new high-resolution observations and modeling of SSTtau J042021+281349, a 400 AU-radius edge-on protoplanetary disk. We have gathered visible and near-infrared scattered light images of the system with the Hubble Space Telescope and Keck adaptive optics system, as well as a 1.3 mm continuum map with CARMA. Compared to the well-known HH 30 disk, this new system is remarkable because of its spectacular bipolar jet and the high degree of lateral symmetry of the disk. Indeed, we argue that this system is a "cleaner" prototype for edge-on disks. In addition, the apparent achromaticity of dust properties (most notably the almost grey opacity law) from the visible to the near-infrared in this disk suggests that it is in an advanced stage of dust evolution.

Keywords. planetary systems: protoplanetary disks; stars: pre–main-sequence

1. Introduction

By naturally occulting the bright central star, edge-on protoplanetary disks offer a vantage viewpoint to the structure of these objects. Specifically, the vertical structure of the disk is directly exposed, allowing analysis of the disk scale height and probing vertical settling of the dust. High-resolution images yield strong constraints on the dust scattering and opacity properties. Coupled with millimeter mapping, full spectral energy distribution (SED) and radiative transfer modeling, it is further possible to constrain the global structure of the disk (e.g., Duchêne *et al.* 2010).

We have conducted an HST/ACS imaging survey of *Spitzer*-selected new candidate edge-on protoplanetary disks to increase the number of such known systems (see contributions by K. Stapelfeldt and S. Wolff in this volume). One of the targets was SST-tau J042021+281349 (hereafter Tau042021), an M0-M1 member of the Taurus star-forming region characterized by a rich emission line spectrum, a very low *apparent* stellar luminosity and a massive infrared excess ($L_\star = 0.004 L_\odot$, $L_{IR}/L_{bol} = 0.44$; Rebull *et al.* 2010). These latter two points are characteristics of edge-on disk systems and, indeed, Tau042021 was first resolved as such in ground-based images (Luhman *et al.* 2009).

2. Observational results and initial modeling results

We have obtained $\leqslant$0.1"-resolution visible and near-infrared images of Tau042021 with HST and Keck adaptive optics, as well as an 0.4"-resolution 1.3 mm map of the system with CARMA (see left panel in Fig. 1). The HST images show a highly symmetric, 400 AU-radius, edge-on disk as well as a collimated Herbig-Haro jet extending up to 8" from the central star with multiple knots along both the jet and counter-jet. In addition to

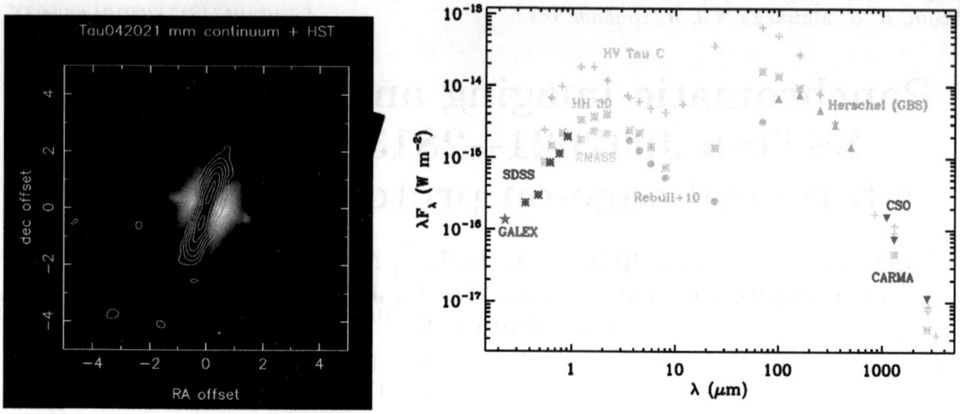

Figure 1. Left: Two-color HST/ACS image (F606W: green; F814W: red) and 1.3 mm continuum contour map of Tau042021. **Right:** Complete SED for Tau042021 (color symbols) compared to those of the HH 30 and HV Tau C edge-on protoplanetary disks (gray asterisks and plus signs).

the typical two parallel nebulae separated by the darklane that defines the disk midplane, our HST images reveal a flat-topped, low surface brightness and markedly blue-colored, nebulosity high up above the disk. The high-resolution millimeter map shows a very flat structure (unresolved along the vertical axis of the disk) with an intriguing double-peak structure. Finally, the SED of Tau042021, with our new CSO and CARMA data, shows a shallow submillimeter spectral slope ($\alpha_{\rm mm} \approx 1.9$, see right panel of Fig. 1).

The system's scattered light images reveal a remarkable lack of chromaticity in the dust scattering and opacity properties (e.g., the dark lane width shrinks by only 20% between 0.8 and 2.2 μm), suggesting that the dust population in the disk has substantially evolved. The weak nebulosity seen above the disk could be an extended disk "atmosphere", for instance related to photoevaporation, or evidence that the disk extends further out radially than suggested by the images, in which case a tapered surface density profile would be required to match all observations (e.g., Isella *et al.* 2009). Extensive simultaneous modeling of the SED and scattered light images is ongoing and will help distinguish between these two possibilities. Preliminary results indicate a "normal" stellar luminosity ($L_\star \approx 0.5 L_\odot$), a massive disk ($M_{\rm dust} \approx 3\ 10^{-4} M_\odot$) and very little foreground extinction ($A_V \leqslant 1$ mag), in line with the ultraviolet detection of the system with GALEX. Finally, the double-peak structure of the CARMA 1.3 mm map is reminiscent of other edge-on protoplanetary disks and this ubiquity suggests that it is more likely an optical depth effect rather than evidence for a central binary source (see also Gräfe *et al.* 2013).

Compared to HH 30, the disk surrounding Tau042021 is larger, observed closer to edge-on ($i \approx 88^{\rm o}$ vs. $\approx 80^{\rm o}$) and does not show pronounced lateral asymmetry. We thus argue that Tau042021 is a better prototype of edge-on protoplanetary disks than HH 30 and anticipate that it will become a new benchmark to compare other similar systems.

References

Duchêne, G., McCabe, C., Pinte, C., *et al.* 2010, *ApJ*, 712, 112
Gräfe, C., Wolf, S., Guilloteau, S., *et al.* 2013, *A&A*, 553, A69
Isella, A., Carpenter, J., & Sargent, 2009, *ApJ*, 701, 260
Luhman, K., Mamajek, E., Allen, P., & Cruz, K., 2009, *ApJ*, 703, 399
Rebull, L., Padgett, D., McCabe, C., *et al.* 2010, *ApJS*, 186, 259

Exploring the Formation and Evolution of Planetary Systems
Proceedings IAU Symposium No. 299, 2013
M. Booth, B. C. Matthews & J. R. Graham, eds.

doi:10.1017/S1743921313008041

Can a planet explain different cavity sizes for small & large dust grains in transition disks?

Antonio Garufi, Henning Avenhaus and Sascha P. Quanz

Institute for Astronomy, ETH Zurich, Wolfgang-Pauli-Strasse 27, Zurich, Switzerland
email: antonio.garufi@phys.ethz.ch

Abstract. Dissimilarities in the spatial distribution of small (μm–size) and large (mm–size) dust grains at the cavity edge of transition disks have been recently pointed out and are now under debate. We obtained VLT/NACO near-IR polarimetric observations of SAO 206462 (HD135344B). The disk around the star shows very complex structures, such as dips and spirals. We also find an inner cavity much smaller than what is inferred from sub-mm images. The interaction between disk and orbiting companion(s) may explain this discrepancy.

Keywords. Stars: individual (SAO 206462, HD 135344B), Techniques: polarimetric

1. Introduction

A small sample of disks, the transition disks, shows a peculiar dip at infrared wavelengths, suggesting a depletion of warm dust around the central star (Strom *et al.* 1989). Disk–companion interaction (Rice *et al.* 2003), photoevaporation (Alexander & Armitage 2007), and particle growth (Dullemond & Dominik 2005) are possible clearing processes.

Polarimetric Differential Imaging (PDI) is allowing high-resolution imaging of circumstellar disks (e.g. Quanz *et al.* 2011, Hashimoto *et al.* 2012) with unprecedented inner working angle (0.1″). Recently, comparisons of PDI images with sub-millimeter images (e.g. Andrews *et al.* 2011) have revealed different spatial distribution for small and large dust grains (see e.g. Dong *et al.* 2012).

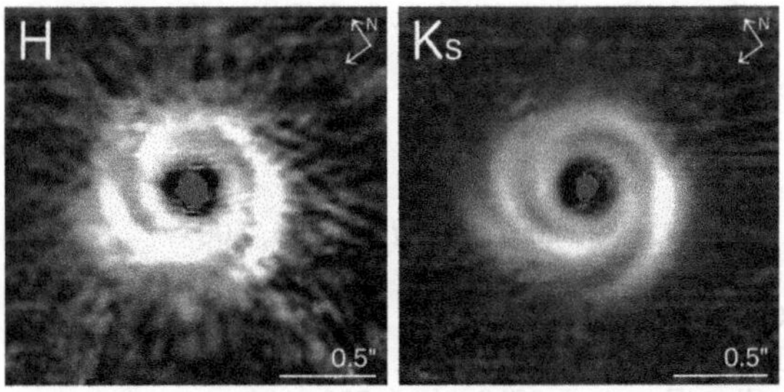

Figure 1. The disk around SAO 206462 from PDI VLT/NACO observations.

2. PDI observations and interpretation

PDI observations of the Herbig Ae/Be SAO 206462 (HD 135344B) were obtained with NACO (Lenzen *et al.* 2003, Rousset *et al.* 2003), the adaptive optics near-IR imager and spectrograph of the VLT, in H and K_S band. The basic principle of PDI is the simultaneous imaging of the linear polarization of the source along two orthogonal directions. The detailed observation setting and data reduction can be found in Quanz *et al.* (2011) and Avenhaus *et al.* (in prep.). The final products are radial Stokes Q_r parameter images of the source described in Garufi *et al.* (in prep.) and shown in Fig. 1. The disk is revealed in scattered light in both bands. The images show three main peculiarities: an inner cavity ($r = 28 \pm 6$ AU) with light depleted by a factor down a few tenths, a quasi-circular rim surrounding the cavity, and two spiral arms extending from the rim outward.

Apart from the spiral structure (probably due to a companion orbiting at large scale, Muto *et al.* 2012), the most tantalizing aspect suggested by these images is the different cavity size of small grains (28 AU) with respect to large grains (39 AU, Brown *et al.* 2009). Similar discrepancies were recently pointed-out by Dong *et al.* (2012).

The observed cavity is probably due to tidal interaction with an orbiting companion. Photoevaporation and dust grain growth are indeed ruled-out: because of high accretion rate (Sitko *et al.* 2012) and sub-AU inner dust belt (Fedele *et al.* 2008) the former, abrupt radial profile (this work) and absence of mm-size grains (Brown *et al.* 2009) the latter.

The scenario with clearing by a companion orbiting inside the cavity may also explain the observed dissimilarity in the cavity sizes. In fact, the pile-up of large grains due to a giant planet can occur at up to 10 tidal radii (Pinilla *et al.* 2012), whereas the outer edge of the gaseous halo cannot exceed 5 tidal radii (Dodson-Robinson & Salyk 2011). We suggest a scenario (see Fig. 2) where a giant planet is generating a pressure bump at 39 AU, which holds back mm-size grains but allows μm-size grains to be dragged inward along with the gas as down as 28 AU. Non-keplerian flows of gas and small dust grains can still be present in the cavity, thus to sustain the inner dust belt and the high accretion rate of the source. We analytically find that a 5 to 15 M_J at 17 to 20 AU is consistent with the observed cavity sizes. However, a multiple-planets system is not ruled-out.

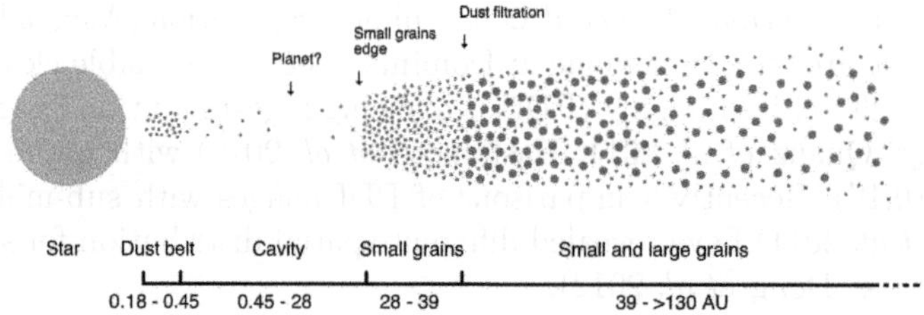

Figure 2. Illustrative sketch showing the dust radial distribution in the disk of SAO 206462

References

Alexander, R. D. & Armitage, P. J. 2007, *MNRAS*, 375, 500
Andrews, S. M., Wilner, D. J., Espaillat, C., *et al.* 2011, *ApJ*, 732, 42
Brown, J. M., Blake, G. A., Qi, C., *et al.* 2009, *ApJ*, 704, 496
Dodson-Robinson, S. E. & Salyk, C. 2011, *ApJ*, 738, 131
Dong, R., Rafikov, R., Zhu, Z., *et al.* 2012, *ApJ*, 750, 161
Dullemond, C. P. & Dominik, C. 2005, *ASPCS*, 434, 971
Fedele, D., van den Ancker, M. E., Acke, B., *et al.* 2008, *A&A*, 491, 809
Hashimoto, J., Dong, R., Kudo, T., *et al.* 2012, *ApJL*, 758, L19
Lenzen, R., Hartung, M., Brandner, W., *et al.* 2003, *SPIE*, 4841, 944
Muto, T., Grady, C. A., Hashimoto, J., *et al.* 2012, *ApJL*, 748, L22
Pinilla, P., Benisty, M., & Birnstiel, T. 2012, *A&A*, 545, A81
Quanz, S. P., Schmid, H. M., Geissler, K., *et al.* 2011, *ApJ*, 738, 23
Rice, W. K. M., Armitage, P. J., Bonnell, I. A., *et al.* 2003, *MNRAS*, 346, L36
Rousset, G., Lacombe, F., Puget, P., *et al.* 2003, *SPIE*, 4839, 140
Sitko, M. L., Day, A. N., Kimes, R. L., *et al.* 2012, *ApJ*, 745, 29
Strom, K. M., Strom, S. E., Edwards, S., Cabrit, S., & Skrutskie, M. F. 1989, *AJ*, 97, 1451

Exploring the Formation and Evolution of Planetary Systems
Proceedings IAU Symposium No. 299, 2013
M. Booth, B. C. Matthews & J. R. Graham, eds.

doi:10.1017/S1743921313008053

Planet gaps in the dust layer of 3D protoplanetary disks: Observability with ALMA

J.-F. Gonzalez[1], **C. Pinte**[2], **S. T. Maddison**[3] **and F. Ménard**[2]

[1]Université de Lyon, Lyon, F-69003, France; Université Lyon 1, Observatoire de Lyon, 9 avenue Charles André, Saint-Genis Laval, F-69230, France; CNRS, UMR 5574, Centre de Recherche Astrophysique de Lyon; École Normale Supérieure de Lyon, Lyon, F-69007, France
email: `Jean-Francois.Gonzalez@ens-lyon.fr`

[2]UJF-Grenoble 1 / CNRS-INSU, Institut de Planétologie et d'Astrophysique de Grenoble, UMR 5274, Grenoble, F-38041, France
email: `christophe.pinte@obs.ujf-grenoble.fr`, `francois.menard@obs.ujf-grenoble.fr`

[3]Centre for Astrophysics and Supercomputing, Swinburne University of Technology, PO Box 218, Hawthorn, VIC 3122, Australia
email: `smaddison@swin.edu.au`

Abstract. Among the numerous known extrasolar planets, only a handful have been imaged directly so far, at large orbital radii and in rather evolved systems. The Atacama Large Millimeter/submillimeter Array (ALMA) will have the capacity to observe these wide planetary systems at a younger age, thus bringing a better understanding of the planet formation process. Here we explore the ability of ALMA to detect the gaps carved by planets on wide orbits.

Keywords. Planetary systems: protoplanetary disks, radiative transfer, methods: numerical, submillimeter

1. Method

In a previous work (Fouchet *et al.* 2010), we ran full 3D, two-fluid Smoothed Particle Hydrodynamics (SPH) simulations of a planet embedded in a gas+dust T Tauri disk for different planet masses and grain sizes. The gas+dust dynamics, where aerodynamic drag leads to the vertical settling and radial migration of grains, is consistently treated. This produces gaps that are more striking and require a smaller planet mass to form in the dust phase than in the gas. In this work, we pass the resulting dust distributions to the Monte Carlo radiative transfer code MCFOST (Pinte *et al.* 2006) to construct synthetic images in the ALMA wavebands. We then use the ALMA simulator from the CASA (Common Astronomy Software Applications, `http://casa.nrao.edu`) package to produce images that include thermal and phase noises for a range of observing parameters. The procedure is detailed in Gonzalez *et al.* (2012).

2. Results

We first look among a wide range of parameters for the best compromise between integration time and required S/N and angular resolution allowing to recover the disk features (Fig. 1, left). We find the optimal observing parameters to be $t = 1$ h and $\theta = 0.1''$ (Fig. 1, center). We also produce images assuming that gas and dust are well mixed (Fig. 1, right) to compare with previous ALMA predictions (e.g. Wolf *et al.* 2002) all made under this hypothesis. This assumption, producing much less defined gaps, with lower contrast, is anyway unrealistic and should not be used.

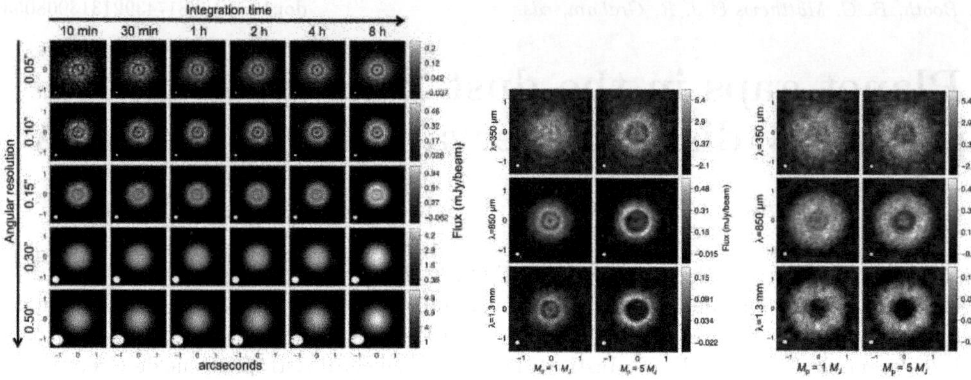

Figure 1. Simulated observations of our reference disk at $d = 140$ pc, $\delta = -23°$ and $i = 18.2°$. Varying integration times and angular resolutions for the $1\,M_J$ planet in the dynamic case at $\lambda = 850$ μm *(left)*. Optimal parameters ($t = 1$ hr, $\theta = 0.1''$) for 1 and $5\,M_J$ planets in the dynamic *(center)* and well-mixed *(right)* cases.

When including phase noise in median sky quality, our images show a substantial deterioration. However, this should not be an issue when observing in dry weather at short wavelengths and using water vapor radiometers providing real-time corrections.

We find that gaps are visible at most disk inclinations: because of dust settling, the disk outer regions are not flared and do not hide them. Planet signatures are fainter in more distant disks but declination has little effect on their detectability thanks to the excellent *uv* coverage of ALMA (Fig. 2). For more details, see Gonzalez *et al.* (2012).

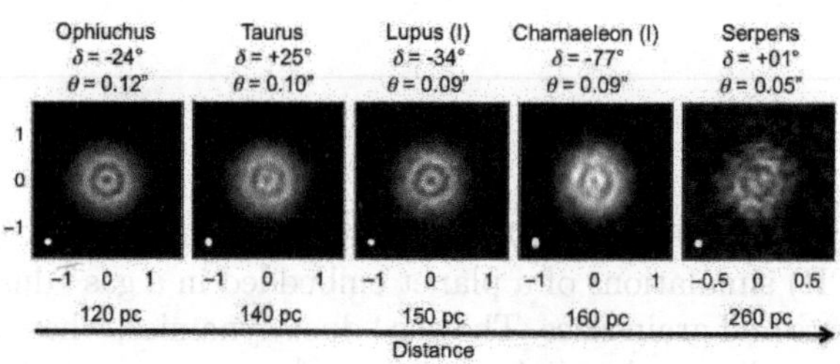

Figure 2. Simulated observations of disks in star forming regions at different distances and declinations for the $1\,M_J$ planet in the dynamic case at $\lambda = 850$ μm.

3. Conclusion

Gaps will be detectable for lighter planets than anticipated from gas-only simulations and a single 1-hour image can be sufficient to detect one and infer the presence of an unseen planet. Characterizing it requires multi-wavelength follow-up and observing at short wavelength in dry weather is crucial to detect the inner disk and discriminate planet gaps from the inner holes of transition disks. ALMA should routinely observe signposts of planets in nearby star forming regions.

References

Fouchet, L., Gonzalez, J.-F, & Maddison, S. T. 2010, *A&A*, 518, A16

Gonzalez, J.-F, Pinte, C., Maddison, S. T., Ménard, F., & Fouchet, L. 2012, *A&A*, 547, A58

Pinte, C., Ménard, F., Duchêne, G., & Bastien, P. 2006, *A&A*, 459, 797

Wolf, S., Gueth, F., Henning, T., & Kley, W. 2002, *ApJ*, 566, L97

Exploring the Formation and Evolution of Planetary Systems
Proceedings IAU Symposium No. 299, 2013
M. Booth, B. C. Matthews & J. R. Graham, eds.

doi:10.1017/S1743921313008065

First images from the PIONIER/VLTI optical interferometry imaging survey of Herbig Ae/Be stars

J. Kluska[1], F. Malbet[1], J.-P. Berger[2], M. Benisty[1], B. Lazareff[1], J.-B. Le Bouquin[1], F. Baron[3,7], C. Dominik[4], A. Isella[5], A. Juhasz[6], S. Kraus[7], R. Lachaume[8], F. Ménard[1,9], R. Millan-Gabet[5], J.D. Monnier[7], C. Pinte[1], W.-F. Thi[1], E. Thiebaut[10] and G. Zins[1]

[1]UJF-Grenoble 1 / CNRS-INSU, Institut de Planétologie et d'Astrophysique de Grenoble (IPAG) UMR 5274, Grenoble, F-38041, France

[2]ESO Santiago de Chile, Chile

[3]Georgia State University, USA

[4]University of Amsterdam, The Netherlands

[5]California Institue of Technology, USA

[6]Leiden Observatory, Leiden, The Netherlands

[7]University of Michigan, USA

[8]Pontificia Univ. Católica de Chile, Chile

[9]UMI-FCA, CNRS y U. de Chile, Chile

[10]Université de Lyon 1, France

Abstract. The close environment of Herbig stars starts to be revealed step by step and it appears to be quite complex. Many physical phenomena interplay : the dust sublimation causing a puffed-up inner rim, a dusty halo, a dusty wind or an inner gaseous component. To investigate more deeply these regions, getting images at the first Astronomical Unit scale is necessary. This has become possible with near infrared instruments on the VLTI. We have developed a new imaging method adapted to young stellar objects where we process separately the stellar component from the rest of the image to reveal the environment by using the spectral differences between these two components. We present the result of this method on the first imaging survey of Herbig stars carried out by PIONIER on the VLTI.

Keywords. techniques: high angular resolution, techniques: image processing, techniques: interferometric, circumstellar matter, planetary systems: protoplanetary disks, stars: pre–main-sequence

1. Introduction

Young stellar objects are forming planets in dusty and gaseous disks surrounding them. The disks are emitting in the infrared. They are detectable in the Spectral Energy Distribution (SED) by an infrared excess to the stellar photosphere (see Fig. 1). These inner parts ($\approx$1500K) create a bump in the SED at NIR wavelengths which is not well explained. Various models can explain the SED shape : a puffed-up inner rim (Isella & Natta, 2005), disk winds (Bans & Königl 2012), dusty halo (Vinkovic *et al.* 2003), etc... The variety of the models comes from the lack of information about the spatial distribution of matter close to the star. Thanks to the high angular resolution provided by optical interferometry, we are able to resolve the geometry of these objects, bring spatial constraints and disentangle between the different scenarios.

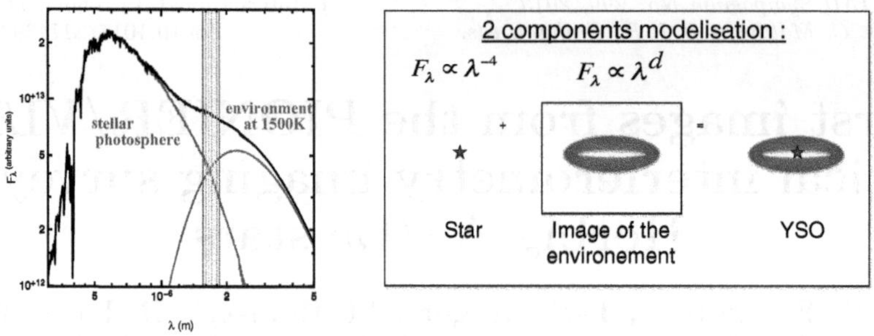

Figure 1. Left : Typical SED of a Young Stellar Object. Vertical lines : the spectral channels of PIONIER in H band. The spectral channels are exactly at the intersection of the spectra of the stellar (shorter wavelengths) and the environment (longer wavelengths) components. Right : In our image reconstruction method, we treat the star and the environment separately.

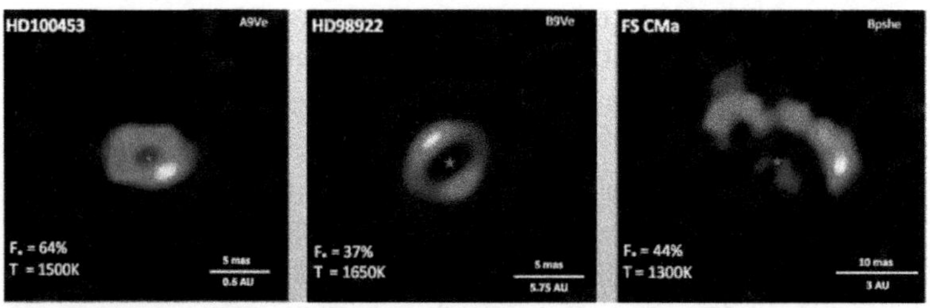

Figure 2. First results of the survey : Reconstructed images of HD100453, HD98922 and FS Cma. The flux ratios and the environment temperature are indicated on each image

2. Image Reconstruction

The main benefit of image reconstruction is to obtain a model-independent image The NIR bands are transition wavelengths between the stellar and environmental fluxes (Fig. 1). We then have to take the star-environement chromatism into account (Kluska *et al.* 2012). We use it to separate the star from the image and reconstruct the environment only (Fig. 1). This method also allows the stellar to environment flux ratio and the environment temperature to be computed. This method was used with the `mira` algorithm (Thiébaut E., 2008) as well as `macim` (Ireland *et al.* 2006).

3. Results

We are carrying out a Herbig Ae/Be survey with VLTi/PIONIER to get statistical information on the inner parts of Young Stellar Objects and to reconstruct a dozen images of the most resolved objects. The first results are shown in Fig. 2. In these images we are able to see inner rims structures with asymmetries due to the inclination and inhomogeneities. We are also able to derive temperatures of the environments which are close to the dust sublimation radius and independently from SED modeling.

The survey is not yet finished and we are expecting more objects to image.

References

Bans, A., Königl, A. 2012, *ApJ*, 758, 100
Ireland, M. J., Monnier, J. D., & Thureau, N. 2006, *SPIE Proceedings*, 6268
Isella, A. & Natta, A. 2005, *A&A*, 438, 899
Kluska, J., Malbet, F., Berger, J.-P., *et al.* 2012, *SPIE Proceedings*, 8445
Le Bouquin, J.-B., Berger, J.-P., Zins, G., *et al.* 2012, *SPIE Proceedings*, 8445
Thiébaut, E. 2008, *SPIE Proceedings*, 7013
Vinković, D., Ivezić, Ž., Miroshnichenko, A. S., & Elitzur, M. 2003, *MNRAS*, 346, 1151

Exploring the Formation and Evolution of Planetary Systems
Proceedings IAU Symposium No. 299, 2013
M. Booth, B. C. Matthews & J. R. Graham, eds.

doi:10.1017/S1743921313008077

Probing protoplanetary disks with Aperture Masking

S. Lacour[1], P. Tuthill[2] and S. Casassus[3]

[1]LESIA, CNRS/UMR-8109, Observatoire de Paris, UPMC, Université Paris Diderot, 5 place Jules Janssen, F-92195, Meudon, France

[2]School of Physics, University of Sydney, NSW 2006, Australia,

[3]Departamento de Astronomía, Universidad de Chile, Camino del Observatorio 1515, Las Condes, Santiago, Chile

Abstract. The interaction between planetary formation and protostellar disks is among the most critical remaining pieces in the puzzle of solar system assembly. Leading theoretical models are constructed around two distinct scenarios: gravitational instabilities and core accretion. The physics of each applies to quite different epochs of formation, and exhibits complex dependencies on parameters like disk density and viscosity. Untangling the effects such processes have on the final planetary statistics necessitates direct observation of exoplanets in their primordial state, prior to orbital migration. Furthermore, detailed study of the environment, such as the way the planets shape the protostellar disk by driving accretion streams across disk gaps, will also constrain formation models. Aperture masking interferometry has demonstrated a unique ability to probe the gaps within stellar disks. It has twin advantages of a higher dynamic range at the diffraction limit (λ/D) than differential imaging, while at the same time giving very extensive UV coverage compared to long baseline interferometry.

Keywords. Disks, planet formation, aperture masking, interferometry

---

Transitional disk are rare, accounting for around 10% of the T Tauri population. They are believed to form the interface between class II pre-main sequence, and class III systems hosting debris disks. They are characterized by a distinctive dip in their infrared SED which suggests that a partially evacuated gap exists in the inner region of the protoplanetary disk (e.g. Calvet *et al.* 2002). The profound implications for studies of planetary formation have become increasingly apparent with the confirmation of the disk-gap architecture by sub-millimetre measurements (e.g. Andrews *et al.* 2011) and optical interferometry (e.g. Olofsson *et al.* 2011). A number of mechanisms have been proposed to cause these gaps, including extensive grain growth (Dullemond & Dominik 2005), photo evaporation (Clarke *et al.* 2001) and binarity (Ireland & Kraus 2008), however most exciting for exoplanetary science is the potential for such systems to arise from the tidal barrier created by dynamical interaction with low-mass disk objects (Bryden *et al.* 1999; Papaloizou 2007). These different mechanisms can be distinguished by studying the distribution of the gas and dust within the gaps. For example, gravitational clearing by a large (stellar) companion or photoevaporation from the primary would almost completely evacuate the inner regions, while a less massive planetary companion would allow gas and small dust grains to exist within its orbit (Lubow *et al.* 1999). Furthermore, in the latter case, the measurement of the size and distribution of this material would allow the orbit and mass of the planetary companion to be constrained (Andrews *et al.* 2011).

Aperture masking on an 8 meter telescope is a key technique to understand the physics and the dynamics of the shaping of such holes in transition disks. This is mainly due to the critical nature of the observational parameter space opened by this method.

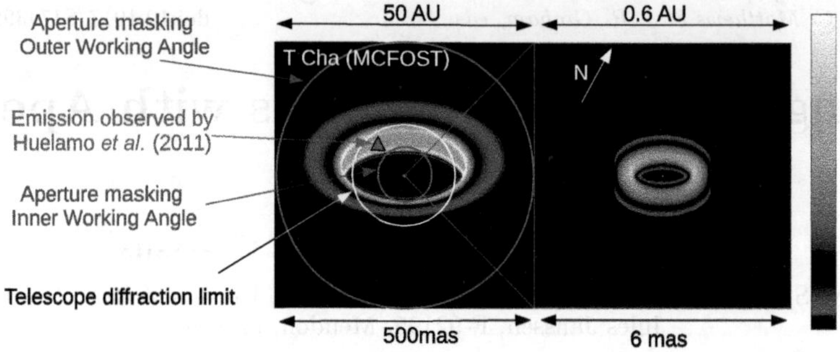

Figure 1. The inner (right) and outer (left) disks from T Cha modeling on which is over-plotted the observational window offered by aperture masking. In the near infrared, on an 8-meter telescope, the inner working angle is 30mas, and the outer working angle around 300mas (at which point differential techniques like ADI become more efficient). At the distance of the Cha I association, it means an IWA at 3AU, and an OWA at 30 AU.

Aperture masking is able to provide information on an object at a resolution between 0.5 λ/D to a few λ/D with a dynamic range of several hundred (Biller *et al.* 2012). At the typical distance of young star-forming clouds ($\approx$150 pc), aperture masking can probe with high dynamic range the immediate vicinity of disk-enshrouded stars at a resolution of a few AUs. As a result, this technique has yielded the key discovery of the protoplanet LkCa 15 b (Kraus & Ireland 2012), which explains the gap in this system. Also, SAM detected structure around a compact source in the inner disk surrounding HD142527 (Biller *et al.* 2012), and yielded resolved observations of the gap in T Cha (Huélamo *et al.* 2011; Olofsson *et al.* 2013) . Thus SAM has unravelled touchstones systems for testing planet formation models. But discoveries are open to interpretation. Although the clear interferometric signals reported by separate teams are hardly open to challenge, could it be that the interpretation as a sub-stellar companion is too simplistic? The problem is twofold: first, aperture masking delivers data streams which have complex observables difficult to understand and model by teams not profoundly involved in the technique. Second, the data interpretation must be done with as wide a variety of tools and constraints as possible. Incisive science will only result from combination of all available observational data combined with advanced modeling to globally interpret the information from different techniques and wavebands.

References

Andrews, S. M., Wilner, D. J., Espaillat, C., *et al.* 2011, ApJ, 732, 42
Biller, B., Lacour, S., Juhász, A., *et al.* 2012, ApJ, 753, L38
Bryden, G., Chen, X., Lin, D. N. C., Nelson, R. P., & Papaloizou, J. C. B. 1999, ApJ, 514, 344
Calvet, N., D'Alessio, P., Hartmann, L., *et al.* 2002, ApJ, 568, 1008
Clarke, C. J., Gendrin, A., & Sotomayor, M. 2001, MNRAS, 328, 485
Dullemond, C. P. & Dominik, C. 2005, A&A, 434, 971
Huélamo, N., Lacour, S., Tuthill, P., *et al.* 2011, A&A, 528, L7
Ireland, M. J. & Kraus, A. L. 2008, ApJ, 678, L59
Kraus, A. L. & Ireland, M. J. 2012, ApJ, 745, 5
Lubow, S. H., Seibert, M., & Artymowicz, P. 1999, ApJ, 526, 1001
Olofsson, J., Benisty, M., Augereau, J.-C., *et al.* 2011, A&A, 528, L6
Olofsson, J., Benisty, M., Le Bouquin, J.-B., *et al.* 2013, A&A, 552, A4
Papaloizou, J. C. B. 2007, A&A, 463, 775

Exploring the Formation and Evolution of Planetary Systems
Proceedings IAU Symposium No. 299, 2013
M. Booth, B. C. Matthews & J. R. Graham, eds.

doi:10.1017/S1743921313008089

The inner disks of EXor-type eruptive stars

Nikoletta Sipos[1] and Ágnes Kóspál[2]

[1]Institute for Astronomy, ETH Zürich, Wolfgang-Pauli-Strasse 27, 8093 Zurich, Switzerland
email: niki@phys.ethz.ch

[2] Research and Scientific Support Department, European Space Agency (ESA, ESTEC, SRE-SA), PO Box 299, 2200 AG, Noordwijk, The Netherlands, email: akospal@rssd.esa.int

Abstract. EX Lupi-type young stars (EXors) show sporadic brightenings of several magnitudes, caused by the episodic increase in the accretion rate of the circumstellar matter onto the young star. As the inner disk plays a crucial role during the onset of the outburst, we examined the quiescent properties of the circumstellar environment of EXors, focusing on the inner regions. We found that in case of three EXors (VY Tau, V1143 Ori and EX Lup) the spectral energy distributions show no or weak excess above the stellar photosphere at NIR-MIR wavelengths, indicative of inner disk clearing. A detailed radiative transfer modeling of the sources revealed that the inner regions of these disks had to go through significant evolution, either the inner radius of the dusty disk is beyond the sublimation radius and/or the inner disks are flattened.

Keywords. stars: formation – stars: protoplanetary disks – stars: individual: VY Tau, V1143 Ori, EX Lup

---

1. Introduction

EXors are young low-mass stars, showing sudden flare-ups separated by intervals of quiescent periods (Herbig 1989). Outbursts typically last for a couple of months or a year, and correspond to an optical brightening of 1–5 magnitudes, while between two eruptions they spend several years in a low-state. This type of eruptive phenomena is associated with a significant enhancement in mass accretion rates (up to 10^{-5} $M_\odot$/yr). Eruptions are important periods for building up stellar mass, and they influence the structure and the composition of the disk (e.g. Mosoni *et al.* 2013, Ábrahám *et al.* 2009). The circumstellar environment and especially the regions closest to the star must play an essential role in driving the eruptions. We examine the dusty circumstellar environment of three EXors (VY Tau, V1143 Ori and EX Lup) in quiescence, looking for features that might explain their eruptive nature and set them apart from normal T Tauri stars.

2. EXors with weak NIR-MIR excess

We compared the spectral energy distributions (SEDs) of the EXors to those of normal T Tauri stars in the NIR-MIR regime. For the comparison we used the T Tauri median SED (Fig. 1). We show here three members of the group, that all have a weak or missing excess over the stellar photosphere compared to typical T Tauri stars at the NIR domain. They run below the lower quartile of the Taurus median at short wavelengths. At longer wavelengths they gradually reach it and EX Lup also exceeds the upper quartile. This characteristic in general might be indicative of an inner disk clearing behind the dust sublimation radius or the settling of the inner disk (Furlan *et al.* 2006, Luhman *et al.* 2010). We also examined five other EXors (XZ Tau, DR Tau, NY Ori, UZ Tau and V1118 Ori), but they have a significant excess above the photosphere at all wavelengths, implying the presence of an optically thick disk already at small radii.

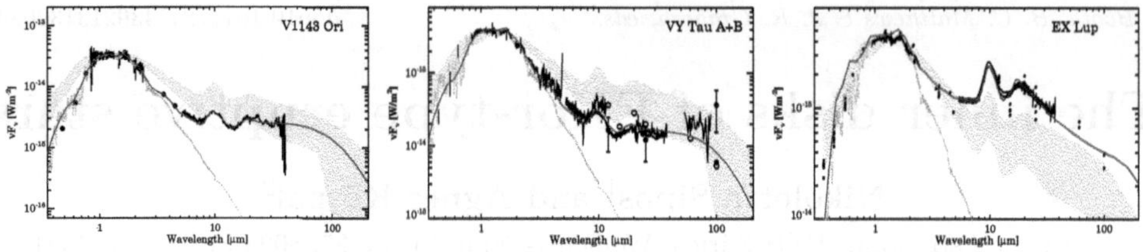

Figure 1. SEDs of V1143 Ori, VY Tau and EX Lup. The light grey band marks the upper and lower quartile of the median SED of Taurus young stars (D'Alessio *et al.* 1999, Furlan *et al.* 2006). The dotted line shows the stellar photosphere of each source. The solid line shows our best fit models (Sipos *et al.* (2009), Sipos & Kóspál (in prep.)).

Table 1. Overview of stellar properties and distances of the sources.

Object	Spectral type	T_{star} (K)	M_{star} ($M_\odot$)	L_{star} ($L_\odot$)	A_V (mag)	Distance(pc)
VY Tau A/B	M0/M2	3800/3370	0.85/0.35	0.26/0.07	0.48	145
V1143 Ori	M2	3500	0.5	0.2	0	400
EX Lup	M0	3800	0.6	0.48	0	155

We tested what kind of geometry may cause the reduced flux at short wavelengths for EX Lup, V1143 Ori and VY Tau by modeling their SEDs with the help of the radiative transfer code *RADMC* (Dullemond & Dominik 2004). The best fit models are shown in Fig. 1, the basic properties of the sources are summarized in Table 1. For the details of the modeling, the applied geometry and the resulting parameters we refer to Sipos *et al.* (2009) and Sipos & Kóspál (in prep.), here we point out our findings in connection with the innermost part of the dusty disks. In the case of EX Lup the dusty disk starts at radii ~ 3.5 times the sublimation radius, beyond that the disk can be characterized with parameters typical of normal T Tauri stars. V1143 Ori also has a dust free inner hole that stretches ~ 3 times behind the sublimation radius, and the scale height of the inner disk is reduced by $\sim 80\%$ compared to a typical value. In order to fit the SED of VY Tau we either had to use an enlarged inner hole or a reduction of the scale height by 60%.

Three out of eight EXors are surrounded by dust free inner holes that stretch further from the central star than the sublimation radius, and in two cases the disks have a reduced scale height, probably due to grain growth or settling. The inner clearing is a surprising result given the high accretion rates during the repetitive eruptions. According to certain outburst scenarios one would expect a massive reservoir in the inner disk that feeds the accretion during the outburst. The enlarged hole can be the result of the eruptions emptying the inner disk, photoevaporation or an unknown small companion.

References

Ábrahám, P., Juhász, A., Dullemond, C. P., *et al.* 2009, *Nature*, 459, 224

D'Alessio, P., Calvet, N., Hartmann, L., Lizano, S., & Cantó, J. 1999, *ApJ*, 527, 893

Dullemond, C. P. & Dominik, C. 2004, *A&A*, 417, 159

Furlan, E., Hartmann, L., Calvet, N., *et al.* 2006, *ApJS*, 165, 568

Herbig, G. H. 1989, ESO Conference and Workshop Proceedings, 33, 233

Luhman, K. L., Allen, P. R., Espaillat, C., Hartmann, L., & Calvet, N. 2010, *ApJS*, 186, 111

Mosoni, L., Sipos, N., Ábrahám, P., *et al.* 2013, *A&A*, 552, A62

Sipos, N., Ábrahám, P., Acosta-Pulido, J., *et al.* 2009, *A&A*, 507, 881

Sipos, N., Kóspál, Á. *in prep.*

Exploring the Formation and Evolution of Planetary Systems
Proceedings IAU Symposium No. 299, 2013
M. Booth, B. C. Matthews & J. R. Graham, eds.

doi:10.1017/S1743921313008090

The Story of Planets: Anchoring Numerics in Reality

Zoë M. Leinhardt

School of Physics, University of Bristol
H.H. Wills Physics Laboratory, Tyndall Avenue, Bristol, BS8 1TL, United Kingdom
email: zoe.leinhardt@bristol.ac.uk

Abstract. Building a complete coherent model of planet formation has proven difficult. There are gaps in the observational record, difficult physical processes that we have yet to fully understand, such as planetesimal formation, and an extensive list of observationally determined constraints that the model must fulfil. For example, the diversity of extrasolar planets detected to date is staggering – from single hot-Jupiters to multiple planet systems with several tightly packed super-Earths. In addition, the characteristics of the host stars are broad from single solar-mass stars to tight binaries and low mass, low metalicity stars. Even more surprising, perhaps, is the frequency of detection and thus, the implied efficiency of the planet formation process. Any theoretical model must not just be able to explain how planets form but must also explain the frequency and diversity of planetary systems. So why is planet formation so prolific? What parameters determine the type of planetary system that will result? How important are the initial parameters of the protoplanetary disk, such as composition, versus stochastic effects, such as gravitational scattering events, that occur during the evolution of the planetary system?

Current observations of extrasolar planets provide snapshots in time of the earliest and latest stages of planet formation but do not show the evolution between the two. It is at this point that we must rely on numerical models to evolve proto-planetary disks into planets. But how can we validate the results of our numerical simulations if the middle stages of planet formation are effectively invisible? Collisions are a core component of planet formation. Planetesimals, the building blocks of planets, collide with one another as they grow and evolve into planets or planetary cores and are viscously stirred by larger protoplanets and fully-formed planets. The range of impact parameters encountered during growth from planetesimals to planets span multiple collision outcome regimes: cratering, merging, disruption, and hit-and-run events. Most of these collisions produce significant debris and dust. If we have a good understanding of the production of collisional debris we can use it as an indirect tracer of on-going planetary evolution even if the planets themselves are not directly detectable.

In this paper I will show how numerical simulations of planet formation including realistic collision modelling can be used to predict, and be constrained by, observations.

Keywords. methods: numerical, solar system: formation, planetary systems: formation, protoplanetary disks

---

1. Introduction

Planet formation is seemingly common around main sequence stars as evidenced by the continual increase in the number of observed planets (889 on 02.06.2013). The sheer number and diversity of the planets drives planet formation theory. Observations of planetary systems provide snapshots of protoplanetary disks or stable planetary systems but they do not provide all the steps in between. A complete self-consistent model of planet formation has eluded astronomers because of gaps in the observational record, incomplete physics in the numerical simulations, and computational constraints. In addition, observations of the earliest phase of planet formation are often indirect. We are not observing the material or objects that we would like to observe, for example, the presence of gas or planets must often be inferred. Thus, the evolution from one state to the other is not fully understood, however, the gaps are being filled in with a combination of observations, numerical simulations, and laboratory experiments.

2. Current Status on our Understanding of Planet Formation

Let us begin by considering where the field stands at the moment both from an observational view point and from a numerical and experimental viewpoint.

2.1. *The Earliest Evidence of Planet Formation: Proplyds, Pre-Transitional, and Transitional Disks*

The Hubble Space Telescope Orion Nebula Proplyd Atlas provides a plethora of beautiful images of protoplanetary disks. These disks are diverse in size, shape, and local environment. In addition, there are an increasing number of resolved pre-transitional and transitional disks that have specific features that indicate planet formation in progress. For example, MWC 758 and SAO 206462 have two clearly identifiable spiral arms (Grady *et al.* 2013, Muto *et al.* 2012). The spiral arms could be caused by perturbing planets. LkCa 15 and SAO 206462 have large dust holes between 10s and 100s of AU wide (Kraus & Ireland 2012). Holes of this size could suggest more than one planet (Dodson-Robinson & Salyk 2011, Zhu *et al.* 2011). LkCa 15 actually shows evidence of an object in the gap, which could be a very young, hot planet.

Pre-transitional objects PDS 70 and UX Tau A also have large gaps 10s of AU wide with inner disks (Hashimoto *et al.* 2012, Tanii *et al.* 2012). Again one of the most convincing ways to create a large gap while maintaining both an inner and outer disk is to evoke at least one embedded planet. A large number of transitional and pre-transitional disks also show evidence of grain growth either large grains have been observed and/or disk size is significantly different when observed at different wavelengths suggesting grain growth or filtration both of which indicate ongoing planet formation and/or the presence of a young planet.

2.2. *Grain Growth*

Protoplanetary disks often show evidence of grain growth. Some disks show the presence of very large grains, for example, UX Tau A has large 30 μm grains and disks in Lupus and Chamaeleon have cm-sized grains (Ubach *et al.* 2012). While other disks seem to have a radial dependence on grain size – the protoplanetary disk AS 209 has a strong radial dependence on maximum grain size. Beyond 70AU sub-mm grains are the largest grains found in the disk while within 70 AU larger mm and cm-sized grains were observed (Perez *et al.* 2012).

Numerical simulations predict disk and gap size should vary with dust size in a transitional disk with an embedded planet (Fouchet *et al.* 2012). In addition, it is often not enough to explain a transitional disk observation by just envoking one embedded planet. GM Aur, for example, an embedded planet plus grain growth are required to explain the large gap and small grain depletion. Small grains should be able to filter through the gap created by the planet although the large grains pile-up at the gap edge (Zhu *et al.* 2012) but the abundance of small grains in the gap is too low, thus, it is possible that the small grains do filter through but also grow over time in the gap.

In evoking grain growth one should also consider how straight forward it is to grow in the local environment. If the environment is strongly perturbed by an embedded object such as a planet, grain growth may be significantly suppressed. Laboratory experiments such as those presented in Jankowski *et al.* (2012) show that aggregates do not always continue to increase in mass and instead may bounce or be disrupted if the impact energy is too high. It is also possible that embedded planets may enhance grain growth in certain locations by concentrating grains at pressure maxima where the increased surface density can lead to more grain collisions.

2.3. *Planetesimal Formation*

Growing grains must eventually make it into the larger building blocks of planets, planetesimals. Planetesimals outside of our own solar system are effectively invisible to observational techniques because they are too small to be directly observable as individual objects. Within our own solar system it is generally assumed that Kuiper Belt Objects (KBOs) are left over planetesimals from our own Solar System's evolution. With the sparse constraints this offers, theorists have come up with many hypotheses to explain planetesimal formation. The three most popular are: 1) *pair-wise accretion* - two planetesimals hit each other and stick. The major argument against this process is that the mean impact speed in a protoplanetary disk can be very high above the disruption threshold. In addition, the aerodynamic drag on metre-sized bodies would cause fast inward migration (Weidenshilling 1977). In recent work by Garaud *et al.* (2013) pair-wise accretion has been shown to work well even in a high-velocity environment as long as there is a distribution of impact speeds amongst the colliders and sweep-up of small material is efficient; 2) *gravitational instability* - direct collapse of dust into planetesimal-mass objects. This process requires a cold, dense environment that does not occur globally in a protoplanetary disk except potentially at very large semi-major axes. Gravitational instability seems to be more successful when used in combination with another instabilities such as the steaming instability or vortices which can collect the solid material to create a locally dense region (for example, Johansen *et al.* 2007); and 3) *dust trapping* - generic term for mechanisms such as vortices and pressure bumps that can collect solid material and increase the local dust density which can then lead to gravitational instability (Dittrich *et al.* 2013, Meheut *et al.* 2012).

Planetesimal formation is particularly difficult to constrain at the moment and the astronomy community has yet to reach a consensus as no planetesimal formation method seems to work well in all conditions. For a more detailed discussion of planetesimal formation please see the review by Eugene Chiang in these proceedings.

2.4. *Planetary Systems*

To date there are about 1000 confirmed planetary systems and thousands of candidates but the astronomy community is still missing a clear observational picture of how planets form, why the formation process is so common, and what parameters determine the type of planetary system will be formed. One of the most surprising characteristics of observed planets is their diversity. There are small planets, large planets, planets in multiple planet systems that are widely spaced, multiple planet systems that are tightly packed, planets that are close to their parent star, planets that are very far away from their parent star, planets around single stars, planets orbiting binary stars, planets orbiting young stars and planets orbiting highly evolved stars. Planet formation is not just common, it is prolific and the process produces objects that are so diverse they stretch the imagination of the most enthusiastic sci-fi follower.

Any coherent, self-consistent planet formation model most be able to explain the observed diversity of extrasolar planets. As mentioned earlier it is observations that are currently driving the field of exoplanet research but often the observations are indirect while numerical and experimental investigations are direct. For significant progress to be made we must now focus on connecting direct numerical simulations with the indirect or inferred results from observational programs. In the next section I will focus on using collisions and the debris produced by collisions as a tracer for planet formation.

3. Collisions within the Solar System

Collisions are fundamental to the evolution of solar systems. Within our own Solar System we see evidence of the final giant impacts that occurred in the last stages of solar system evolution. The Earth-Moon system and Pluto-Charon could have been formed by large impacts (Cuk & Stewart 2012, Canup 2005). In the asteroid belt there are many dynamical families with similar orbital and spectral parameters. Objects within one dynamical family are believed to have originated from the same parent body which was disrupted by a high speed impact (Nesvorny *et al.* 2006). There is also one dynamical family in the Kuiper Belt which appears to be the result of a collision event (Leinhardt *et al.* 2010).

Regardless of how the building blocks of planets formed, it is generally understood that in order for planetesimals to grow into planets they must grow via collisions (gravitational instability may indeed avoid this process for large giant planets at large semi-major axis). Thus, the evolution of planetesimals is dominated by a series of individual collisions with other planetesimals. The collision outcome: merging, erosion, cratering, hit-and-run, depends on the specific impact conditions and although the planetesimals may be observationally undetectable the debris that is produced as a result of collisions may indeed be the key to connecting numerical simulations of early planetary growth with observations of evolving protoplanetary disks.

Interpreting the collisional evidence requires that we understand the collisions and have a model to describe them. Previous collision models were not developed to be able to determine the underlying parent population. Now we need a model that can be used to do just that. The work that I am going to present here is motivated by explaining the diversity of extrasolar planets and identifying tracers of the formation process that have to date been unobservable.

3.1. *Numerical Simulations of Collisions*

In order to develop the model we constructed a series of numerical simulations of planetesimal collisions in isolation and fit scaling-laws to the results. The collisions ranged from slow sub-sonic impacts to fast shock generating super-sonic impacts. Using results from previous studies and our own simulations we investigated collisions with a range of impact parameters and bulk compositions (Leinhardt & Stewart 2012).

In this work we found that there is no one velocity that will determine disruption of a particular target. In other words the catastrophic disruption threshold for gravity dominated objects,

$$Q^{\star}_{RD\,grav} = q_g(\rho_1 G)^{3\bar{\mu}/2} R_{C1} V^{\star(2-3\bar{\mu})}, \qquad (3.1)$$

where ρ_1 is 1 g cm^{-3}, G is the gravitational constant, q_g and $\bar{\mu}$ are material parameters fit to the empirical results, R_{C1} is the radius of the combined mass of the target and projectile with a bulk density of ρ_1, and $V^{\star}$ is the critical impact speed, is velocity dependent. It is the momentum of the impact that determines the outcome not the impact energy as has often been assumed in the literature. The catastrophic disruption threshold in the gravity regime has a velocity dependence that scales with momentum, $\bar{\mu} \sim 0.35$. Equation 3.1 fits the empirical data well over five orders of magnitude even though the data is very diverse and includes multiple numerical techniques including an

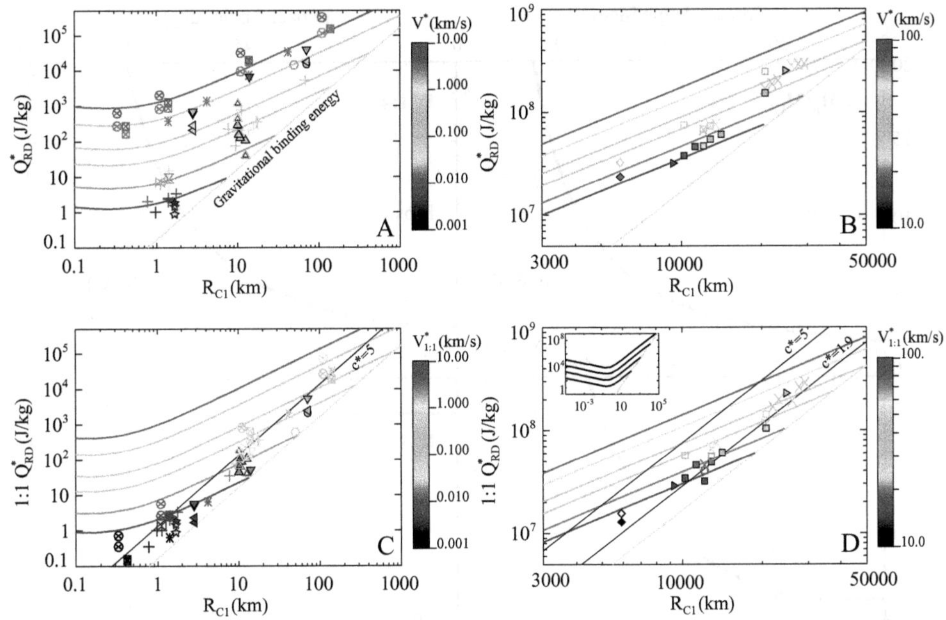

Figure 1. Collision outcome maps from Leinhardt & Stewart (2012). Each region indicates the type of collision outcome from a two body collision given the impact parameter and impact speed for four example mass ratios. The solid curves indicate the transition between different collision outcomes. The thick solid black curve identifies the catastrophic disruption threshold where 50% of the total system mass is permanently removed. The dashed and dotted lines indicate other interesting transitions such as projectile erosion and projectile disruption.

N-body gravity code, Eulerian grid codes, and Lagrangian smoothed particle hydrocode techniques from multiple authors (for example, Benz & Asphaug 1999, Leinhardt *et al.* 2000, Leinhardt & Stewart 2009, and Jutzi *et al.* 2010). In addition to the catastrophic disruption threshold we have also fit the data with scaling laws for the mass of the largest remnant, the size distribution of the remnant tail, and the velocity dispersion of the remnants. The mass of the largest remnant scales linearly with $Q^\star_{RD}$ for impacts within the disruption regime (impact energy less than twice $Q^\star_{RD}$) but is fit better with a power law for more energetic super-catastrophic impacts ($Q_{RD} > 2Q^\star_{RD}$). The results give us detailed collision outcome maps which identify the type of collision that will occur for a given mass ratio, impact parameter, and impact speed (Fig. 1). The collision outcome maps have several interesting characteristics - as projectile and target reach similar masses hit-and-run impacts become more prevalent. Hit-and-run impacts could be an efficient mechanism for stripping mantle material from projectiles and increase the diversity of densities within one planetary system. Satellite forming giant impacts such as the Earth-Moon or Pluto-Charon tend to occur on the transition between accretion and hit-and-run suggesting that hit-and-run impacts occurred within our solar system.

The final result is a model that can predict the collision outcome taking into account most parameters including impact parameter, mass ratio, and impact speed. For a more detailed discussion of this collision model see Leinhardt & Stewart (2012) and Stewart & Leinhardt (2012).

3.2. *Implications of Collision Model*

To check the long term implications of our collision model we retrospectively applied our new model to the N-body simulations of the last stage of planet formation. The original

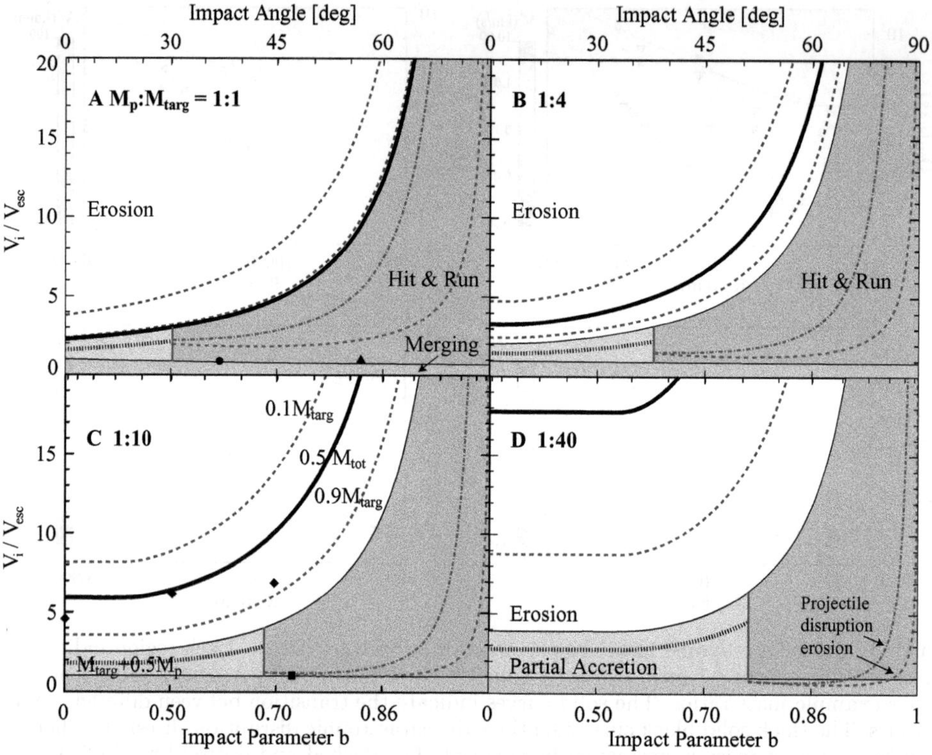

Figure 2. Synthetic dust images of pre-transitional disks with embedded Jupiter-mass planet. Figure 3 from Dobinson *et al.* (submitted) reprinted here with permission from the author. The first row shows the instantaneous surface density of dust calculated from direct N-body simulations of a protoplanetary gas-free disk with an embedded Jupiter-mass planet. The second row shows the flux density using RADMC-3D (Dullemond 2012) to calculate the full radiative transfer. The first column shows the results from circular embedded planet using the rubble collision model from Leinhardt & Richardson (2005), which allows planetesimals to both grow and erode do to collisions. The next three columns show results from merging simulations in which the underlying planetesimal population is forced to merge as the result of a collision. These three columns have embedded planets with varying degrees of eccentricity: 0.0, 0.1, and 0.2, respectively.

simulations (O'Brien *et al.* 2006, Raymond *et al.* 2009) assumed that all collisions resulted in perfect merging events. We reanalysed these simulations in Stewart & Leinhardt (2012) and found that outcomes of giant impacts span all possible collision regimes including hit-and-run, accretion, erosion and catastrophic disruption. Fragmentation during giant impacts is also significant. The majority of the ejected material is mantle from partial accretion events or energetic hit-and-runs that result in the disruption of the projectile. If the ejected material is not totally re-accreted giant impacts can create planets depleted in volatiles and mantle (including water and atmosphere) compared to initial embryos.

In summary, we found less than 10% of the collisions between protoplanets would result in perfect merging events and a large fraction (1/5 to 1/3) would result in hit-and-run events. Depending on the assumptions made about the reaccretion of mantle material the high fraction of stripping events could lead to a significant increase in the range of densities of the final planetary embryos.

4. Current Work on Early Indications of Planet Formation

4.1. *Transitional Disks: Collisional Dust from an Embedded Planet*

Let us now return to early indications of planet formation. With the ever increasing observational capabilities of ground and space based facilities such as ALMA and Kepler there is no time better than the present to definitively link the dynamic process of planet formation which is simulated numerically with a young system protoplanetary disk in the midst of planetary formation. As a result, we have begun focusing our numerical efforts on transitional and pre-transitional disks. Assuming that some transitional and pre-transitional disks are the result of embedded planets we have constructed a series of numerical simulations with 10^6 self-interacting planetesimals and one embedded Jupiter-mass planet. Assuming that the planet formed via core accretion there should be planetesimals that have evolved on the same timescale. These planetesimals will collide with each other due to perturbations from the planet and the planetesimals themselves. The collisions will generate observable dust (see Fig. 2 from Dobinson *et al.* submitted). When considered in isolation this second generation dust looks remarkably similar to primordial dust trapped in the gas disk of a transitional disk with an embedded planet (Gonzalez *et al.* 2012). However, it is clear that collisions are fundamentally important to the evolution of planetary systems and cannot be ignored. We hope that as this work progresses, multiple planets will be embedded in the planetesimal disk, and the primordial dust signature will also be taken into account. We will then be more confident about identifying young systems that are likely to be in the midst of planet formation from the dust signatures.

5. Summary and Conclusions

Planet formation seems to occur around most stars producing a large diversity of objects and systems. There are large gaps in our observational record and numerical methods are not powerful enough to fill in all of the gaps. Progress is being made observationally, numerically, and experimentally but we are at a point where these methods must now be used together and not in isolation in order for significant progress to be made. Interpretation of observations is difficult and often non-unique. For example, in most cases the presence of planets must be inferred. The planets are not directly observed.

Collisions are important in the formation of planets and should produce observable signatures, however, the signature from the collisions is complicated by the presence of primordial and growing dust grains. In order to interpret the observational results from young protoplanetary systems more work needs to be done including collisional evolution and grain growth in the presence of a gas disk. However, it is clear that planets within planetesimal disks do produce clear observational signatures and with a bit more work the community should be able to confidently identify unique signatures of planet formation within transitional and pre-transitional disks.

References

Benz, W. & Asphaug E. 1999, *Icarus*, 142, 5
Canup, R. 2005, *Science*, 307, 546
Ćuk, M. & Stewart S. T. 2012, *Science*, 338, 1047
Dittrich, K., Klahr, H., & Johansen, A. 2013, *ApJ*, 763, 117
Dodson-Robinson, S. E., Salyk, C. 2011, *ApJ*, 738, 131
Dullemond, C. P. 2012, *ASCL*, 1202.015
Fouchet, L., Gonzalez, J.-F., & Maddison, S. T. 2010, *A&A*, 518, A16

Garaud, P., Meru, F., Galvagni, M., & Olczak, C. 2013, *ApJ*, 764, 146
Gonzalez, J.-F., Pinte, C., Maddison, S. T., Ménard, F. & Fouchet, L. 2012, *A&A*, 547, A58
Grady, C. A., *et al.* 2013, *ApJ*, 762, 48
Hashimoto, J., *et al.* 2012, *ApJL*, 758, L19
Jankowski, T., Wurm, G., Kelling, T., Teiser, J., Sabolo, W., Gutiérrez, P. J. & Bertini, I. 2012, *A&A*, 542, A80
Johansen, A., Oishi, J. S., Mac Low, M.-M., Klahr, H., Henning, T. & Youdin, A. 2007, *Nature*, 448, 1022
Jutzi, M., Michel, P., Benz, W., Richardson, D. C. 2010, *Icarus*, 207, 54
Kraus, A. L. & Ireland, M. J. 2012, *ApJ*, 745, 5
Leinhardt, Z. M., Richardson, D. C., Quinn, T. 2000, *Icarus*, 146, 133
Leinhardt, Z. M., Richardson, D. C. 2005, *ApJ*, 625, 427
Leinhardt, Z. M. & Stewart, S. T. 2009, *MNRAS*, 199, 542
Leinhardt, Z. M., Marcus, R. A. & Stewart, S. T. 2010, *ApJ*, 714, 1789
Leinhardt, Z. M. & Stewart, S. T. 2012, *ApJ*, 745, 79
Meheut, H., Meliani, Z., Varniere, P. & Benz, W. 2012, *A&A*, 545, A134
Muto, T., *et al.* 2012, *ApJL*, 748, L22
Nesvorný, D., Enke, B. L., Bottke, W. F., Durda, D. D., Asphaug, E. & Richardson, D. C. 2006, *Icarus*, 183, 296
Pérez, L. M., *et al.* 2012, *ApJL*, 760, L17
Stewart, S. T. & Leinhardt, Z. M. 2012, *ApJ*, 751, 32
Tanii, R., *et al.* 2012, *PASJ*, 64, 124
Ubach, C., Maddison, S. T., Wright, C. M., Wilner, D. J., Lommen, D. J. P., & Koribalski, B. 2012, *MNRAS*, 425, 3137
Weidenschilling, S. J. 1977, *MNRAS*, 180, 57
Zhu, Z., Nelson, R. P., Hartmann, L., Espaillat, C., Calvet, N. 2011, *ApJ*, 729, 47
Zhu, Z., Nelson, R. P., Dong, R., Espaillat, C. & Hartmann, L. 2012, 755, 6

Exploring the Formation and Evolution of Planetary Systems
Proceedings IAU Symposium No. 299, 2013
M. Booth, B. C. Matthews & J. R. Graham, eds.

doi:10.1017/S1743921313008107

Zooming in on the Formation of Protoplanetary Disks

Åke Nordlund[1,2], Troels Haugbølle[1], Michael Küffmeier[1,2], Paolo Padoan[3] and Aris Vasileiades[1,2]

[1]Centre for Star and Planet Formation, University of Copenhagen, Øster Voldgade 5, DK-2100 Copenhagen, Denmark
email: aake@nbi.dk

[2]Niels Bohr Institute, University of Copenhagen
Juliane Maries Vej 30, DK-2100 Copenhagen, Denmark

[3]ICREA & ICC, University of Barcelona, Marti i Franqus 1, E-08028 Barcelona, Spain

Abstract. We use the adaptive mesh refinement code RAMSES to model the formation of protoplanetary disks in realistic star formation environments. The resolution scales over up to 29 powers of two ($\sim$ 9 orders of magnitude) covering a range from outer scales of 40 pc to inner scales of 0.015 AU. The accretion rate from a 1.5 solar mass envelope peaks near 10^{-4} $M_\odot\,yr^{-1}$ about 6 kyr after sink particle formation and then decays approximately exponentially, reaching 10^{-6} $M_\odot\,yr^{-1}$ in 100 kyr. The models suggest universal scalings of physical properties with radius during the main accretion phase, with kinetic and / or magnetic energy in approximate balance with gravitational energy. Efficient accretion is made possible by the braking action of the magnetic field, which nevertheless allows a near-Keplerian disk to grow to a 100 AU size. The magnetic field strength ranges from more than 10 G at 0.1 AU to less than 1 mG at 100 AU, and drives a time dependent bipolar outflow, with a collimated jet and a broader disk wind.

Keywords. ISM: magnetic fields, jets and outflows; stars: formation; planetary systems: formation, accretion disks

1. Introduction

We present first-of-a-kind *ab initio* simulations of the formation of protoplanetary disks, using a locally developed version of the public domain AMR code RAMSES (Teyssier 2002; Commercon *et al.* 2011) to zoom in from giant molecular cloud (GMC) scales to the scales of circumstellar disks. The purpose of this work is to characterize the typical properties of accretion onto solar mass protostars with as few free parameters as possible, and (in the near future) to gather a statistical sample of such conditions, to quantify the extent of statistical variation of properties. This is a vast improvement over models where initial and boundary conditions have to be parametrized and a multi-dimensional parameter space has to be sampled. Here the initial and boundary conditions follow instead from the statistical properties of the interstellar medium (ISM), which are reasonably well established, as per for example Larson (1981); Solomon *et al.* (1987) and Crutcher (2011). We anticipate that since accretion properties are expected to have unavoidable statistical variation from system to system, uncertainties in the properties of the ISM on GMC scales may in fact be of secondary importance. In this contribution we report on the results of a single case study, where a prestellar core in a filament collapses and forms a relatively isolated 1.5 solar mass star. We expect that due to mass loss by bipolar outflows, stars ending up with exactly one solar mass form from systems that accrete more than one solar mass — indeed, one of the future goals of this line of work is to quantify the amount of mass lost in outflows during the accretion process.

2. Methods

The simulations proceed in three steps, with the first step covering 16 levels of (factor two) refinement (smallest cell size $\sim$ 125 AU), following individual star formation in a 40 pc GMC model for $\sim$ 10 Myrs with velocity dispersion and density statistics consistent with Larson's relations (Larson 1981), and the kinetic energy maintained by supernova explosions rather than artificial forcing. In the second step, which uses 22 levels of refinement, the neighborhoods of individual stars with a final system mass of 1-2 solar masses are followed during their accretion process, with a smallest mesh size of 2 AU, sufficient to follow the development of the large scale structure of their accretion disks during most of the accretion process. Finally a selection of these disks are studied over time intervals of the order 100 yr at three different epochs, using 29 levels of refinement and cell sizes ranging down to 0.015 AU, sufficient to study and compare the local dynamics of the disks at these different epochs. As is customary for RAMSES simulations, we count levels of refinement compared to the full box size. The root grid is at level 7.

To be able to carry out these simulations we have enhanced the public domain version of the RAMSES code (Teyssier 2002; Commercon *et al.* 2011) in several ways. In addition to the MPI parallelization we have added OpenMP parallelization of all routines that contribute significantly to the computing time, thus improving load balancing by allowing a larger number of grids to be handled by each MPI thread. The handling of Hilbert-Peano keys has been modified to allow more levels of refinement to be used. A new sink particle recipe allows the (symplectic) tracking of stellar (sink particle) dynamics, with massive stars ending their lives in supernova explosions after mass dependent life times interpolated from tables. Utilizing a galilean transformed variant of the built-in 'geometric refinement', which follows individual sinks, we are able to zoom in to concentric sets of domains around selected sink particles. A combination of adaptive mesh refinement criteria, including Jeans length, and density, pressure and velocity gradients are used, with a roughly constant number of cells per level, resulting in an approximately linear dependence of the computing time on the number of levels.

3. Results

Perusal of the accretion histories of different $\sim$ solar mass stars shows that the accretion time (defined for example as the time needed to accrete 90% of the final mass) can vary from less than 30 kyr to more than 300 kyr, depending on circumstances. Here we report on a case where the 90% accretion time is about 75 kyr. Figure 1 shows the accretion history and distribution of mass with radius. Within 6 kyr of the sink particle creation the accretion rate reaches a peak at $\approx 6\,10^{-5}$ $M_\odot\,yr^{-1}$. The mass of the sink particle at that time is about 10% of the final mass. Decreasing approximately exponentially with time after that, the accretion rate reaches 10^{-6} $M_\odot\,yr^{-1}$ after about 100 kyr. As illustrated by the right hand side panel of Fig. 1, during the main accretion phase the radial mass distribution follows approximately the same $r^{3/2}$ distribution that one would expect from a free fall situation, with radial velocities scaling as $r^{-1/2}$ and volume densities $\rho \sim r^{-3/2}$.

Three epochs during this evolution were singled out for study at the highest spatial resolution, using 29 levels of refinement; the first one at $t \approx 2$ kyr, during the initial phase of rising accretion, the second one at $t \approx 50$ kyr, while the accretion rate is still of the order 10^{-5} $M_\odot\,yr^{-1}$, and the third one at $t \approx 100$ kyr, when the accretion rate

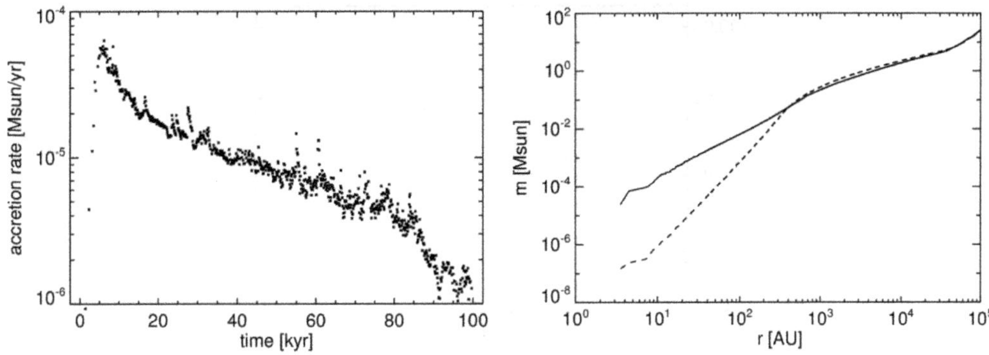

Figure 1. Mass accretion rate to the central sink particle as a function of time for the model with 22 levels of refinement (left), and radial profiles of mass, integrated over radii (right), initially (dashed) and during the main accretion phase (full drawn).

Figure 2. Visualizations of the mass density and magnetic field lines, at three epochs after the formation of a sink particle that ended up accreting ≈ 1.5 solar masses: 2 kyr after formation (left), 50 kyr after formation (center), and 100 kyr after formation (right). The point of view is essentially face-on in the left hand side panel.

has dropped to about 10^{-6} $M_\odot\,yr^{-1}$. Figure 2 shows 3-D VAPOR (Clyne & Rast 2005) renderings of the mass density and magnetic field lines at the three epochs, illustrating striking differences. As shown also by Fig. 3, the extent of the rotationally supported disk is initially very small–of the order of 1 AU–and the magnetic field lines are hour-glass shaped. At the second epoch the extent of the rotationally supported disk is ≈ 10 AU, the magnetic field lines are wound into spiral shapes, supporting a central, well-collimated (but essentially one-sided) jet and a broader disk wind outflow. Finally, during the third epoch, the rotationally supported disk extends out to ≈ 100 AU, the magnetic field lines are wound much tighter than in the 2nd case, and there is only a weak disk wind outflow.

Figure 3 shows the (mass weighted) average speed of rotation in shells of constant radius, as well as the Alfvén speed (based on the rms magnetic field and the average mass density), and the velocity dispersion relative to the average speed of rotation. The line with constant slope shows the final Keplerian speeds of rotation, illustrating the near-equipartion that apparently exist between gravitational and kinetic energy at small radii and between gravitational and magnetic energy at large radii.

4. Discussions and Conclusions

One of the most important results of this work is that it illustrates the self-consistent development of a situation that satisfies two of the critical requirements necessary to form realistic protoplanetary disks: 1) The initial very efficient shedding of angular momentum, allowing half of the stellar mass to be accreted in less than 25 kyr, while 2) nevertheless allowing an extended rotationally supported protoplanetary disk to develop and grow in

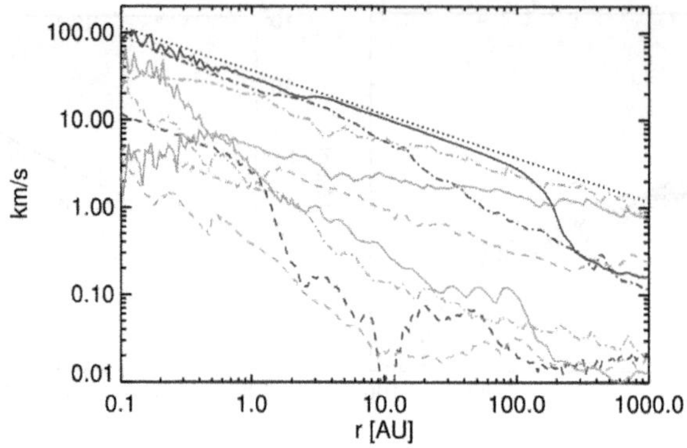

Figure 3. Radial profiles of characteristic velocities at three epochs; 2 kyr (dashed), 50 kyr (dot-dashed), and 100 kyr (full drawn). From top to bottom the three curves (blue, green and orange in the on-line version) for each epoch are: the mass weighted rotational velocity, the average Alfvén speed (see the text), and the root mean square velocity dispersion. The dotted straight line (slope -1/2) shows the Keplerian rotation speed for the fully accreted star.

size towards the end of the accretion phase. The presence of a strong azimuthal magnetic field in the disk could suppress shear driven turbulence and allow the Goldreich & Ward (1973) mechanism to operate.

At larger radii than those where a rotationally supported disk exists there is instead approximate balance between gravitational and magnetic energy density, while the kinetic energy density is much smaller. This may be seen as a balance between gravity trying to pull in and compress the magnetic field and magnetic forces resisting the pull. This is supported by visualizations (available as part of the on-line copy of the symposium presentation) illustrating the chaotic nature of the mass density and magnetic field, extending over scales with widely varying time scales (essentially following a free-fall time scaling $t_{ff} \sim \rho^{-1/2}$). The widely different time scales at different radii presumably make an ordered collapse impossible, resulting instead in a generic, chaotic dynamical state, which allows accretion to proceed at some significant fraction of the free fall speed.

As a byproduct of this type of modeling, which starts out from a supernova driven interstellar medium, we can follow the transport of short-lived radionuclides (SLRs), from the time of ejection from supernovae and until they become part of the proto-planetary disks. As shown by Vasileiades *et al.* (2013) the transport time is on average short enough to be consistent with initial abundance of 26Al in the Solar System derived from cosmochemistry. Of particular interest in the near future is to characterize the amount of variation with time of the SLR abundance during the life time of PP-disks surrounding solar mass stars.

Acknowledgements: Research at Centre for Star and Planet Formation is funded by the Danish National Research Foundation. PP is supported by the FP7-PEOPLE-2010-RG grant PIRG07-GA-2010- 261359. Supercomputing resources at Leibniz Rechenzentrum (PRACE project pr86li) and at DeIC/KU in Copenhagen are gratefully acknowledged.

References

Clyne, J. & Rast, M. 2005, in Proceedings of Visualization and Data Analysis, pp. 284-294
Commercon, B., Teyssier, R., Audit, E., Hennebelle, P., & Chabrier, G. 2011, *A&A*, 539, 35
Crutcher, R. M. 2011, *ARAA*, 50, 29
Goldreich, P. & Ward, W. R. 1973, *ApJ*, 183, 1051

Larson, R. 1981, *MNRAS*, 194, 809
Solomon, P. M., Rivolo, A. R., Barrett, J., & Yahil, A. 1987, *ApJ*, 319, 730
Teyssier, R. 2002, *A&A*, 385, 337
Vasileiades, A., Nordlund, A., & Bizzarro, M. 2013, *ApJL*, 768, 8

Discussion

PUDRITZ: As you know, many MHD simulations with evolved fields find that disks cannot form - magnetic braking is too extreme and disks do not form. We (Seifreid *et al.* 2012) find that Kepler disks can form in turbulent media. Have you observed these effects in your simulations? Are disks Keplerian?

NORDLUND: Yes, disks are Keplerian at sufficiently small radii. The region dominated by near-Keplerian motions grows secularly with time, as the mass of the central object grows and the accretion rate drops.

Exploring the Formation and Evolution of Planetary Systems
Proceedings IAU Symposium No. 299, 2013
M. Booth, B. C. Matthews and *J. R. Graham, eds.*

doi:10.1017/S1743921313008119

Problems and Prospects in Planetesimal Formation

Eugene Chiang[1], Ruth Murray-Clay[2] and Ji-Ming Shi[1]

[1]Department of Astronomy, UC Berkeley
Hearst Field Annex B-20, Berkeley CA 94720, USA
email: echiang@astro.berkeley.edu, jmshi@berkeley.edu

[2]Harvard-Smithsonian Center for Astrophysics
60 Garden Street, Cambridge MA 02138, USA
email: rmurray-clay@cfa.harvard.edu

Abstract. On the long road from dust to planets there are planetesimals. We review and criticize proposed mechanisms for planetesimal formation, and point out topics for future work.

Keywords. planets and satellites: formation, solar system: formation, instabilities, turbulence, Kuiper Belt, minor planets, asteroids, accretion, accretion disks

---

1. Introduction

A popular topic in astronomy classes is tidal disruption: the shredding of a body that comes too close to another. Around black holes, this is called spaghettification. But the reverse process can also occur. Gravitational instability can form stars, planets, and perhaps even planetesimals: the building blocks of planets. The hope is that "spaghettified" dust particles, each no larger than a cm, can collect into regions so dense that they gravitationally collapse into the first-generation planetesimals, having sizes $\gtrsim$ kilometers.

A benchmark density is the Roche density, which is that required for a one-fluid body to resist tidal shear (cf. §4), and which we will approximate by $\rho_{\rm Roche} \sim M_*/a^3$, where M_* is the mass of the central star and a is the orbital distance. The Roche density is much higher than the gas densities usually attributed to protoplanetary disks; for example, $\rho_{\rm Roche}$ exceeds the gas density in the minimum-mass solar nebula by $\sim 10^2$. The situation for dust is still more daunting. For dust to reach Roche density, the dust-to-gas ratio $\mu \equiv \rho_{\rm d}/\rho_{\rm g}$ must increase by 10^4, from the diffuse interstellar medium value of 10^{-2} to $\mu_{\rm Roche} \sim 10^2$. Comparing $\mu_{\rm Roche}$ to the highest dust-to-gas ratios on Earth (e.g., smoke-filled rooms; Saharan dust storms) is an exercise left for the reader.

This article comments on some recent efforts to understand how such extremely dusty environments might be achieved.

2. The Streaming Instability and Particle-Particle Sticking

One way to concentrate particles is through the streaming instability (SI), a two-fluid instability afflicting gas and dust in rotating disks. Youdin & Goodman (2005) investigated linear wave disturbances in these two frictionally coupled fluids and discovered that there is always an unstable mode that clumps particles aerodynamically. See Jacquet *et al.* (2011) for a mechanistic explanation.

The SI drives turbulence, and dust clumps within "streaming turbulence" have μ as large as $\sim$3000 (Bai & Stone 2010), well exceeding $\mu_{\rm Roche}$. Indeed gravitational collapse into bound clumps has been observed in simulations that combine streaming turbulence with self-gravity (e.g., Johansen *et al.* 2012). Unfortunately the degree of clumping is sensitive to particle size, or more accurately, particle stopping time $t_{\rm stop}$: the time a particle takes to "come up to speed" in a gas flow. Small particles, easily entrained in gas, have short $t_{\rm stop}$. The highest values of μ occur when $t_{\rm stop}$ is of order the orbital time Ω^{-1}, i.e., the turnover time for the largest eddies in streaming turbulence. Bai & Stone (2010) found that when $\tau_{\rm stop} \equiv \Omega t_{\rm stop} \approx 0.1$–1, streaming turbulence drives $\max \mu \sim 3000$, but when $\tau_{\rm stop} \sim 0.01$–0.1, $\max \mu \sim 30$. Decreasing $\tau_{\rm stop}$ decreases $\max \mu$.†

Unfortunately, models of particle-particle sticking have yet to robustly produce $\tau_{\rm stop} \sim 1$ particles in abundance. As particles grow, their relative velocities increase to the point where collisions result in bouncing or destruction. Standard calculations find $\max \tau_{\rm stop} \sim 10^{-3}$ (mm-cm) at 1 AU (Zsom *et al.* 2011). Greater growth is achieved by accounting for the full distribution of relative velocities rather than using rms values; the idea here is that a few lucky bodies can grow to large sizes through a series of low-velocity collisions. Figure 6 of Garaud *et al.* (2013) impresses: at 1 AU, particles stick to sizes of 10^2 m (see also Meru *et al.*, this volume)! What remains to be shown is whether such growth (which is slow because it relies on an improbable sequence of events) proceeds fast enough to overcome the "meter-size barrier": the loss of $\tau_{\rm stop} \sim 1$ (m-size) bodies to the star from inward drift.

In addition to relying on initial particle sizes that may be unrealistically large, the SI might spawn planetesimals that are too big. Johansen *et al.* (2012) argued that planetesimals having compact-equivalent sizes of several hundred km would form from the SI at ~30 AU.‡ Schlichting *et al.* (this volume) cannot reproduce the size distribution of Kuiper belt objects observed today if the primordial belt were to comprise such large objects—but for a contrary view, see Campo Bagatin & Benavidez (2012).

3. Revisiting Goldreich-Ward

A path to gravitational instability that does not rely on $\tau_{\rm stop} \sim 1$ particles is vertical settling of grains into a thin dense sheet at the midplane (Goldreich & Ward 1973). The usual objection is that vertical shear across this "sublayer" triggers the Kelvin-Helmholtz instability (KHI), driving turbulence and halting sedimentation (Weidenschilling 1980). But direct numerical simulations show that the KHI limits itself to the top and bottom surfaces of the sublayer—naturally, since density/velocity gradients are steepest there—leaving dust in the heart of the sublayer free to settle into dense cores (Lee *et al.* 2010ab; Figure 1). In disks whose bulk (height-integrated) metallicities are enriched over solar by a few, μ approaches 30 in these cores. Still larger μ could be attained by adding self-gravity and better resolving the cores spatially.

Thus the classic Goldreich-Ward scenario appears viable insofar as the danger posed by the KHI can be neutralized in disks enriched in metals (see also Weidenschilling 2006; Chiang 2008). Nevertheless, other forms of turbulence threaten. The sublayer tolerates diffusivities no larger than $\alpha \lesssim 3 \times 10^{-7} (\tau_{\rm stop}/10^{-3})(a/{\rm AU})^{4/7}$ (Chiang & Youdin 2010, p509). Streaming turbulence undercuts particle concentration by preventing settling and lowering the average particle density (turbulence extends the wings of the density PDF but lowers the mean).

† Vortices also strongly prefer to concentrate $\tau_{\rm stop} \sim 1$ particles (Meheut *et al.*, this volume).

‡ Our own naive estimate of the mass of bound objects formed by the SI using $\tau_{\rm stop} \sim 1$ particles is $\sim\rho_{\rm Roche}(\eta a)^3$, with $\eta a \sim (c_{\rm gas}/\Omega a)^2 a$ the largest eddy size, gas sound speed $c_{\rm gas}$, and Kepler speed Ωa. Our estimate yields a clump mass of $0.05 M_\oplus$ at 30 AU, of order the mass of the entire Kuiper belt interior to 50 AU.

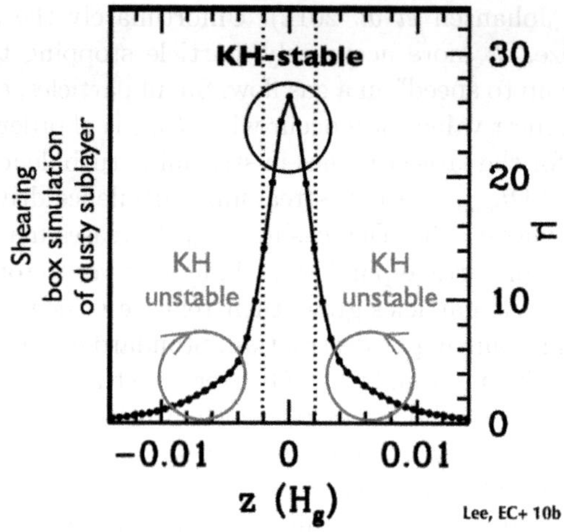

Figure 1. Dust-to-gas volume density ratio μ vs. height z (in units of gas scale height H_g) in a settled "sublayer" (figure adapted from Lee *et al.* 2010b). The bulk disk metallicity (dust-to-gas surface density ratio) is $Z_{\rm disk} = \Sigma_{\rm d}/\Sigma_{\rm g} = 0.06 = 4\times$ solar. Only the top and bottom surfaces of the sublayer were observed to be Kelvin-Helmholtz (KH) unstable; the core settled freely, and would have settled to still larger μ than is shown here if the grid resolution were finer.

4. So You've Reached Roche Density: Now What?

We have used the Roche density as a benchmark, but it is not much more than that. Attaining Roche density does not guarantee that dust can collapse against gas pressure or rotational shear. Consider a clump of dimension ℓ so large that its sound-crossing time $\ell/c > t_{\rm stop}$. In this "large-clump" regime, the dust-gas mixture acts as a tightly coupled suspension (like milk). Its effective sound speed is $c = c_{\rm gas}/\sqrt{1+\mu}$, where the square root accounts for how dust adds to the mixture's "mean molecular weight." In this regime, $\ell/c \ll 1/\sqrt{G\rho_{\rm Roche}}$: large Roche-density clumps are Jeans-stable (stabilized by gas pressure; Cuzzi *et al.* 2008; Shi & Chiang 2013).

"Small" clumps ($\ell/c \ll t_{\rm stop}$) do not face this difficulty; short-wavelength/ high-frequency sound waves in gas do not couple with dust. Thus small clumps are not supported by gas pressure—but they can still resist collapse because of their finite angular momentum. Although pressure is no longer an impediment, particle collapse speeds are limited by drag to terminal speeds of $v_{\rm term} \sim f t_{\rm stop}$, where the self-gravitational acceleration $f \sim G(\rho\ell^3)/\ell^2 \sim G\rho\ell$. The terminal-speed limit, frustratingly slow, is generic to secular (i.e., dissipative) gravitational instability (GI), and renders collapsing clumps vulnerable to disruption by turbulence (Youdin 2011, see his Figure 4). For self-gravity to defeat rotational shear, $v_{\rm term} \gtrsim \Omega\ell$ or $\rho \gtrsim \rho_{\rm Roche}/\tau_{\rm stop}$: a truly high bar to clear for small $\tau_{\rm stop}$. Recent studies of secular GI punt on the obstacle of rotational shear by studying only axisymmetric disturbances (Youdin 2011; Michikoshi *et al.* 2012). The good news is that axisymmetric modes grow by secular GI no matter what the gas pressure and for any ρ, even $\rho \ll \rho_{\rm Roche}$, but the bad news is that growth plods along at terminal speeds.

5. Future Directions

(*a*) All the problems of planetesimal formation arise because of gas. Perhaps planetesimals form only where gas is severely depleted. The idea seems heretical because gas is required to form giant planets like Jupiter. But maybe the acquisition of gas envelopes occurs late in disk evolution, by rocky cores intercepting the last dregs of disk gas transported by viscous accretion across large distances.

(*b*) The new collisional kernels adopted by Garaud *et al.* (2013) are promising, but need to be incorporated into a global model that includes radial drift of solids.

(*c*) How dead are MRI-dead zones, and to what extent can particles settle in such environments (Bai & Stone 2013; Simon *et al.,* this volume)?

(*d*) The threat posed by the KHI to Goldreich-Ward can be circumvented for metal-enriched disks, but what about the danger presented by the SI? For $\tau_{\rm stop} \sim 10^{-3}$ and supersolar bulk metallicities, does the SI prevent vertical settling to $\mu \gtrsim 30$?

(*e*) How does secular GI proceed non-axisymmetrically?

We thank P. Hopkins, C. Ormel, R. Rafikov, and A. Youdin for discussions.

References

Bai, X.-N. & Stone, J. M. 2010, *ApJ*, 722, 1437
Bai, X.-N. & Stone, J. M. 2013, *ApJ*, 769, 76
Campo Bagatin, A. & Benavidez, P. G. 2012, *MNRAS*, 423, 1254
Chiang, E. 2008, *ApJ*, 675, 1549
Chiang, E. & Youdin, A. 2010, *AREPS*, 38, 493
Cuzzi, J. N., Hogan, R. C., & Shariff, K. 2008, *ApJ*, 687, 1432
Garaud, P., Meru, F., Galvagni, M., & Olczak, C. 2013, *ApJ*, 764, 146
Goldreich, P. & Ward, W. 1973, *ApJ*, 183, 1051
Jacquet, E., Balbus, S., & Latter, H. 2011, *MNRAS*, 415, 3591
Johansen, A., Youdin, A. N., & Lithwick, Y. 2012, *A&A*, 537, A125
Lee, A. T., Chiang, E. I., Asay-Davis, X., & Barranco, J. 2010a, *ApJ*, 718, 1367
Lee, A. T., Chiang, E. I., Asay-Davis, X., & Barranco, J. 2010b, *ApJ*, 725, 1938
Michikoshi, S., Kokubo, E., & Inutsuka, S. 2012, *ApJ*, 746, 35
Shi, J.-M. & Chiang, E. 2013, *ApJ*, 764, 20
Weidenschilling, S. J. 1980, *Icarus*, 44, 172
Weidenschilling, S. J. 2006, *Icarus*, 181, 572
Youdin, A. 2011, *ApJ*, 731, 99
Youdin, A. N. & Goodman, J. 2005, *ApJ*, 620, 459
Zsom, A., Ormel, C. W., Dullemond, C. P., & Henning, T. 2011, *A&A*, 534, A73

Discussion

NORDLUND: If the mean magnetic field is strong enough, it can stabilize the flow against the Kelvin-Helmholtz instability.

CHIANG: Yes, magnetic tension can stabilize the flow against the KHI. But midplanes are poorly ionized and expected to be magnetically inactive.

FRASER: As clumps collapse by self-gravity, how much do particles grow by sticking, and can such growth protect the clump against turbulent disruption?

CHIANG: This needs to be further investigated. Relative velocities may be slower inside the clump, enabling growth beyond the usual limits.

GAIDOS: Some planets are known to be formed around metal-*poor* stars.

CHIANG: Yes, but stellar metallicity and disk metallicity are distinct. What is relevant is the disk metallicity $Z_{\rm disk}$ local to a given radius. Because solids drift radially (Andrews *et al.,* this volume), $Z_{\rm disk}$ can differ substantially from the host star metallicity.

Exploring the Formation and Evolution of Planetary Systems
Proceedings IAU Symposium No. 299, 2013
M. Booth, B. C. Matthews & J. R. Graham, eds.

doi:10.1017/S1743921313008120

A Herschel View of Dust Evolution in Protoplanetary Disks

Catherine Espaillat

Boston University
Department of Astronomy
725 Commonwealth Avenue, Boston, MA 02215, USA
email: cce@bu.edu

Abstract. The details of how protoplanetary disks evolve from initially well-mixed distributions of gas and dust to systems composed mostly of rocky planets and gas giants like our own solar system is a fundamental question in astronomy. It is widely accepted that the first step in planet formation is dust grain growth and settling to the disk midplane. This dust evolution in disks can be studied in greater detail with far-infrared and submillimeter wavelength observations, which offer us unique access to the outer disk's deeper layers. Here we present *Herschel* far-infrared and submillimeter spectra of GM Aur taken with PACS and SPIRE. GM Aur is a transitional disk, whose inner disk hole is proposed to have been cleared by yet unseen planets. By utilizing *Herschel* data, we can potentially link the properties of dust evolution in the outer disk to dust clearing in the inner disk. In particular, preliminary SED modeling presented here suggests that GM Aur may have a lower gas-to-dust mass ratio than typically assumed for disks, which may be linked to disk clearing by planets. With further study, such *Herschel* data may provide insight for theoretical modeling of dust evolution and planet formation.

Keywords. (stars:) planetary systems: protoplanetary disks, stars: formation, accretion, accretion disks

1. Introduction

Many questions exist concerning how the gas and dust in a disk around a typical T Tauri star (TTS) evolves into a planetary system and observations of TTS may provide clues. Some objects in particular that have gained increasing attention in this regard are those dubbed "transitional disks" (Strom *et al.* 1989). Their spectral energy distributions (SEDs) lack excess emission above the stellar photosphere at infrared (IR) wavelengths and they were first identified using near-infrared (NIR) ground-based photometry and IRAS mid-IR (MIR) photometry Strom *et al.* 1989; Skrutskie *et al.* 1990). These "dips" were thought to indicate a hole in the dust distribution of the disk.

IR dips in the SEDs of several transitional disks were later confirmed with the *Spitzer Space Telescope* (Werner *et al.* 2004). The Infrared Spectrograph (IRS; Houck *et al.* 2004) provided unprecedented, simultaneous wavelength coverage between ~5 μm and 38 μm, uncovering new details regarding these IR dips (Furlan *et al.* 2006). Many transitional disk SEDs had nearly photospheric NIR (1–5 μm) and MIR (5–20 μm) emission, coupled with substantial emission at wavelengths beyond ~20 μm.

Detailed modeling of many transitional disk SEDs incorporating *Spitzer* data has been performed. SEDs of transitional disks with little or no NIR and MIR emission have been fit with models of inwardly truncated optically thick disks (Calvet *et al.* 2002; Calvet *et al.* 2005; Espaillat *et al.* 2007; Espaillat *et al.* 2008). The inner edge or "wall" of the outer disk is frontally illuminated by the star, dominating most of the emission seen in the IRS spectrum (e.g., GM Aur; Figure 1, left). Some of the holes in transitional disks

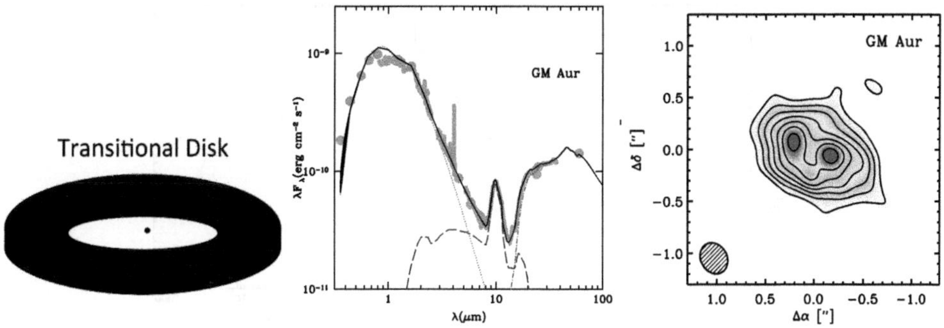

Figure 1. ***Left:*** Schematic of transitional disk structure. Progressing outward from the star (black) is the disk hole (white), outer wall (light gray), and outer disk (dark gray). ***Center:*** Observed (symbols) and simulated (black line) SED of the transitional disk of GM Aur taken from Espaillat *et al.* (2011). The transitional disk model consists of the stellar photosphere (dotted line) and an outer wall and outer disk (dot-short-dash line). Within the cleared dust region there is $\sim 10^{-11}$ $M_\odot$ of submicron-sized optically thin dust (long-dash line) which creates the strong 10 μm silicate feature. ***Right:*** 345 GHz (850 μm) submillimeter continuum image of GM Aur from Andrews *et al.* (2011), confirming the presence of a clearing with a ~20 AU radius.

are relatively dust-free (e.g., DM Tau; Calvet *et al.* 2005; Espaillat *et al.* 2010) while SED model fitting indicates that others with strong 10 μm silicate emission have a small amount of optically thin dust in their disk holes to explain this feature (e.g., GM Aur; Figure 1, center; Calvet *et al.* 2005; Espaillat *et al.* 2010). For many transitional disks, the truncation of the outer disk has been confirmed with submillimeter and millimeter interferometric imaging (e.g., GM Aur; Figure 1, right; Hughes *et al.* 2007; Hughes *et al.* 2009; Andrews *et al.* 2009; Andrews *et al.* 2010; Andrews *et al.* 2011).

The distinct SEDs of transitional disks most likely signify that these objects are in an important phase of disk evolution. Many researchers have proposed that these disks are forming planets on the basis that cleared regions in disks are predicted by planet formation simulations (e.g., Paardekooper & Mellema 2004; Zhu *et al.* 2012). To explore the potential link between transitional disks and planet formation, we present *Herschel* far-infrared and submillimeter spectra of GM Aur. Data at these wavelengths trace regions deeper in the disk, where dust evolution is expected to occur. By utilizing *Herschel* data of transitional disks, we can look for new evidence supporting the presence of planets in disks. In turn, the results can be used to inform models of dust evolution and planet formation.

2. Overview

The pre-*Herschel* SED of GM Aur (Figure 2, left) lacks substantial coverage in the far-infrared (FIR) and submillimeter wavelengths, even though this object is one of the best-studied in the literature from the U-band to millimeter wavelengths. Adding in *Herschel* PACS and SPIRE spectra significantly expands the wavelength coverage in the SED (Figure 2, right). Preliminary modeling analysis of the updated SED of GM Aur has yielded an interesting result. We find that the best-fitting model to the pre-*Herschel* SED of GM Aur, which was presented by Espaillat *et al.* (2011), does not fit the observed emission at ~100–500 μm (Figure 3, left). Extra emission at these wavelengths is seen from the disk that the previous model does not predict.

In our new preliminary fit to the updated SED of GM Aur, we can account for the extra emission in the FIR and submillimeter with a disk that is more massive (Figure 3,

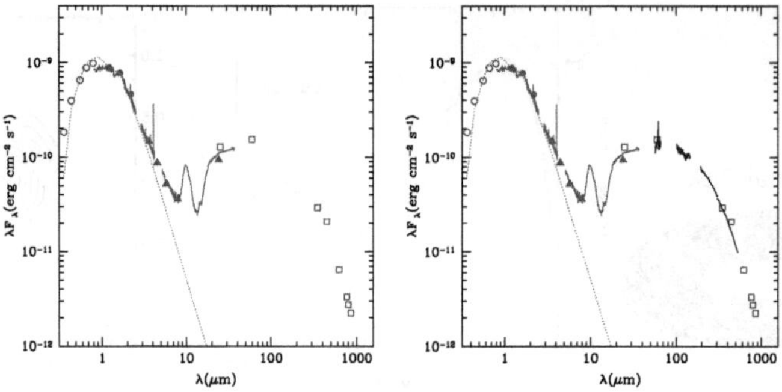

Figure 2. ***Left:*** Observed SED of GM Aur taken from Espaillat *et al.* (2011). We show SpeX and IRS spectra from Espaillat *et al.* (2010, 2011) as well as the stellar photosphere (dotted line). Open circles correspond to ground-based (UBVRIL) photometry taken from Kenyon & Hartmann (1995). Closed circles are 2MASS photometry from Skrutskie *et al.* (2006). Triangles are *Spitzer* photometry from Luhman *et al.* (2010). Squares correspond to IRAS and millimeter fluxes. IRAS data are taken from Weaver *et al.* (1992). Millimeter data were obtained from Andrews & Williams (2005), Weintraub *et al.* (1989), Beckwith & Sargent (1991), and Hughes *et al.* (2009). ***Right:*** Observed SED of GM Aur including preliminary *Herschel* PACS and SPIRE spectra. *Herschel* significantly expands our wavelength coverage at FIR and submillimeter wavelengths. Note that the processing of the *Herschel* data may be improved with future pipeline releases and we refer the reader to Espaillat *et al.* (in preparation) for a more recent data reduction and relevant details.

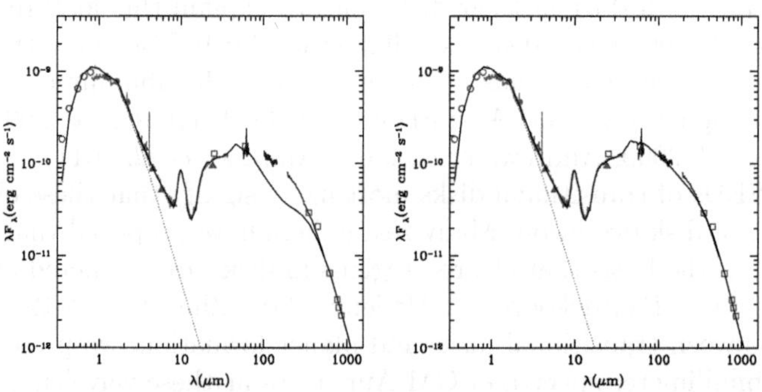

Figure 3. ***Left:*** Model SED (black line) presented in Figure 1 from Espaillat *et al.* (2011) overlaid on the updated SED of GM Aur from Figure 2 (right). The model is a good fit to the pre-*Herschel* SED presented in the left panel of Figure 2. It does not fit the updated SED at 100–500 μm, however, highlighting the importance of including *Herschel* data in SED modeling. ***Right:*** Preliminary model fit (black line) to the updated SED of GM Aur. This is a better fit to the *Herschel* data at 100–500 μm; more details can be found in Section 2. We stress that this is a preliminary fit and that analysis is still underway. We refer the reader to Espaillat *et al.* (in preparation) for the most recent SED modeling analysis results.

right). However, this disk is Toomre unstable. To reconcile this dilemma, GM Aur may have a lower gas-to-dust mass ratio in its disk than what is typically assumed (i.e., 100). A disk model with a lower gas-to-dust mass ratio can produce similar emission levels but with a less massive disk since $F_{mm} \propto M_{dust} \propto$(dust-to-gas ratio)$\times M_{gas}$. Therefore, with a lower gas-to-dust ratio (i.e., a higher dust-to-gas ratio), a similar millimeter flux could be obtained with less gas (which is the bulk of the disk mass). We stress that this result is preliminary and other possibilities that may be more important need to be explored before this initial result is confirmed. We refer the reader to Espaillat *et al.* (in preparation) for more recent analysis results.

3. Implications

Lower gas-to-dust mass ratios in disks have been proposed previously. One example is the transitional disk of TW Hya where Thi *et al.* (2010) inferred a gas-to-dust ratio of ∼30 using PACS observations of the [OI] line. This result, however, has been recently disputed by Bergin *et al.* (2013) who find the gas-to-dust ratio of TW Hya is ∼100 using the HD line. More work needs to be done to decipher disk gas-to-dust mass ratios.

In the meantime, we can speculate that if indeed the gas-to-dust mass ratios of transitional disks is lower, this difference may be related to the mechanism that is potentially causing the holes in these disks, namely planets. Recently, there have been steps to more closely tie theory to observations by simulating disk clearing due to dust filtration. The dust filtration model predicts substantial holes in the large dust grain distribution of the disk, while the gas and small dust grain distribution have undetectable gaps (Zhu *et al.* 2012). This mechanism could potentially lead to lower gas-to-dust ratios in disks if indeed gas is being depleted in the outer disk as it filters through into the inner disk and is accreted onto the star while large dust grains are being held back in the outer disk. Still, dust filtration does not explain all aspects of transitional disk observations (e.g., MIR deficits), indicating that we are still missing important physics in these objects. More theoretical and observational work needs to be done and the particular case of GM Aur will be explored in more detail by Espaillat *et al.* (in preparation).

References

Andrews, S. M. & Williams, J. P. 2005, *ApJ*, 631, 1134
Andrews, S. M., Wilner, D. J., Hughes, A. M., Qi, C., & Dullemond, C. P. 2009, *ApJ*, 700, 1502
Andrews, Wilner, Hughes, Qi, & Dullemond, 2010, *ApJ*, 723, 1241
Andrews, S. M., *et al.* 2011, *ApJ*, 732, 42
Beckwith, S. V. W. & Sargent, A. I. 1991, *ApJ*, 381, 250
Bergin, E. A., Cleeves, L. I., Gorti, U., *et al.* 2013, *Nature*, 493, 644
Calvet, N., *et al.* 2002, *ApJ*, 568, 1008
Calvet, N., *et al.* 2005, *ApJL*, 630, L185
Espaillat, C., *et al.* 2007a, *ApJL*, 664, L111
Espaillat, C., *et al.* 2008, *ApJL*, 689, L145
Espaillat, C., *et al.* 2010, *ApJ*, 717, 441
Espaillat, C., Furlan, E., D'Alessio, P., *et al.* 2011, *ApJ*, 728, 49
Furlan, E., *et al.* 2006, *ApJS*, 165, 568
Houck, J. R., *et al.* 2004, *ApJS*, 154, 18
Hughes, A. M., *et al.* 2007, *ApJ*, 664, 536
Hughes, A. M., *et al.* 2009, *ApJ*, 698, 131
Kenyon, S. J. & Hartmann, L. 1995, *ApJS*, 101, 117
Luhman, K. L., Allen, P. R., Espaillat, C., Hartmann, L., & Calvet, N. 2010, *ApJS*, 186, 111
Paardekooper, S.-J. & Mellema, G. 2004, *A&A*, 425, L9
Skrutskie, M. F., *et al.* 1990, *AJ*, 99, 1187
Skrutskie, M. F., *et al.* 2006, *AJ*, 131, 1163
Strom, K. M., Strom, S. E., Edwards, S., Cabrit, S., & Skrutskie, M. F. 1989, *AJ*, 97, 1451
Thi, W.-F., Mathews, G., Ménard, F., *et al.* 2010, *A&A*, 518, L125
Weaver, W. B. & Jones, G. 1992, *ApJS*, 78, 239
Weintraub, D. A., Sandell, G., & Duncan, W. D. 1989, *ApJL*, 340, L69
Werner, M. W., *et al.* 2004, *ApJS*, 154, 1
Zhu, Z., Nelson, R. P., Dong, R., Espaillat, C., and Hartmann, L. 2012, *ApJ*, 755, 6

Discussion

RAFIKOV: Your argument in favour of high dust/gas was that otherwise the gas disk mass is $\approx 1 M_*$. But that means that initially, in the non-transitional phase, the gas disk still had to be very massive, $\approx 1 M_*$. Is this a problem?

ESPAILLAT: If the dust is not accreted or photoevaporated at the same rate as the gas, then the dust per unit mass can increase with time. In this case, the initial gas disk did not need to be much more massive.

LEINHARDT: What is the radial dependence of the gas to dust ratio in transitional disks? Do you expect a significant radial dependence of gas to dust ratio in the outer disk due to dust filtration at the wall?

ESPAILLAT: We currently do not have a good observational constraint on the radial dependence of the gas to dust ratio in any disk. For the transitional disks, if dust filtration is occurring, it is expected that there is an enhancement of dust near the wall and ALMA will be able to test this.

Exploring the Formation and Evolution of Planetary Systems
Proceedings IAU Symposium No. 299, 2013
M. Booth, B. C. Matthews & J. R. Graham, eds.

doi:10.1017/S1743921313008132

Substructure and Signs of Planet Formation in the Disk of HD 169142

M. Osorio[1], G. Anglada[1], C. Carrasco-González[2,3], J. M. Torrelles[4], P. D'Alessio[3], L. F. Rodríguez[3], N. Calvet[5], J. F. Gómez[1], J. M. Mayen-Gijon[1] and W. R. F. Dent[6]

[1]Instituto de Astrofísica de Andalucía (CSIC), Camino Bajo de Huétor 50, E-18008 Granada, Spain email: osorio@iaa.es

[2]Max-Planck-Institut für Radioastronomie, Auf dem Hügel 69, 53121 Bonn, Germany

[3]Centro de Radioastronomía y Astrofísica, UNAM, 58089 Morelia, Michoacán, Mexico

[4]Instituto de Ciencias del Espacio (CSIC) - UB/IEEC, Martí i Franquès 1, E-08028 Barcelona, Spain

[5]Department of Astronomy, University of Michigan, 825 Dennison Building, 500 Church St, Ann Arbor, MI 48109, USA

[6]ALMA SCO, Alonso de Córdova 3107, Vitacura, Santiago, Chile

Abstract. We carried out 7 mm VLA observations at very high angular resolution that reveal substructure and evidence of planet formation in the disk of HD 169142. Our observations, along with near-infrared polarimetric imaging, show that this disk has a ring of enhanced, asymmetric emission at a radius of ~25 AU from the central star. This ring, whose inner region appears devoid of emission, is surrounded by an annular gap in surface density in the ~30-70 AU range of radii. Several mechanisms have been invoked in the literature to explain this kind of gaps and cavities. Among them, one of the most interesting is the possibility that one or more planets in formation are creating these cavities. Since our 7 mm observations show a compact source lying in the 30-70 AU gap, we speculate that this compact source could be tracing dust emission associated with a possible protoplanet. We model the broad-band spectral energy distribution of the disk and we infer its physical structure. From this modeling we infer the presence of a small ($r \sim 0.7$ AU) disk inside the central cavity, suggesting that the HD 169142 disk is in the pre-transitional disk phase.

Keywords. stars: formation, planetary systems: formation, protoplanetary disks

1. Introduction

Transitional disks are protoplanetary disks that have developed central cavities of several tens AU in size almost devoid of dust but full of gas since the central star is still accreting (Calvet *et al.* 2005). A subclass of these disks, the so-called pre-transitional disks, with a remaining inner disk inside the central cavity, was introduced to account for significant near-infrared excesses observed in a number of objects (Espaillat *et al.* 2008). Several mechanisms have been invoked to explain these central cavities such as photoevaporation (Alexander *et al.* 2006), grain-growth (Dullemond & Dominik 2004), or planetary formation (Augereau & Papaloizou 2004). If planets were actually opening these cavities, then transitional and pre-transitional disks would be the best sites to search for planets in the verge of their formation.

The presence of most of these cavities has been indirectly inferred through the modeling of the Spitzer spectra. But in the last years an increasing number have been angularly resolved by interferometers like the Submillimeter Array (SMA; see Andrews *et al.* 2012)

and the Atacama Large Millimeter/submillimeter Array (ALMA; see Casassus *et al.* 2013).

An interesting case of study is the protoplanetary disk around the Herbig Ae/Be star HD 169142. HD 169142 is an intermediate-mass star ($\sim 2\ M_{\odot}$) located at a distance of 145 pc. Scattered light images as well as molecular observations reveal a face-on disk extending up to 250 AU. Dent *et al.* (2006) detected relatively strong, angularly unresolved 7 mm emission through Very Large Array (VLA) observations. Since centimeter emission (Dent *et al.* 2006; Osorio *et al.* 2013) is not detected towards this object, it is expected that the observed 7 mm emission traces the dust emission from the disk, without free-free emission contamination from a possible radio jet. The lack of detectable free-free emission also suggests that this object is in a relatively advanced evolutionary stage where the accretion and outflow processes are almost halted. On the other hand, the observed SED does not exhibit a silicate feature at 10 microns, a result that is suggestive of significant grain growth, a process required for planetary formation.

The presence of an inner cavity in the HD 169142 disk was inferred from the modeling of the observed spectral energy distribution (SED). From these models, estimates of ~40 AU (Grady *et al.* 2007) and ~20 AU (Meeus *et al.* 2010; Honda *et al.* 2012) were obtained for the radius of this central cavity. Confirmation of a central cavity has been obtained recently through NACO/VLT polarized light images by Quanz *et al.* (2013) (gray-scale image in Fig. 1). In addittion to a central cavity with a radius of ~20 AU, these observations reveal a bright ring located at a radius of ~ 25 AU and an annular gap extending from ~30 to 70 AU.

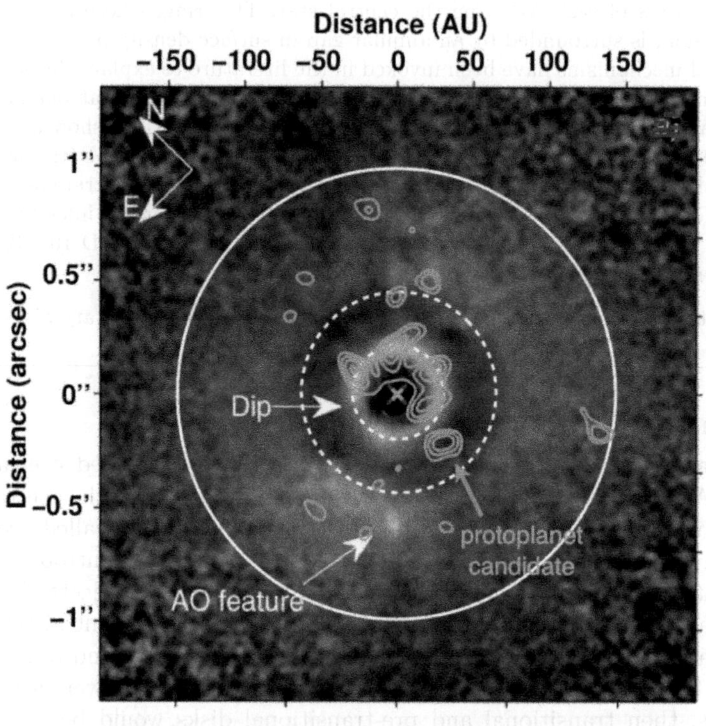

Figure 1. VLA map at 7 mm (contours; Osorio *et al.* 2013) overlaid on the VLT/NACO polarized light image in the H band (gray-scale; from Quanz *et al.* 2013). The size of the VLA beam is $\sim 0.2''$. Saturated pixels in the central region of the H-band image have been masked out. The cross marks the 2MASS position of the HD 169142 star, after correction for proper motions to the epoch of the observations (2012).

2. VLA Observational Results

VLA observations at 7 mm and 6 cm were carried out in the CnB and A configurations along 2012 (Osorio *et al.* 2013). No emission was detected at 6 cm, indicating that there is no significant free-free emission associated with this object. In Figure 1 (contours) we show a preliminary map of the combined CnB+A data. The 7 mm map, with an angular resolution of $\sim 0.2''$, shows a (partial) ring-like structure that coincides pretty well with the ring observed in the polarized infrared image obtained by Quanz *et al.* (2013). The ring is asymmetric, with the weakest part coinciding with a dip observed in the polarized light image. Outside the ring, at a radius of 43 AU, there is an unresolved 7 mm source (0.15 mJy) that falls within the annular gap observed by Quanz *et al.* (2013). We speculate that this 7 mm source could be tracing circumplanetary dust associated with a young protoplanet. Assuming a temperature of 100-300 K, a 7 mm opacity of 0.9-5×10^{-3} cm^{-2} g^{-1}, and a gas-to-dust mass ratio of 100, a mass of 0.2-2 M_J is obtained. If the 7 mm source traces the protoplanet that has created the 30-70 AU annular gap, its orbital period would be 186 yr, and we expect significant proper motions in a few years.

3. Modeling of the Disk

We modeled the Spectral Energy Distribution (SED) of the circumstellar disk of HD 169142 using the accretion disk models developed by D'Alessio *et al.* (1999, 2001, 2006). Following the observations, we have truncated the disk at an inner radius of 20 AU, and we considered that the disk is extending out to 250 AU. In this model the annular gap located between 30 and 70 AU is not considered because it is expected to have a small effect on the SED, as it appears to be partially filled (see Quanz *et al.* 2013) and the temperatures expected at these distances are relatively low. We included the emission of an inner wall at a radius of 20 AU with a temperature of 200 K, calculated from the equilibrium equation, assuming a stellar luminosity of 17 $L_\odot$. However, this model does not reproduce the near-infrared range of the observed SED (dotted line in Fig. 2). To fit this wavelength range, we included a small central disk of 0.7 AU radius, with a wall heated at a temperature of 1200 K (the dust sublimation temperature) located at 0.4 AU, its inner radius. By adding these components to the previous model we are able to explain the broad-band SED and the 7 mm image. This model (thick solid line in Fig. 2) is in agreement with the constrains set by the VLA observations. It predicts weak 7 mm emission (undetectable with the sensitivity achieved with the VLA) inside a radius of 20 AU, because the small central disk and its wall radiate most of the emission in the 2-10 μm wavelength range (see Fig. 2). Note that the central region was masked out in the H-band observations of Quanz *et al.* (2013); so, these infrared observations cannot be used to constrain the size of the small inner disk.

4. Summary and Proposed Scenario for HD 169142

We obtained a high angular resolution ($0.2''$) VLA image at 7 mm of the inner structure of the HD 169142 protoplanetary disk. The morphology of the 7 mm emission is well correlated with the VLT infrared scattered light image obtained by Quanz *et al.* (2013), showing a ring of radius $\sim$ 25 AU surrounded by an annular gap from $\sim$ 30 to 70 AU.

The SED fitting and the VLA/VLT data indicate that the HD 169142 disk is a pre-transitional disk, with a small inner disk, from 0.4 to 0.7 AU, and an outer disk from 20 to 250 AU. There is, therefore, a gap from 0.7-20 AU, and a second, probably less developed, gap extending from 30-70 AU.

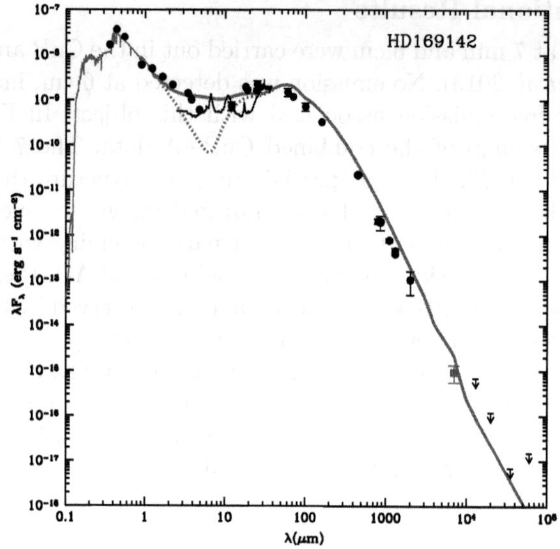

Figure 2. Observed (points and thin line) and model (dotted and thick solid lines) SEDs of the HD 169142 protoplanetary disk. The thick solid line corresponds to a model including the small ($\sim$ 0.7 AU) central disk and its wall, while the dotted line corresponds to a model without this central disk (see text).

Inside the 30-70 AU annular gap we detect a 7 mm unresolved source, that we tentatively interpret as tracing the dust associated with a protoplanet candidate. We estimate a total mass of 0.2-2 M_J for this proposed circumplanetary structure.

We interpret these results as evidence of planetary formation in the disk of HD 169142. This hypothesis can be further tested with new VLA, VLT, and ALMA observations.

Acknowledgements

GA, CC-G, JFG, JMM-G, MO, and JMT acknowledge support from MINECO (Spain) AYA2008-06189-C03 and AYA2011-30228-C03 grants, co-funded with FEDER funds.

References

Alexander, R. D., Clarke, C. J., & Pringle, J. E. 2006, *MNRAS*, 369, 216
Andrews, S. M., Rosenfeld, K. A., Wilner, D. J., & Bremer, M. 2011, *ApJ*, 742, L5
Augereau, J. C. & Papaloizou, J. C. B. 2004, *A&A*, 414, 1153
Calvet, N., D'Alessio, P., Watson, D. M., *et al.* 2005, *ApJ* (Letters), 630, L185
Casassus, S., van der Plas, G., M, S. P., *et al.* 2013, *Nature*, 423, 191
D'Alessio, P., Calvet, N., Hartmann, L., Lizano, S., & Cantó, J. 1999, *ApJ*, 527, 893
D'Alessio, P., Calvet, N., & Hartmann, L. 2001, *ApJ*, 553, 321
D'Alessio, P., Calvet, N., Hartmann, L., Franco-Hernández, R., & Servín, H. 2006, *ApJ*, 638, 314
Dullemond, C. P. & Dominik, C. 2004, *A&A*, 421, 1075
Espaillat, C., Calvet, N., Luhman, K. L., Muzerolle, J., & D'Alessio, P. 2008, *ApJ* (Letters), 682, L125
Grady, C. A., Schneider, G., Hamaguchi, K., *et al.* 2007, *ApJ*, 665, 1391
Honda, M., Maaskant, K., Okamoto, Y. K., *et al.* 2012, *ApJ*, 752, 143
Meeus, G., Pinte, C., Woitke, P., *et al.* 2010, *A&A*, 518, L124
Osorio, M., *et al.* 2013, in preparation
Quanz, S. P., Avenhaus, H., Buenzli, E., *et al.* 2013, *ApJ* (Letters), 766, L2

Exploring the Formation and Evolution of Planetary Systems
Proceedings IAU Symposium No. 299, 2013
M. Booth, B. C. Matthews & J. R. Graham, eds.

doi:10.1017/S1743921313008144

Simultaneous Visible and Near-Infrared Variability of Classical T Tauri Stars

Yukako Aimi[1], Misato Fukagawa[1], Tomonori Yasuda[2], Takuya Yamashita[3], Kouji Kawabata[2], Makoto Uemura[2], Akira Arai[4], Mahito Sasada[5], Takashi Ohsugi[2], Michitoshi Yoshida[2] and Hiroshi Shibai[1]

[1]Dept. of Earth and Space Science, Graduate School of Science, Osaka University, 1-1, Machikaneyama, Toyonaka, Osaka 560-0043, Japan
email: `aimi@iral.ess.sci.osaka-u.ac.jp`

[2]Hiroshima Astrophysical Science Center, Hiroshima University, 1-3-1 Kagamiyama, Higashi-Hiroshima, Hiroshima 739-8526, Japan

[3]National Astronomical Observatory of Japan 2-21-1, Osawa, Mitaka, Tokyo 181-8588, Japan

[4]Nishi-Harima Astronomical Observatory, Center for Astronomy, University of Hyogo, 407-2, Nishigaichi, Sayo-cho, Sayo, Hyogo 679-5313, Japn

[5]Dept. of Astronomy, Kyoto University, Kitashirakawa-Oiwake-cho, Sakyo-ku, Kyoto 606-8502, Japan

Abstract. Temporal structural changes of protoplanetary disks surrounding T Tauri stars (TTSs) can cause magnitude variations of TTSs. On the other hand, variability is also expected due to cool spots and/or hot spots on the surface of the star, thus it is important to distinguish the causes of the observed variability. Our sample consists of 23 TTSs (22 classical T Tauri stars, 1 weak-lined T Tauri star) and 4 Herbig Ae/Be stars. The observations were performed over a period of about 3 months in the V, J, and K_S band, simultaneously. We detected variability for all stars in the three bands (>0.05 mag in V, >0.09 mag in J, >0.09 mag in K_S). Color-magnitude relations obtained between V, J, and K_S bands suggest that stellar spots are not the only cause of variability for most of our targets. In addition, the data implies that six stellar systems contain larger grains than in the interstellar medium if the variability is only caused by extinction due to circumstellar matter.

Keywords. stars: pre-main-sequence, (stars:) planetary systems: protoplanetary disks, stars: variables: other.

1. Sample, Observations, and Data Reduction

Observations were carried out using TRISPEC, a simultaneous optical and near-infrared imager mounted on the *Kanata* telescope of Hiroshima University. We obtained photometric data in the V, J, and K_S band for 22 CTTSs, 1 WTTS, and 4 Herbig Ae/Be stars between November 2007 and February 2008, with a coverage of at least one observation every five days.

Magnitudes were measured by aperture photometry relative to the comparison stars located in the same field of view ($7'.0 \times 7'.0$). The photometric accuracy spans from 0.03 to 0.05 mag in the three bands.

2. Results

We detected significant variability in all three bands. The amplitude of the variability ranged 0.05–3.14 mag in V, 0.09–1.58 mag in J, and 0.09–0.75 mag in K_S.

Periodicity analysis. Fourteen out of 27 stars were analyzed for periodicity in the V band. Six stars showed quasi-periodic variability with periods between 3.6 and 10 days.

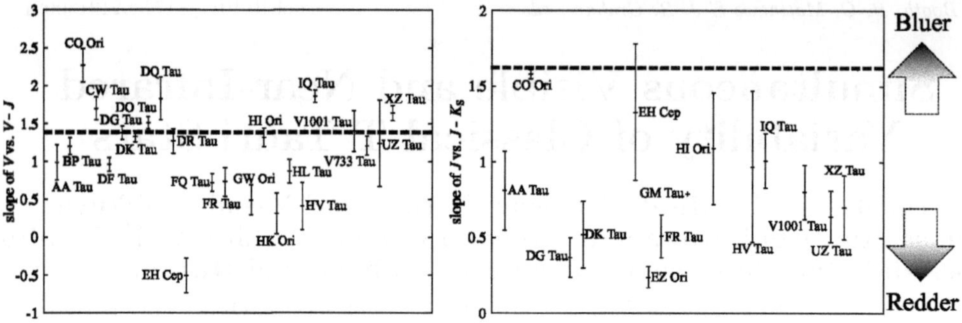

Figure 1. Slopes of variation in the color-magnitude diagrams. Left: V vs. $V - J$, right: J vs. $J - K_S$. The dashed lines represent extinction by dust grains of $R_V = 3.1$. (Some stars are not included in this plot because they did not show the relation expressed by a linear function in the color-magnitude diagrams.)

Color-magnitude diagram. Targets appear to be redder when they are dimmer in V (J). Excluding data with errors larger than 0.2 mag, we performed least-squares fitting for the color-magnitude variations (Fig. 1). First, we calculated the slope of variation on J vs. $J - K_S$ based on a scenario that the cause of the variability is a cool spot (T_{spot} = 1000–3500 K, $Area_{spot}/Area_{star}$ = 0.05–0.5) or a hot spot (T_{spot} = 4500–10000 K, $Area_{spot}/Area_{star}$ = 0.01–0.05), and found that it is larger than 2.0. This is inconsistent with our observations (Fig. 1 right), suggesting that stellar spots are not the only cause of variability for most of our targets. Another possible cause is variable extinction by circumstellar matter. Slopes of V vs. $V - J$ for CO Ori, CW Tau, DO Tau, DQ Tau, IQ Tau, and XZ Tau are larger than 1.39 ($R_V = 3.1$, Mathis, 1990), implying grain growth occurred in these systems. However the slope of J vs. $J - K_S$ is lower than in the case of pure extinction. While extinction plays a role in determining the K_S flux, the dominant factor in this wavelength range is the near-infrared excess from the disk. Taking this into account it is possible to explain the observations by assuming that the near-infrared excess itself is variant over time.

Disk Modeling. Assuming a protoplanetary disk with a warp, we used the Monte Carlo radiative transfer code of Whitney *et al.* (2003a, 2003b) for investigating the relation between extinction and the thermal radiation from the disk. The magnitude varies due to disk rotation. Stellar mass, radius (= R_{star}), and temperature, as well as the mass, inner and outer radius of the disk were taken from Bouvier *et al.* (1999). The scale height of the warp (= w) was varied between 0.5 and 2.7 R_{star}. We found that a larger w yields a larger slope of V vs. $V - J$ for a disk with an inclination of 73 degrees. A higher warp leads to a shift toward bluer colors which can be explained by a shadow cast onto the region where J flux is dominant. On the other hand, the slope of J vs. $J - K_S$ becomes smaller, because infrared excess from the warp increased. However, slopes predicted from any of the warp models were not consistent with our observations which showed smaller slopes. We therefore conclude that additional near-infrared emission sources are needed.

References

Mathis, J. S. 1995, *ARAA*, 38, 37

Whitney, Barbara A., Wood, Kenneth, Bjorkman, J. E., Wolff, Michael J. 2003a, *ApJ*, 591, 1049

Whitney, Barbara A., Wood, Kenneth, Bjorkman, J. E., Cohen, Martin 2003b, *ApJ*, 598, 1079

Bouvier, J., *et al.* 1999, *A&A*, 349, 619

Exploring the Formation and Evolution of Planetary Systems
Proceedings IAU Symposium No. 299, 2013
M. Booth, B. C. Matthews & J. R. Graham, eds.

doi:10.1017/S1743921313008156

A First Look at the Disk Population in the Auriga-California Molecular Cloud

Hannah Broekhoven-Fiene[1], Brenda C. Matthews[1,2], Paul M. Harvey[3] and members of the *Spitzer* Gould Belt Survey

[1]Department of Physics and Astronomy, University of Victoria,
PO Box 3055 STN CSC, Victoria, BC, V8W3P6, Canada
email: broekhov@uvic.ca
[2]National Research Council of Canada,
Victoria, BC, V9E 2E7, Canada
[3]Astronomy Department, University of Texas at Austin,
1 University Station C1400, Austin, TX 78712-0259, USA

Abstract. The Auriga-California Molecular Cloud (AMC) is one of two nearby (within 500 pc) giant molecular clouds, the other being the Orion A Molecular Cloud (OMC). We aim to study the properties of circumstellar disks in the AMC to compare the planet formation potential and processes within the AMC to those for other clouds. A first look with measurements from *Spitzer* observations suggests that AMC disk properties, such as the distribution of disk luminosities and the evolution of the mid-IR excesses, are not vastly different from those in other regions. Follow-up observations in the submm, mm and cm can be used to measure disk masses and the degree of grain growth from spectral slopes to more completely characterize the disk population.

Keywords. stars: formation, planetary systems: protoplanetary disks, ISM: clouds

1. Introduction

The AMC was only recently identified as being one contiguous cloud (Lada *et al.* 2009). Consequently, it is relatively unstudied compared to other star-forming regions in the Gould Belt. Although the AMC is similar in mass (10^5 $M_\odot$), size (80 pc) and distance (~450 pc) to the OMC, its star formation properties are quite different (Lada *et al.* 2009). The AMC is forming 20x fewer stars than the OMC as well as forming less massive stars. The OMC has 20 OB stars whereas the AMC has only one B star, LkHα 101. This raises the question of how the reservoir of material in the cloud relates to that used for the formation of star systems.

Spitzer observations of the AMC have been used to identify 166 young stellar objects (YSOs) in the cloud (Broekhoven-Fiene *et al.*, submitted). A substantial number of these (91/166) are Class II sources, i.e. likely disk hosts with no remnant circumstellar envelope. The Class II sources define our target sample of potential protoplanetary disk hosts in the AMC.

2. First Look at Disk Properties with *Spitzer*

Some disk properties can be measured from the *Spitzer* observations by modelling the spectral energy distributions of Class II and Class III YSOs. This is done by fitting a model stellar spectrum to the observed short wavelength photometry (optical and near-infrared). The excess emission that is observed in the infrared is attributed to dust emission from the disk. By determining the wavelength range of excess emission, the ratio of the disk and stellar luminosities is estimated (see Fig. 1). The distribution of disk

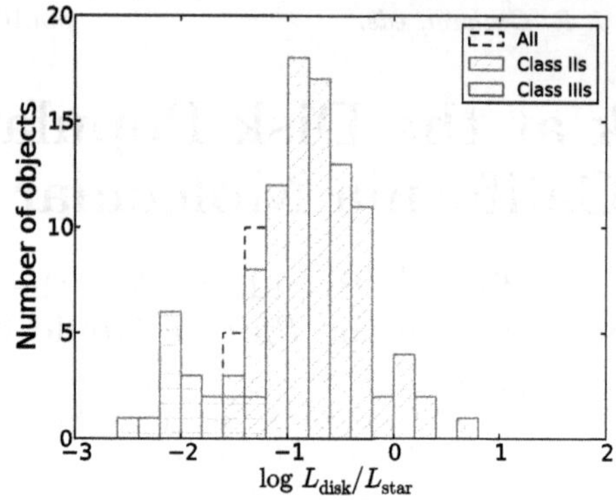

Figure 1. The ratio of the disk luminosity to the stellar luminosity for Class II and Class III sources in the AMC. (Figure adapted from Broekhoven-Fiene *et al.*, submitted.)

luminosities is similar for disk populations in other clouds observed by the *Spitzer* Gould Belt Survey suggesting, at first glance, the disk populations appear similar regardless of the cloud properties (Broekhoven-Fiene *et al.*, submitted).

3. Current questions/future work

We want to investigate how the disk population in the AMC compares to disk populations in other Gould Belt clouds and how the disks in the environment of LkHα 101 compare to those further away along the adjoining filament. Measuring the disk masses and the degree of grain growth by measuring the spectral slope is the best way to target the total amount of material in the disk available for planet formation. The disk mass distribution in the AMC can then be directly compared to other disk populations, such as Taurus (Andrews & Williams 2005), ρ Ophiucus (Andrews & Williams 2007) and Orion (Mann & Williams 2009).

The AMC was mapped by *Herschel* at 70, 160, 250, 350, and 500 μm (see Harvey *et al.* 2013) and part of the cloud is also being mapped with SCUBA-2 by the JCMT Gould Belt Legacy Survey (PI Jenny Hatchell) at 450 and 850 μm. Submillimeter fluxes from these observatories provide preliminary measurements of the disk masses for a handful of sources. Detections at the longer wavelengths by both *Herschel* and SCUBA-2 identify the targets that are likely the brightest sources for potential follow-up observations at longer wavelengths.

ALMA and the VLA are optimal observatories to use to both measure disk masses in the AMC and estimate the degree of grain growth, while constraining the contribution from free-free emission. Furthermore, the sensitivity of the these observatories is advantageous given the distance to the AMC ($\sim$ 450 pc). ALMA Cycle 0 observations of protoplanetary disks in the OMC (PI Rita Mann) demonstrate ALMA's ability to observe disks at the distance of the OMC and the AMC.

References

Andrews, S. M. & Williams, J. P. 2005, *ApJ*, 631, 1134
Andrews, S. M. & Williams, J. P. 2007, *ApJ*, 671, 1800
Harvey, P. M., Fallscheer, C., Ginsburg, A., *et al.* 2013, *ApJ*, 764, 133
Lada, C. J., Lombardi, M., & Alves, J. F. 2009, *ApJ*, 703, 52
Mann, R. K. & Williams, J. P. 2009, *ApJ*, 694, L36

Exploring the Formation and Evolution of Planetary Systems
Proceedings IAU Symposium No. 299, 2013
M. Booth, B. C. Matthews & J. R. Graham, eds.

doi:10.1017/S1743921313008168

Ice Lines in Circumbinary Protoplanetary Disks

Christian D. Clanton

Department of Astronomy, The Ohio State University,
140 W. 18th Ave., Columbus, OH 43210, USA
email: clanton@astronomy.ohio-state.edu

Abstract. I find the location of the ice line in circumbinary disks heated by steady mass accretion and stellar irradiation, comparing the position with the minimum stable semimajor axis, interior to which planetary orbits are unstable. I show that there is a critical binary separation for which binaries with separations larger than this critical value have ice lines that lie interior to the boundary of stability. The critical separation for an equal-mass binary of 1 $M_\odot$ stars is ≈ 1.04 AU, scaling weakly with mass accretion rate and Rosseland mean opacity of the disk. For a steady mass accretion rate of $\dot{M} \sim 10^{-8}\ M_\odot\ \mathrm{yr}^{-1}$ and a Rosseland mean opacity of $\kappa_\mathrm{R} \sim 1\ \mathrm{cm}^2\ \mathrm{g}^{-1}$, I show that $\gtrsim 80\%$ of all binary systems with component masses $M_\star \lesssim 2.0\ M_\odot$ have ice lines interior to the boundary of stability. This suggests that rocky planets should not be common in these systems. Searching for planets around binaries with separations larger than the critical separation with *Kepler* or microlensing will provide a test of this prediction.

Keywords. binaries: general, planets and satellites: formation, stars: pre–main-sequence, planetary systems: protoplanetary disks

---

I adopt a simple disk model with the flared geometry described in Chiang & Goldreich (1997), with a surface density distribution similar to the minimum mass Solar nebula (MMSN), $\Sigma = \Sigma_0\,(a/\mathrm{AU})^{-3/2}$, where $\Sigma_0 = 10^3\ \mathrm{g\ cm}^{-2}$ (Weidenschilling 1977). Although I consider heating from both stellar irradiation and mass accretion, it turns out that the temperature of the disk midplane in regions of the disk where the ice line is expected to lie is dominated by mass accretion for binaries with component masses $M_\star \lesssim 2.0\ M_\odot$. Fig. 1 plots the location of the ice line relative to the minimum stable semimajor axis as a function of binary separation for both equal-mass and unequal-mass binaries.

In the mass accretion-dominated regime, there is an analytic solution for the location of the ice line,

$$a_\mathrm{ice}^\mathrm{acc} \approx 2.13\ \mathrm{AU}\left(\frac{M_\mathrm{tot}}{M_\odot}\right)^{2/9}\left(\frac{\kappa_\mathrm{R}}{1\ \mathrm{cm}^2\ \mathrm{g}^{-1}}\right)^{2/9}\left(\frac{\dot{M}}{10^{-8}\ M_\odot\ \mathrm{yr}^{-1}}\right)^{2/9}\left(\frac{T_\mathrm{ice}}{160\ \mathrm{K}}\right)^{-8/9} \tag{0.1}$$

where M_tot is the total mass of the binary, κ_R is the Rosseland mean opacity of the disk, $\dot{M}$ is the mass accretion rate, and T_ice is the temperature of ice condensation.

Comparing this with the minimum stable semimajor axis found by Holman & Wiegert (1999), I find a critical binary separation, for which binaries with separations larger than this critical value have ice lines interior to the boundary of stability. For equal-mass binaries on circular orbits, this critical separation is $a_\mathrm{bin}^\mathrm{crit} \approx 0.89$ AU, with the same scalings as (Eq. 0.1).

I compute the binary periods corresponding to the critical separations for equal-mass binaries and integrate over the observed binary period distribution of Raghavan *et al.* (2010) to estimate the fraction of binaries with ice lines interior to their inner disk edge. I find that ice lines lie interior to the minimum stable semimajor axis for $\gtrsim 80\%$ of

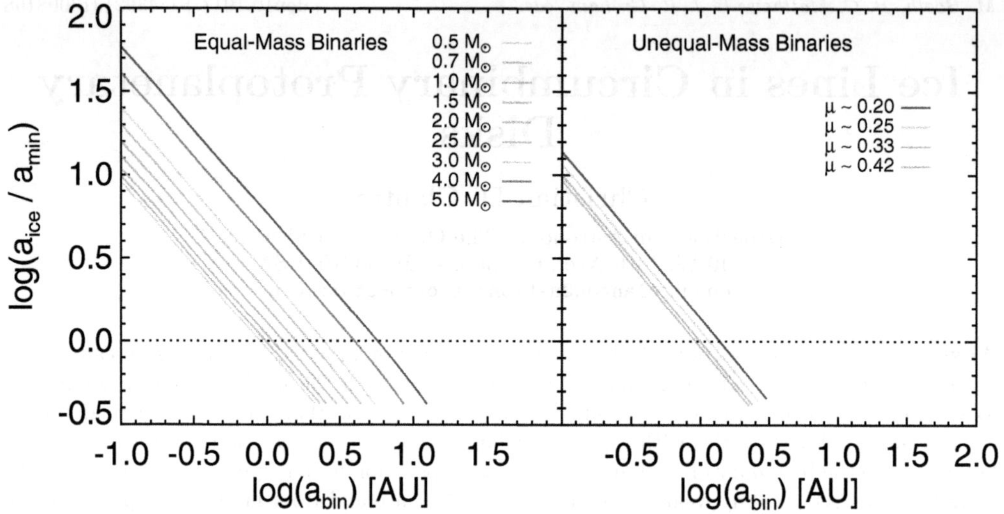

Figure 1. Ice line location relative to the minimum stable semimajor axis as a function of binary separation for both equal-mass (left) and unequal-mass binaries (right). Adapted from Clanton (2013).

all equal and unequal-mass binaries with components less massive than $M_\star \lesssim 2\ M_\odot$. This suggests that rocky planets should not form in these systems. For more details, see Clanton (2013).

References

Chiang, E. & Goldreich, P. 1997, *ApJ* 490, 368

Clanton, C. 2013, *ApJ* (Letters) 768, L15

Holman, M. & Wiegert, P. 1999, *AJ* 117, 621

Raghavan, D., McAlister, H., Henry, T., Latham, D., Marcy, G., Mason, B., Gies, D., White, R., ten Brummelaar, T. 2010, *ApJS* 190, 1

Weidenschilling, S. 1977, *Ap&SS* 51, 153

Exploring the Formation and Evolution of Planetary Systems
Proceedings IAU Symposium No. 299, 2013
M. Booth, B. C. Matthews & J. R. Graham, eds.

doi:10.1017/S174392131300817X

The different faces of transitional discs

M. de Juan Ovelar[1], M. Min[2], C. Dominik[2,3], C. Thalmann[2], P. Pinilla[4], M. Benisty[5] and T. Birnstiel[6]

[1]Leiden Observatory, Leiden University,
2300RA Leiden, The Netherlands
email: mjovelar@strw.leidenuniv.nl

[2]Astronomical Institute Anton Pannekoek, University of Amsterdam,
1090 GE Amsterdam, The Netherlands

[3]Department of Astrophysics/IMAPP, Radboud University Nijmegen,
6500 GL Nijmegen, The Netherlands

[4]Institut für Theoretische Astrophysik, Universität Heidelberg,
Albert-Ueberle-Straße 2, 69120 Heidelberg, Germany

[5]Institut de Planètologie et Astrophysique Grenoble,
414 rue de la Piscine, 38400 st. Martin d'Heres, France

[6]Harvard-Smithsonian Center for Astrophysics,
60 Garden Street, Cambridge, MA 02138, USA

Abstract. Recent imaging observations of transitional discs have revealed discrepancies between the structure observed at different wavelengths. In some targets, the gap measured using sub-mm observations disappears when observed using near infrared polarimetry, suggesting that the empty region is actually filled with small particles of dust (Dong *et al.,* 2012). Assuming the gapped structure observed in transitional discs is caused by the presence of a planet, we try to explain such discrepancies simulating observations of physical models of disc/planet systems with VLT/SPHERE-ZIMPOL, Subaru/HiCIAO, VLT/VISIR and ALMA.

Keywords. Transition Discs – Protoplanetary discs – (Stars:) circumstellar matter

---

1. Method

In order to generate realistic images of a transitional disc, we use similar models as those presented in Pinilla *et al.* (2012a). These combine 2-D hydrodynamical and 1-D dust evolution simulations including coagulation, fragmentation and radial drift, to self consistently reproduce the evolution of the gas and dust present in the disc. We focus on the cases of a disc with a planet of masses $M_{\rm p} = [1, 9, 15]$ $\rm M_{jup}$. after 3 Myr of evolution. The resulting distribution of dust and gas for these three cases, after 3 Myr of evolution, is input into the Monte-Carlo radiative transfer code MCMax (Min *et al.*, 2009), to produce theoretical intensity and polarised intensity images of the emitted and scattered flux of the disc. We then simulate realistic observations with VLT/SPHERE, Subaru/HiCIAO and ALMA at wavelengths $\lambda = [0.65, 1.6, 850]$ μm using specific instrument simulators (SPHERE-ZIMPOL (Thalmann *et al.* 2008) or convolving the full resolution images with an either measured or simulated point-spread-function (PSF).

2. Results and Conclusions

Figure 1 shows the synthetic observations of the disc-planet systems studied.

(*a*) The trapping and filtering mechanisms triggered by the presence of a > 1 Mjup mass planet in the disc yield a distinction in the radial distribution of the dust particles

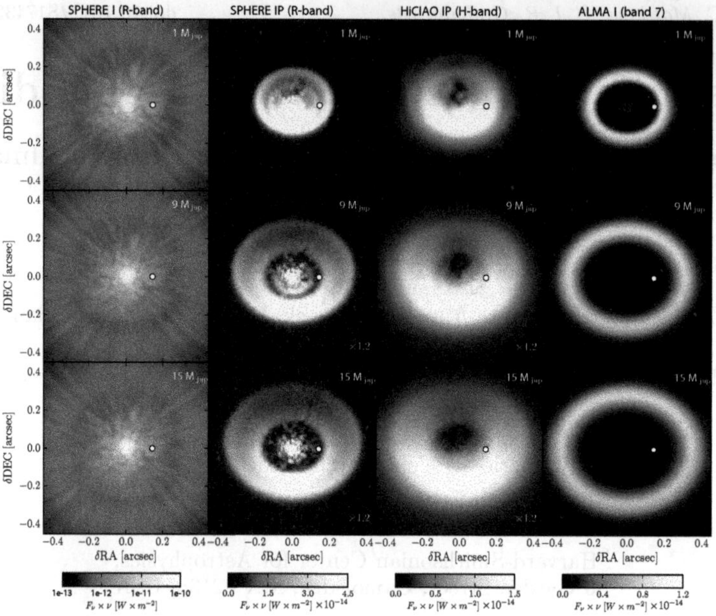

Figure 1. Simulated images of a 3 Myr disc model with a planet orbiting at 20 AU (white dot). Columns show images of intensity (I) and polarised intensity (PI) in the R-band (0.65 μm) with SPHERE-ZIMPOL, PI in the H-band (1.6 μm) with HiCIAO and intensity with ALMA at 850 μm. Top, middle and bottom rows show images obtained for planet masses of $M_p = [1, 9, 15]$ $\mathrm{M_{jup}}$ respectively. The system has an inclination of $i = 35°$

according to their size. This causes observations at different wavelengths to show different structures. Particles with larger sizes tend to be trapped in the region outside the planet's orbit, while smaller particles are allowed to move to the inner radii. The former show up in sub-mm emission measurements while the latter are detected in polarised scattering flux.

(*b*) The position of the pressure *"trap"*, the amount of particles trapped and filtered, and the particle size threshold of the filtering process depend strongly on the mass of the planet. This effect makes possible to constrain planet mass and position by combining multi-wavelength observations.

(*c*) Combination of sub-mm and polarimetric images at visible wavelengths, is the best imaging "strategy" to characterise the main features of this dust grain size distribution. An instrument like SPHERE-ZIMPOL in the R-band ($\lambda = 0.65$ μm) is able to differentiate between the three planet mass cases studied in polarised intensity, particularly if the system is inclined.

References

Dong, R., Rafikov, R., Zhu, Z., *et al.* 2012, *ApJ*, 750, 161
Min, M., Dullemond, C. P., Dominik, C., Koter, A. D., & Hovenier, J. W. 2009, *A&A*, 497, 155
Pinilla, P., Benisty, M., & Birnstiel, T. 2012a, *A&A*, 545, A81
Thalmann, C., Schmid, H. M., Boccaletti, A., et al. 2008, *SPIE*, Vol. 7014

Exploring the Formation and Evolution of Planetary Systems
Proceedings IAU Symposium No. 299, 2013
M. Booth, B. C. Matthews & J. R. Graham, eds.

doi:10.1017/S1743921313008181

Dynamics of the inner edge of the dead zone in protoplanetary disks

Julien Faure[1,2], Sebastien Fromang[1,2] and Henrik Latter[3]

[1]CEA, Irfu, SAp, Centre de Saclay, F-91191 Gif-sur-Yvette, France; [2]UMR AIM, CEA-CNRS-Univ. Paris VII, Centre de Saclay, F-91191 Gif-sur-Yvette, France; [3]Department of Applied Mathematics and Theoretical Physics, University of Cambridge, Centre for Mathematical Sciences, Wilberforce Road, Cambridge, CB3 0WA, UK

Abstract. In protoplanetary disks, the inner boundary between an MRI active and inactive region has recently been suggested to be a promising site for planet formation. A set of numerical simulations has indeed shown that vortex formation mediated by the Rossby wave instability is a natural consequence of the disk dynamics at that location. However, such models have so far considered only the case of an isothermal equation of state, while the complex thermodynamics is at the heart of how this region works. Using the Godunov code Ramses, we have performed 3D global numerical simulations of protoplanetary disks that relax the isothermal hypothesis. We find that, at the interface, the disk thermodynamics and the turbulent dynamics are intimately entwined, because of the importance of turbulent dissipation and thermal ionisation.

1. Introduction

We aim to investigate the interplay of thermodynamics and dynamics in the inner regions of protoplanetary disks. We then simulate the feedback of temperature on turbulence via the thermal heating with a set of numerical experiments of MRI turbulence. To complete the thermodynamical feedback loop, we link the resistivity to the temperature using, for simplicity, a step function. Such a form capture the strong sensitivity of resistivity in Saha's law around the threshold T_{MRI}.

2. Results

Initially, the gas is in a turbulent state from $R = 1.5$ to $R = 3.5$ and a dead zone extends from $R = 3.5$ to the outer boundary of the simulation domain.

We first consider the case of a resistivity that is a function of position only. 900 orbits after the beginning of the simulation, the temperature has reached a quasi steady state. We check that the flow is laminar in the dead zone region. However, we note that, at this location, the temperature (Fig. 1) does not drop to the minimum value, as one might assume in absence of turbulent heating. The curve on the (Fig. 1) establishes that the increased temperature is mostly due to density waves shocking inside the dead zone.

We then move to the more realistic case in which the resistivity is self-consistently calculated as a function of temperature. We restart the simulation after closing the thermodynamical feedback loop. The left panel of Fig. 2 shows the radial profile of the turbulent activity throughout the simulation. We see that the interface between the active and the dead zone is not static: a *MRI front* propagates outward over several disk scale heights. This dynamical behavior had been anticipated by previous exploratory works which modelled the complicated interaction between turbulence and temperature via a crude diffusive process (Latter & Balbus, 2012). On the right panel of Fig. 2, we show the

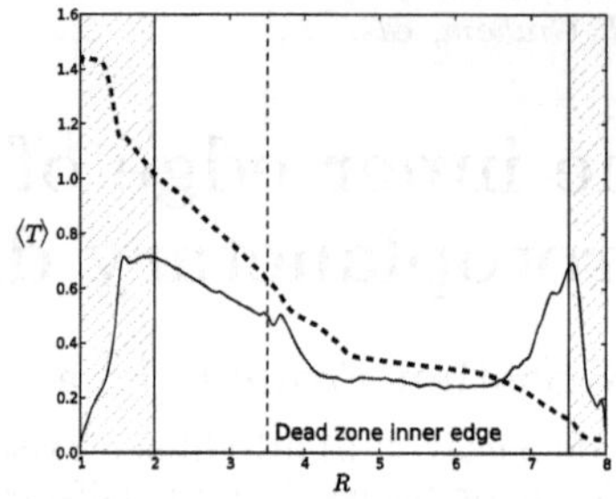

Figure 1. Temperature radial profile in the MHD simulation (black dashed line). The equilibrium temperature profile balancing the wave heating is shown by the plain black line.

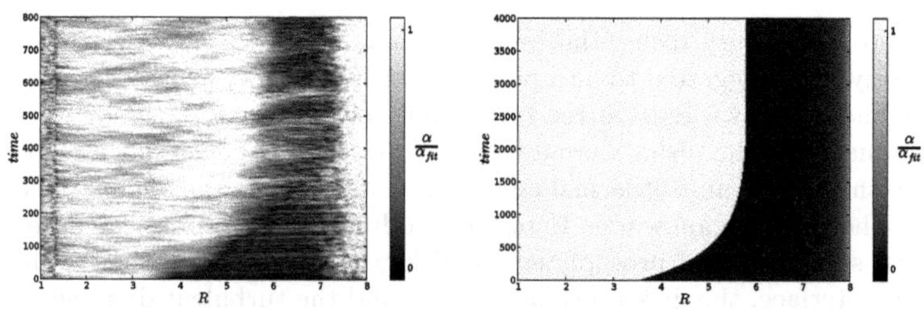

Figure 2. Turbulent activity evolution, in the MHD simulation (left panel), in the mean field simulation (right panel)

result obtained using the *mean field model* of Latter & Balbus (2012) initialized similarly to our MHD simulation. The two fronts propagate outward and stop at the same *critical radius*. The two y-axis scales indicate the front seen in the MHD simulation propagates much faster than the other one.

3. Conclusions

We have shown that the active zone controls the dead zone thermodynamics via waves excited at the interface. These waves transport energy, heating up the gas deeply inside the dead zone. The good agrement between the two final front position demonstates the mean field model can be used to understand the front propagation engine: A front propagates outward (inward) if the thermal energy reservoir before the interface is (not) sufficient to support dead zone thermal feeding until temperature exceed the ionization threshold. Hence, the final front position can be predicted once knowing the disk thermal structure only. By contrast, the velocity of the front is sensitive to the thermal flux connecting the active zone to the dead zone. A diffusive description of that flux fails at explaining the front velocities as it induces much too slow fronts. Instead, fronts move rapidly because waves efficiently transport energy across the interface. This emphasizes their importance for the dead zone dynamics as well.

For near future, we plan to investigate how vortices, known to form at the interface, are sensitive to the thermodynamical interplay we described here.

References

Latter, H. N., & Balbus, S. 2012, *MNRAS*, 424, 1977–1990

Exploring the Formation and Evolution of Planetary Systems
Proceedings IAU Symposium No. 299, 2013
M. Booth, B. C. Matthews & J. R. Graham, eds.

doi:10.1017/S1743921313008193

Visible Light Adaptive Optics Imaging of the Orion 218-354 Silhouette Disk

Katherine B. Follette[1,5], Laird M. Close[1], Jared R. Males[1,4], Derek Kopon[2], Ya-Lin Wu[1], Katie M. Morzinski[1,4,5], Philip Hinz.[1], Timothy J. Rodigas[1], Alfio Puglisi[3], Simone Esposito[3], Armando Riccardi[3], Enrico Pinna[3], Marco Xompero[3], & Runa Briguglio[3]

[1]Steward Observatory, The University of Arizona, 933 N Cherry Ave, Tucson, AZ 85721, USA
[2]Max Planck Institute for Astronomy, Königstuhl 17, 69117 Heidelberg, Germany
[3]INAF - Osservatorio Astrofisico di Arcetri, Largo E. Fermi 5, I-50125, Firenze, Italy
[4]NASA Sagan Fellow
[5]kfollette@as.arizona.edu

Abstract. We present the first ground-based adaptive optics images of a silhouette disk. This disk, Orion 218-354, is seen in silhouette against the bright nebular background of Orion, and was resolved using the new Magellan Adaptive Secondary AO system and its VisAO camera in Simultaneous Differential Imaging (SDI) mode. PSF subtraction of Orion 218-354 reveals a disk ~1" (400 AU) in radius, with the degree of absorption increasing steadily towards the center of the disk. By virtue of the central star being unsaturated, these data probe inward to a much smaller radius than previous HST images. Our data present a different picture than previous observers had hypothesized, namely that the disk is likely optically thin at Hα at least as far inward as ~20AU. In addition to being among the first high-resolution AO images taken in the optical on a large telescope, these data reveal the power of SDI imaging to illuminate disk structure, and speak to a bright future for visible AO imaging. Analysis of the results described briefly here can be found in full detail in Follette *et al.* (2013).

Keywords. Instrumentation: Adaptive Optics, Planetary Systems: Protoplanetary Disks

---

1. Introduction

Silhouette disks have long been of interest to the astronomical community as they provide perhaps the most unequivocal direct evidence of the existence of dusty circumstellar disks around young stars, as well as some of the most robust measurements of disk extents and inclinations. To date, all observations of these disks in silhouette have been done from space with HST. Similar observations have been precluded from the ground by a number of factors, chief among them the inability of most AO systems to achieve them. In particular, the wavelengths of the nebular lines necessary to see these disks in silhouette all lie shortward of 1μm, well outside the operating wavelengths of most AO systems. The central stars in these disks are also fainter than is manageable with traditional wavefront sensors. MagAO is the first AO system capable of imaging these disks from the ground. It is able to do so by virtue of its adaptive secondary mirror, which allows for correction on the necessary spatial scales to achieve AO imaging in the visible, as well as its pyramid wavefront sensor, which allows for binning of the pyramid pupils to lock on fainter natural guide stars.

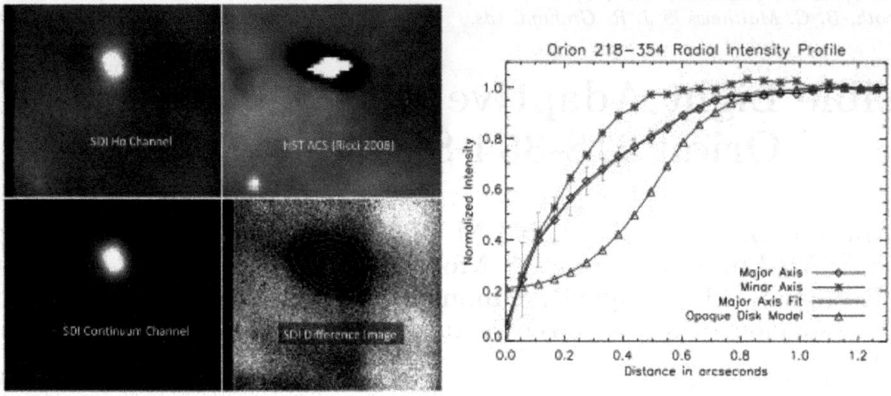

Figure 1. Left panel: Upper left: MagAO Hα image, showing the central star and the outer regions of the disk seen in silhouette. Lower left: MagAO SDI continuum (643 nm) image. Lower right: SDI difference image (Hα minus scaled continuum), revealing the entire disk in silhouette. Contours denote levels of decreasing flux moving towards the center of the disk. Upper right: HST ACS image of the same disk on the same physical scale from Ricci (2003).Right panel: Radial profile of MagAO ADI difference image, as well as model fits and the convolved best-fit optically thick disk model of McCaughrean & O'dell (1996), which is clearly a poor match to the data

2. Observations

We observed the silhouette disk Orion 218-354 with MagAO's visible light science camera, VisAO, on December 6, 2012 as part of the commissioning effort of the system (Close *et al.* (2013)). The disk was observed in simultaneous differential imaging (SDI) mode, in which a Wollaston prism splits the beam into two beams of equal brightness (for unpolarized sources). Each of these is passed through a different narrowband filter. One filter is centered on a spectral line of interest (Hα, [OI] or [SII]), and one on the neighboring continuum. This allows for simple, robust PSF subtraction through the nearly identical simultaneous continuum channel PSF.

3. Conclusions

From the Orion 218-354 data, we conclude that the radial intensity profile of this disk is suggestive of a small optical depth at Hα, a surprising result given previous HST work suggesting that all of the silhouette disks are optically thick (MacCaughrean (1996)). We convolve the best fit optically thick disk model of earlier observers with our continuum PSF and find that it is inconsistent with our data both in an absolute sense and in its general shape and character. In Follette *et al.* (2013), we use the observed profile to estimate the disk extent, geometry, small grain dust mass and radial mass density profile, and use these as inputs to the Whitney Monte Carlo radiative transfer code to generate the first multiwavelength model of the Orion 218-354 SED.

References

Close, L. M., *et al.*, 2013, *ApJ*, 774, 94
Follette, K. B., *et al.*, 2013, *ApJL*, 775, 113
McCaughrean, M. J. & O'dell, C. R. 1996, *AJ*, 111, 1977
Ricci, L., Robberto, M., & Soderblom, D. R. 2008, *AJ*, 136, 2136

Exploring the Formation and Evolution of Planetary Systems
Proceedings IAU Symposium No. 299, 2013
M. Booth, B. C. Matthews & J. R. Graham, eds.

doi:10.1017/S174392131300820X

A gas-rich disk around DX Cha

A. S. Hales[1,2], I. De Gregorio-Monsalvo[1,3], B. Montesinos[4], S. Casassus, W. R. F. Dent, C. Eiroa, A. M. Hughes, G. Garay, D. Mardones, F. Ménard, A. Palau, S. Pérez, N. Phillips, J. M. Torrelles and D. Wilner

[1]Atacama Large Millimeter/Submillimeter Array, Joint ALMA Observatory, Alonso de Córdova 3107, Vitacura 763-0355, Santiago - Chile
email: ahales@alma.cl

[2]National Radio Astronomy Observatory, 520 Edgemont Road, Charlottesville, Virginia, 22903-2475, United States

[3]European Southern Observatory, Karl-Schwarzschild-Str. 2, 85748, Garching bei Mnchen, Germany

[4]Department of Astrophysics, Centre for Astrobiology (CAB, CSIC-INTA), ESAC Campus, P.O. Box 78, 28691 Villanueva de la Cañada, Madrid, Spain

Abstract. DX Cha (HD 104237) is a southern, optically bright Herbig Ae star with an infrared excess, that is part of a small stellar group younger than 5 Myr. We used the APEX and ASTE submillimeter telescopes in Chile to search for continuum and gas emission around this system. Using LABOCA on APEX we detect strong continuum emission around HD104237-A and system component HD104237-E. Our ASTE spectrum detects a double-peaked ^{12}CO(3-2) line profile towards the system, typical of a rotating disk. The new data are used as constraints with MCFOST to produce a disk model that fits the entire SED as well as the observed CO line profile.

---

1. Introduction

DX Cha (HD 104237) is the optically brightest Herbig Ae star (A4-A8Ve) in the sky. At 115 pc, it is part of the young ϵ Cha stellar group (3-5 Myr; Feigelson *et al.* 2003). At least five stars are found within 15 arcseconds, and of these two are lower-mass stars that have IR excesses and emission lines, associated to T Tauri type stars (Grady *et al.* 2004). The brightest member, DX Cha (or HD 104237A), is actually a binary system, where the secondary is a K3 star (HD104237b) at 0.2 AU separation (Böhm *et al.* 2004). Recent modeling of the optical to mid-infrared SED by Fang *et al.* (2013) supports the earlier Meeus *et al.* (2001) interpretation of HD104237's SED as a star being surrounded by a self-shadowed (i.e., non-flared) dust disk. Here we present submillimeter data of the dust and gas disk surrounding DX Cha together with radiative transfer modeling efforts to fit the data.

2. LABOCA Imaging

The LABOCA 870 microns continuum map of HD 104237 (Figure 1) shows a clear detection of the primary disk, but also significant extension to the southeast that is consistent with emission from a second disk around the D or E components in the system. Based on the mid-IR detection of a disk around the E component by Grady *et al.* (2004), we have extracted fluxes using a model of two point sources separated by the 14.9" A-E separation vector (Figure 1). From this method we obtain 870 μm fluxes of 154 mJy and 91 mJy for the A and E components. Due to the system configuration, any flux from the

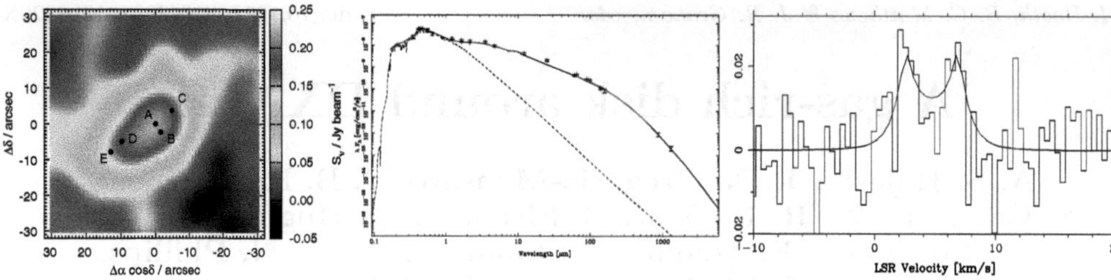

Figure 1. *Left:* LABOCA map of HD 104237 smoothed with half-beam Gaussian (FWHM = 9.3″), with stellar component positions plotted. *Middle:* Observed SED of HD 104237 (black points with error bars) plotted against the MCFOST disk model that fits the optical to submillimeter data (solid line). *Right:* MCFOST CO model of HD 104237 (solid line) fits well the ASTE line, data. Units are in antenna temperatures (T_A^*).

Table 1. HD104237 disk model

Physical parameter	Value	Reference
Disk total mass: ($M_\odot$)	$100\times M_d$	
Disk dust mass M_d ($M_\odot$)	4×10^{-4}	
Inner Rim: R_{in} (AU)	0.45	Tatulli *et al.* (2007)
Characteristic radius R_C (AU)	90	
Characteristic height h_C (AU)	6	
Surface density exponent: γ	0.8	
Flaring exponent: ψ	0.0	
Inclination angle: i (deg)	31	
Systemic Velocity: v_{lsr} (km/s)	4.2	

D component will mostly be incorporated in the E component flux, and any emission from the B and C components will be included in the A component flux.

3. Results and Conclusions

The 3D radiative transfer code MCFOST (Pinte *et al.* 2006) was used to model the SED and the molecular emission of HD104237A, assuming that both the infrared excess and $^{12}CO(3\text{-}2)$ emission stem from a circumstellar disk (Figure 1). We find the SED and CO line can be well fit by a non-flared disk 90 AU in radius and of 4×10^{-2} $M_\odot$ total mass. The parameters of this representative model are shown in Table 1. The CO surface brightness distributions predicted by the standard model over predicts the disk intensity by factors of 2-3 (assuming CO/H_2 ratios of 10^{-4}). By also incorporating the freeze-out of ^{12}CO molecules into dust grains at $T < 20$ K, however, the model fits well the line data. HD 104237 is a unique laboratory to study the extreme of SED-predicted flat disks. Resolved submillimeter images with existing interferometers (e.g. ALMA) are required to further constrain the physical and chemical structures of the disk.

References

Böhm, T., Catala, C., Balona, L., & Carter, B. 2004, *A&A*, 427, 907
Fang, M., van Boekel, R., Bouwman, J., *et al.* 2013, *A&A*, 549, A15
Feigelson, E. D., Lawson, W. A., & Garmire, G. P. 2003, *ApJ*, 599, 1207
Grady, C. A., Woodgate, B., Torres, C. A. O., *et al.* 2004, *ApJ*, 608, 809
Meeus, G., Waters, L. B. F. M., Bouwman, J., *et al.* 2001, *A&A*, 365, 476
Pinte, C., Ménard, F., Duchêne, G., & Bastien, P. 2006, *A&A*, 459, 797
Tatulli, E., Isella, A., Natta, A., *et al.* 2007, *A&A*, 464, 55

Exploring the Formation and Evolution of Planetary Systems
Proceedings IAU Symposium No. 299, 2013
M. Booth, B. C. Matthews & J. R. Graham, eds.

doi:10.1017/S1743921313008211

The Early Era: How do protostellar discs form?

Marc Joos[1,2], Patrick Hennebelle[1,2], Andrea Ciardi[2] and Sébastien Fromang[1]

[1]Service d'Astrophysique (CEA/DSM/Irfu/SAp)

[2]Laboratoire de Radioastronomie, ENS - LERMA, Observatoire de Paris, Universit Pierre et Marie Curie

Abstract. Discs are a key element in star and planet formation; however, magnetic fields can efficiently transport angular momentum away from the central region of the collapsing core during the dense core collapse, preventing disc formation. We perform numerical simulations of magnetically supercritical collapsing cores with a misalignment between the rotation axis and the magnetic field (Joos *et al.* 2012) and in a turbulent environment (Joos *et al.* 2013). The early formation of massive discs can take place at moderate magnetic intensities if the rotation axis is tilted or in a turbulent environment, because of misalignment and turbulent diffusion.

Keywords. star formation, low mass stars, disc formation, magnetohydrodynamics, turbulence, numerical simulations

---

1. Introduction

Discs are a key element in star and planet formation: protostars probably grow by accreting material from protostellar accretion discs, which at later stages are the natural progenitors of planets. The formation of discs is a challenging theoretical question, and the conditions for their presence remain unclear.

Magnetic fields certainly play an important role, and they can efficiently transport angular momentum away from the central regions of the collapsing core, thereby preventing the early formation of massive discs. This occurs even for relatively low magnetic field intensities (Mellon & Li 2008; Price & Bate 2007; Hennebelle & Fromang 2008), at a level largely in accordance with the observational measurement of core magnetizations (Crutcher 1999; Falgarone *et al.* 2008). However, most previous simulations were performed in a particular configuration, where the magnetic field and the rotation axis are initially aligned and without turbulence.

We therefore perform three-dimensional, adaptive mesh, numerical simulations of magnetically supercritical collapsing dense cores in both non-turbulent and turbulent environments, using the magneto-hydrodynamic code RAMSES (Teyssier 2002; Fromang *et al.* 2006).

2. Orientation matters

We perform a first set of simulations with an initial mass of 1 $M_\odot$ and an angle α between the rotation axis and the magnetic field varying from 0 to 90°.

We show that it affects the transport of angular momentum: specific angular momentum in the inner part of the collapsing core increases with the angle α. More momentum is therefore available to form discs.

Indeed, transport processes are less efficient when α increases. First, the strong magnetic braking acting in the aligned configuration is due to a strong lever arm, due to the peculiar geometry of the problem (with the typical hourglass shape of the magnetic field lines). In misaligned configurations (i.e. in other geometric configurations), the strengh of this lever arm decreases, and, therefore, so does the magnetic braking. Second, the outflows launched by the growth of a magnetic tower transport an significant amount of angular momentum. The suppression of the outflows with increasing α is responsible for a weaker transport of angular momentum.

More angular momentum is therefore available to build more massive discs. Even a small angle (of 20°) is sufficient, in the intermediate magnetisation case, to form disc 3 times larger than in the aligned configuration.

3. The role of turbulence

The interstellar medium is known to be turbulent, and turbulent energy is known to cascade down to small scales. Star formation is therefore occuring in a turbulent environment. To investigate the effects of turbulence, we perform simulations with an initial core mass of 5 $M_\odot$ since more massive clouds are expected to be more turbulent (Larson 1981).

Turbulence effects are twofold: first, the angular momentum included in the turbulent velocity field imposed as initial conditions induces a misalignment between the rotation axis and the magnetic field. We already show how this orientation matters and help to form more massive discs. Second, turbulence is responsible for an efficient diffusion of the magnetic flux; magnetic flux is diffused out of the inner part of the collapsing core, decreasing the strengh of the magnetic braking. It also favors the formation of more massive discs.

4. Conclusions

In contrast to earlier analyses, we show that the transport of angular momentum acts less efficiently in collapsing cores with non-aligned rotation and magnetic fields. We also show that the turbulence is responsible for a misalignment between the rotation axis and the magnetic field and can diffuse out of the magnetic field of the inner regions efficiently. The magnetic braking is therefore reduced, and massive discs can be built.

The early formation of massive discs can take place at moderate magnetic intensities if the rotation axis is tilted or in a turbulent environment, because of misalignment and turbulent diffusion.

Bibliography

Crutcher, R. M. 1999, *ApJ*, 520, 706.
Falgarone, E., Troland, T. H., Crutcher, R. M., *et al.* 2008, *A&A*, 487, 247.
Fromang, S., Hennebelle, P., & Teyssier, R. 2006, *A&A*, 457, 371.
Hennebelle, P. & Fromang, S. 2008, *A&A*, 477, 9.
Joos, M., Hennebelle, P., & Ciardi, A. 2012, *A&A*, 543, 128.
Joos, M., Hennebelle, P., Ciardi, A., & Fromang, S. 2013, *A&A*, 554, 17.
Larson, R. B. 1981, *MNRAS*, 319, 340.
Mellon, R. R. & Li, Z.-Y., 2008, *ApJ*, 681, 1356.
Price, D. J. & Bate, M. R., 2007, *MNRAS*, 377, 77.
Teyssier, R. 2002, *A&A*, 385, 337.

Exploring the Formation and Evolution of Planetary Systems
Proceedings IAU Symposium No. 299, 2013
M. Booth, B. C. Matthews & J. R. Graham, eds.

doi:10.1017/S1743921313008223

Circumstellar Disks in Very Young Embedded Clusters

Naibí Mariñas[1], Elizabeth A. Lada[1], Paula S. Teixeira[2] and Charles J. Lada[3]

[1]Astronomy Department, University of Florida,Gainesville, FL 32611
[2]University of Vienna, Tuerkenschanzstrasse 17, 1180, Vienna, Austria
[3]Harvard-Smithsonian CFA, Cambridge, MA

Abstract. We used FLAMINGOS near-IR photometry and spectroscopy and Spitzer mid-IR photometry to study disk fractions in the 1 to 2 Myr old NGC2264 clusters. We find that stars with masses < 0.3 solar masses have lower disk fractions than stars of solar mass or higher at these early ages. We also find that most disks disappear within the first 4 Myr, which is consistent with previous studies of disk lifetimes. Our study suggests that either some very low mass stars form without disks or that their disks are less massive and/or colder than predicted from models and not detected with Spitzer/Flamingos sensitivities.

Keywords. stars: pre-main-sequence, stars: circumstellar matter

1. Introduction

The standard model of star formation predicts that all stars are formed by accretion and are surrounded by circumstellar disks. Planets will form in these disks and this process is deeply connected to the amount of gas and dust available. However, do all stars really begin their lives surrounded by disks?

Near- and mid-IR studies of the frequency distribution of disks in young stellar clusters show that disks dissipate quickly, within 3 to 5 Myr, (Haisch, Lada & Lada 2001; Hernandez *et al.* 2008) and that their evolution seems to depend on the mass of the central star, intermediate mass stars have shorter disk lifetimes than solar and low mass stars (e.g. Haisch, Lada & Lada 2000; Carpenter *et al.* 2006; Lada *et al.* 2006). These short disk lifetimes imply that even when we consider clusters with young ages, between 3 to 5 Myr old, this mass dependent disk evolution already dominates the observed disk fractions. It is the stellar population with ages < 2 Myr that offers a window to investigate the initial frequency distribution of disks.

To address this issue, we have recently investigated the disk fractions of the young (1 to 2 Myr) population of the NGC2264 northern and southern clusters (Marinas *et al.* 2013, Marinas *et al.* in preparation).

2. Overview

We used FLAMINGOS near-IR spectra (R 1000 and S/N $>$ 50) to obtain spectral types of cluster members by comparing the spectral features and continuum shape of the stars with those of standards. For sources later than M2, the broad H band absorption feature was used, for K3-M2 stars we used line ratios of MgI (1.50 μm) and SiI (1.59 μm) to OH (1.69 μm), which are sensitive to effective temperatures. For the earlier sources, HI (1.68 μm) was used as well. Using this spectral classification and JHK FLAMINGOS photometry we estimated luminosities for all the sources. Spectral types were converted to temperature by using the temperature scale of Kenyon and Hartmann (1995) and Luhman *et al.* (2003). Placing the sources on the H-R diagram, and using theoretical evolutionary pre-main-sequence (PMS) tracks (Baraffe *et al.* 1998, Chabrier *et al.* 2000, Siess *et al.* 2000) we derived median ages of 1.4 Myr for the NGC2264 southern cluster and 1.8 Myr for the NGC2264 northern cluster. Given the uncertainties associated with theoretical PMS model ages, the two clusters seem to be coeval.

To obtain infrared excess for the sources we compared the expected SED slope from the photosphere alone over the wavelength range 1.68 to 8 μm, estimated by using NextGen models given each star effective temperature and surface gravity, with the slope of the SED from measured fluxes corrected for interstellar extinction. The excess for each source was correlated with stellar mass and age. A more detailed description of the data reduction and analysis can be found in Marinas *et al.* 2013.

The final sample included 226 stars from the two NGC2264 clusters.

3. Results

Disk Lifetimes Infrared excess in this sample decreases rapidly with age. While the disk fraction in stars younger than 2 Myr is 56 $\pm$ 4 %, the disk fraction for stars 2 to 4 Myr is 40 $\pm$ 5 % and the disk fraction for stars older than 4 Myr is only 24 $\pm$ 6 %.

Initial Disk Fractions and Stellar Mass Disk fractions in the young NGC2264 increase with stellar mass, resembling the disk distributions found in lower stellar density regions like ChaI and Taurus (Luhman *et al.* 2010). We find a disk fraction of 68 $\pm$ 8% for 1 to 3 solar mass stars, a disk fraction of 51 $\pm$5% for 0.3 to 1 solar mass stars and a disk fraction of 42 $\pm$ 7% for 0.1 to 0.3 solar mass stars. Since we expect stars in very young clusters and those in low density environments to be less evolved, this result suggest that stellar mass might play a role in the initial distribution of disks.

Disk Evolution and Stellar Mass In the 3 Myr old cluster IC348, disk fraction peaks for the very low mass stars at about 50 %, and decreases for the more massive stars (Luhman *et al.* 2010). This is the opposite of what we find in NGC2264. The difference in the disk fraction dependance on stellar mass between these two clusters of similar stellar densities, but different ages, could be the result of mass dependent disk evolution. While a very young region like NGC2264 would still be more representative of the original disk distribution, a faster disk evolution for more massive stars would deplete disks around these higher mass stars faster and result in the distribution found in the older IC348.

Studies of other nearby clusters of similar ages are needed to investigate the role of stellar mass in the formation and early retention of circumstellar disks.

References

Baraffe, I., Chabrier, G., Allard, F., & Hauschildt, P. H. 1998, *A&A*, 337, 403
Carpenter, J. M., Mamajek, E. E., Hillenbrand, L. A. *et al.*, 2006, *ApJL*, 651, 49
Chabrier, G., Baraffe, I., Allard, F., & Hauschildt, P. 2000, *ApJ*, 542, 464
Haisch, K. E. Jr, Lada, E. A., & Lada, C. J. 2000, *AJ*, 120, 1396
Haisch, K. E. Jr., Lada, E. A., & Lada, C. J. 2001, *AJ*, 121, 2065
Hernandez, J., Hartmann, L., Calvet, N. *et al.*, 2008, *ApJ*, 686, 119
Kenyon, S., J. & Hartmann, L. 1995, *ApJS*, 101, 117
Lada, C. J., Muench, A. A., Luhman, K. L. *et al.*, 2006, *AJ*, 131,1574
Luhman, K. L, Stauffer, J. R., Muench, A. A., Rieke, G. H. *et al.*, 2003, *ApJL*, 593, 1093
Luhman, K. L., Allen, P. R., Espaillat, C., Hartmann, L., & Calvet, N. 2010, *ApJS*, 186, 111
Marinas, N., Lada, E. A., Teixeira, P. S., & Lada, C. J. 2013, *ApJ*, 772
Siess L., Dufour, E., & Forestini, M. 2000, *A&A*, 358, 593

Exploring the Formation and Evolution of Planetary Systems
Proceedings IAU Symposium No. 299, 2013
M. Booth, B. C. Matthews & J. R. Graham, eds.

doi:10.1017/S1743921313008235

On the evolution of the Snow Line in Protoplanetary Discs

Rebecca G. Martin[1] and Mario Livio[2]

[1]NASA Sagan Fellow, JILA, University of Colorado, Boulder, CO 80309, USA
email: rebecca.martin@jila.colorado.edu

[2]Space Telescope Science Institute, 3700 San Martin Drive, Baltimore, MD 21218, USA

Abstract. We examine the evolution of the snow line in a protoplanetary disc. If the magneto-rotational instability (MRI) drives turbulence throughout the disc, there is a unique snow line outside of which the disc is icy. The snow line moves closer to the star as the infall accretion rate drops. Because the snow line moves inside the radius of the Earth's orbit, the formation of our water-devoid planet is difficult with this model. However, protoplanetary discs are not likely to be sufficiently ionised to be fully turbulent. A dead zone at the mid-plane slows the flow of material through the disc and a global steady state cannot be achieved. We model the evolution of the snow line also in a disc with a dead zone. As the mass is accumulating, the outer parts of the dead zone become self gravitating, heat the massive disc and thus the outer snow line does not come inside the radius of the Earth's orbit. With this model there is sufficient time and mass in the disc for the Earth to form from water-devoid planetesimals at a radius of 1AU. Furthermore, the additional inner icy region within the dead zone predicted by this model may allow for the formation of giant planets close to their host star without the need for much migration.

Keywords. Earth – planetary systems: protoplanetary discs – planets and satellites: formation

1. Introduction

The inner terrestrial planets have very little water compared to the outer giant planets. The current water content of the Earth is 0.023% by weight compared to the outer planets that have greater than 40%. In the solar nebula, ice forms beyond the snow line radius, $R_{\rm snow}$. The snow line in our solar system is currently observed to be within the asteroid belt. Thus, at the time of planetesimal formation, the snow line was at a radius of $R = 2.7\,$AU from the Sun. Outside of the snow line, planetesimal formation is much easier because of the increased solid density. Thus, we expect massive planets to form outside of the snow line. Fully turbulent disc models predict a minimum snow line radius that is inside the orbital radius of the Earth. With this model it is hard to explain the formation of our water devoid Earth. However, protoplanetary discs are not likely to be fully ionised. They contain a dead zone about the mid-plane where the MRI cannot drive turbulence. We examine this disc model here.

2. Protoplanetary Discs with Dead Zones

Dead zones form when the ionisation fraction is too low for the magnetic field to be well coupled to the disc (Gammie 1996). The inner parts of the disc have a temperature higher than the critical required for complete thermal ionisation, $T_{\rm c} > T_{\rm crit}$. Further out, the magneto-rotational instability (MRI) is suppressed except in the surface layers of the disc that are ionised by cosmic rays or X-rays. In the low surface density outer parts of the disc, where $\Sigma < \Sigma_{\rm crit}$, cosmic rays can penetrate and ionise the whole disc.

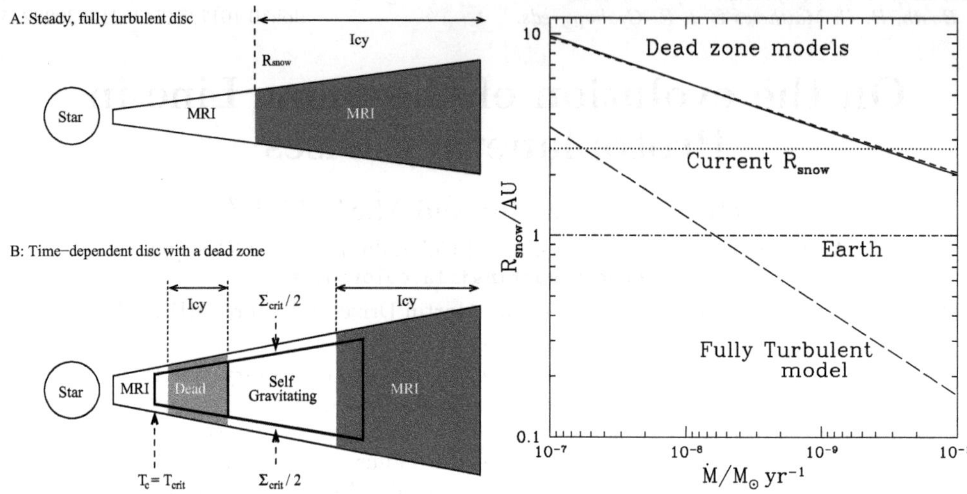

Figure 1. Left: Sketches of the protoplanetary disc structure with and without a dead zone. Right: The snow line radius as a function of infall accretion rate.

Material builds up in the dead zone until the outer parts become self gravitating (see Fig. 1) and this drives some extra turbulence and heating within the disc. In the early stages of disc evolution, this can lead to transitions between gravitationally and magnetically produced turbulence and outbursts on to the young star (e.g. Armitage *et al.* 2001, Martin & Lubow 2011). In the later stages when the infall accretion rate has dropped, the outbursts cease but a dead zone may still be present. The dead zone changes the temperature profile of a disc significantly from the fully turbulent model.

We find an analytic solution for the outermost snow line radius in a disc with a dead zone (Martin & Livio 2013). We find good agreement with the one-dimensional time-dependent layered accretion disc models (Martin & Livio 2012, Martin *et al.* 2012). In the disc with a dead zone, the extra gravitational turbulence heats the disc and the outer icy region remains outside the orbital radius of the Earth. Thus, the formation of a water devoid Earth at $R = 1$ AU is possible.

3. Conclusions

In a fully turbulent disc, the unique snow line passes inside the radius of the Earth's orbit causing problems for the formation of our water-devoid planet Earth. When a dead zone forms in the disc, it prevents mass flow through the disc and the outer regions become self-gravitating. The turbulence driven by self-gravity increases the temperature of the outer parts of the dead zone and thus the outer icy region is much farther from the star. On a more speculative note, the inner icy region within the dead zone, not predicted by the turbulent disc model, could allow for the formation of icy planets or gas giants in the inner regions of exo-solar systems if there is water present there.

References

Armitage P. J., Livio M., & Pringle J. E., 2001, *MNRAS*, 324, 705
Gammie C. F., 1996, *ApJ*, 457, 355
Martin R. G. & Lubow S. H., 2011, *ApJ*, 740, 6
Martin R. G., Lubow S. H., Livio M., & Pringle J. E., 2012, *MNRAS*, 423, 2718
Martin R. G. & Livio M., 2012, *MNRAS*, 425, L6
Martin R. G. & Livio M., 2013, *MNRAS*, 434, 633

Exploring the Formation and Evolution of Planetary Systems
Proceedings IAU Symposium No. 299, 2013
M. Booth, B. C. Matthews & J. R. Graham, eds.

doi:10.1017/S1743921313008247

Large grains can grow in circumstellar discs

Farzana Meru[1], Marina Galvagni[2], Christoph Olczak[3,4,5] and Pascale Garaud[6]

[1]Institut für Astronomie, ETH Zürich, Wolfgang-Pauli-Strasse 27, 8093 Zürich
email: farzana.meru@phys.ethz.ch

[2]Institute of Theoretical Physics, Universität Zürich, Winterthurerstrasse 190, 8057 Zürich
[3]Astronomisches Rechen-Institut (ARI), Zentrum für Astronomie Universität Heidelberg, Mönchhofstrasse 12-14, 69120 Heidelberg
[4]Max-Planck-Institut für Astronomie (MPIA), Königstuhl 17, 69117 Heidelberg
[5]National Astronomical Observatories of China, 20A Datun Lu, Beijing 100012
[6]Department of Applied Mathematics and Statistics, UCSC, 1156 High Street Santa Cruz

Abstract. We perform coagulation & fragmentation simulations to understand grain growth in T Tauri & brown dwarf discs. We present a physically-motivated approach using a probability distribution function for the collision velocities and separating the deterministic & stochastic velocities. We find growth to larger sizes compared to other models. Furthermore, if brown dwarf discs are scaled-down versions of T Tauri discs (in terms of stellar & disc mass, and disc radius), growth at the same location with respect to the outer edge occurs to similar sizes in both discs.

Keywords. methods: analytical; methods: numerical; planetary systems: protoplanetary disks

1. Introduction

Previous models of coagulation and fragmentation of an ensemble of dust aggregates often assumed that a collision between particles of particular sizes occurs with a single velocity. We develop a more physically-motivated model for the collision velocities.

2. Principles behind the new physically-motivated model

The key features are: (i) a particle's velocity is *not* single-valued but is described by a probability distribution function (PDF); (ii) a particle's velocity in any direction is a Gaussian with a mean given by the deterministic velocity in that direction (i.e. radial drift, azimuthal drift or vertical settling) and a standard deviation given by the stochastic velocities (i.e. turbulence & Brownian motion). The analytical 1D PDF of *relative velocities* in each direction is produced for each pair of particles, and combined to give a 3D collision velocity PDF.

We assume that collisions with velocities lower than the bouncing velocity, $v_{\rm b}$, lead to growth while those higher than the fragmentation velocity, $v_{\rm f}$, break apart if their mass ratio is smaller than the *mass transfer* parameter and stick if it is larger. Physically this means that collisions between unequal-sized aggregates are likely to lead to growth while equal-sized aggregates are likely to fragment - a result shown by experiments (e.g. Teiser & Wurm 2009) and simulations (Meru *et al.* 2013b). At all other velocities the aggregates bounce. Fig. 1 shows that the final 3D PDF may cover any of these collisional outcomes - these are used to simulate the local size evolution of an ensemble of particles in a disc.

3. Application to T Tauri and brown dwarf discs

We simulate grain growth locally in a disc with mass $M_{\rm d} = 0.05 M_\star$ around a $0.75 {\rm M}_\odot$ star (see Garaud *et al.* 2013 for parameters) and find growth to larger sizes (Fig. 2, left).

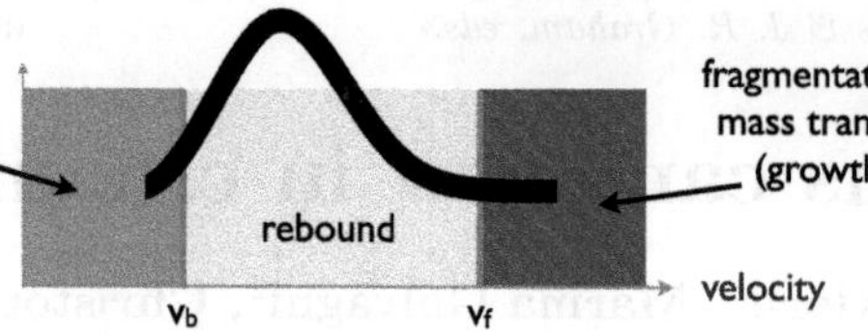

Figure 1. Schematic diagram showing the collisional outcomes of sticking and bouncing at low velocities, and fragmentation and mass transfer at high velocities. The PDF has an expectation in one zone but extends into others.

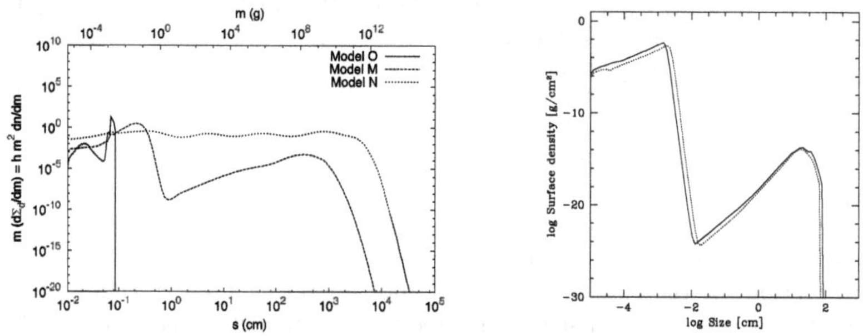

Figure 2. Surface mass density distribution of particles. Left: at 1au for a model (i) without a velocity PDF (Model O), (ii) where the stochastic velocities dominate over the deterministic ones (Model M), and (iii) the new model (Model N). Right: for the brown dwarf disc (solid line) and T Tauri disc (dotted line) at 2/3 of the distance to the outer radius.

We also model growth in brown dwarf & T Tauri discs where the latter is a scaled-up version of the former, i.e. the same disc to star mass ratio, simulated at the same location with respect to the truncation radius, R_t, and where M_d & R_t are set using the observed relation $M_d \propto R_t^{1.6}$ (Andrews *et al.* 2010). We simulate growth at 10au in a brown dwarf disc with $M_d = 4 \times 10^{-4} M_\odot$ and $R_t = 15$au around a $60 M_{Jup}$ brown dwarf, and at 60au in a T Tauri disc with $M_d = 7 \times 10^{-3} M_\odot$ and $R_t = 90$au around a $1 M_\odot$ star. Fig. 2 (right) shows that growth at the equivalent location in both discs occurs to the same size (Meru *et al.* 2013a).

4. Conclusions

We present a model for growth and fragmentation that considers a particle's velocity PDF and separates the deterministic and stochastic velocities. We find growth to large sizes and the emergence of two particle populations. In addition if brown dwarf discs are scaled-down versions of T Tauri discs, growth occurs to similar sizes. Our model may potentially explain the large ($\approx$ mm-sized) grains observed in brown dwarf discs (e.g. Bouy *et al.* 2008) and the long-standing problem of grain growth in discs: growing dust while maintaining a small-size population.

References

Andrews, S. M., *et al.* 2010, *ApJ*, 723, 1241
Bouy, H., *et al.* 2008, *A&A*, 486, 877
Garaud, P., Meru, F., Galvagni, M., & Olczak, C. 2013 *ApJ*, 764, 146
Meru, F. and Galvagni, M. & Olczak, C., 2013, *ApJ*, 774, L4
Meru F., Geretshauser R. J., Schaefer C., Speith R., & Kley W., 2013, *MNRAS*, 435, 2371
Teiser, J. & Wurm, G., 2009, *MNRAS*, 393, 1584

Exploring the Formation and Evolution of Planetary Systems
Proceedings IAU Symposium No. 299, 2013
M. Booth, B. C. Matthews & J. R. Graham, eds.

doi:10.1017/S1743921313008259

LIPAD Simulations of Giant Planet Core Formation

Henry Ngo[1,2], Martin J. Duncan[2] and Harold F. Levison[3]

[1]Division of Geological and Planetary Science, California Institute of Technology, MC 150-21, Pasadena, California, USA 91101
email: hngo@caltech.edu

[2]Department of Physics, Engineering Physics & Astronomy, Queen's University, 99 University Avenue, Kingston, Ontario, Canada K7L 3N6

[3]Department of Space Studies, Southwest Research Institute, 1050 Walnut Street, Suite 300, Boulder, Colorado, USA 80302

Abstract. We present some preliminary results from our investigation of giant planetary core formation using numerical simulations with the Lagrangian Integrator for Planetary Accretion and Dynamics (LIPAD) by Levison *et al.* (2012). LIPAD couples dynamics with collisional evolution, including fragmentation. We start with a cold planetesimal disk using particles of a few kilometres in size. Our simulations show growth from kilometre-sized planetesimals to several Earth-mass sized embryos (tens of thousands of kilometers) can occur. However, these embryos may not be large enough to start runaway gas accretion necessary to build the envelopes of gas giant planets.

Keywords. accretion, methods: n-body simulations, methods: numerical, planets and satellites: formation

1. Introduction

Gas giant planet cores must reach a critical mass (about 10 Earth masses; Mizuno *et al.* 1978) to initiate runaway accretion of their gaseous envelopes from the solar nebula. Observations of stellar disks (e.g. Hernández *et al.* 2009) show that the gas disk lifetime is only 1-10 Myr. The formation of a massive core on such a short timescale is a major challenge for the core accretion model of planet formation. In this work, we seek to determine dynamical processes that could allow for planetary cores to grow this quickly.

2. Methods

We have performed numerical simulations of giant planet core formation from a disk of planetesimals using a new Lagrangian integrator ("LIPAD") by Levison *et al.* (2012). LIPAD is built on top of the Symplectic Massive Body Algorithm (SyMBA) by Duncan *et al.* (1998). A key feature of LIPAD is the ability to couple dynamics with collisional evolution. For example, unlike other numerical integrators, LIPAD allows planetesimals to grow or fragment in size due to interactions with each other. Thus, entire planets or planetary cores can grow from a planetesimal disk. LIPAD handles the large number of planetesimals and collisions statistically. In addition to the planetesimal tracers (a tracer represents a large number of planetesimals on similar orbits), there are also embryo particles that have full N-body interactions with planetesimals and each other. Finally, LIPAD imposes a minimum size on the planetesimals to keep the number of computed collisions to a tractable amount. When planetesimals fragment to a size below this limit, which was set to 0.5 km in this work, these smallest particles are put into a new class that do not have collisional interactions at all. Instead, a constant inward drift rate (due to aerodynamic drag) is imposed.

3. Results

We start with an initial disk of 10,000 kilometre-sized planetesimal tracers from 4 to 10 AU and integrate for 4.5 Myr and include the drag effects of a decaying gas disk. This is a follow-up on previous work by Levison *et al.* (2010) with the SyMBA integrator.

In our work, we notice that there are three main classes out of outcomes, depending on the imposed drift rate on the smallest non-colliding particle class. When the drift rate is very high, these particles sweep by the embryos too quickly for accretion. When the drift rate is very low, the particles become locked into mean motion resonances with the embryos and this causes both the embryos and particles to drift inwards towards the Sun within a million years. The maximum growth of embryos occur for an intermediate drift rate, where the largest embryos form are about 6 Earth masses in size.

These simulations are best viewed in animated form and examples of the above three cases can be found at: `http://gps.caltech.edu/~hngo/IAUS299`.

4. Conclusions

Our work is the first time this type of integrator has been used to study planet formation in the outer Solar System. We found that the smallest particles in the system played an important role in determining the growth and migration of the embryos, an idea also suggested by Wetherill & Stewart (1993) and by Rafikov (2004). In the previous work by Levison *et al.* (2010), the size of the planetesimals (which were not allowed to change in SyMBA) determined the growth and migration of the embryos. Here, with LIPAD, the planetesimals now can vary in size but now it appears that the resulting growth and migration is dependent on the choice of drift rate for the smallest particles. Future work would involve a more realistic treatment of these smallest particles.

Acknowledgements

The authors are grateful for funding from NSERC and the NSF. Partial travel support for H.N. came from the Foster and Coco Stanback Fund. Computations were performed on the GPC supercomputer at the SciNet HPC Consortium. SciNet is funded by: the Canada Foundation for Innovation under the auspices of Compute Canada; the Government of Ontario; Ontario Research Fund - Research Excellence; and the University of Toronto.

References

Duncan, M., Levison, H., & Lee, M. 1998, *AJ*, 116, 2067

Hernández, J., Calvet, N., Hartmann, L., Muzerolle, J., Gutermuth, R., Stauffer, J. 1993, *ApJ*, 707, 705

Levison, H., Thommes, E., & Duncan, M. 2010, *AJ*, 139, 1297

Levison, H., Duncan, M., & Thommes, E. 2012, *AJ*, 144, 119

Mizuno, H., Nakazawa, K., & Hayashi, C. 1978, *Progress of Theoretical Physics*, 60, 699

Rafikov, R. 2004, *AJ*, 128, 1348

Wetherill, G. W., & Stewart, G. R. 1993, *Icarus*, 106, 190

Exploring the Formation and Evolution of Planetary Systems
Proceedings IAU Symposium No. 299, 2013
M. Booth, B. C. Matthews & J. R. Graham, eds.

doi:10.1017/S1743921313008260

Inviscid and viscous flow past embedded planets: implications for planet formation

Chris W. Ormel and Ji-Ming Shi

Astronomy Department, University of California, Berkeley
Berkeley CA 94720
email: [ormel,jmshi]@astro.berkeley.edu

Abstract. We have investigated the flow pattern that arises past (proto)planets embedded in the nebula disk during its formation. We consider the regime where the planet mass is large enough to gravitationally perturb the gas ($\gtrsim$0.01 Earth masses), but not too large to open a gap or accrete unlimited amounts of gas from the disk. We consider both inviscid and viscid flows and aim to understand the flow pattern on the scales of the Bondi radius. Having described the flow pattern of the gas, we integrate trajectories of small, solid particles. In agreement with previous findings, we show that the ensuing accretion rates can be high—although, due to radial drift motions, pebble accretion is not necessarily an efficient mechanism.

Keywords. hydrodynamics – methods: analytical – planets and satellites: formation – protoplanetary discs

---

1. Introduction

Quite generally, planets can be characterized into three groups: rocky planets, gas giants, and the intermediate-mass planets. The latter are dominated by a rock/icy core, yet contain appreciable quantities of gas—gas that has been acquired during the nebula phase. Planets like Neptune and the ubiquitously discovered super-Earths (mini Neptunes) fall in this intermediate category.

We aim to describe the interaction between such planets and the nebular gas. In particular, we aim to solve for the flow pattern (gas density and velocity) of the gas on the scales of the Bondi radius of the planet—the radius where the thermal motions of the gas become similar to the escape velocity from the planet, $R_{\rm bondi} = GM_{\rm planet}/c_s^2$ (where G is Newton's constant, $M_{\rm planet}$ the planet mass, and c_s the thermal velocity of the gas). We consider a configuration in which the flow is steady in a frame corotating with the planet, which is assumed to move in a circular orbit.

2. Inviscid and viscid flows

Figure 1 shows the flow pattern that arises in the vicinity of the planet. The flow is barotropic, subsonic, 2D, but compressible. In that case a stream function (Ψ) formulation is feasible (Korycansky & Papaloizou 1996; Ormel 2013). We solve for Ψ numerically and have also derived analytical approximations that closely match the numerical ones.

In Fig. 1 one clearly identifies the horseshoe region where the flow makes a characteristic 'U-turn'. Furthermore, one notes circulating flow in the counterclockwise (prograde) direction near the center on scales of the Bondi radius. This region is the atmosphere of the planet. These two regions are separated by a critical streamline: the separatix (the thick curve that makes an 'X'). Material within the atmosphere is bound to the planet.

We have also conducted hydro simulations with Athena (Stone *et al.* 2008). These simulations have a similar setup as the inviscid simulations, but we do account for a

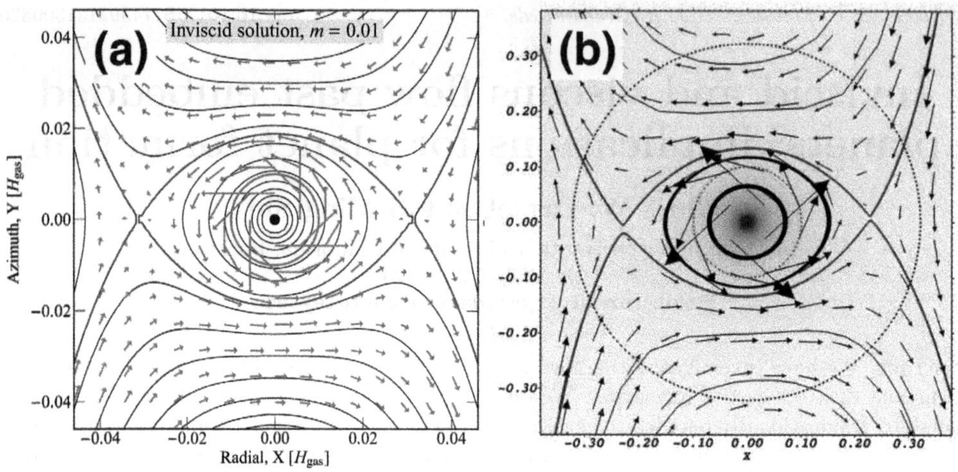

Figure 1. Inviscid (*left*) and viscous (*right*) flow pattern around a planet. The flow pattern is described by velocity vectors (arrows) and streamlines (solid lines). The thick streamline denotes the separatix. Length units are the scaleheights of the disk ($H_{\rm gas}$). The planet mass is nondimensionalized as $m = R_{\rm bondi}/H_{\rm gas}$. The masses in the two panels is not the same: $m = 0.01$ (left) and $m = 0.1$ (right). The flow in both configurations is steady.

viscosity. By choosing a gravitational smoothing parameter on the order of the grid spacing, we can, for the first time, resolve the flow close to the planet, that is, within the Bondi radius (Fig. 1b). We adopt a dimensionless mass parameter $m = R_{\rm Bondi}/H_{\rm gas} = 0.1$, where $H_{\rm gas}$ is the scaleheight of the gas and a kinematic viscosity $\nu = 10^{-3} c_s H_{\rm gas}$.

As the critical streamline provides the interface between the disk and planet atmosphere, we can in this way solve for the size of the planet.

3. Efficient accretion of small particles?

A new model for the formation of planets is by accretion of small particles (Ormel & Klahr 2010; Lambrechts & Johansen 2012). Using the flow solutions of the gas we can integrate trajectories of these particles, for which gas drag is a critical agent determining their fate: accretion on the planet or drift past the planet. With the new flow solution we find that very small particles (dust) couple too strongly with the gas and thereby tend to avoid accretion. For larger particles (those that couple more loosely) the accretion cross section is enhanced. However these particles also tend to drift to the star very quickly (Weidenschilling 1977). Altogether, the accretion efficiency of particles depends sensitively on the local properties of the gas (turbulent strength, gas density, and pressure gradients).

References

Lambrechts M. & Johansen A. 2012, *A&A*, 544, A32

Korycansky, D.G & Papaloizou, E. 1996, *ApJS*, 105, 181

Ormel, C. W. 2013, *MNRAS*, 428, 3526

Ormel C. W. & Klahr H. H. 2010, *A&A*, 520, A43

Stone, J. M., Gardiner, T. A., Teuben, P., Hawley, J. F., Simon J. B., 2008, *ApJS*, 178, 137

Weidenschilling, S. J. 1977, *MNRAS*, 180, 57

Exploring the Formation and Evolution of Planetary Systems
Proceedings IAU Symposium No. 299, 2013
M. Booth, B. C. Matthews & J. R. Graham, eds.

doi:10.1017/S1743921313008272

Gas signatures of Herbig Ae/Be disks probed with Herschel SPIRE spectroscopy

Matthijs H. D. van der Wiel[1], David A. Naylor[1], Giambattista Aresu[2] and Göran Olofsson[3]

[1]Institute for Space Imaging Science, Department of Physics and Astronomy, University of Lethbridge, Lethbridge, AB, Canada; email: matthijs.vanderwiel@uleth.ca

[2]Kapteyn Astronomical Institute, University of Groningen, Groningen, The Netherlands

[3]Stockholm University, Stockholm, Sweden

Abstract. Herbig Ae/Be objects, like their lower mass counterparts T Tauri stars, are seen to form a stable circumstellar disk which is initially gas-rich and could ultimately form a planetary system. We present *Herschel* SPIRE 460–1540 GHz spectra of five targets out of a sample of 13 young disk sources, showing line detections mainly due to warm CO gas.

Keywords. circumstellar matter, planetary systems: protoplanetary disks, ISM: molecules

1. Observations, processing, and line detections

In a 16-hour guaranteed time project (P.I. Göran Olofsson) using the SPIRE Fourier Transform Spectrometer (FTS), spectra in the 460–1540 GHz range were obtained of 13 protoplanetary disks around Herbig stars: AB Aur, HD 100546, HD 97048, HD 163296, T Tau, HD 142527, HD 144432, RY Tau, HD 104237, HD 36112, HD 169142, HD 100453 and TW Hya. The data were processed in HIPE 9, followed by subtraction of a median background signal obtained from the off-center detectors. All spectra show a smoothly rising dust continuum, while only the first five of the above list show detectable line signal. The disk emission lines, presented in Fig. 1, are unresolved both spatially (the SPIRE FTS beam is 42–17″) and spectrally ($R \sim 400$–1300). Spectral lines are fitted using a dedicated Fourier Transform line fitter tool (available online at `www.uleth.ca/phy/naylor/`). Table 1 lists lines detected in at least one target besides T Tau.

2. Interpretation and analysis plans

First, it is evident from the observations that the continuum and CO lines toward T Tau originate in an extended protostellar envelope rather than in a much smaller disk. This is likely also the case for the cold H_2O vapor. Second, the N^+ line toward HD 163296 is 10^4–10^5 times brighter than what is predicted by current (X-ray) irradiated disk models (e.g., Aresu *et al.* 2011), even when invoking X-ray luminosities much higher than appropriate for HD 163296. Since N^+ is also detected in all off-center detectors up to 1′ away from the star, we hypothesize that a single (external) source is responsible for ionizing nitrogen both in the disk/jet of HD 163296 and in the surrounding gaseous medium.

Finally, we plan to use the unique set of 4–3 to 13–12 lines of CO and ^{13}CO ($E_{\rm up}/k \sim 50$–500 K) to characterize warm gas in the Herbig protoplanetary disks.

Acknowledgements: This work is supported by CSA and NSERC. SPIRE has been developed by a consortium of institutes led by Cardiff University (UK) and including contributions from Canada, China, France, Italy, Spain, Sweden, UK, and USA.

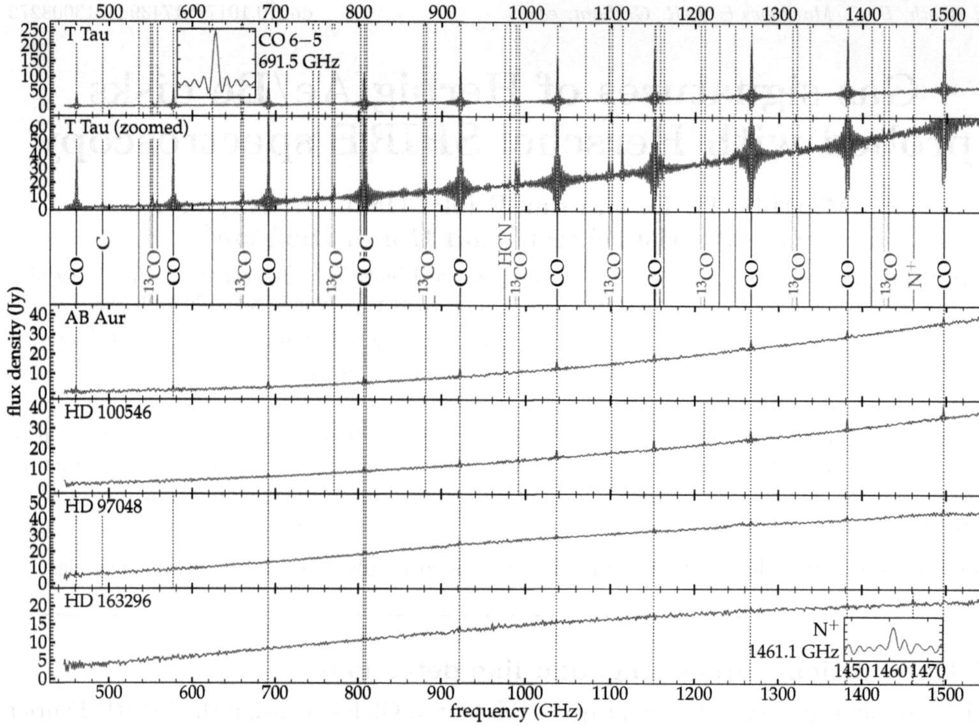

Figure 1. *Herschel* SPIRE-FTS spectra of T Tau (with the second panel zoomed in on the flux density axis to highlight weaker lines), AB Aur, HD 100546, HD 97048 and HD 163296. Detected gas lines are indicated by vertical dotted lines. The ringing of the sinc-shaped line profile typical of an FTS (see inset in top panel) is particularly noticeable near the ^{12}CO lines in T Tau.

Table 1. Spectral lines identified in the SPIRE spectra of at least one target besides T Tau.

line transition$^{(1)}$	$E_{\rm up}/k$ **(K)**	**frequency**$^{(2)}$ **(GHz)**	**line flux**$^{(3)}$ $(10^{-18}$ W m$^{-2})$ **[uncertainty]** T Tau	AB Aur	HD 100546	HD 97048	HD 163296
CO 4–3	55	461.0	448 [5]	35 [2]	-	37 [3]	-
C ^{3}P$_1$–^{3}P$_0$	24	492.2	35 [5]	-	-	26 [3]	-
CO 5–4	83	576.3	599 [5]	25 [2]	-	13 [3]	-
CO 6–5	116	691.5	740 [5]	38 [2]	17 [2]	17 [3]	8 [2]
^{13}CO 7–6	148	771.2	100 [5]	6 [2]	7 [2]	-	-
CO 7–6	155	806.7	1141 [5]	38 [2]	22 [2]	17 [3]	10 [2]
C ^{3}P$_2$–^{3}P$_1$	62	809.3	93 [5]	14 [2]	8 [2]	17 [3]	5 [2]
CO 8–7	199	921.8	1531 [5]	48 [2]	36 [2]	31 [3]	16 [2]
HCN 11–10	281	974.5	10 [4]	11 [2]	-	-	-
^{13}CO 9–8	238	991.3	75 [4]	5 [2]	14 [2]	-	-
CO 9–8	249	1036.9	1106 [4]	47 [2]	45 [2]	25 [3]	9 [2]
CO 10–9	304	1152.0	1202 [4]	39 [2]	53 [2]	29 [3]	12 [2]
^{13}CO 11–10	349	1211.3	50 [4]	-	14 [2]	-	-
CO 11–10	365	1267.0	1225 [4]	46 [2]	51 [2]	32 [3]	12 [2]
CO 12–11	431	1382.0	1224 [4]	38 [2]	59 [2]	24 [3]	10 [2]
N$^+$ ^{3}P$_1$–^{3}P$_0$	70	1461.1	-	-	-	-	17 [2]
CO 13–12	503	1496.9	1262 [4]	44 [2]	62 [2]	41 [3]	11 [2]

Notes: $^{(1)}$ In addition to the lines listed here, in T Tau we detect six H_2O lines and ∼25 others identified as ^{13}CO, C^{18}O, HCO$^+$, HCN and N$_2$H$^+$. $^{(2)}$ Rest frequency from JPL, Pickett *et al.* (1998), except N$^+$ from SLAIM. $^{(3)}$ A '-' indicates a non-detection. Uncertainties include only the formal fitting error.

References

Aresu, G., Kamp, I., Meijerink, R., Woitke, P., Thi, W.-F., & Spaans, M. 2011, *A&A*, 526, A163

Pickett, H. M., Poynter, R. L., Cohen, E. A., *et al.* 1998, *J. Quant. Spec. Radiat. Transf.*, 60, 883

Exploring the Formation and Evolution of Planetary Systems
Proceedings IAU Symposium No. 299, 2013
M. Booth, B. C. Matthews & J. R. Graham, eds.

doi:10.1017/S1743921313008284

Large-scale planetesimal formation by streaming instability

Chao-Chin Yang and Anders Johansen

Lund Observatory, Department of Astronomy and Theoretical Physics, Lund University
Box 43, SE-22100 Lund, Sweden
email: ccyang@astro.lu.se

Abstract. In current theory of planet formation, streaming instability is one of the most promising mechanisms to overcome the meter-barrier in the course of core accretion. Almost all previous works, however, were focused on a local region of protoplanetary disks with a limited size of about 0.2 gas scale heights. Only one radial filamentary particle concentration was seen in these studies. To address this, we conduct the largest-scale simulations of this kind to date, up to 0.8 gas scale heights both horizontally and vertically. We demonstrate that streaming instability remains robust on large scale and multiple radial particle concentrations exist in large enough boxes. This result may be important in characterising the feeding zone of planetesimal formation.

Keywords. Hydrodynamics, instabilities, methods: numerical, planetary systems: formation

---

It has been a long standing problem in core accretion scenario of planet formation for meter/decimetre-sized solid objects to remain in the gaseous protoplanetary disks due to constant head wind (Weidenschilling 1977). One of the most promising mechanisms to overcome this meter-barrier is the streaming instability (Youdin & Goodman 2005), in which viscous interaction between solids and gas can concentrate the solids to the extent that gravitational collapse takes over to produce Ceres-sized planetesimals (Johansen *et al.* 2007). Systemic study of this mechanism seems warranted to explore the preferable conditions and the properties of the resulting solid concentration.

We conduct the largest-scale hydrodynamical simulations of this kind to date, up to $0.8H$ both horizontally and vertically, where H is the gas scale height. We use the Pencil Code† to model a local shearing box of gas and solid particles. The vertical component of the stellar gravity is included so that the particles quickly settle into a thin mid-plane layer. The gas and particles interact by viscous drag with a Stokes number of 0.3, where the Stokes number is defined as the product of local orbital frequency and particle stopping time due to gas drag. The total solid-to-gas ratio is fixed at 0.02, which is approximately over the critical solid-to-gas ratio required to trigger strong clumping of these particles (Johansen, Youdin, & Mac Low 2009).

The top panel of Figure 1a demonstrates the process of particle sedimentation and turbulent stirring. The particle scale height first decreases to a minimum within about one local orbital period and then increases due to stirring by the streaming instability. There exists a significant increase in particle scale height when increasing the vertical size of the simulation box from $0.2H$ to $0.4H$, while the results are similar for runs of vertical sizes $0.4H$ and $0.8H$. This indicates there exists a minimum vertical size of about $0.4H$ to properly capture the turbulent stirring of particles by streaming instability.

Furthermore, the concentration of particles by streaming instability seems also dependent on the dimensions of the simulation box, as shown in the bottom panel of Figure 1a.

† The code is publicly available at `https://code.google.com/p/pencil-code/`.

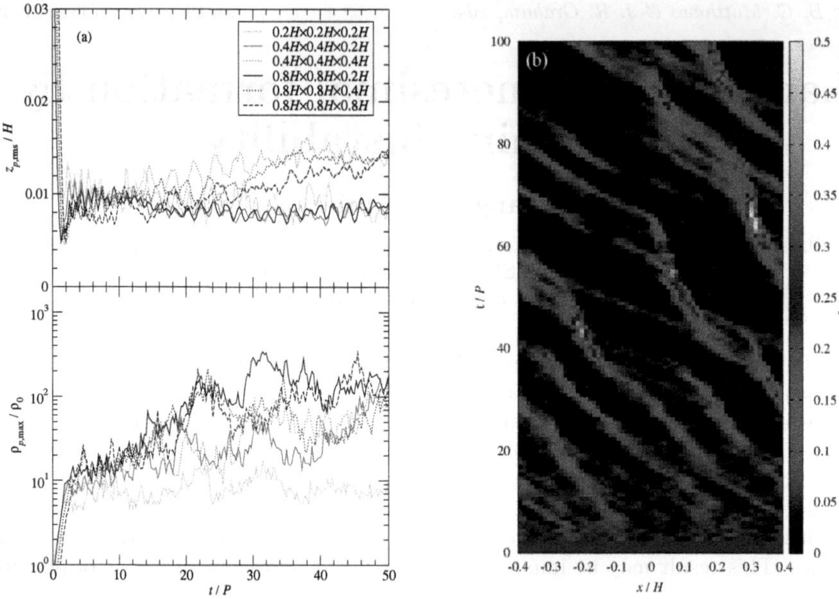

Figure 1. (a) Average particle scale height (top) and maximum particle density (bottom) as a function of time in terms of the local orbital period. Different lines denote simulation boxes of different dimensions, where H is the gas scale height. (b) Space-time diagram of the particle density integrated over the vertical direction and averaged over the azimuthal direction from a $0.8H \times 0.8H \times 0.8H$ simulation box. The horizontal axis represents the radial direction, while the vertical axis denotes the time.

The larger the horizontal dimensions, the higher the peak particle density. On the other hand, the peak particle density seems little affected by the vertical dimension of the simulation box. This phenomenon could be understood by the difference in the size of the particle feeding zone in the azimuthal direction.

Finally, Figure 1b shows the particle density, integrated over the vertical direction and averaged over the azimuthal direction, as a function of time and radial position. In contrast to previous works, where only one radial filamentary structure in particle density was seen, we find multiple concentrations at different radii given large enough horizontal dimension. For a $0.8H \times 0.8H \times 0.8H$ simulation box, at least two particle concentrations exist simultaneously. This indicates that we might have captured the characteristic wavelength of particle concentrations in the non-linear stage of the streaming instability, which helps characterising the feeding zone of planetesimal formation.

In summary, we have systematically investigated the streaming instability of viscously interacting gas and solid particles in a protoplanetary disk with various box dimensions. We find convergence in final particle scale height with vertical dimension. We have also demonstrated multiple radial concentrations of solid particles do exist in large enough simulation boxes, which may have important implications in planet formation scenarios.

References

Johansen, A., Oishi, J. S., Mac Low, M.-M., *et al.* 2007, *Nature*, 448, 1022
Johansen, A., Youdin, A., & Mac Low, M.-M. 2009, *ApJ*, 704, L75
Weidenschilling, S. J. 1977, *MNRAS*, 180, 57
Youdin, A. N. & Goodman, J. 2005, *ApJ*, 620, 459

Exploring the Formation and Evolution of Planetary Systems
Proceedings IAU Symposium No. 299, 2013
M. Booth, B. C. Matthews & J. R. Graham, eds.

doi:10.1017/S1743921313008296

Planet formation in evolving protoplanetary discs

Richard Alexander

Department of Physics & Astronomy, University of Leicester, Leicester, LE1 7RH, UK
email: richard.alexander@leicester.ac.uk

Abstract. I attempt to summarize our knowledge of planet formation in evolving protoplanetary discs. I first review the physics of disc evolution and dispersal. For most of the disc lifetime evolution is driven by accretion and photoevaporation, and I discuss how the interplay between these processes shapes protoplanetary discs. I also discuss the observations that we use to test these models, and the major uncertainties that remain. I will then move on to consider planet formation and migration in evolving discs, and discuss how observations of both discs and planets can be used to inform our understanding of protoplanetary disc evolution.

Keywords. protoplanetary discs, planet–disc interactions, planetary systems

1. Introduction

We now know that many, if not most, stars host planetary systems, but their origins remain a mystery. Understanding the evolution of protoplanetary discs is a critical ingredient of any predictive theory of planet formation. Protoplanetary discs are the sites of planet formation, but planet formation is an inefficient process (only a small fraction of the initial disc mass ends up in planets) and consequently discs dominate the dynamics of forming planetary systems. Here I review our understanding of the physics of protoplanetary disc evolution and dispersal, and its consequences for the architecture of planetary systems.

1.1. *Disc evolution*

The existence of gas-rich protoplanetary discs was first suggested as long ago as the work of Kant & Laplace, but such discs were not directly observed until the late 1980s (e.g., Sargent & Beckwith 1987). Since then observations of discs have progressed at a startling rate, and we now have a census of thousands of these objects. Discs are observed across a wide range of wavelengths, using both gas and dust tracers, and this has allowed us to build up a detailed picture of their properties. Observations of young clusters find that the fraction of stars with an infrared excess (which traces warm dust at $\sim$AU radii) declines from $\sim$100% at ages $\lesssim$1Myr to $\lesssim$10% at ages of 5–10Myr (e.g., Haisch *et al.* 2001; Mamajek 2009). A similar trend is seen in signatures of accretion on to the stellar surface (e.g., Fedele *et al.* 2010), which trace gas in the inner disc, and in (sub-)mm continuum emission, which traces cold dust at large ($\sim$100AU) radii (e.g., Andrews & Williams 2005). Although age estimates for young stars, which are typically derived from comparing observations to pre-main-sequence evolutionary tracks, are subject to significant uncertainties (Hillenbrand 2009; Bell *et al.* 2013), from these data we can conclude that typical disc lifetimes are a few Myr, with order-of-magnitude scatter.

In addition to statistical studies of discs in clusters, these observations also allow us to determine global properties of individual discs. By making standard assumptions about the dust opacity and dust-to-gas ratio we can estimate disc masses from the (sub-)mm luminosity: typical disc masses range from $M_{\rm d} \sim 0.1{\rm M}_\odot$ to $\lesssim 0.001{\rm M}_\odot$ (e.g., Andrews & Williams 2005). We can also derive instantaneous accretion rates, by measuring the excess short-wavelength emission produced by the accretion shock on the stellar surface; if we know the stellar mass and radius, we can convert an observed accretion luminosity into a mass accretion rate (e.g., Calvet & Gullbring 1998). Accretion rates measured in this manner (for $\sim$solar-mass stars) range from $\dot{M}_{\rm acc} \gtrsim 10^{-7}\,{\rm M}_\odot{\rm yr}^{-1}$ to $\lesssim 10^{-10}{\rm M}_\odot{\rm yr}^{-1}$ (e.g., Gullbring *et al.* 1998). We can therefore estimate an empirical accretion timescale $t_{\rm acc} \sim M_{\rm d}/\dot{M}_{\rm acc}$, and find that for most sources $t_{\rm acc}$ is of order Myr, again with large scatter (e.g., Jones *et al.* 2012). The accretion time-scale is approximately the time it will take all of the observed disc mass to be accreted (at the observed accretion rate), so we expect protoplanetary discs to evolve substantially due to accretion during their ~Myr lifetimes. This in turn implies that there is no such thing as "typical conditions" for planet formation; planets form in evolving protoplanetary discs, and the ubiquity of planets suggests that they form under a wide range of different conditions.

These observations show that protoplanetary discs evolve, at least in part due to disc accretion, but shed little light on the physical processes driving this evolution. The decline in observed accretion rates with stellar age is broadly consistent with classical viscous accretion disc evolution (e.g., Lynden-Bell & Pringle 1974; Hartmann *et al.* 1998), but the origin of the required angular momentum transport remains unclear (see, e.g., the review by Armitage 2011). At early times, when the disc is still embedded in its parent envelope, disc evolution is probably dominated by gravitational instabilities (e.g., Lodato 2007). Magnetically-driven jets and winds can also remove angular momentum and drive accretion in the disc, particularly at small radii, and may remain important throughout the disc's evolution (e.g., Königl & Salmeron 2011). However, for most of the disc lifetime the dominant source of angular momentum transport is thought to be magnetohydrodynamic (MHD) turbulence, driven by the magnetorotational instability (MRI; Balbus & Hawley 1991; Balbus 2011). It remains unclear how much of the disc is sufficiently ionized to sustain the MRI, and it seems likely that substantial "dead zones", with little or no turbulence, exist in the majority of discs (e.g., Gammie 1996; Bai 2011). Recent work also suggests that MRI-driven winds may play a role in disc evolution, potentially driving significant mass- and angular momentum-loss over a wide radial extent (e.g., Fromang *et al.* 2013; Bai & Stone 2013). Thus, although simple viscous disc models are sufficient to explain most observations of protoplanetary disc accretion, our ignorance of how discs accrete remains a major obstacle to a complete theory of planet formation.

1.2. *Disc dispersal*

Observations of young stars also allow us to infer a great deal about disc dispersal. Different disc tracers are strongly correlated, despite originating in very different regions of the disc, and tend to vanish together; few objects are seen with "partial" discs (e.g., Andrews & Williams 2005). This suggests that disc dispersal is almost simultaneous across a wide range of radii, from $\lesssim 0.1$AU to > 100AU. Moreover, the relative lack of objects with properties between those of disc-bearing and disc-less stars suggests that dispersal is rapid: statistical estimates suggest that the dispersal time-scale is at least an order of magnitude shorter than the disc lifetime (e.g., Simon & Prato 1995; Koepferl *et al.* 2013). Finally, non-detections of warm or hot gas around disc-less weak-lined T Tauri stars set very strict upper limits on the gas surface density (orders of magnitude below canonical protoplanetary disc values), implying that disc dispersal is very efficient (Pascucci *et al.* 2006; Ingleby *et al.* 2011). Disc evolution therefore has two distinct time-scales (the ~Myr disc lifetime and the much shorter disc dispersal time), and the process(es) which drive final disc clearing must operate very efficiently over the entire radial extent of the disc.

These observational constraints rule out several obvious candidates as mechanisms for disc dispersal. Resolved observations of disc structure suggest that most of the disc mass resides at large radii ($\gtrsim 50$AU; Andrews *et al.* 2009), but the characteristic viscous time-scale is $\gtrsim$Myr at these radii. Discs therefore live for no more than a few viscous time-scales in their outer regions, and consequently accretion cannot play a major role in outer disc clearing. Similarly, winds or jets driven by the stellar magnetic field cannot be the dominant disc clearing mechanism at large radii, though MRI-driven winds may play a role. However, in most discs the primary driver of final disc dispersal is thought to be disc photoevaporation (see, e.g., Clarke 2011; Alexander *et al.* 2014).

2. Disc evolution models

2.1. *Disc Photoevaporation*

Photoevaporation occurs when high-energy (UV and/or X-ray) radiation is incident on protoplanetary discs. This irradiation creates a hot ($\sim 10^3$–10^4K) layer on the disc surface, and beyond a certain radius the heated gas is unbound. The result is a pressure-driven flow, which is referred to as a photoevaporative wind. The characteristic critical radius depends on the temperature of the hot gas, and is approximately

$$R_c \simeq \frac{0.2GM_*}{c_{\rm hot}^2} \simeq 1.8 \left(\frac{M_*}{1{\rm M}_\odot}\right) \left(\frac{T_{\rm hot}}{10^4{\rm K}}\right)^{-1} {\rm AU}\,, \tag{2.1}$$

where M_* is the stellar mass, and $c_{\rm hot}$ and $T_{\rm hot}$ are the sound speed and temperature in the heated surface layer, respectively (Hollenbach *et al.* 1994; Liffman 2003; Font *et al.* 2004). The irradiation can be both "internal" (i.e., from the central star) or "external" (from nearby massive stars); external heating dominates in some cases (e.g., the propldys in the centre of the Orion Nebula Cluster; Johnstone *et al.* 1998), but for the majority of discs we expect photoevaporation to be dominated by irradiation from the central star.

The temperature of the surface layers depends primarily on the irradiation, and three regimes are important for protoplanetary discs: ionizing, "extreme" UV (EUV; 13.6–100 eV); non-ionizing far-UV (FUV; 6–13.6 eV) and X-rays (0.1–10 keV):

EUV: In the EUV case heating and cooling is dominated by ionization and recombination of atomic hydrogen, and for an optically thick disc ionizing photons from recombinations in the disc atmosphere dominate the heating at the ionization front (Hollenbach *et al.* 1994). The resulting disc atmosphere is close to isothermal at $T_{\rm hot} \simeq 10^4$K (akin to an H II region on the disc surface), and most of the mass-loss originates at 1–2AU. As the flow is recombination-limited the integrated mass-loss rate $\dot{M}_{\rm w}$ scales as the square-root of the ionizing luminosity, and for fiducial parameters we find $\dot{M}_{\rm w} \sim 10^{-10}{\rm M}_\odot{\rm yr}^{-1}$ (Hollenbach *et al.* 1994; Font *et al.* 2004). However, if the inner disc is optically thin to ionizing photons (as expected during disc dispersal) then direct irradiation of the inner disc edge dominates. In this case most of the mass-loss comes from close to the inner edge, and $\dot{M}_{\rm w}$ increases by a factor of $\sim$10 (Alexander *et al.* 2006a).

FUV: FUV irradiation is analogous to the well-studied problem of photodissociation regions (PDRs): photons are predominantly absorbed by either dust grains or polycyclic aromatic hydrocarbons (PAHs), and the gas is heated by a combination of collisions with grains and FUV-pumping of H_2. The resulting gas temperatures vary substantially with both radius and height above the disc midplane, from $\sim$100K at large radii to $\gtrsim$1000K close to the star (Adams *et al.* 2004; Gorti & Hollenbach 2008, 2009). The

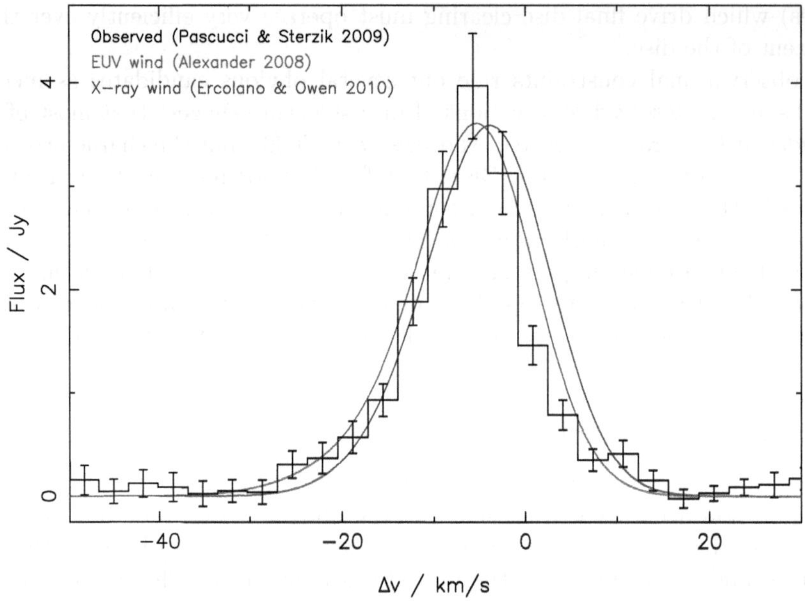

Figure 1. TW Hya [Ne II] 12.81μm line profile: the black line shows the observed line profile from Pascucci & Sterzik (2009), while the red and blue curves show the predictions from photoevaporative wind models driven by stellar EUV (Alexander 2008b) or X-ray (Ercolano & Owen 2010) photons, respectively. Although the data do not yet distinguish between these models, the blue-shifted line represents a clear detection of a slow, ionized wind, and provides unambiguous evidence of on-going photoevaporation of the TW Hya disc.

resulting mass-loss profile has a peak at 5–10AU, and also rises towards the disc outer edge ($\gtrsim$100AU); in some cases the mass-loss at large radii dominates. The integrated wind rate depends mainly on the FUV luminosity incident on the disc, and for fiducial parameters $\dot{M}_w \sim 10^{-8} M_\odot yr^{-1}$ (Gorti & Hollenbach 2009).

X-rays: Young solar-mass stars have long been known to be bright X-ray sources, with typical luminosities $L_X \sim 10^{30} erg\,s^{-1}$ (e.g., Feigelson *et al.* 2007). X-rays are primarily absorbed by K-shell ionization of heavy elements, and the resulting ~keV photoelectrons collisionally ionize and/or heat the disc gas. As in the FUV case, the disc atmosphere is not well characterised by a single value of T_{hot}, but hydrodynamic models find that most of the resulting wind originates in the atomic layer, at temperatures of 3000–5000K (Owen *et al.* 2010), and consequently the mass-loss profile peaks at around 3AU. The mass-loss rate scales close to linearly with L_X, and for typical parameters we again find $\dot{M}_w \sim 10^{-8} M_\odot yr^{-1}$ (Owen *et al.* 2011, 2012).

These different "flavours" of photoevaporation models all predict mass-loss at rates that are large enough to play a significant role in disc evolution, but it is not clear which heating mechanism dominates in real discs. Some theoretical issues remain, but the dominant uncertainty in these models is the irradiating flux which reaches the disc surface (which is an input parameter in the models). Recent work has attempted to resolve this question observationally, by looking for emission line diagnostics which trace the wind structure. Photoevaporative winds are characterised by relatively low-density gas which is at least partially ionized, and are therefore expected to produce copious forbidden line emission (Hollenbach & Gorti 2009). Hydrodynamic models show that the characteristic signature of such a wind is a small blue-shift in the emission from face-on discs (Font *et al.* 2004),

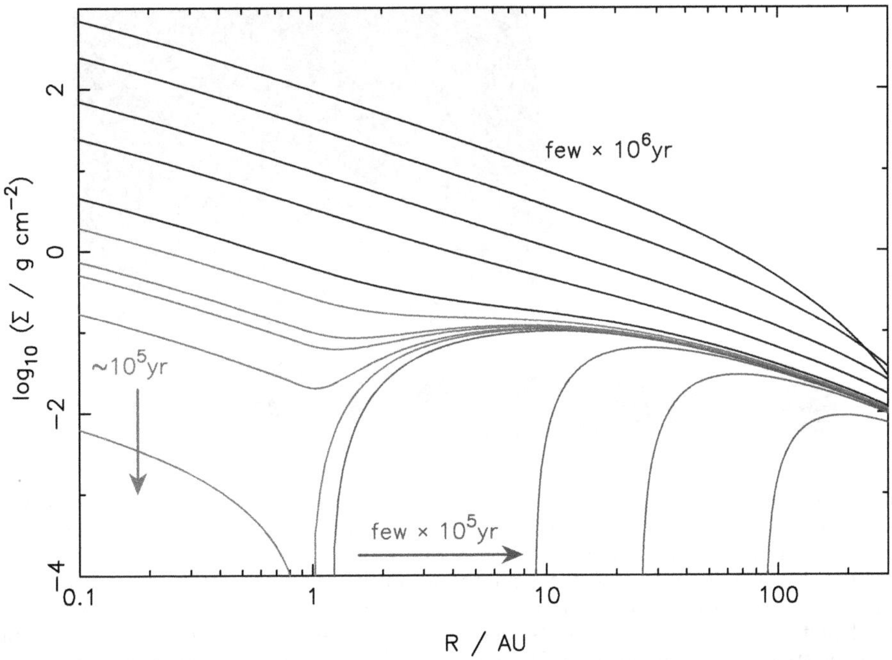

Figure 2. Surface density evolution of a characteristic viscous/photoevaporation disc model, colour-coded to denote the three stages of the disc's evolution. For most of the disc's lifetime (few Myr, blue curves) photoevaporation is negligible and the disc follows the behaviour of a classical viscous accretion disc. However, once the disc accretion rate drops sufficiently photoevaporation is able to open a gap in the disc at $\simeq R_c$, after which the inner disc rapidly accretes on to the star ($\sim 10^5$yr, green curves). Once the inner disc becomes optically thin, photoevaporation clears the disc from inside out (few $\times 10^5$yr, red curves). [Figure adapted from Alexander (2008a), using the median disc model of Alexander & Armitage (2009).]

and the [Ne II] 12.81μm line is arguably the most promising such diagnostic (Alexander 2008b). In this case the line originates close to the base of the wind, and models predict small blue-shifts ($\Delta v \sim 5$–10 km s^{-1}) for both EUV & X-ray heated winds (Alexander 2008b; Ercolano & Owen 2010). This blue-shift was first detected in the TW Hya disc (Pascucci & Sterzik 2009, see Fig. 1), and has subsequently been seen in around a dozen other sources (Pascucci *et al.* 2011; Sacco *et al.* 2012; Baldovin-Saavedra *et al.* 2012). The observed line profiles do not yet distinguish between the different photoevaporation models (see Fig. 1), but these observations represent an unambiguous detection of a slow wind which is at least partially ionized. When combined with line profiles from other species (such as [O I], e.g., Rigliaco *et al.* 2013), or other wind diagnostics such as free-free emission (Pascucci *et al.* 2012), these observations offer the prospect of measuring photoevaporation rates empirically.

2.2. *Secular evolution*

These photoevaporative mass-loss rates, inferred from both theory and observations, are comparable to the observed accretion rates for many T Tauri stars, and are clearly large enough to play a role in the later stages of disc evolution. However, in order to understand how protoplanetary discs evolve we must consider accretion and photoevaporation simultaneously. This was first done by Clarke *et al.* (2001), who coupled EUV photoevaporation to a 1-D viscous accretion model. Models of this type have subsequently been refined and improved by a number of different groups (e.g., Alexander *et al.* 2006b; Gorti

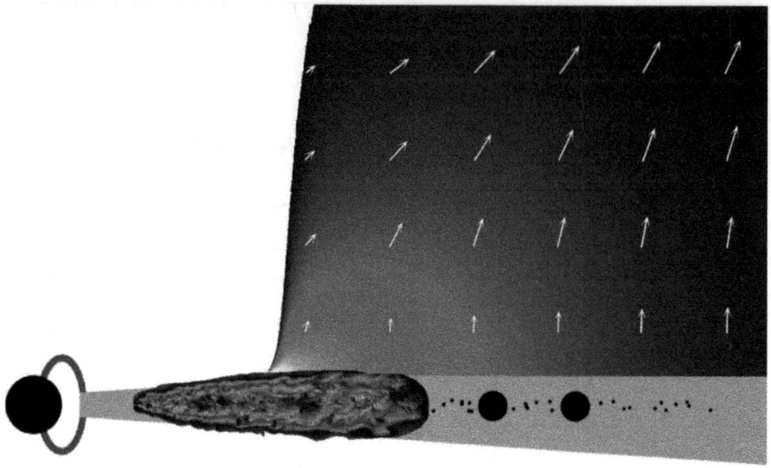

Figure 3. Schematic picture of an evolving protoplanetary disc. MHD turbulence drives accretion, while planetesimals and the cores of gas-giant planets form from the agglomeration of dust particles at the disc midplane. UV and X-ray photons from the star irradiate the upper layers of the disc, driving mass loss via a photoevaporative wind. Accretion dominates for most of the disc's lifetime, before photoevaporation takes over at late times and clears the disc. This simplified picture neglects many details (such as dead zones and MHD-driven winds), but is broadly consistent with observations of T Tauri stars. [Turbulent disc simulation from Beckwith *et al.* (2011); photoevaporative wind simulation from Alexander (2008b).]

et al. 2009; Owen *et al.* 2010; Bae *et al.* 2013), and although the details depend on the choice of model parameters and input physics, all photoevaporation/viscous evolution models display the same characteristic behaviour. Initially the accretion rate is much higher than the photoevaporation rate and the disc evolves viscously, accreting mass inwards and spreading at large radii. However, as the disc mass is finite the disc accretion rate declines with time, and after a few Myr the accretion rate becomes comparable to the rate of photoevaporative mass-loss. At this point the wind opens a gap in the disc at $\simeq R_c$, and once photoevaporation overcomes accretion the inner disc (inside R_c) is cut-off from re-supply and drains on to the star on its (short) viscous time-scale ($\sim 10^5$yr). Once the inner disc has been drained photoevaporation rapidly clears the outer disc, and the entire disc is dispersed on a time-scale of a few $\times\, 10^5$yr. This three-stage evolution scenario (illustrated in Fig. 2) is consistent with disc observations across a wide range in wavelength (e.g., Alexander *et al.* 2006b), and successfully reproduces the two-time-scale behaviour demanded by observations. We note also that this type of inside-out clearing inevitably gives rise to a short "inner hole" phase, and represents a possible mechanism for producing some of the observed transitional discs (e.g., Cieza *et al.* 2008).

Although different photoevaporation models result in qualitatively similar behaviour, the different mass-loss rates and profiles do give rise to significant quantitative differences in how discs evolve. In particular, the large wind rates predicted for X-ray- and FUV-dominated photoevaporation imply that mass-loss is significant even at early times, and that the total mass "lost" to photoevaporation over the disc's lifetime may be larger than that accreted on to the star. These models also result in distinctly different properties during the clearing phase (most notably higher disc masses and accretion rates than in the EUV case), and therefore make very different predictions for the properties of transition discs. However, the dominant uncertainties in this picture of disc evolution remain our lack of understanding of angular momentum transport, and also of protoplanetary disc initial conditions (in particular the initial angular momentum distribution).

Despite these uncertainties, this evolutionary scenario is now supported by mature theoretical models and a wealth of observational evidence. For much of the protoplanetary disc lifetime the evolution of the gas disc is essentially governed by the competition between accretion and photoevaporation. During this time solid material in the disc is agglomerating into progressively larger bodies, and planets are forming and migrating through the disc; this scenario is illustrated schematically in Fig. 3. The processes driving disc evolution and dispersal essentially compete with those of planet formation, depleting the disc and ultimately terminating the epoch of (giant) planet formation.

3. Implications for planetary systems

Having reviewed the physics of protoplanetary disc evolution, I now consider the implications of these results for the formation and evolution of planetary systems. Disc evolution and dispersal influences planet formation in a number of different ways, by altering the disc's chemistry, changing the dust-to-gas ratio, and ultimately by starving forming planets of the gas reservoir from which they accrete (e.g., Shu *et al.* 1993; Throop & Bally 2005; Guillot & Hueso 2006). Disc evolution also influences the architectures of young planetary systems: the migration of low-mass planets is very sensitive to the local disc structure, and disc clearing halts gas-driven planet migration. These issues are discussed in detail in Alexander *et al.* (2014); for reasons of length, I focus here on the effects of disc clearing on giant planet migration.

3.1. *Halting planet migration*

It was recognised more than 30 years ago that planets in discs migrate (Goldreich & Tremaine 1979, 1980), and when large numbers of exoplanets were discovered in relatively short-period orbits ($P \ll 1$yr), migration offered a natural explanation. However, the predicted migration time-scales are invariably much shorter than observed disc lifetimes (see, e.g., Kley & Nelson 2012), so planet migration must be slowed or stopped in order to produce planets at their observed locations. Type I migration, which applies to low-mass planets ($\lesssim 0.1\mathrm{M_{Jup}}$), can be halted or even reversed by changes in the local disc structure (Paardekooper & Papaloizou 2009), but Type II migration, which applies to giant planets ($\gtrsim 0.5\mathrm{M_{Jup}}$) is driven by the viscous accretion flow and continues as long as the disc accretes. Disc dispersal offers a obvious means of stopping Type II migration, but the manner and time-scale of disc clearing can have a strong influence on the dynamics of migrating planets.

Armitage *et al.* (2002) showed that the distribution of then-known exoplanets was consistent with a simple model, where giant planets form outside the snow-line (at 5–10AU) and undergo Type II migration before being stranded when the disc is dispersed. Subsequent population synthesis calculations have considered the processes of planet formation in much more detail (e.g., Ida & Lin 2004; Mordasini *et al.* 2009), but have generally adopted highly simplified models of disc dispersal (such as exponential depletion at all radii). Alexander & Armitage (2009) considered the migration of giant planets in a viscous/photoevaporation disc model, and ran large numbers of models to generate statistical samples. They tested their model against observations of both discs and exoplanets, and showed that this simple scenario successfully reproduces the observed properties of protoplanetary discs and the distribution of giant planets at ~AU radii. However, as disc dispersal by photoevaporation is not scale-free, it has the potential to leave a signature on the resulting planet distribution. Alexander & Pascucci (2012) revisited this issue, and found that photoevaporative gap-opening alters planet migration in a manner that is rather sensitive to both the planet mass and the efficiency of accretion

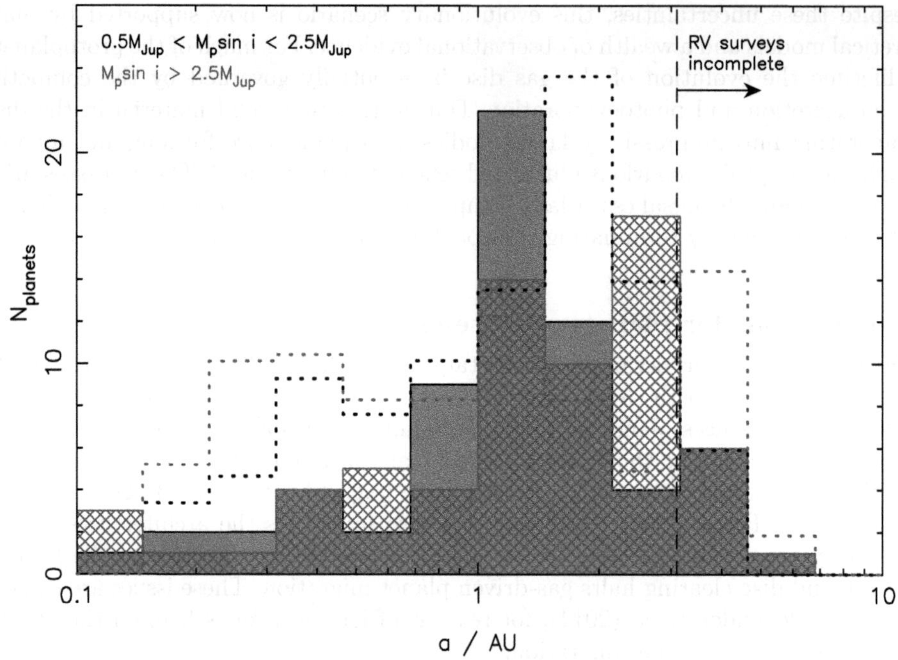

Figure 4. Exoplanet semi-major axes: the shaded histograms show current exoplanet data (taken from `exoplanets.org`; Wright *et al.* 2011); the dashed histograms show model predictions from Alexander & Pascucci (2012). The "pile-up" of ~Jupiter-mass planets is plausibly explained by the interaction between migrating planets and gap-opening due to photoevaporation. [Figure adapted from Alexander & Pascucci (2012).]

across the planet's orbit. The migration rate of planets close to R_c changes when the gap opens, and this results in mass-dependent "deserts" and "pile-ups" in the distribution of planet semi-major axes (see Fig. 4). Alexander & Pascucci (2012) tentatively associated the observed pile-up of ~Jupiter-mass planets at ~1AU with this effect, and suggested that we may be able to link features in the observed exoplanet distribution to the physics of protoplanetary disc evolution. Other mechanisms, such as planet traps (e.g., Hasegawa & Pudritz 2012), can also lead to similar planet pile-ups, and it remains to be seen whether such features in the initial planet distribution are preserved over Gyr time-scales. However, these results suggest that it may be possible to use the observed properties of exoplanets to inform our understanding of both disc evolution and planetary accretion.

4. Summary

In summary, we now understand that planets form in evolving protoplanetary discs, and that disc evolution plays a major role in both the formation of planets and the subsequent dynamical evolution of planetary systems. The main processes driving disc evolution on Myr time-scales are accretion, driven by MHD turbuence, and photoevaporation by high-energy radiation from the central star (though magnetically-driven winds may also play a role). Both accretion and photoevaporation are now observed directly, and there is good agreement between these observations and theoretical predictions. However, while our picture of protoplanetary disc evolution and dispersal is now relatively mature, its implications for planet formation have yet to be explored in detail. Disc evolution alters the conditions under which planets form, while disc clearing halts planet migration and may leave footprints in the observed distribution of exoplanets. Future observations and models of both discs and planets will explore these issues further, and allow us to build up a comprehensive picture of the early evolution of planetary systems.

Acknowledgements. I am grateful to many colleagues and collaborators, especially Ilaria Pascucci, Sean Andrews, Phil Armitage & Lucas Cieza, for a number of useful and interesting discussions. I also thank Ilaria Pascucci and James Owen for providing some of the data used in Fig. 1, and Phil Armitage for the turbulent disc image used in Fig. 3. My research is supported by the Science & Technology Facilities Council (STFC), through an Advanced Fellowship (ST/G00711X/1) and a Consolidated Grant (ST/K001000/1).

References

Adams, F. C., Hollenbach, D., Laughlin, G., & Gorti, U. 2004, *ApJ*, 611, 360

Alexander, R. 2008a, *New Astron. Revs.*, 52, 60

Alexander, R. D. 2008b, *MNRAS*, 391, L64

Alexander, R. D. & Armitage, P. J. 2009, *ApJ*, 704, 989

Alexander, R. D., Clarke, C. J., & Pringle, J. E. 2006a, *MNRAS*, 369, 216

Alexander, R. D., Clarke, C. J., & Pringle, J. E. 2006b, *MNRAS*, 369, 229

Alexander, R. D. & Pascucci, I. 2012, *MNRAS*, 422, L82

Alexander, R., Pascucci, I., Andrews, S., Armitage, P., & Cieza, L. 2014, Protostars & Planets VI, in press, arXiv:1311.1819

Andrews, S. M. & Williams, J. P. 2005, *ApJ*, 631, 1134

Andrews, S. M., Wilner, D. J., Hughes, A. M., Qi, C., & Dullemond, C. P. 2009, *ApJ*, 700, 1502

Armitage, P. J. 2011, *ARAA*, 49, 195

Armitage, P. J., Livio, M., Lubow, S. H., & Pringle, J. E. 2002, *MNRAS*, 334, 248

Bae, J., Hartmann, L., Zhu, Z., & Gammie, C. 2013, *ApJ*, 774, 57

Bai, X.-N. 2011, *ApJ*, 739, 50

Bai, X.-N. & Stone, J. M. 2013, *ApJ*, 769, 76

Balbus, S. A. 2011, in Physical Processes in Circumstellar Disks around Young Stars, ed. P. J. V. Garcia, 237–282

Balbus, S. A. & Hawley, J. F. 1991, *ApJ*, 376, 214

Baldovin-Saavedra, C., Audard, M., Carmona, A., *et al.* 2012, *A&A*, 543, A30

Beckwith, K., Armitage, P. J., & Simon, J. B. 2011, *MNRAS*, 416, 361

Bell, C. P. M., Naylor, T., Mayne, N. J., Jeffries, R. D., & Littlefair, S. P. 2013, *MNRAS*, 434, 806

Calvet, N. & Gullbring, E. 1998, *ApJ*, 509, 802

Cieza, L. A., Swift, J. J., Mathews, G. S., & Williams, J. P. 2008, *ApJL*, 686, L115

Clarke, C. 2011, in Physical Processes in Circumstellar Disks around Young Stars, ed. P. J. V. Garcia, 355–418

Clarke, C. J., Gendrin, A., & Sotomayor, M. 2001, *MNRAS*, 328, 485

Ercolano, B. & Owen, J. E. 2010, *MNRAS*, 406, 1553

Fedele, D., van den Ancker, M. E., Henning, T., Jayawardhana, R., & Oliveira, J. M. 2010, *A&A*, 510, A72

Feigelson, E., Townsley, L., Güdel, M., & Stassun, K. 2007, Protostars and Planets V, 313

Font, A. S., McCarthy, I. G., Johnstone, D., & Ballantyne, D. R. 2004, *ApJ*, 607, 890

Fromang, S., Latter, H., Lesur, G., & Ogilvie, G. I. 2013, *A&A*, 552, A71

Gammie, C. F. 1996, *ApJ*, 457, 355

Goldreich, P. & Tremaine, S. 1979, *ApJ*, 233, 857

Goldreich, P. & Tremaine, S. 1980, *ApJ*, 241, 425

Gorti, U., Dullemond, C. P., & Hollenbach, D. 2009, *ApJ*, 705, 1237

Gorti, U. & Hollenbach, D. 2008, *ApJ*, 683, 287
Gorti, U. & Hollenbach, D. 2009, *ApJ*, 690, 1539
Guillot, T. & Hueso, R. 2006, *MNRAS*, 367, L47
Gullbring, E., Hartmann, L., Briceno, C., & Calvet, N. 1998, *ApJ*, 492, 323
Haisch, Jr., K. E., Lada, E. A., & Lada, C. J. 2001, *ApJL*, 553, L153
Hartmann, L., Calvet, N., Gullbring, E., & D'Alessio, P. 1998, *ApJ*, 495, 385
Hasegawa, Y. & Pudritz, R. E. 2012, *ApJ*, 760, 117
Hillenbrand, L. A. 2009, in IAU Symposium, Vol. 258, ed. E. E. Mamajek, D. R. Soderblom, & R. F. G. Wyse, 81–94
Hollenbach, D., Johnstone, D., Lizano, S., & Shu, F. 1994, *ApJ*, 428, 654
Hollenbach, D. & Gorti, U. 2009, *ApJ*, 703, 1203
Ida, S. & Lin, D. N. C. 2004, *ApJ*, 604, 388
Ingleby, L., Calvet, N., Hernández, J., *et al.* 2011, *AJ*, 141, 127
Johnstone, D., Hollenbach, D., & Bally, J. 1998, *ApJ*, 499, 758
Jones, M. G., Pringle, J. E., & Alexander, R. D. 2012, *MNRAS*, 419, 925
Kley, W. & Nelson, R. P. 2012, *ARAA*, 50, 211
Koepferl, C. M., Ercolano, B., Dale, J., *et al.* 2013, *MNRAS*, 428, 3327
Königl, A. & Salmeron, R. 2011, in Physical Processes in Circumstellar Disks around Young Stars, ed. P. J. V. Garcia, 283–352
Liffman, K. 2003, *PASA*, 20, 337
Lodato, G. 2007, Nuovo Cimento Rivista Serie, 30, 293
Lynden-Bell, D. & Pringle, J. E. 1974, *MNRAS*, 168, 603
Mamajek, E. E. 2009, in American Institute of Physics Conference Series, ed. T. Usuda, M. Tamura, & M. Ishii, Vol. 1158, 3–10
Mordasini, C., Alibert, Y., & Benz, W. 2009, *A&A*, 501, 1139
Owen, J. E., Clarke, C. J., & Ercolano, B. 2012, *MNRAS*, 422, 1880
Owen, J. E., Ercolano, B., & Clarke, C. J. 2011, *MNRAS*, 412, 13
Owen, J. E., Ercolano, B., Clarke, C. J., & Alexander, R. D. 2010, *MNRAS*, 401, 1415
Paardekooper, S.-J. & Papaloizou, J. C. B. 2009, *MNRAS*, 394, 2283
Pascucci, I., Gorti, U., & Hollenbach, D. 2012, *ApJL*, 751, L42
Pascucci, I. & Sterzik, M. 2009, *ApJ*, 702, 724
Pascucci, I., Gorti, U., Hollenbach, D., *et al.* 2006, *ApJ*, 651, 1177
Pascucci, I., Sterzik, M., Alexander, R. D., *et al.* 2011, *ApJ*, 736, 13
Rigliaco, E., Pascucci, I., Gorti, U., Edwards, S., & Hollenbach, D. 2013, *ApJ*, 772, 60
Sacco, G. G., Flaccomio, E., Pascucci, I., *et al.* 2012, *ApJ*, 747, 142
Sargent, A. I. & Beckwith, S. 1987, *ApJ*, 323, 294
Shu, F. H., Johnstone, D., & Hollenbach, D. 1993, *Icarus*, 106, 92
Simon, M. & Prato, L. 1995, *ApJ*, 450, 824
Throop, H. B. & Bally, J. 2005, *ApJL*, 623, L149
Wright, J. T., Fakhouri, O., Marcy, G. W., *et al.* 2011, *PASP*, 123, 412

Discussion

CURRIE: I wondered if you could comment in more detail on how you expect disc dispersal driven by photoevaporation to change with stellar mass? The radiation field of a G-star is very different from that of an M-star, and we see some differences in the frequency of transitional discs at different stellar masses.

ALEXANDER: I'm largely agnostic about how the models scale with stellar mass, as there are a number of input parameters (radiation fields, initial disc properties, angular momentum transport, etc.) which are not well constrained. In essence, the models have sufficiently many free parameters that predictions about how they scale with stellar mass are speculative at best.

GRAHAM: The region [in Fig. 4] marked "RV surveys incomplete" is where next-generation direct imaging surveys will operate; what would your expectation be in this region?

ALEXANDER: The region from 5–20AU is particularly interesting, as planets at those semi-major axes may not have migrated far from their formation location. Model predictions in this region are are very sensitive to the assumptions we make about when and when planets form, and my hope is that these new direct imaging surveys will offer a way to break some of these degeneracies.

PUDRITZ: Large-scale MHD winds from discs have been predicted for a long time. How would you discriminate observationally between these and photoevaporative winds?

ALEXANDER: MHD and photoevaporative winds have very different temperature and ionization structures, so by observing the wind in a variety of different diagnostic lines we should be able to distinguish between the two.

???: Following on from that, how would you expect an MHD wind close to the star to influence photoevaporation?

ALEXANDER: A jet or wind very close to the star shields the disc from stellar irradiation, which in turn dramatically reduces the photoevaporation rate. However, we expect the optical depth of these winds to decline as the disc evolves (as jets typically scale with the disc accretion rate), and allow photoevaporation to proceed at late stages in the disc's evolution.

Exploring the Formation and Evolution of Planetary Systems
Proceedings IAU Symposium No. 299, 2013
M. Booth, B. C. Matthews & J. R. Graham, eds.

doi:10.1017/S1743921313008302

Disk Inhomogeneities and the Origins of Planetary System Architectures and Observational Properties

Yasuhiro Hasegawa[1] **and Ralph E. Pudritz**[2,3]

[1]EACOA fellow, Institute of Astronomy and Astrophysics, Academia Sinica (ASIAA), Taipei 10641, Taiwan
email: yasu@asiaa.sinica.edu.tw

[2]Department of Physics and Astronomy, McMaster University, Hamilton, ON L8S 4M1, Canada

[3]Origins Institute, McMaster University, Hamilton, ON L8S 4M1, Canada
email: pudritz@physics.mcmaster.ca

Abstract. Recent high-resolution observations show that protoplanetary disks have various kinds of structural properties or inhomogeneities. These are the consequence of a mixture of a number of physical and chemical processes taking place in the disks. Here, we discuss the results of our comprehensive investigations on how disk inhomogeneities affect planetary migration. We demonstrate that disk inhomogeneities give rise to planet traps - specific sites in protoplanetary disks at which rapid type I migration is halted. We show that up to three types of traps (heat transitions, ice lines and dead zones) can exist in a single disk, and that they move differently as the disk accretion rate decreases with time. We also demonstrate that the position of planet traps strongly depends on stellar masses and disk accretion rates. This indicates that host stars establish preferred (initial) scales of their planetary systems. Finally, we discuss the possible observational signatures of disk inhomogeneities.

Keywords. accretion, accretion disks, radiative transfer, turbulence, (stars:) planetary systems: formation, (stars:) planetary systems: protoplanetary disks.

---

1. Introduction

The advent of "next generation" telescopes has revealed that protoplanetary disks are rich in their structures (Tamura 2009). For instance, SMA observations have recently inferred that there is a jump in the column density of ^{13}CO at CO-ice lines (Qi *et al* 2011). The evidence suggests that the jump occurs due to the condensation of CO onto dust grains at the CO-ice line, where the disk temperature declines to about 160 K.

One of the most serious problems in theories of planet formation is planetary migration (Kley & Nelson 2012). It arises from tidal, resonant interactions between gaseous disks and protoplants that form in the disks. The most advanced studies show that the timescale of type I migration, that is applicable for low-mass planets such as terrestrial planets and cores of gas giants, is very rapid ($\sim 10^5$ years for planets with $\sim 1 M_{\oplus}$ at $\sim$ 1 AU), and that its direction is highly coupled with disk properties such as the surface density and the temperature of disks. Here, we report on our investigations as to how disk structures or "disk inhomogeneities" slow planetary migration and explore their consequences for planetary system architectures. We also discuss how important the inhomogeneities are by bridging between the observations of exoplanets and protoplanetary disks.

2. Disk Inhomogeneities and the Resultant Planet Traps

We first discuss dead zones, which are predicted to be present in the inner region of disks (Gammie 1996). In general, protoplanetary disks are ionized by X-rays from the central stars and cosmic rays. In the inner region of disks, however, the ionization is not so efficient due to the high column density there. Magnetorotational instabilities (MRIs) will be largely suppressed there, because MRIs require a good coupling between charged fluids in the gas disks and magnetic fields threading the disks. As a result, protoplanetary disks are characterized mainly by two zones: the inner, high density region is the so-called MRI "dead" zone and is weakly turbulent whereas the outer, low density region is a MRI active zone and is highly turbulent.

We examine the effects of dead zones on the thermal structure of disks and investigate how planetary migration is affected by the presence of dead zones. Figure 1 (panels (a) and (b)) shows the results of the density and thermal structures of disks with dead zones (Hasegawa & Pudritz 2010a). The dust distribution (panel (a)) shows that a dusty "wall" is present at the outer boundary of a dead zone that is 6 AU in size. This arises because dust settling in the dead zone is significantly enhanced due to the low value of turbulence whereas dust in the active zone is still mixed well with the gas. Adopting the dust distribution, we computed the disk temperature. In order to make our calculations reliable, we performed Monte Carlo based, radiative transfer simulations, assuming the main heat source is stellar irradiation (also see Hasegawa & Pudritz 2010b). Once we obtain the

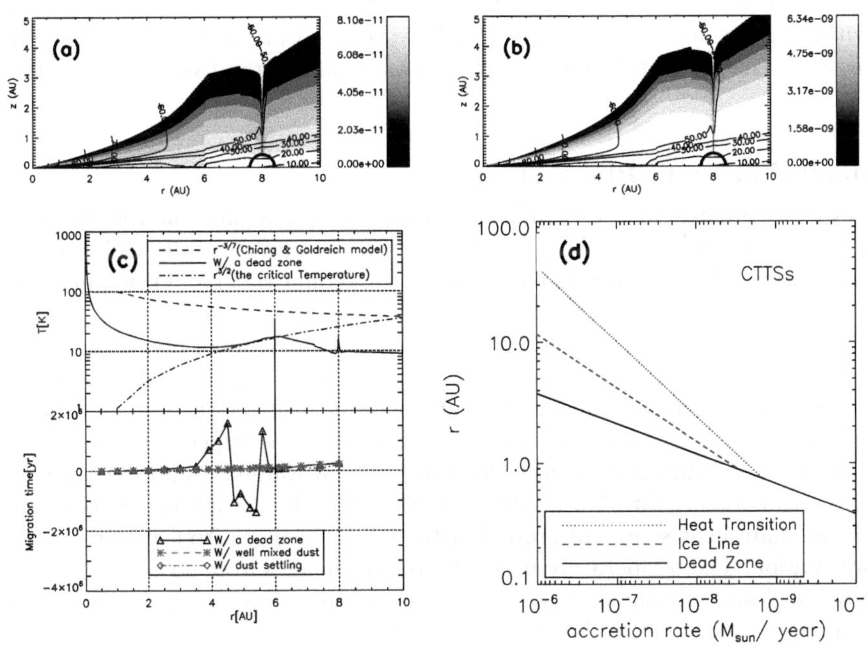

Figure 1. Radiatively heated protoplanetary disks with dead zones (in (a) and (b); adapted from Hasegawa & Pudritz 2010a). The dust and gas density distribution, that is denoted by black-and-white in g cm^{-3}, is shown with the temperature contours in Kelvin, on panel a) and b) respectively. Planetary migration in radiatively heated disks with dead zones which is 6 AU in size (in (c); adapted from Hasegawa & Pudritz 2010a). In c) (top), the temperature profile of the mid-plane region as a function of the distance from the central star. In c) (bottom), the timescale of type I migration as a function of position of a planet from the central star. Position of planet traps (in (d); adapted from Hasegawa & Pudritz 2011). The position of planet traps as a function of disk accretion rate for classical T Tauri stars (CTTSs).

temperature structure of disks, we re-calculate the gas distribution using the computed disk temperature, assuming the vertical hydrostatic equilibrium. It is interesting that wall-like structures do not appear in the gas distribution (see Panel (b)).

We investigate how the dusty wall affects the thermal structure of disks. In order to proceed, we plotted the temperature of the midplane as a function of the distance from the central star. Figure 1 (panel (c); top) shows that the disk temperature has a *positive* gradient in the dead zone. This is the major finding in Hasegawa & Pudritz (2010b) and can be understood as what follows: in general, the central stars irradiate the surface layer of disks directly, so that the midplane region of disks is *indirectly* heated by such surface layers. When a disk has a dead zone, a dusty wall is left at the outer boundary of the dead zone. Then, the wall is heated efficiently by stellar irradiation or the surface layers. As a result, the wall becomes thermally hot and leads to the back heating of the dead zone at smaller disk radii, which produces a positive temperature gradient in the dead zone.

We then use these results to examine how this thermal structure affects planetary migration. More specifically, we analytically calculate the timescale of type I migration, adopting the density and thermal structure of the disk with the dead zone. Figure 1 (panel (c); bottom) shows that the timescale of type I migration becomes *negative* in the dead zone where the positive temperature gradient is established - planets migrate *outward.* This occurs because such a temperature profile makes the inner resonances closer to a planet and the outer ones further away from the planet. As a result, the balance of the torque, that drives planetary migration, reverses there. Thus, when planets migrate either within or beyond dead zones, they will be trapped at the location where the net torque is zero. The location is often referred to as a planet trap (Masset *et al.* 2006).

3. Unified Picture of Planet Traps

As shown in this example, disk inhomogeneities play a significant role in planetary migration through the creation of planet traps. Since a number of disk inhomogeneities are already discussed in the literature, we take into account them and generalize how the disk inhomogeneities act as planet traps.

We adopt an analytical approach for comprehensively examining disk inhomogeneities and the resultant planet traps (Hasegawa & Pudritz 2011). We utilized analytical modeling of disk inhomogeneities and investigated how the direction of type I migration switches from inward to outward there. We consider three types of inhomogeneities: dead zones, ice lines, and heat transitions. We focus on water-ice lines, because the ice lines are most important for planet formation. Heat transitions are disk inhomogeneities at which the main heat source of protoplanetary disks changes from viscous heating to stellar irradiation. The temperature profile therefore changes from steep to shallow ones.

We confirmed analytically that the net torque becomes zero at all the three types of disk inhomogeneities, so that they can act as planet traps for rapid type I migration. Also, we showed that the position of planet traps is sensitive to the surface density of disks, or equivalently the disk accretion rate. Figure 1 (panel (d)) shows how the position of planet traps evolves with the disk accretion rate. It is important that single disks have up to three types of planet traps and that different traps move inward at different rates.

Table 1 summarizes the position of planet traps for various types of stars. The position of planet traps vary largely, depending on the mass of the central stars. Thus, our results indicate that stellar masses and disk accretion rates play a significant role in establishing a preferred (initial) scale of planetary systems.

Table 1. The position of planet traps. Adapted from Hasegawa & Pudritz (2011)

	Herbig Ae/Be stars r_{edge} (AU)	r_{il} (AU)	r_{ht} (AU)
$\dot{M} = 10^{-4}(M_{\odot}/\text{year})$	3.9	155	830
$\dot{M} = 10^{-6}(M_{\odot}/\text{year})$	1.2	20	47
	CTTSs r_{edge} (AU)	r_{il} (AU)	r_{ht} (AU)
$\dot{M} = 10^{-6}(M_{\odot}/\text{year})$	3.7	11.7	42.4
$\dot{M} = 10^{-8}(M_{\odot}/\text{year})$	1.2	1.5	2.4
	M stars r_{edge} (AU)	r_{il} (AU)	r_{ht} (AU)
$\dot{M} = 10^{-8}(M_{\odot}/\text{year})$	4.2	N/A	N/A
$\dot{M} = 10^{-9}(M_{\odot}/\text{year})$	2.3	1.5	N/A

N/A denotes the case that planet traps excited by the disk inhomogeneities are not active due to convergence with a dead zone trap.

4. Implications of Planet Traps for Observations

W have discussed how useful disk inhomogeneities and the resultant planet traps are for resolving the problem of rapid type I migration and the origin of planetary system architectures. The next fundamental question is how such effects are tested against observations. In order to address such a question, we have recently computed evolutionary tracks of planets that grow at planet traps. We have found that, when planet traps are coupled with the core accretion scenario, the observations of exoplanets done by radial velocity techniques can be reproduced very well (Hasegawa & Pudritz 2012; and Pudritz & Hasegawa, these Proceedings). As discussed above, the presence of dead zones alters the disk structure significantly, so that some kind of features may be shown in the observables. One possibility is in the SEDs/images of disks with inhomogeneities. We are currently investigating this for further supporting the importance of disk inhomogeneities and the resultant planet traps.

References

Gammie, C. F. 1996, *ApJ*, 457, 355

Hasegawa, Y. & Pudritz, R. E. 2010a, *ApJ*, 710, L167

Hasegawa, Y. & Pudritz, R. E. 2010b, *MNRAS*, 401, 143

Hasegawa, Y. & Pudritz, R. E. 2011, *MNRAS*, 417, 1236

Hasegawa, Y. & Pudritz, R. E. 2012, *ApJ*, 760, 117

Kley, W. & Nelson, R. P. 2012, *ARAA*, 50, 211

Masset, F. S., *et al.* 2006, *ApJ*, 642, 478

Qi, C., *et al.* 2011, *ApJ*, 740, 84

Tamura, M. 2009, *Proceedings of American Institute of Physics Conference Series*, 1158, 11

Exploring the Formation and Evolution of Planetary Systems
Proceedings IAU Symposium No. 299, 2013
M. Booth, B. C. Matthews & J. R. Graham, eds.

doi:10.1017/S1743921313008314

The bright end of the exo-Zodi luminosity function: Disk evolution and implications for exo-Earth detectability

G. M. Kennedy and M. C. Wyatt

Institute of Astronomy, University of Cambridge, Madingley Road, Cambridge CB3 0HA, UK
email: gkennedy@ast.cam.ac.uk

Abstract. This contribution summarises the first characterisation of the $12\,\mu$m warm dust ("exo-Zodi") luminosity function around Sun-like stars, focussing on the dustiest systems that can be identified by the WISE mission (Kennedy & Wyatt 2013). We use the sample of main-sequence stars observed by Hipparcos within 150pc as an unbiased sample, and report the detection of six new warm dust candidates. The ages of five of these new sources are unknown, meaning that they may be sites of terrestrial planet formation or rare analogues of other old warm dust systems. We show that the dustiest old (> Gyr) systems such as BD+20 307 are 1 in 10,000 occurrences. Bright warm dust is much more common around young (<120 Myr) systems, with a ~1% occurrence rate. We show that a two component *in situ* model where all stars have initially massive warm disks and in which warm debris is also generated at some random time along the stars' main-sequence lifetime, perhaps due to a collision, can explain the observations. However, if all stars only have initially massive warm disks these would not be visible at Gyr ages, and random collisions on the main-sequence are too infrequent to explain the high disk occurrence rate for young stars. That is, neither component can explain the observations on their own. Despite these conclusions, we cannot rule out an alternative dynamical model in which comets are scattered in from outer regions because the distribution of systems with the appropriate dynamics is unknown. Our *in situ* model predicts that the fraction of stars with exo-Zodi bright enough to cause problems for future exo-Earth imaging attempts is at least roughly 10%, and is higher for populations of stars younger than a few Gyr. This prediction of roughly 10% also applies to old stars because bright systems like BD+20 307 imply a population of fainter systems that were once bright, but are now decaying through fainter levels. Our prediction should be strongly tested by the Large Binocular Telescope Interferometer, which will provide valuable constraints and input for more detailed evolution models. A detection fraction lower than our prediction could indicate that the hot dust in systems like BD+20 307 has a cometary origin due to the quirks of the planetary dynamics. Population models of comet delivery need to be developed to help distinguish between different possible origins of warm dust.

Keywords. circumstellar matter, planetary systems

1. Introduction

Perhaps the most important long term goal of humankind is to find and communicate with aliens, the inhabitants of extra-Solar Systems. This goal is a difficult one, with many challenges to overcome before it is technically possible to even image an Earth-like planet around another star. When it is eventually attempted, imaging detection of these planets could still be thwarted by the presence of exo-Zodiacal dust located in or near the terrestrial zone, which with sufficient dust levels can mask the presence of exo-Earths (e.g. Beichman *et al.* 2006; Absil *et al.* 2010; Roberge *et al.* 2012). Therefore, in preparation for future missions that will target specific stars, a critical piece of information is whether those stars harbour dust at a level that precludes the detection of an Earth-like planet.

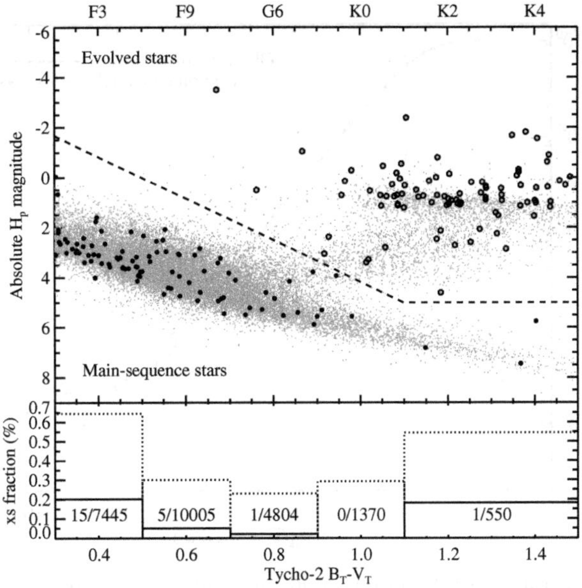

Figure 1. *Top:* HR diagram showing stars with $0.3 < B_T - V_T < 1.5$, split into Sun-like main-sequence stars and evolved stars by the dashed line. The top axis shows spectral types calculated using synthetic photometry of PHOENIX stellar atmosphere models (Brott & Hauschildt 2005). Stars with $W1 - W3$ greater than 0.1mag are shown as circles; open for evolved stars and filled for main-sequence stars. *Bottom:* Fraction of main-sequence stars with $W1 - W3 > 0.1$. The dotted bars show all stars with $W1 - W3 > 0.1$ (i.e. filled dots in the top panel), while the solid bars and numbers show those with plausible warm dust emission after individual checking.

This contribution briefly describes our study of the exo-Zodi luminosity function (Kennedy & Wyatt 2013), which had several goals. First, the frequency of exo-Zodi detected as infra-red excesses at 12μm as a function of brightness (the so-called "luminosity function") was derived using data from the WISE mission. Second, these detections, which were necessarily at the bright end of the luminosity function, were used in conjunction with a simple evolution model to make predictions for the frequency of much fainter, but much more common exo-Zodi. In the interests of brevity, here we have omitted many details, and refer the reader to Kennedy & Wyatt (2013) for a full description.

2. Sample and warm dust candidates

Stars are selected from the Hipparcos catalogue using the HR diagram shown in Figure 1. We use a fairly broad definition of Sun-like; $0.3 < B_T - V_T < 1.5$. We exclude evolved stars, which may show excesses due to circumstellar material associated with mass loss. Because they are selected solely using optical photometry and parallax, the 24,174 stars selected by these criteria are unbiased with respect to the presence of warm dust. Figure 1 shows all 27,333 stars within the $B_T - V_T$ range, the cut used to exclude evolved stars (dashed line), and those found to have 12μm excesses (large dots and circles).

Not all stars with apparent 12μm emission represent a true indication of warm dust, so all 96 excess systems were vetted by inspecting the WISE images and constructing spectral energy distributions to check for consistency among the photometry (notes on individual systems are given in the appendix of Kennedy & Wyatt 2013). Most were found to be spurious due to image contamination and/or poor flux measurement, leaving 22 warm dust candidate systems, all but seven of which were previously known.

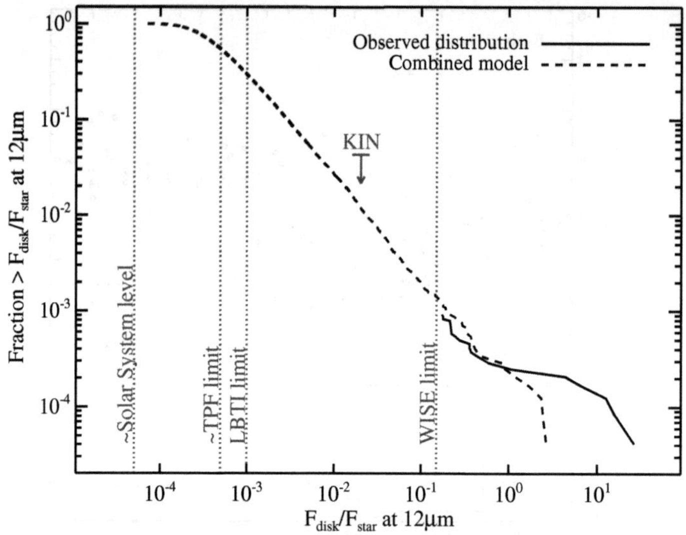

Figure 2. Observed luminosity function (solid line) and a simple prediction for faint exo-Zodi for stars between 0 and 13 Gyr with a single realisation of a Monte-Carlo population model of evolution. Grey dotted lines show the limits for LBTI and WISE sensitivity, as well as approximate levels for the Solar System Zodiacal cloud and for which TPF-like missions will be compromised by exo-zodiacal dust. The upper limit from the 23-star KIN survey is also shown.

3. Excess selection and warm dust candidates

If warm dust is sufficiently bright relative to the host star, it will manifest as a detectable mid-infrared (i.e. 10-20 μm) excess relative to the stellar photosphere, which for ~300 K dust peaks near 12 μm. The WISE bands are well suited for this task, and because it has about an order of magnitude better sensitivity to 300K dust compared to W4 (see Fig 2 of Kennedy & Wyatt 2012) we use the 12 μm $W3$ band.†

Our luminosity function is therefore formally the distribution of 12 μm disk-to-star flux density ratios, which we will use in cumulative form. We use the term "luminosity function" largely for conciseness. The disk luminosity of course not only depends on the 12 μm flux ratio, but also the disk temperature and stellar properties.

Figure 2 shows the 12 μm luminosity function derived using the WISE data and one realisation of a Monte-Carlo population model of evolution. The model under-predicts the number of brightest excesses, but varies considerably due to the small number of such disks. Below $F_{\rm disk}/F_\star \sim 0.1$ all realisations, and hence the predictions for fainter disks, are very similar.

The right side of the figure is covered by photometric surveys like the one presented here, whereas fainter disks (left side) can only be detected with more sophisticated methods, such as those employed by the KIN and LBTI. We have taken the approximate Solar System 12 μm dust level as $F_{\rm disk}/F_\star = 5 \times 10^{-5}$, and the LBTI sensitivity as 20 times this level (Hinz 2009). The predicted limit for exo-Earth detection by a Terrestrial Planet Finder (TPF)-like mission is also shown (at ten times the Solar System level).

The predicted disk fraction is about 50% at the LBTI sensitivity. An unbiased LBTI survey of just a few tens of stars would therefore result in a strong test of our prediction, and has the potential for detection of a dozen or so disks if the evolution predicted by

† The WISE bands are known as $W1$, $W2$, $W3$, and $W4$, with wavelengths of 3.4, 4.6, 12, and 22 μm respectively (Wright *et al.* 2010; Jarrett *et al.* 2011).

our model is correct. Such a result would be a great success for the instrument, but less promising for the study of exo-Earths; another prediction of our model is that roughly only a few tens of percent of stars are amenable to exo-Earth detection if our combined model is correct.

4. Summary and Conclusions

We have presented an analysis of the bright end of the 12 μm warm dust luminosity function as seen by WISE around nearby Sun-like stars. We report the possible detection of 6 new warm dust candidates, HD 19257, HD 23586, HD 94893, HD 154593, HD 165439, and HD 194931.

There is a clear preference for excess emission around young stars, with at least 14/22 systems younger than 120 Myr, despite the ~10 Gyr stellar lifetime. However, two of the dustiest systems are ~1-2 Gyr old, and another may be ~ Gyr old, suggesting that while youth is an important factor for setting the brightness of warm dust, it is not the sole factor.

Using a simple *in situ* evolution model to interpret the observed luminosity function, we show that neither a picture where all systems have initially massive disks, nor one where all excesses are due to randomly timed collisions, can explain the fact that warm dust is observed around both young and old stars. A simple combination of these two *in situ* scenarios reproduces the observed luminosity function. However, we cannot rule out that warm dust is in fact dominated by comet delivery. In this scenario comets are scattered from a more distant reservoir, and reliable luminosity function predictions cannot be made because they depend on the distribution of systems with the appropriate dynamics.

We have made simple predictions for the number of exo-Zodi that might be detected by an unbiased LBTI survey. For the expected sensitivity we anticipate a handful of detections and a strong test of our evolution model. Such tests are a crucial step towards understanding the origins and evolution of exo-Zodi, and ultimately the imaging of Earth-like planets.

References

Absil, O. *et al.* 2010, in Astronomical Society of the Pacific Conference Series, Vol. 430, Pathways Towards Habitable Planets, ed. V. Coudé Du Foresto, D. M. Gelino, & I. Ribas, 293

Beichman, C. A. *et al.* 2006, *ApJ*, 652, 1674

Brott, I. & Hauschildt, P. H. 2005, in ESA Special Publication, Vol. 576, The Three-Dimensional Universe with Gaia, ed. C. Turon, K. S. O'Flaherty, & M. A. C. Perryman, 565

Hinz, P. M. 2009, in American Institute of Physics Conference Series, Vol. 1158, American Institute of Physics Conference Series, ed. T. Usuda, M. Tamura, & M. Ishii, 313–317

Jarrett, T. H. *et al.* 2011, *ApJ*, 735, 112

Kennedy, G. M. & Wyatt, M. C. 2012, *MNRAS*, 426, 91

—. 2013, *MNRAS*

Roberge, A. *et al.* 2012, *PASP*, 124, 799

Wright, E. L. *et al.* 2010, *AJ*, 140, 1868

Discussion

S. Metchev: In our search for W3 and W4 excesses in WISE (see Patel & Metchev poster) we found that 75% of our W3 excesses have even stronger W4 excesses. That is, they do not seem to be exo-Zodi. Do you have estimates of temperatures for your detections?

KENNEDY: Yes, most of our stars have W4 detections so we can determine temperatures. These are given in Kennedy & Wyatt (2013) and are typically a few hundred Kelvin. We also have some discussion of temperatures in Section 4.2.

C. STARK: You began your talk by listing the different mechanisms that could produce exo-Zodiacal dust. It's not unreasonable to think that different mechanisms may dominate at different ages and/or dust levels. So is it valid to extrapolate measurements of dense disks associated with young stars to lower mass disks around older stars?

KENNEDY: This is a very good point. The model we used for the prediction is based on the assumption that disks decay collisionally, and there is no specification of what the mechanism is that starts the evolution. So in this sense the model is OK because the mechanism doesn't matter as long as the subsequent evolution is something like our model. However, as we discuss in our paper, comet delivery is an equally viable cause of exo-Zodi, and it could be that this works preferentially at young ages and is less important at late times. If this were the main cause of exo-Zodi then the extrapolation could be very different. Making any kind of prediction for this scenario is extremely dependent on the assumptions made, so we chose to only model the collisional scenario as it is easier to define.

S. ERTEL: In the context of the model random collisions are needed. Is it possible to constrain how frequent these collisions have to be in order to reproduce the detection rate you presented (e.g. once in each system per lifetime, more often)?

KENNEDY: In a crude way yes we can. There is a degeneracy between the number of collisions during a stars' lifetime and the decay rate after that collision. But if the decay rate is the same as needed for the disks that start decaying early on, the number of collions per stellar lifetime is just one. It's not a great constraint but is probably indicative if this scenario is correct.

Exploring the Formation and Evolution of Planetary Systems
Proceedings IAU Symposium No. 299, 2013
M. Booth, B. C. Matthews & J. R. Graham, eds.

doi:10.1017/S1743921313008326

Orbital Motion and Multi-Wavelength Monitoring of LkCa15 b

Michael J. Ireland[1,2] **and Adam L. Kraus**[3,4]

[1]Australian Astronomical Observatory, PO Box 296, Epping NSW 1710, Australia
email: mireland@aao.gov.au

[2]Department of Physics & Astronomy, Macquarie University, NSW 2109 Australia

[3]Department of Astronomy, University of Texas at Austin,
2515 Speedway, Stop C1400 Austin, Texas 78712-1205, USA
email: alk@astro.as.utexas.edu

[4]Harvard-Smithsonian Center for Astrophysics, 60 Garden St, Cambridge, MA 02138, USA

Abstract. As part of a deep multi-year non-redundant aperture mask infrared imaging campaign observing transition disks, we present multi-epoch monitoring of the resolved emission seen within the disk gap of LkCa 15. Orbital motion of both the central source and extended lobes as presented in Kraus and Ireland (2012) is clearly detected at the level of ~4 degrees/year (deprojected), in both K and L'-bands. Based on these data as well as single-epoch H and M bands epochs, we present two models for the central source - thermal emission as a planetary accretion signature and scattering. The thermal emission model is preferred.

Keywords. planets and satellites: formation, stars: individual: LkCa 15, instrumentation: high angular resolution, techniques: image processing

1. Introduction

Since the discovery of the first hot jupiter planet (Mayor and Queloz 1995) it has been clear that orbital evolution plays an important role in the configuration of mature planetary systems. Understanding planetary formation therefore requires not only a census of mature systems, but also direct detection of planets as they are forming. Luckily, the bulk of the gravitational potential energy of a forming giant planet is released early in its lifetime. This is especially true in the core-accretion model, where the release of energy in an accretion-shock can in principle radiate with a luminosity approaching $0.1\,L_\odot$ for $\sim 10^5$ years (Marley *et al.* 2007).

The greatest difficulty in directly detecting newly-formed exoplanets with current-generation telescopes is the relatively large distances to nearby star forming regions. In Taurus in the northern hemisphere or ρ Ophiucus in the southern hemisphere, a linear separation of 5 AU corresponds to only 0.035 arc sec, which is within the formal diffraction limit of the world's largest telescopes for all bandpasses longer than the H-band (1.55 μm).

Maximizing contrast at the diffraction-limit requires more than just an adaptive optics system. Fundamental noise sources such as photon-noise play only a small role in detecting faint companions. For example, a K = 20 object buried in the Airy ring of a K = 10 star, taken with a high-Strehl, 25% efficiency adaptive optics system on a 10m telescope would be detected with a photon-limited signal-to-noise of more than 5 in only 10 s. At this contrast level of 10^4, phase noise is the dominant error source – in particular, phase noise that manifests itself as quasi-static speckles. Techniques such as LOCI (Lafrenière *et al.* 2007), KL image projection (Soummer *et al.* 2012) and POISE (Ireland 2013) help to distinguish the signal of a companion from a phase-noise signal, but it

Table 1. Summary of LkCa 15 Observations with Keck.

Date	Filter	Observation Time	Detection?
2008 Dec	K'	0.5 hours	N
2009 Nov	L'	4 hours	Y
2010 Aug	L'	2 hours	Y
2010 Nov	K'	3.5 hours	Y
2010 Nov	L'	1.5 hours	Y
2012 Jan	H[1]	6 hours	N
2012 Jan	Ms	2 hours	Y
2012 Aug	L'	1.5 hours	N
2012 Dec	K'	3 hours	Y
2012 Dec	L'	1 hour	Y

Notes:
[1] Actually the CH4-short filter of NIRC2, comprising the short-wavelength half of the H-band.

also helps to define an observable that is least-affected by phase-noise. A coronagraph removes the diffracted electric field in the image-plane, making quasi-static speckles 2nd-order (rather than 1st-order) in pupil-plane phase errors. A technique even less sensitive to pupil-plane phase errors is *kernel-phase*, where detection thresholds are 3rd order in pupil-plane phase (Martinache 2012, Ireland 2012, Ireland 2013). In order for kernel-phase to be effective at moderate Strehl ratios, the pupil geometry also must be modified with an aperture-mask (Lloyd *et al.* 2006). It is the observable of kernel-phase, with or without an aperture-mask that we have used to maximise our sensitivity to exoplanets at the diffraction-limit of the world's largest telescopes (Keck and VLT).

2. Observations of Transition Disks

Detecting young exoplanets at solar-system scales requires not only the best technique but also a carefully constructed target list. We have been systematically targeting the so-called "transition" disks, which have a cleared inner-hole. For more massive disks where photo-evaporation can not yet play a significant role in disk shaping, the key clearing possible mechanisms are significant grain growth and dynamical interactions with forming protoplanets (Dodson-Robinson and Salyk 2011) or companions (Ireland and Kraus 2008). Significant grain growth is itself a signature that a disk is at a planet forming stage, so for disks that are not cleared by a stellar companion, a transition disk spectrum can be seen as a signpost for planetary formation.

We have carried out the deepest observations for transition disks with the L' filter at the Keck telescope, achieving typical contrast limits of 6 magnitudes. Our target list included UX Tau, GM Aur, LkCa 15, LkHa 330 and SR 21. Only one target showed an unambiguous sign of substructure within its disk gap, LkCa 15. This detection was published in Kraus and Ireland 2012.

3. New Observations of LkCa 15

The full set of Keck, NIRC2 observations of LkCa15 are given in Table 1. All 2012 observations are unpublished, with preliminary results first presented at this conference. The 2012 K-band epoch is shown in Figure 1, with a similar observing technique and image reconstruction technique to Kraus and Ireland (2012). The white circles in the image show the positions of the NE lobe, central source and SW lobe as presented in Kraus and Ireland (2012). The central source has approximately the same brightness between epochs, but the lobes are clearly brighter in 2012. L-band images (not shown) are consistent with similar levels of orbital motion and variability. When all data are taken

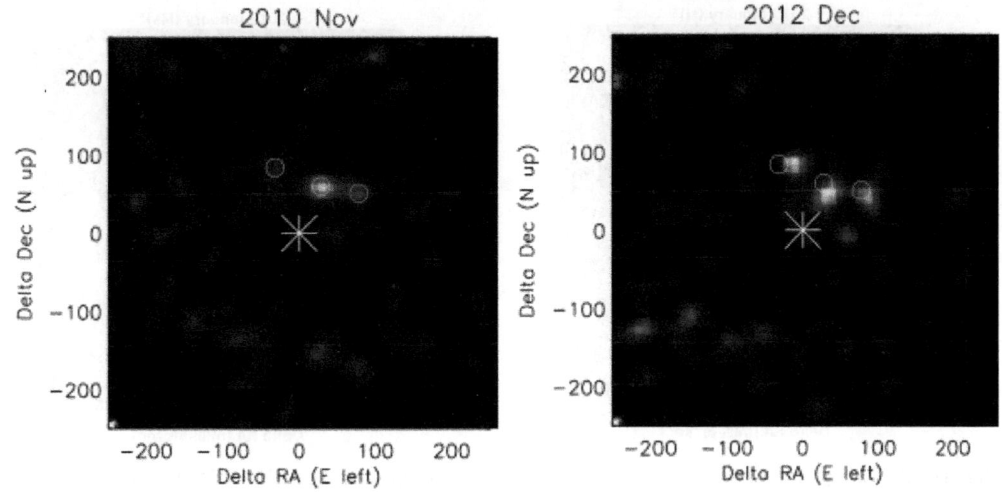

Figure 1. Image reconstructions from 2010 and 2012 data sets in K-band, showing clockwise orbital motion. Each image was reconstructed from closure-phase data only using the MACIM algorithm Ireland *et al.* 2006. Variability is also seen in the resolved structure to the NW of the central star (modelled as a point source).

together, we see orbital motion of 6.0 ± 1.5 degrees per year, which is approximately 4 degrees per year deprojected for a circular orbit once the system inclination of 49 degrees is taken into account. This is consistent with Keplarian rotation for at 20 AU orbit.

During 2012 Jan, we attempted to detect emission from LkCa 15b in H-band, using the 9-hole aperture mask of NIRC2 and the CH4-short filter. Including overheads and calibrations, 6 hours was spent on target, with no detections down to a contrast of 7 magnitudes. Figure 2 shows the reconstructed image from these observations: the noise peak in the NW corner would correspond to a companion with contrast of 7.2 magnitudes.

In the Ms filter, aperture-masking data has poor signal-to-noise because of both the light loss at the aperture-mask, and the sensitivity to thermal background due to the aperture-mask technique spreading information over a sky area ~50 times larger than data with an unobstructed pupil. However, Strehl ratios are much higher in the Ms filter than K' and L', meaning that it is possible to use the Kernel-phase technique on data taken with an unobstructed pupil under most seeing conditions. The reconstructed image (Figure 2) has resolved structure at the same location as the K' and the L' filters, and a integrated contrast of 3.5 magnitudes.

4. Discussion

When taken together with the resolved velocity CO observations of Simon *et al.* (2000), the clockwise orbital motion means that the resolved emission is geometrically behind LkCa 15. In turn, this means that forward scattering can not be evoked as an emission mechanism to give an apparent albedo greater than unity. Given the L' contrast of the resolved structure of ~4.5 magnitudes and the Ms contrast of the structure of 3.5 magnitudes, this in turn means that it is extremely difficult to explain the amount of emission from the resolved structure as scattering. The very red colors (K'-Ms color more than ~1.5 magnitudes redder than the star) are also inconsistent with typical scattering processes, which typically have a blue or neutral colour.

This leaves thermal emission as the obvious explanation for the resolved structure. The K'-Ms color implies a temperature of 900K for small, optically-thin grains, or an even

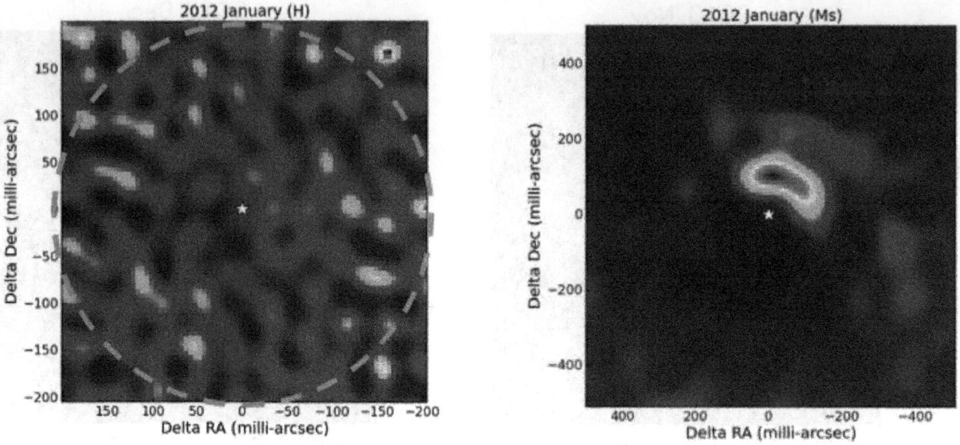

Figure 2. Image reconstructions from H-band (1.55 μm, left, no detection), and Ms-band (4.7 μm, right), with the star not shown (modelled as a central point source). The Maximum Entropy image reconstruction technique as described in Ireland(2013) was used, which images from both aperture-masking data (H-band) and kernel-phase data taken with an unobstructed telescope pupil (Ms filter). The dashed line shows the rough window size for the aperture-mask analysis in H-band. In M-band, the resolved structure has an integrated brightness approximately 3.5 magnitudes fainter than the central star.

higher temperature for optically-thick material. This is much higher than the $\sim$250 K maximum temperature for small grains at 20 AU separations in equilibrium with the stellar radiation field. The integrated luminosity of $3\times10^{-3}\,L_\odot$ between 2 and 5 microns is also very high, and only consistent with the luminosity of the Marley *et al.* (2007) models for 1–3 M_J masses for a period of $\sim10^5$ years. This luminosity is therefore remains consistent with planetary formation, but the mechanism for producing the emission remains unknown.

References

Dodson-Robinson, S. E. & Salyk, C. 2011, *ApJ*, 738, 131

Ireland, M. 2012, *in Proc. SPIE* Vol. 8447, 79

Ireland, M. J. 2013, *MNRAS*, 433, 1718

Ireland, M. J. & Kraus, A. L. 2008, *ApJ*, 678, L59

Ireland, M. J., Monnier, J. D., & Thureau, N. 2006, *in Proc. SPIE* Vol. 6268, 58I

Kraus, A. L. & Ireland, M. J. 2012, *ApJ*, 745, 5

Lafrenière, D., Doyon, R., Marois, C., Nadeau, D., Oppenheimer, B. R., Roche, P. F., Rigaut, F., Graham, J. R., Jayawardhana, R., Johnstone, D., Kalas, P. G., Macintosh, B., & Racine, R. 2007, *ApJ*, 670, 1367

Marley, M. S., Fortney, J. J., Hubickyj, O., Bodenheimer, P., & Lissauer, J. J. 2007, *ApJ*, 655, 541

Martinache, F. 2010, *ApJ*, 724, 464

Mayor, M, & Queloz, D. 1995, *Nature*, 378, 355

Simon, M., Dutrey, A., & Guilloteau, S. 2000, *ApJ*, 545, 1034

Soummer, R., Pueyo, L., & Larkin, J. 2012, *ApJ*, 755, L28

Discussion

CHIANG: Why can't reflection explain the near-infrared photometry?

IRELAND: It isn't possible for reflection to give $\Delta M = 3.5$, even with an albedo of 1.0. However, the strongest argument is due to red colours: I have tried to create fake dust that would reproduce the colours, but would need the imaginary refractive index to increase by a factor of 10 between e.g. L' and K' filters. I don't think there is a physical dust type that can do this.

CHIANG: Is there evidence for the time-variability in the photometry?

IRELAND: Yes, especially in the K' filter. The "lobes" vary by more than a factor of 2 between epochs.

GRAHAM: What is the speckle coherence lifetime needed?

IRELAND: The exposure times are 10 to 20 seconds, much longer than any coherence time. See Ireland (2013) for details.

GRAHAM: Comment: The median age of the GPI input sample is ≈125 Myr and therefore sensitive to the heat of formation and will distinguish between hot start/cold start, but not see the accretion spike.

Exploring the Formation and Evolution of Planetary Systems
Proceedings IAU Symposium No. 299, 2013
M. Booth, B. C. Matthews & J. R. Graham, eds.

doi:10.1017/S1743921313008338

HST/STIS imaging of Fomalhaut: New main belt structure and confirmation of Fomalhaut *b*'s eccentric orbit

P. Kalas[1], J. R. Graham[1], M. P. Fitzgerald[2] and M. Clampin[3]

[1]Astronomy Department, University of California, Berkeley, CA 94720
email: kalas@berkeley.edu

[2]Dept. of Physics and Astronomy, UCLA, Los Angeles, CA

[3]NASA Goddard Space Flight Center, Greenbelt, MD

Abstract. High contrast imaging observations with the Hubble Space Telescope show that the nearby star Fomalhaut is surrounded by a dusty debris belt and a candidate planet, Fomalhaut *b*, located just inside the edge of the belt. Fomalhaut *b* has unexpected characteristics, such as a relatively blue spectrum, leading to the hypothesis that the detected object is a low-mass planet hosting a giant planetary dust ring or cloud seen in reflected light. Here we present new HST/STIS observations made in 2010 and 2012 that authenticate the existence of Fomalhaut *b*. Our MCMC analysis of four epochs of astrometry spread over eight years indicate that the orbit has a ∼170 AU and $e \sim 0.85$. Fomalhaut *b*'s orbit is apsidally aligned with the main belt, and periapse is located approximately ∼30 AU south of the star. We also show the existence of a ∼50 AU wide azimuthal dust depletion in the dust belt. These new findings provide a revised picture of Fomalhaut as a dynamically complex system, where the orbit of Fomalhaut *b* and the belt structure signify the presence of additional massive planets orbiting the star.

Keywords. planets & satellites: individual: Fomalhaut *b*

Background: As of this writing, the existence of Fomalhaut *b* has been confirmed by three epochs of observations with the Hubble Space Telescope (HST) between 2004 and 2010, in work published by three independent research groups (Kalas *et al.* 2008, Kalas *et al.* 2010, Kalas 2011, Currie *et al.* 2012, Galicher *et al.* 2013). Though confirmed as a real, low-mass, common proper motion companion to the A3V star Fomalhaut (α PsA; HD 216956), Fomalhaut *b* has surprising observational properties. One of these surprises was that the third epoch of observation made in 2010 showed that Fomalhaut *b*'s sky plane motion was directed toward instead of parallel to the dust belt's inner edge. This motion implied a highly eccentric orbit. However, Kalas *et al.* (2010) & Kalas (2011) argued that this required observational confirmation because the third epoch represented a change of instrumentation from HST/ACS to HST/STIS. Thus there were potential problems concerning systematic errors in the relative astrometry that could be introduced by the changeover in imaging cameras. Here we report that the fourth epoch detection of Fomalhaut *b*, made with HST/STIS in 2012, *confirms* that Fomalhaut *b* has a highly eccentric orbit. For detailed information we refer the reader to Kalas *et al.* (2013).

HST/STIS Observations & Results: Figure 1 shows our 2012 detection of Fomalhaut's dust belt and Fomalhaut *b* using HST/STIS. STIS consists of a 1024×1024 pixel (0.0407″ / pixel) CCD operating between 0.20 μm and 1.03 μm, with non-transparent focal plane, wedge-shaped masks to prevent saturation of bright stars, but without a pupil plane mask to suppress diffraction around the secondary support spiders. The bright PSF halo and diffraction spikes are subtracted by angular differential imaging – the telescope is rotated to 12 different orientations each separated by 2° about the optical axis in 12 different orbits. One of the key problems for relative astrometry is ascertaining the position of Fomalhaut A (the central star) behind the occulting wedges. We used a variety of techniques to estimate this position, such as by using the diffraction spikes to provide one estimate of the position. We also re-analysed the astrometry for the 2004 and 2006 observations made with HST/ACS, and the 2010 observations made with HST/STIS. Other sources of astrometric error, such as uncorrected geometric distortion and guide star tracking drift, were also estimated using a variety of techniques. A detailed list of astrometric error sources, and the four epochs of astrometry with their associated error bars, are provided in Kalas *et al.* (2013).

Figure 1 illustrates three of the key new findings made with the HST/STIS images. First, the astrometry confirms that the sky-plane motion of Fomalhaut *b* is directed to the

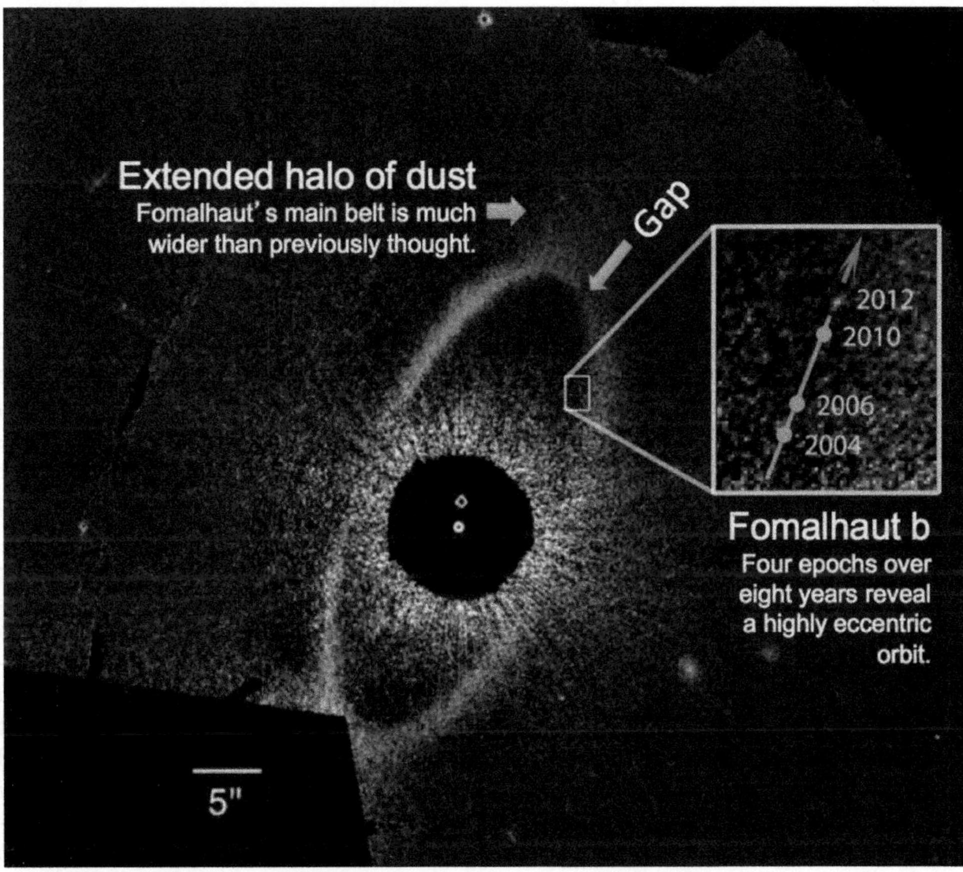

Figure 1. Detections of Fomalhaut *b* and structure in Fomalhaut's dust belt made in 2010 & 2012 with optical HST/STIS observations (Kalas *et al.* 2013). North is up, east is left, and 5″ corresponds to 39 AU. The box inset illustrates the positions of Fomalhaut *b* over eight years, with a trajectory that is northwestward and will appear to cross the belt over the next few decades. Periastron is south of the star at 30 ± 24 AU. The circle marks the position of the star, and the diamond marks the geometric center of the belt that is offset from the star by ~13 AU in the sky plane. The main belt has a tenuous halo of nebulosity detected to the northwest suggesting that dust grains extend to at least 209 AU radius. An additional dust belt feature discovered in the 2010 STIS observations and confirmed in the 2012 data is an azimuthal gap at PA=331°. In the deprojected reference frame, the gap can be characterized as a 50% decrease in scattered light with FWHM ≈ 50 AU.

northwest in a path that *in projection* crosses the dust belt. Second, the northwest side of the belt shows a tenuous nebulosity indicating that the dust seen in optical scattered light is extended to at least 209 AU radius. Third, the dust belt surface brightness in the azimuthal direction diminishes at PA=331°. In a deprojected view, the azimuthal belt gap has FWHM~50 AU.

The 1-σ errors on the astrometry of Fomalhaut *b* relative to the star are approximately 20 mas for the 2004 and 2006 HST/ACS data, and 40 mas for the 2010 and 2012 HST/STIS data. We developed an MCMC method to sample the posterior probability distribution for orbital elements based on the astrometry, yielding median and rms values of $a = 170 \pm 30$ AU, $e = 0.85 \pm 0.07$ (assuming $M_\star = 1.92 \pm M_\odot$; Mamajek 2012).

The orbit of Fomalhaut *b* is confirmed to be highly eccentric. This puts periastron slightly over a century ago in the region a few tens of AU from the star, and in the sky plane projection Fomalhaut *b* appears to cross the belt over the next few decades. The median inclination of the orbital plane (−56°) is close to that of the belt plane (−66°) and apsidal alignment between Fomalhaut *b* and the belt is indicated, within the uncertainties. Overall, additional astrometric points in future work are required to determine if Fomalhaut *b* is on a collision course with the main belt, or if it will pass above or below the belt over the next few decades.

Implications of Fomalhaut *b*'s highly eccentric orbit: The dynamical history of the Fomalhaut system is clearly more complex than previously thought. Kalas *et al.* (2005) presented detailed measurements of the dust belt's stellocentric offset and sharp inner edge consistent with the hypothetical secular perturbations of an undetected planet. When the existence of Fomalhaut *b* was subsequently discovered as lying just 18 AU interior to the dust belt, it naturally became the top candidate for explaining these belt properties (Kalas *et al.* 2008, Chiang *et al.* 2009). However, Fomalhaut *b*'s high eccentricity now complicates the dynamical picture because: **(1)** a mechanism, such as planet-planet scattering by a more massive Fomalhaut *c*, is required to explain Fomalhaut *b*'s high eccentricity, and **(2)** the stellocentric and sharp inner edge of the dust belt also require a yet-to-be-detected planet on a relatively low eccentricity orbit.

Fomalhaut *b*'s origin and physical nature also continue to be open topics for additional research. Kalas *et al.* (2008) argued that because Fomalhaut *b* is detected in the optical and not in the infrared, the optical flux could represent light scattered from a circumplanetary ring system. Kennedy & Wyatt (2011) studied the possibility that instead of a circumplanetary dust ring, a circumplanetary dust cloud could be generated by the collisional erosion of irregular satellites. The third possibility, that Fomalhaut *b* is a dust clump resulting from a collision between two planetesimals, was considered by Kalas *et al.* (2008), Currie *et al.* (2012) and Galicher *et al.* (2013) and generally thought to be unlikely, though not ruled out. Kalas *et al.* (2013) argue that the periastron distance of ~30 AU decreases the previously assumed Hill radius for Fomalhaut *b* ($R_H = a_{pl}(1-e)(m_{pl}/3M_\star)^{1/3}$). They establish through a numerical experiment that Fomalhaut *b* needs a few times a Ceres mass to maintain a long lived cloud of dust from irregular satellites, as in the Kennedy & Wyatt (2011) model.

For the maximum mass of Fomalhaut *b*, the infrared non-detection suggest $\lesssim$1 Jupiter mass based on models of planet luminosity evolution (Kalas *et al.* 2008, Janson *et al.* 2012, Kenworthy *et al.* 2013). A complementary, model-dependent mass constraint derives from the dynamics of the system. Kalas *et al.* (2013) numerically studied cases where a Jupiter mass Fomalhaut *c* is nested within and coplanar with the dust belt, with $a = 120$ AU, $e = 0.1$, and Fomalhaut *b* is a planet with masses between Earth and Jupiter that has a highly eccentric coplanar or non-coplanar orbit (relative to Fomalhaut *c* and the dust belt). The coplanar cases are relatively short-lived (i.e. the orbit of Fomalhaut b changes on $\sim 10^6$ yr timescales). A coplanar, Saturn-mass Fomalhaut b erodes the belt edges on 10^5 year timescales (the belt spreads radially), but a coplanar, Neptune-mass has an insignificant effect on that timescale. In the non-coplanar simulations, the ascending and descending nodes of Fomalhaut b may lie significantly within or beyond the perimeter of the belt and Fomalhaut c, with negligible dynamical effect on $\sim 10^6$ yr timescales. However, secular perturbations are possible on 10^7-10^8 yr timescales that modify the orbit of Fomalhaut b. Secular perturbations can *reduce* Fomalhaut b's a and e such that its apastron occurs within the main belt. Fomalhaut b therefore transitions into a significant perturber of the belt. Alternately, secular perturbations could *increase* Fomalhaut b's a and e such that its perturbations on the system transition to longer timescales.

To summarize, a variety of factors and assumptions underly any determination of Fomalhaut b's mass and origin from dynamics alone. Future work should be dedicated to detecting the hypothetical Fomalhaut c. Moreover, Fomalhaut b could be directly detected at sub-mm wavelengths with ALMA if in fact the circumplanetary dust cloud or ring hypothesis is correct. The azimuthal gap shown in Fig. 1 is also unexplained as of this writing – Fomalhaut b's orbit does not cross this region. Over a decade timescale, the 0.4″ sky plane rotation of the gap is potentially observable with ALMA.

Acknowledgements: We thank GO-11818 & GO-12576 provided by NASA from STScI; NSF AST-0909188; NASA Origins 09-SSO09-0124; and Univ. Calif. LFRP-118057.

References

Chiang, E., Kite, E., Kalas, P., Graham, J. R. & Clampin, M. 2009, *ApJ*, 693, 734
Currie, T., Debes, J., Rodigas, T. J., *et al.* 2012, *ApJL*, 760, L32
Galicher, R., Marois, C., Zuckerman, B., & Macintosh, B. 2013, *ApJ*, 769, 42
Janson, M., Carson, J. C., Lafrenière, D., *et al.* 2012, *ApJ*, 747, 116
Kalas, P., Graham, J. R., & Clampin 2005, *Nature*, 435, 1067
Kalas, P., Graham, J. R., Chiang, E., *et al.* 2008, *Science*, 322, 1345
Kalas, P., Graham, J. R., Fitzgerald, M., & Clampin, M. 2010, in *In the Spirit of Lyot 2010*, ed. A. Boccaletti.
Kalas, P. 2011, in *The Astrophysics of Planetary Systems*, IAU Symposium, Vol. 276, p. 279.
Kalas, P., Graham, J. R., Fitzgerald, M., & Clampin, M. 2013, in *ApJ*, 775, 56
Kennedy, G. M. & Wyatt, M. C. 2011, *MNRAS*, 412, 2137
Kenworthy, M. A., Meshkat, T., Quanz, S. P., *et al.* 2013, *ApJ*, 764, 7
Mamajek, E. E. 2012, *ApJL*, 754, L20

Discussion

Eric Gaidos: Photometric variability was reported in the discovery paper of Fomalhaut b. What is the status of this with the new observations?

Paul Kalas: Between 2010 and 2012 we do not detect photometric variability in the STIS detections of Fomalhaut b. In our tests using injected point sources, we find that the photometric uncertainty due to speckle noise is $\sim 35\%$. These tests also show that an injected point source may gain an artificial, extended morphology due to the noise. With respect to the first two epochs of discovery images made with HST/ACS, other groups report somewhat conflicting photometry that illustrate the significant role of using different data reduction techniques and measurement methods. The issue of photometric variability and extended morphology therefore require additional future work.

Exploring the Formation and Evolution of Planetary Systems
Proceedings IAU Symposium No. 299, 2013
M. Booth, B. C. Matthews & J. R. Graham, eds.

doi:10.1017/S174392131300834X

The multiple spirals in the disk of HD100546

A. Boccaletti[1], A.-M. Lagrange[2], E. Pantin[3], J.-C. Augereau[2]
S. P. Quanz[4] and H. Meheut[3]

[1]LESIA, Observatoire de Paris, 5 place Jules Janssen, 92195 Meudon, France
email: anthony.boccaletti@obspm.fr

[2]IPAG, Université Joseph Fourier, CNRS, BP 53, 38041 Grenoble, France

[3]SAp, CEA, France

[4]Institute for Astronomy, ETH Zurich, Wolfgang-Pauli-Strasse 27, 8093 Zurich, Switzerland

Abstract. We report the detection of multi-armed spirals in the environment of HD 100546 using NICI at Gemini South in the Ks band. These data feature a better angular resolution and higher contrast than previous HST images, which allows to resolve the former known spiral into a multiple pattern. An analytic model with a gravitational perturber is used to fit the spiral pattern. We derived limit of detections which set constraints on the discovered forming planet.

Keywords. stars: individual (HD 100546) - protoplanetary disks - high angular resolution

1. Introduction

The Herbig Be star HD 100546 (5–10 Myr, 97 pc) is known to harbour a pre-transitional protoplanetary disks already imaged with other instruments. Using NICMOS on HST the disk was seen as an elliptical pattern without obvious structure in the near IR (Augereau *et al.* (2001)), but the image from Ardila *et al.* (2007), owing to an improved angular resolution at visible wavelengths, revealed a much more complex morphology. The disk region between 1.5 – 3" is of particular interest since it contains two trailing spiral arms developing towards the south-east and the north-west respectively. The innermost part of the disk intensively studied by Benisty *et al.* (2010) with the VLTI suggests an inner hole between 4 and 13 AU indicative of an inner giant planet. Recently, Quanz *et al.* (2013) reported the discovery of a planet forming candidate.

2. The spirals of HD 100546

We retrieved archival data obtained with the Near Infrared Coronagraphic Imager (NICI), installed at Gemini South. The observation were done on March 6th, 2010, using a single channel configuration in the Ks band (2.15 μm) with the 0.32" coronagraphic mask and with the telescope pupil at a constant orientation to allow for angular differential imaging (ADI). We followed the data reduction procedure described in Boccaletti *et al.* (2013) using various ADI flavours like KLIP and LOCI. While the large scale component of the disk is removed by ADI, a complex set of spiral patterns is detected in the southeast part of the field in the place where Ardila *et al.* (2007) reported the detection of a single spiral. The spirals are labeled in Fig. 1 from S1 to S6. S1 is the brightest and longest spiral, it starts from $\rho = 1.40$", $PA = 374°$ to $\rho = 2.27$", $PA = 329°$. The equivalent projected length is about 155 AU. Both S1 and S2 are seen in the ACS/HST image as a single broken spiral. The image was un-projected assuming an inclination of 42° and the spines of the spirals were fitted with the Muto *et al.* (2012) model, which assumes a gravitational perturber launching a single spiral pattern in a gaseous disk. On average for all spirals,

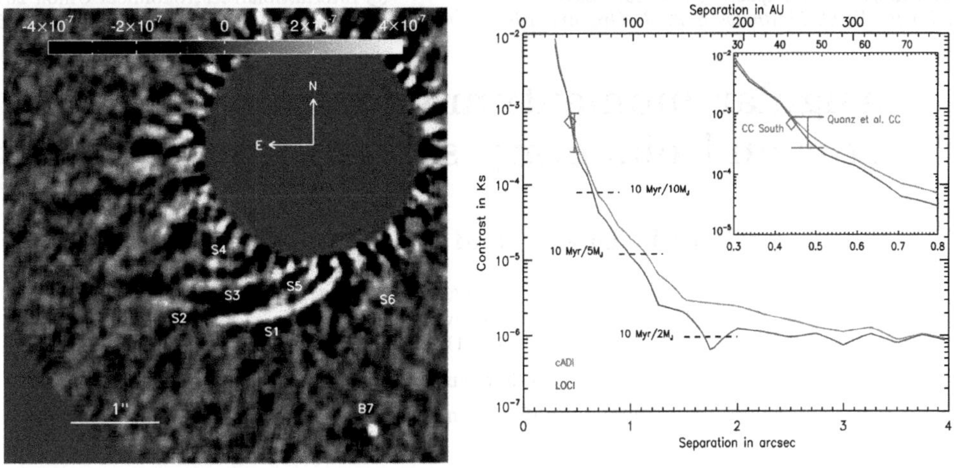

Figure 1. Left: Image of the HD 100546 environment processed with LOCI. The spirals are labeled S1 to S6. Right: Limits of point source detection estimated with ADI and LOCI using fake planets to calibrate the ADI self-subtraction. The expected intensity of the forming planet (Quanz *et al.* CC) is reported as well as a new candidate identified in NICI image (CC south).

we obtained $r_c = 0.23 \pm 0.04$", $\theta_0 = 230 \pm 55°$, and $hc = 0.13 \pm 0.04$, with r_c and θ_0 the polar coordinates of the perturber and hc the disk aspect ratio. However, we note that the parameters are highly degenerated. If one or several perturbers are responsible for these spirals, the model of Muto *et al.* (2012) derives a physical separation of $20 - 30$ AU from the star, which is out of the gap detected from interferometry and closer than the location of the forming planet. Using a fake spiral injected at the northwest, we were able to measure the width and photometry of S1 as well as a lower limit of the anisotropic scattering factor ($g > 0.48 - 0.61$) which is derived from the non detection of the northwest spiral. In addition, we did not detect in Ks any counter part of the Quanz *et al.* (2013) forming planet. As a result, the contrast (Fig. 1, right) achieved in the NICI data at a separation of 0.48" allows to rule out a self-luminous planet more massive than $16 - 18\,M_J$ according to the age and whatever the evolutionary model. Quanz *et al.* (2013) suggested that if the planet is still forming the flux measured in Lp may not entirely come from its photosphere

Any large perturbation in a disk of gas tends to launch sound waves resulting in a spiral shape due to the differential rotation of the disk. Several scenarios can be considered to account for the presence of spirals. A protoplanetary disk can also undergo gravitational instabilities. But in the case of HD 100546, the estimation of the Toomre parameter, based on currently estimated disk's mass as well as the disk aspect ratio, is well above unity. We concluded that gravitational instabilities are unlikely to account for the presence of spirals and the effect of gravitational perturbers is preferred. More detailed analysis will be presented in a forthcoming paper.

References

Augereau, J. C., Lagrange, A. M., Mouillet, D., & Menard, F., 2001, *A&A*, 365, 78
Ardila, D. R., Golimowski, D. A., Krist, J. E., *et al.*, 2007, *ApJ*, 365, 78
Benisty, M., Tatulli, E., Menard, F., & Swain, M. R., 2010, *A&A*, 511, 75
Boccaletti, A., Lagrange, A. M., Bonnefoy, M., Galicher, R., & Chauvin, G., 2013 *A&A*, 551, 14
Muto, T., Grady, C. A., Hashimoto, J., Fukagawa, M. *et al.*, 2012, *ApJ*, 748, 22
Quanz, S. P., Amara, A., Meyer, M. R., *et al.*, 2013 *ApJL*, 766, 1

Exploring the Formation and Evolution of Planetary Systems
Proceedings IAU Symposium No. 299, 2013
M. Booth, B. C. Matthews & J. R. Graham, eds.

doi:10.1017/S1743921313008351

Angular momentum evolution during star and planetary system formation

Claire L. Davies[1] **and Jane S. Greaves**[2]

[1,2]SUPA School of Physics & Astronomy,
University of St Andrews, North Haugh,
St Andrews, Fife, UK, KY16 9SS
[1]email: cd54@st-andrews.ac.uk
[2]email: jsg5@st-andrews.ac.uk

Abstract. We focused on analysing the role played by protoplanetary disks in the evolution of angular momentum during star formation. If all the angular momentum contained within collapsing pre-stellar cores was conserved during their formation, proto-stars would reach rotation rates exceeding their break-up velocities before they reached the main sequence (Bodenheimer 1995). In order to avoid this occuring, methods by which proto-stars can lose angular momentum must exist. Angular momentum can be transferred from star to disk via stellar magnetic field lines through a process called magnetic braking (Camenzind 1990; Königl 1991). Alternatively, the stellar angular momentum can be lost from the star-disk system entirely via stellar- or disk-winds (e.g. Pelletier & Pudritz 1992; Matt & Pudritz 2005).

The proportion of lost stellar angular momentum retained within the protoplanetary disk is important to studies of planetary system formation. If the bulk motion within the disk remains Keplerian, any increase of angular momentum in the disk causes an outward migration of disk material and an expansion of the disk. Therefore, an increase in disk angular momentum may cause a reduction in the disk surface density, often used to indicate the disk's ability to form planets.

We made use of multi-wavelength data available in the literature to directly calculate the stellar and disk angular momenta for two nearby regions of star formation. Namely, these were the densely populated and highly irradiated Orion Nebula Cluster (ONC) and the comparitively sparse Taurus-Auriga region. Due to the limited size of the ONC dataset, we produced an average surface density profile for the region. We modelled the stars as solid body rotators due to their fully convective nature (Krishnamurthi *et al.* 1997) and assumed the disks are flat and undergo Keplerian rotation about the same rotation axis as the star.

We observed the older disks within each of the two star forming regions to be preferentially larger and less massive, consistent with viscous accretion theories and disk dispersal. However, when drawing comparisons between the two regions, the ONC sample appeared to have less massive disks than the Taurus-Auriga sample, even though the population of Taurus-Auriga is older. This may suggest an influence of the birth cloud environment on protoplanetary disk evolution.

Finally, the older stars within the ONC were observed to harbour disks that contained more angular momentum than their younger counterparts whereas, in the Taurus-Auriga sample, the amount of angular momentum contained in the older and younger samples was consistent. We suggest that the missing disk angular momentum in the older Taurus-Auriga disks may be contained within yet-undetected planets.

Keywords. stars: formation, planetary systems: formation, planetary systems: protoplanetary disks, stars: rotation

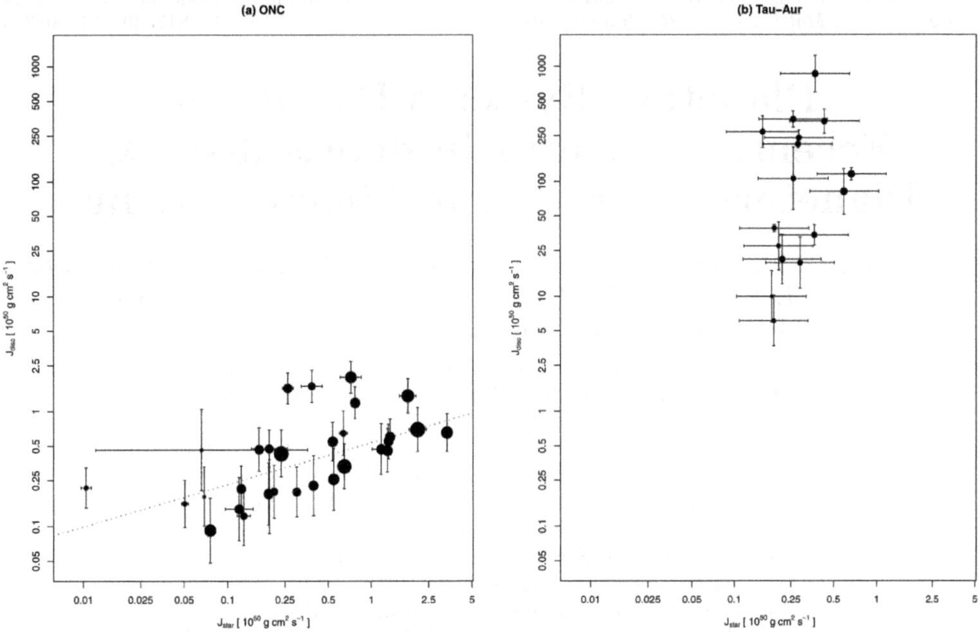

Figure 1. Comparison of stellar and disk angular momenta in (a) ONC and (b) Taurus-Auriga. The data points are scaled according to their stellar radii. The result of a χ^2 fitting procedure is shown by the dashed line for the ONC but no such correlation was found in the case of the Taurus-Auriga. Within the ONC sample, the older stars appear to harbour disks containing higher angular momentum. However, the disk angular momenta of the older and younger samples within Taurus-Auriga remains consistent. The missing angular momentum in the older Taurus-Auriga disks may have formed as yet-undetected planets.

References

Bodenheimer, P. 1995, *ARAA*, 33, 199
Camenzind, M. 1990, *Rev. Mexicana AyA*, 3, 234
Königl, K. 1991, *ApJ*, 370, 37
Krishnamurthi, A., Pinsonneault, M. H., Barnes, S., & Sofia, S. 1997 *ApJ*, 480, 303
Matt, S. & Pudritz, R. E. 2005, *ApJ*, 632, 135
Pelletier, G. & Pudritz, R. E. 1992, *ApJ*, 394, 117

Exploring the Formation and Evolution of Planetary Systems
Proceedings IAU Symposium No. 299, 2013
M. Booth, B. C. Matthews & J. R. Graham, eds.

doi:10.1017/S1743921313008363

Planetary Systems Dynamics Eccentric patterns in debris disks & Planetary migration in binary systems

V. Faramaz[1], H. Beust[1], J.-C. Augereau[1], A. Bonsor[1], P. Thébault[2], Y. Wu[3], J. P. Marshall[4], C. del Burgo[5], S. Ertel[1], C. Eiroa[4], B. Montesinos[6], A. Mora[7], and the DUNES team

[1]UJF-Grenoble 1 / CNRS-INSU, Institut de Planétologie et d'Astrophysique de Grenoble (IPAG) UMR 5274, Grenoble, F-38041, France
[2]LESIA, Observatoire de Paris, 92195 Meudon France
[3]Dpt. of Astronomy and Astrophysics, University of Toronto, ON M5S 3H4, Canada
[4]Dpt. Fìsica Teòrica, Facultad de Ciencias, Universidad Autònoma de Madrid, Cantoblanco, 28049 Madrid, Spain
[5]Instituto Nacional de Astrofísica, Óptica y Electrónica, Luis Enrique Erro 1, Sta. Ma. Tonantzintla, Puebla, Mexico
[6]Dpt. de Astrofisica, Centro de Astrobiologìa (INTA-CSIC), ESAC Campus, P.O.Box 78, E-28691 Villanueva de la Cañada, Madrid, Spain
[7]Aurora Technology B.V., ESA-ESAC Gaia SOC. P.O. Box 78, E28691 Villanueva de la Cañada, Madrid, Spain

Abstract. We present some highlights of two ongoing investigations that deal with the dynamics of planetary systems. Firstly, until recently, observed eccentric patterns in debris disks were found in young systems. However recent observations of Gyr-old eccentric debris disks leads to question the survival timescale of this type of asymmetry. One such disk was recently observed in the far-IR by the *Herschel Space Observatory* around ζ^2 Reticuli. Secondly, as a binary companion orbits a circumprimary disk, it creates regions where planet formation is strongly handicapped. However, some planets were detected in this zone in tight binary systems (γ Cep, HD 196885). We aim to determine whether a binary companion can affect migration such that planets are brought in these regions and focus in particular on the planetesimal-driven migration mechanism.

Keywords. Circumstellar matter – Methods: N-body Simulations – Celestial mechanics – Stars: ζ^2 Reticuli, planetary systems

---

Can eccentric debris disks be long-lived? Recent *Herschel* far-IR observations of the debris disk surrounding the 2-3 Gyr old star ζ^2 Reticuli, obtained as part of the DUNES key program (see Eiroa *et al.*, 2013), reveal an asymmetric double-lobed circumstellar feature (Fig. 1, left panel), interpreted as a ring like structure seen almost edge-on with an elliptical shape and minimum eccentricity of 0.3, extending from 70 to 120 AU (Eiroa *et al.*, 2010). So far, eccentric patterns in debris disks were observed in young systems (HR 4796: 10 Myr ; Fomalhaut: 440 Myr), and attributed to the dynamical shaping of the disk by massive companion on an eccentric orbit (see e.g. Wyatt, 1999). We discuss here the origin of this eccentric pattern in the ζ^2 Reticuli system, and the ability to maintain the disk asymmetry over Gyrs. Through a modeling method combining analytical predictions, numerical N-body treatment (Levison & Duncan, 1994) and radiative transfer, we can put constraints on the orbital parameters of hypothetical perturbers that could induce a disk structure compatible with observational constraints. The synthetic images support the hypothesis that the observed double-lobe feature is the signature of an eccentric ring (see Fig. 1, right panel), and our synthetic model supports the presence

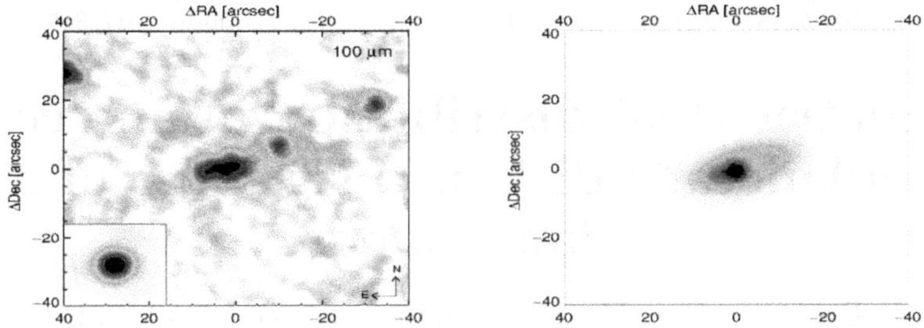

Figure 1. *Left:* Herschel/PACS Observations of ζ^2 Reticuli at 100 microns. *Right:* Synthetic image at 100 microns.

of a planetary mass companion orbiting at several tens of AU from its host star, and on a significantly eccentric orbit (Faramaz *et al.* 2013, in prep.).

Planetary migration in binary systems Planetesimal-driven migration mechanism leads a planet to exchange angular momentum with planetesimals during close encounters. Angular momentum passes from outer to inner material and components experience a mutual gravitational attraction. In a system with a single planet, timescales favour interactions with inner material, and thus inward migration (see Fig. 1 of Kirsh *et al.* 2009). Our goal is to investigate this process in binary systems: preliminary results show that in the presence of a binary companion, the direction of migration could be reversed: interactions with outer material seems favoured, which would move planets out rather than in (see Fig. 2). This may provide an explanation for the presence of planets in regions of binary systems where models fail to explain in situ formation (see the review by Thébault P., 2011).

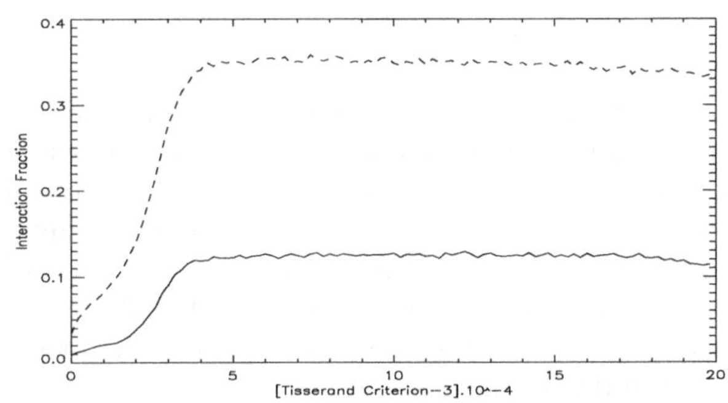

Figure 2. Probability for a planetesimal to suffer a close encounter as a function of its Tisserand parameter, and thus as a function of its distance to a planet of mass $4.5M_{\oplus}$ at 10 AU on a circular orbit in the presence of a $1M_{\odot}$ binary companion on a circular orbit at 100 AU, within $3R_{\rm Hill}$, and for inner (solid line) and outer (dashed line) material.

References

Eiroa, C., *et al.* 2010, *A&A*, 518, L131

Eiroa, C., *et al.* 2013, *A&A*, 555, A11

Faramaz, V., *et al.*, In prep.

Kirsh, D. R., Duncan, M., Brasser, R., & Levison, H. F. 2009, *Icarus*, 199, 197–209

Levison, H. F. & Duncan, M. J 1994, *Icarus*, vol. 108, no. 1, p. 18–36

Thébault, P. 2011, *Celestial Mechanics and Dynamical Astronomy*, vol. 111, p. 29–49

Wyatt, M. C. 1999, *ApJ*, 527:918–944

Computations presented in this paper were performed at the Service Commun de Calcul Intensif de l'Observatoire de Grenoble (SCCI) on the super-computer funded by Agence Nationale pour la Recherche under contracts ANR-07-BLAN-0221, ANR-2010-JCJC-0504-01 and ANR-2010-JCJC-0501-01.

Funding for this project was provided by a grant from la Région Rhône-Alpes.

C. Eiroa, J.P. Marshall, and B. Montesinos are supported partly by Spanish grant AYA 2011-26202.

Exploring the Formation and Evolution of Planetary Systems
Proceedings IAU Symposium No. 299, 2013
M. Booth, B. C. Matthews & J. R. Graham, eds.

doi:10.1017/S1743921313008375

Asymmetric dust distribution in an eccentric protoplanetary disk as a signpost of a gas giant planet

Pin-Gao Gu, He-Feng Hsieh and Hsi-Wei Yeh

Instituter of Astronomy & Astrophysics, Academia Sinica,
Taipei 10617, Taiwan
email: gu@asiaa.sinica.edu.tw

Abstract. The presence of a cavity in a protoplanetary disk revealed by dust continuum emissions is sometimes postulated as a signpost of an embedded gas giant planet. More peculiarly, dust emissions exterior to the cavity are often observed to be asymmetric. We explore the possibility of the asymmetry as a result of the asymmetric distribution of dust in an eccentric protoplanetary disk under the secular gravitational perturbation of an embedded massive gas giant planet. We find that the surface density of the dust well coupled to the disk gas is enhanced around the apocenter of the disk. In addition, the azimuthal distributions of particles of various sizes can deviate significantly due to different coupling to the gas. Overall, the asymmetric structure exhibits a phase correlation between the gas velocity field and dust density distribution. A Doppler map for an eccentric disk is also presented based on Cycle 1 ALMA observations. Our study potentially provides a reality check as to whether an asymmetric disk gap detected at sub-mm and cm wavelengths is a signpost of a massive gas giant planet.

1. Introduction

The presence of a cavity and/or non-axisymmetric features in a protoplanetary disk as revealed by dust continuum emissions has at times been postulated as a potential signpost of an embedded gas giant planet (Andrews *et al.* 2011). If the planet is sufficiently massive, the disk exterior to the planet's orbit can become moderately eccentric (thus we only show the results for the outer disk. See below) and start precessing on the secular timescale, depending on the disk mass and viscosity (Dunhill *et al.* 2012). In such an eccentric protoplanetary disk harboring a massive planet, the orbits of dust particles would be eccentric as well due to the gas drag.

We run the FARGO code (Masset 2000) to obtain the gas profile in a two-dimensional disk with a massive planet, and then employ the secular perturbation theory and incorporate aerodynamic drag (Paardekooper *et al.* 2008) to estimate the dust velocity and density that will in turn reveal the asymmetric structures inherent to an eccentric disk. The purpose of this study is to focus only on the secular behaviors of dust associated with an eccentric protoplanetary disk in the presence of a massive giant planet. It should be noted that although the vortex model produces a much larger dust density contrast than that in our model (van der Marel *et al.* 2013), the disk conditions required for vortex generation in general allow for the generation of disk eccentricity.

2. Results

Fig. 1 shows a snapshot of the spatial distribution of the dust surface density for 0.01 cm and 1 m particles in both the protoplanetary and anemic (low-density transitional) disks. The distributions exhibit $m = 1$ structures with the density excess about

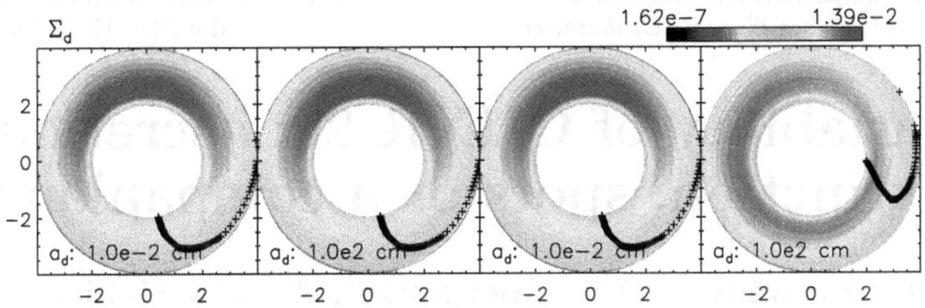

Figure 1. Spatial distribution of the dust surface density for 0.01 cm and 1 m particles in the cases of the protoplanetary (left 2 panels) and anemic disks (right 2 panels). The units of the surface density are about 0.356 g/cm^2 and 9×10^{-4} g/cm^2 for the protoplanetary and anemic disks, respectively. The plus signs in the plot mark the azimuthally averaged longitude of pericenter of particles.

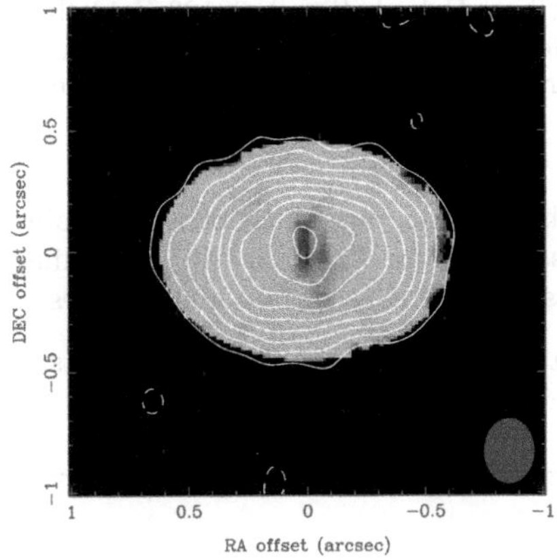

Figure 2. Doppler image of an eccentric disk. Green denotes the Keplerian circular rotation. Red indicates an excess of redshifted velocities elongated along the minor axis of the disk.

5-10% around the apocenter. The dust enhancement is caused by the gas streamlines converge near the apocenter of the disk. All the $m = 1$ dust distributions look quite similar except for that for the 1 m particles in the anemic disk shown in the lower bottom panel. For those large particles in the anemic disk, the dimensionless secular timescale $\tau_{s,sec} \approx$ 1–10 and thus the particles are weakly coupled. As a result, the 1 m particles precess with the gas disk but with a phase lag $\approx \arctan \tau_{s,sec} \approx 80°$, as illustrated in Fig. 1.

Fig. 2 presents a Doppler image of the simulated protoplanetary disk subtracted from the Keplerian circular motion (colors), overlaid on the total integrated intensity map at 345.8 GHz corresponding to ^{12}CO 3-2 line (contours). A planet of 5 M_j is placed at 20 AU on the disk plane. The disk has the radius of 80 AU at 140 pc away and is inclined by 45 degrees from the line of sight. This simulated observational map is constructed using Cycle 1 ALMA observations with 6 hour integration, an $0.27'' \times 0.2''$ angular resolution, and an 0.11 km/s velocity resolution.

References

Andrews, S. M., *et al.* 2011, *ApJ*, 732, 42

Dunhill, A., Alexander, R., & Armitage, P. 2013, *MNRAS*, 428, 3072

Masset, F. 2000, *A&AS*, 141, 165

Paardekooper, S.-J., Thébault, P., & Mellema, G. 2008, *MNRAS*, 386, 973

van der Marel, N., *et al.* 2013, *Science*, 340, 1199

Exploring the Formation and Evolution of Planetary Systems
Proceedings IAU Symposium No. 299, 2013
M. Booth, B. C. Matthews & J. R. Graham, eds.

doi:10.1017/S1743921313008387

Variability of CHXR 20: accretion, extinction, spots or a companion?

Taisiya G. Kopytova[1,2] Viki Joergens[1,3], Aurora Sicilia-Aguilar[4], Maria V. Rodríguez-Ledesma[1,5] and Reinhard Mundt[1]

[1]Max-Planck Institut für Astronomie, Königstuhl 17, 69117 Heidelberg, Germany
email: kopytova@mpia.de

[2]International Max-Planck Research School for Astronomy and Cosmic Physics at the University of Heidelberg, IMPRS-HD, Germany [3]Zentrum für Astronomie, Institut für Theoretische Astrophysik, Albert-Ueberle-Str. 2, 69120 Heidelberg, Germany [4]Departamento de Física Teórica, Facultad de Ciencias, Universidad Autónoma de Madrid Cantoblanco, 28049 Madrid, Spain [5]Institut für Astrophysik, Georg-August-Universität, Friedrich-Hund-Platz, 37077 Göttingen, Germany

Abstract. We study the photometric variability of a pre-main sequence star (K6, 0.9$M_\odot$) CHXR 20. We test several scenarios for the variability including variable accretion, variable extinction, cool and hot spots on the stellar surface and the presence of a potential companion.

Keywords. stars: activity, stars: pre-main sequence, stars: spots, stars: planetary systems: protoplanetary disks

1. Introduction

CHXR 20 (K6, Luhman 2004, 2007; 0.9 $M_\odot$, based on models by Siess *et al.* 2000) is a pre-main sequence member of the Chameleon I dark cloud. Its SED shows an infrared excess (Luhman *et al.* 2008) pointing to the presence of a circumstellar disk. Lawson & Crause (2009) report a V-band variability of CHXR 20 suggesting that it is a candidate for an eclipsing binary system with an orbital period of 7.7 days.

In this work, we study the variability of CHXR 20 using available 2MASS (Cutri *et al.* 2003), WISE (Wright *et al.* 2010) and Spitzer (Luhman *et al.* 2008) data, and additionally, our own observations obtained with WFI and FEROS at the 2.2m telescope in La Silla, Chile.

2. Observations

The I_c-band lightcurve obtained with WFI has a quasi-periodic character. The period search analysis determines a period of the variations of 7.0 days, which is consistent with the study of Lawson & Crause (2009).

Available JHKs data from 2MASS point to the strong variability in the near-infrared, though the mid-infrared fluxes (3.6, 4.5, 5.8, 8.0, 24.0 μm, Spitzer; 3.4, 4.6, 12, 22.0 μm, WISE) do not change significantly (Fig. 1).

High-resolution FEROS spectra show activity indicators: Ca II infrared lines and a highly variable Hα feature. The amplitude of radial velocity variations is 3.4 km s^{-1}, though the available data (eight observational epochs) does not allow us to conclude if these variations are periodic or not.

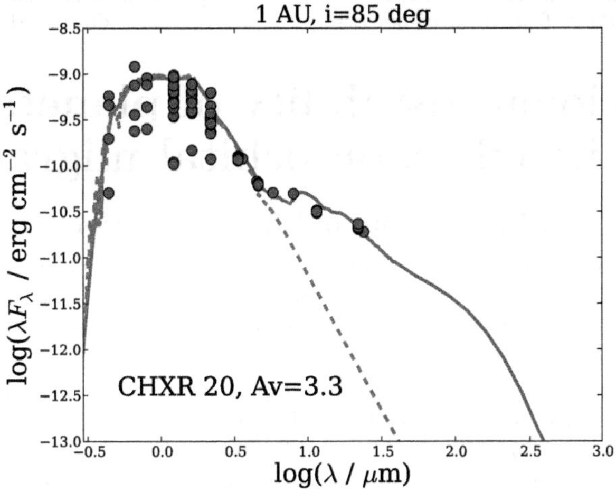

Figure 1. The spectral energy distribution of CHXR 20. The dashed line is a model photosphere. The disk modeling (solid line) with RADMC points to the presence of an inner gap in the disk.

3. Discussion and conclusions

Using the relation between color and magnitudes in the near-infrared, we notice that CHXR 20 gets redder when fainter. This allows us to exclude variable accretion from the potential sources of the photometric variability. On the other hand, the same color-magnitude relation is consistent with the reddening vector for CHXR 20 which was calculated following Cardelli *et al.* (1989). The latter favors the variable extinction origin of the near-infrared variability of CHXR 20.

Using the method proposed by Carpenter *et al.* (2001), we estimate that cool spots would not cause any photometric variability. However, hot spots with temperature 9000–10000 K covering 20–30% of the stellar surface could cause the variability in the optical but not in the near-infrared.

The SED modeling using RADMC (Dullemond & Domonik 2004) suggests a possibility of the presence of an inner gap in the disk (Fig. 1). Given the radial velocity variations of 3.4 km s^{-1} and the period of photometric variability of 7 days, we tentatively suggest that the gap could be cleared-out by a companion with the mass of 15 M_{Jup} at a maximum separation of $\sim$ 0.1 AU. The presence of the companion would also explain the observed photometric variability.

References

Cardelli, J. A., Clayton, G. C., & Mathis, J. S. 1989, *ApJ*, 345, 245

Carpenter, J. M., Hillenbrand, L. A., & Skrutskie, M. F. 2001, *AJ*, 121, 3160

Cutri, R. M., Skrutskie, M. F., van Dyk, S., Beichman, C. A., *et al.* 2000, *VizieR Online Data Catalog*

Dullemond, C. P. & Dominik, C. 2004, *A&A*, 417, 159

Lawson, W. A. & Crause, L. A. 2009, *PASA*, 26, 31

Luhman, K. L. 2004, *ApJ*, 602, 816

Luhman, K. L. 2007, *ApJS*, 173, 104

Luhman, K. L., Allen, L. E., Allen, P. R., Gutermuth, R. A., Hartmann, L., Mamajek, E. E., Megeath, S. T., Myers, P. C., & Fazio, G. G. 2008, *ApJ*, 675, 1375

Siess, L., Dufour, E., & Forestini, M. 2000, *A&A*, 358, 593

Wright, E. L., Eisenhardt, P. R. M., Mainzer, *et al.* 2010, *AJ*, 140, 1868

Exploring the Formation and Evolution of Planetary Systems
Proceedings IAU Symposium No. 299, 2013
M. Booth, B. C. Matthews & J. R. Graham, eds.

doi:10.1017/S1743921313008399

Gravitational instability of planetary gaps and its effect on orbital migration

Min-Kai Lin and Ryan Cloutier

Canadian Institute for Theoretical Astrophysics, 60 St. George Street, Toronto, ON, M5S 3H8, Canada
email: mklin924@cita.utoronto.ca, cloutier@cita.utoronto.ca

Abstract. Gap formation by giant planets in self-gravitating disks may lead to a *gravitational edge instability* (GEI). We demonstrate this GEI with global 3D and 2D self-gravitating disk-planet simulations using the ZEUS, PLUTO and FARGO hydrodynamic codes. High resolution 2D simulations show that an unstable outer gap edge can lead to outward migration. Our results have important implications for theories of giant planet formation in massive disks.

Keywords. hydrodynamics, instabilities, methods: numerical, planetary systems: protoplanetary disks, planets and satellites: formation

1. Introduction: the gravitational edge instability

Models of giant planet formation in self-gravitating protoplanetary disks may involve gap-opening in massive disks. However, gaps or edges in self-gravitating disks are potentially unstable to a gravitational edge instability (GEI) associated with local potential vorticity maxima (Meschiari & Laughlin 2008; Lin & Papaloizou 2011).

We confirm the GEI for planet-induced gaps with global disk-planet simulations. We consider a locally isothermal disk with constant aspect-ratio $h = 0.05$ and Keplerian Toomre parameter $Q_k = 1.5$ at the outer disk boundary $r_{\rm out} = 2.5r_p$, where r_p is the fixed orbital radius of a planet with mass $M_p = 10^{-3}M_*$ and M_* is the central star mass. We simulated this self-gravitating disk-planet system in 3D using the ZEUS-MP code (Hayes *et al.* 2006) and PLUTO code (Mignone *et al.* 2007); and in 2D using the FARGO code (Baruteau & Masset 2008). The 3D simulations are performed in spherical polar co-ordinates covering two vertical scale-heights.

Fig. 1 shows the outer gap edge develops a spiral instability with azimuthal wavenumber $m = 2$ (for 3D runs one of the spirals coincides with the planet wake in the snapshot). The spirals extend close to r_p. In fact, the GEI co-rotation radius is just inside a giant planet's co-orbital region (Lin & Papaloizou 2011). So the GEI supplies fluid onto horseshoe orbits ahead of the planet, which provides a positive disk-on-planet torque (Lin & Papaloizou 2012).

2. Instability-induced outward migration and implications

We also performed 2D FARGO simulations with different M_p where the planet is allowed to migrate, so that $r_p = r_p(t)$. Fig. 2 shows that for $t \lesssim 100$ orbits, outward migration proceeds at a faster rate with increasing M_p, because the GEI becomes stronger. Eventually, a GEI spiral-planet scattering event triggers rapid outward type III migration. In the context of the disk fragmentation model for wide-orbit giant planet formation (Vorobyov 2013), this suggests that gap opening may not be sufficient to maintain giant planets on a fixed orbital radius beyond a few tens of orbits.

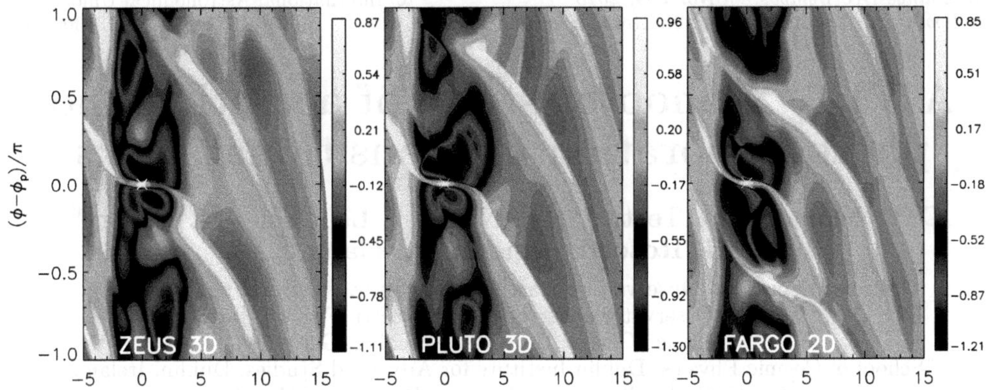

Figure 1. Gravitational edge instability of a planetary gap. The logarithmic midplane density perturbation is shown for 3D runs (left, middle) and the logarithmic surface density perturbation is shown for the 2D run (right). Snapshots are taken at 15 (10) orbits after the planet is introduced into the 3D (2D) disk. Horizontal axis are in units of Hill radii. The planet is at the origin. The left and right simulations are taken from Lin (2012) and Cloutier & Lin (2013), respectively.

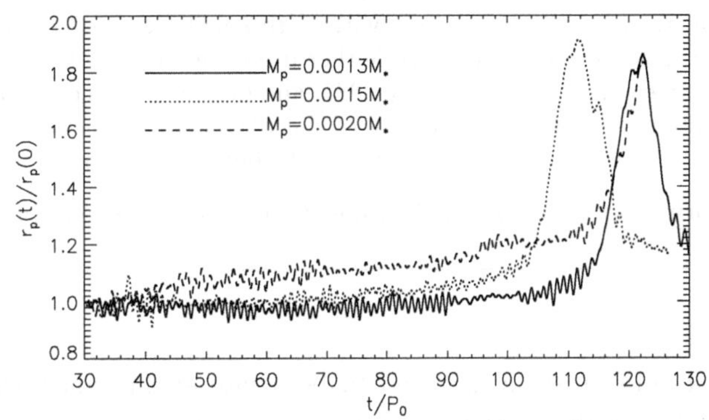

Figure 2. Orbital migration of giant planets when the outer gap edge is gravitationally unstable. Here, P_0 is the orbital period at the planet's initial orbital radius. Adapted from 2D FARGO simulations presented in Cloutier & Lin (2013).

Acknowledgments. Several computations were performed on the GPC supercomputer at the SciNet HPC Consortium. SciNet is funded by: the Canada Foundation for Innovation under the auspices of Compute Canada; the Government of Ontario; Ontario Research Fund - Research Excellence; and the University of Toronto.

References

Baruteau, C. & Masset, F. 2008, ApJ, 678, 483

Cloutier, R. & Lin, M.-K. 2013, MNRAS, 434, 621

Hayes, J. C., Norman, M. L., Fiedler, R. A., Bordner, J. O., Li, P. S., Clark, S. E., ud-Doula, A., & Mac Low, M. 2006, ApJS, 165, 188

Lin, M.-K. 2012, MNRAS, 426, 3211

Lin, M.-K. & Papaloizou, J. C. B. 2011, MNRAS, 415, 1445

Lin, M.-K. & Papaloizou, J. C. B. 2012, MNRAS, 421, 780

Meschiari, S. & Laughlin, G. 2008, ApJL, 679, L135

Mignone, A., Bodo, G., Massaglia, S., Matsakos, T., Tesileanu, O., Zanni, C., & Ferrari, A. 2007, ApJS, 170, 228

Vorobyov, E. I. 2013, A&A, 552, A129

Exploring the Formation and Evolution of Planetary Systems
Proceedings IAU Symposium No. 299, 2013
M. Booth, B. C. Matthews & J. R. Graham, eds.

doi:10.1017/S1743921313008405

A VLT/X-Shooter study of accretion and photoevaporation in Transitional Disks

C. F. Manara[1], L. Testi[1,2,3], A. Natta[2,4], L. Ricci[5], M.Benisty[6], G. Rosotti[3,7] and B. Ercolano[3,7]

[1]European Southern Observatory, Garching, Germany, email: cmanara@eso.org
[2]INAF - Osservatorio Astrofisico di Arcetri, Firenze, Italy
[3]Excellence Cluster Universe, Garching, Germany
[4]School of Cosmic Physics, Dublin Institute for Advanced Studies, Dublin, Ireland
[5]California Institute of Technology, Pasadena, CA, USA
[6]Laboratoire d'Astrophysique, Observatoire de Grenoble, Grenoble, France
[7]Universitats-Sternwarte Munchen, Munich, Germany

Abstract. We present preliminary results of a detailed study of the accretion, stellar, and wind properties of transitional disks (TDs) carried out with the X-Shooter spectrograph. Combining new and archival spectra, we collected a sample of more than 20 TDs from different nearby star-forming regions. Our sample includes objects with both small (<5-15 AU) and large (>20–30 AU) known inner hole size from the literature (either from mm-observations or IR SED fitting). We check their stellar parameters ($T_{\rm eff}$, L_*, A_V, M_*) and derive their accretion properties ($L_{\rm acc}$, $\dot{M}_{\rm acc}$) in a self-consistent way, which makes use of the wide wavelength coverage of X-Shooter, and study their wind properties by mean of different forbidden emission lines analysis.

Keywords. stars: formation, stars: pre–main-sequence, planetary systems: protoplanetary disks

---

1. Introduction

Transitional Disks (TDs) are considered to be a late evolutionary stage of optically thick massive disks whose inner regions are being evacuated, leaving behind large holes that can be detected both by modeling the infrared spectral energy distribution (SED) or, in some cases, by mm-interferometry. These holes could be produced by processes of photoevaporation, grain growth, or planet formation. Still, none of these processes alone has been shown to be sufficient to explain all observations. In this context, the combination of inner hole size, mass accretion rate ($\dot{M}_{\rm acc}$) and wind properties is a powerful observational diagnostic of disk evolution models, but the current measurements of $\dot{M}_{\rm acc}$ for TDs are mostly based on secondary indicators (such as the 10% Hα width), and very few data on the wind properties for these objects are available.

2. Method

Accretion and spectral type fitter: We fit the observed spectra of our targets with a grid of models that include a range of photospheric template spectra (Class III YSOs from Manara *et al.* 2013a, augmented with some earlier spectral type (SpT) templates), a range of possible values for the extinction and for the excess spectrum produced by the disk accretion process, modeled as an isothermal hydrogen slab emission (for a detailed description of the method see Manara *et al.* 2013b). With the best fitting model (see Fig. 1a) we derive self-consistently from the complete X-Shooter spectrum SpT, A_V, and $L_{\rm acc}$ for the input target. In general, but with some notable exceptions (see Sect. 3), our findings confirm or slightly modify previous values available in the literature.

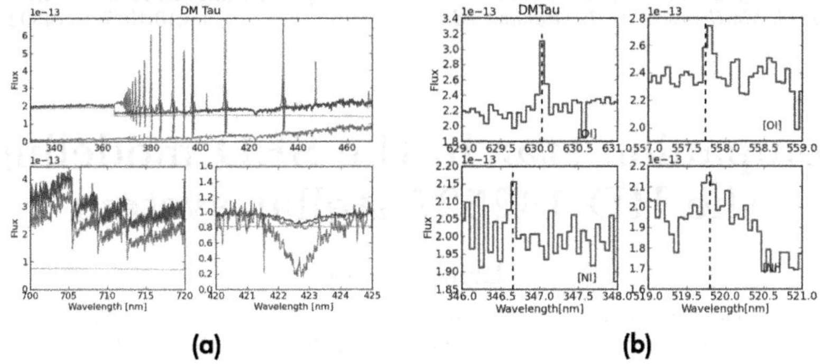

Figure 1. *(a)*: Example of fitting procedure described in Sect. 2. We show the Balmer jump fit, the normalization region, and the veiled Ca I line. *(b)*: Example of forbidden emission lines.

Forbidden lines: We search for emission in various forbidden lines to check for the presence of winds and disk photoevaporation in each source. In particular, we check various [OI] lines ($\lambda\lambda = 630, 636.4, 557.7$ nm), the [NI] lines ($\lambda\lambda = 519.8, 346.6$ nm), and the [SII] line at $\lambda = 673.1$ nm (see Fig. 1b). A clear detection of all these lines is usually evidence of outflows, possibly caused by disk photoevaporation. In order to asses that, we have to study in detail the kinematics and intensity of these lines.

3. Results and discussion

We confirm that the TD SR21 ($R_{in} = 36$ AU) is not accreting. The X-Shooter spectrum of this object does not show any signature of accretion, neither in the UV excess or in the emission lines. Another notable exception is Ser29, (SSTc2d J182911.5 + 002039, $R_{in} = 8$ AU), for which we find a value of $\dot{M}_{acc} = 1.6 \cdot 10^{-9} M_{\odot}/\text{yr}$ instead of the one reported in the literature, which is $\dot{M}_{acc} \sim 10^{-7} M_{\odot}/\text{yr}$ (Merìn *et al.* 2010). Most of the objects with large inner holes ($R_{in} > 30$ AU, e.g. SR21, LkHα330, RX J1615) do not show the signatures expected from photoevaporating winds (forbidden line emission). On the contrary, most of the objects with $R_{in} < 25$ AU show forbidden line emission.

Disk accretion signatures have been used to exclude the possibility that photoevaporation plays a dominant role in the disk clearing process (e.g. Espaillat *et al.* 2010), because gas should not be anymore present in the inner-hole if photoevaporation is occuring. As opposite to that, planet induced gaps may still allow for large values of $\dot{M}_{acc}$. On the other hand, photoevaporation models predict ongoing accretion for $\sim 10^{-5}$ years after hole opening, while the hole is still small (Owen *et al.* 2012). In our sample we see that some objects with very large inner holes, i.e. $R_{in} > 30$ AU, in particular SR21, have low or negligible accretion, and their inner holes could be possibly originated by photoevaporation. At the same time, there are some TDs with large inner holes and not negligible accretion rates ($\dot{M}_{acc} \gtrsim 10^{-9} M_{\odot}/\text{yr}$) for their masses ($M_* \gtrsim 1 M_{\odot}$) that, within the current theoretical framework, could be explained only with planet formation models. More analysis on their wind and stellar properties will help to understand their real origin.

References

Espaillat, C., D'Alessio, P., Hernandez, J., Nagel, E., *et al.* 2010, *ApJ*, 717, 441
Manara, C. F., Testi, L., Rigliaco, E., Alcalà, J. M., *et al.* 2013, *A&A*, 551, A107
Manara, C. F., Beccari, G., Da Rio, N., De Marchi, G., *et al.* 2013, *A&A*, 558, A114
Merìn, B., Brown, J. M., Oliveira, I., Herczeg, G. J., *et al.* 2010, *ApJ*, 718, 1200
Owen, J. E., Clarke, C. J., & Ercolano, B. 2012, *MNRAS*, 422, 1880

Exploring the Formation and Evolution of Planetary Systems
Proceedings IAU Symposium No. 299, 2013
M. Booth, B. C. Matthews & J. R. Graham, eds.

doi:10.1017/S1743921313008417

A companion star in the SED modeling of the HD 142527 stellar system

Erick Nagel

Departamento de Astronomía, Universidad de Guanajuato,
Callejón de Jalisco S/N, Valenciana, 36240, Guanajuato, México
email: erick@astro.ugto.mx

Abstract. The discovery of a likely companion of the Herbig Ae/Be star HD 142527 motivates the study of the effect that it produces on the SED. The main change on the system configuration is the formation of a gap in the disk in which the companion is embedded, following the orbit of the secondary star. This results in the formation of a wall (at the outer gap edge), which is illuminated by stellar radiation. We present a model for the SED, taking into account all the components of the system: the two stars, the disk with two gaps (one produced by the stellar companion and the other by potential planets), three walls (two associated with the gaps and the other with dust sublimation), optically thin material in the gaps and in a halo. The size of the modeled spherical halo is smaller than found in a previous study.

Keywords. binaries:general, planetary systems: protoplanetary disks

1. The HD 142527 stellar system

An image in the near-infrared by Fukagawa *et al.* (2006) shows two bright opposing arcs and a spiral arm. Mid-infrared images by Fujiwara *et al.* (2006) also show arc-like emission in the outer disk. Fukagawa *et al.* (2006) suggest that an eccentric companion is responsible for these features, while Fujiwara *et al.* (2006) suggest that the far side rim of the outer inclined disk is brighter because it is exposed, and the inner side is brighter due to forward scattering by the small grains, which explains their observation of the opposite arcs. Biller *et al.* (2012) used NIR observations and detected an emission peak which they interpret as a stellar companion about 14 *au* from the central star.

As a reference, we use the model of Verhoeff *et al.* (2011), which consists of three main components: (i) a small inner disk with inner and outer radii of 0.3 and 30 *au*, respectively; (ii) an optically thin spherical halo with radius 30 *au*; and (iii) an outer massive disk between 130 *au* and 200 *au* with a very high inner wall. For our model, the inner disk has a gap produced by the likely stellar companion. There is an outer gap between the inner and outer disk. Both gaps have optically thin dust.

2. Results

The SED of the disk is calculated using the model of a passive irradiated circumstellar disk around a Herbig Ae star by Dullemond, Dominik & Natta (2001). In the gaps, there is optically thin material that connects the outer with the inner parts of the disk, its emission is calculated as in Nagel *et al.* (2012). The emission of the outer rim is included using the formalism described by D'Alessio *et al.* (2005). The optically thin dust inside the gap is consistent with a planet at 90 *au* (Casassus *et al.* 2012). The importance of including this material is highlighted by Verhoeff *et al.* (2011), looking at a radial brightness profile which they were unable to fit. We consider that the innermost wall is

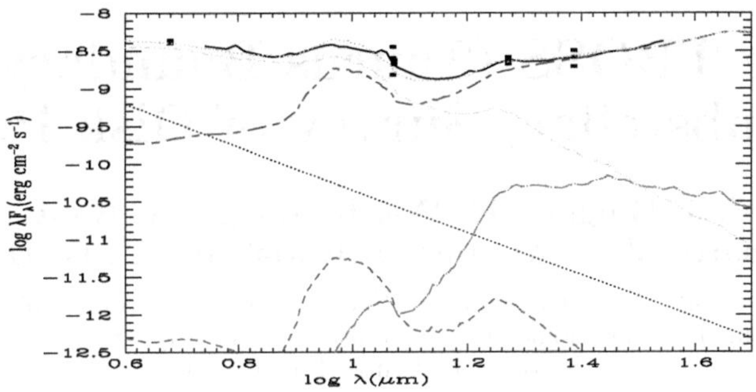

Figure 1. Spectrum of the binary+disk system HD 142527 with a $0.1\,M_\odot$ secondary star. The Spitzer spectrum is represented with a black line. The squares define the photometric fluxes (taken with errors from Verhoeff *et al.* 2011). The stellar spectrum is shown as a black pointed line. The red line represents the inner gap emission, the magenta line corresponds to the outer gap emission, the cyan line is the halo emission, the blue line is the inner and outer disk SED and finally in green is represented the best fit.

curved due to the dependency of sublimation temperature on density and its emission is calculated as in Nagel *et al.* (2013). From the fitting of the SED, the best model for a $0.1\,M_\odot$ companion star is shown in Fig. 1. The mass in dust located in the inner gap and outer gap is $3.52 \times 10^{-12}\,M_\odot$ and $8.88 \times 10^{-8}\,M_\odot$, respectively. The radius of the halo is $5\,au$, smaller than the value found in Verhoeff *et al.* (2011).

3. Conclusions

The SED fitting with the new distribution of optically thin dust leads to a halo that is six times smaller than previous models (i.e. $5\,au$ rather than $30\,au$). As pointed out by Verhoeff *et al.* (2011), a spherical halo of dust is not easy to obtain. A possible mechanism is dynamical excitation of planetesimals by an inwardly migrating planet (Krijt & Dominik 2011). For a 3-planet system, the interaction with an outer disk of planetesimals results in spherical clouds $100 - 1000\,au$ in size (Raymond & Armitage 2013). Stirring of material up to a height of $5\,au$ (this model) requires a weaker interaction with planets than the excitation required to produce a $30\,au$ halo (Verhoeff's model).

References

Biller, B., *et al.* 2012, *ApJ*, 753, L38

Casassus, S., Perez, M. S., Jordán, A., Ménard, F., Cuadra, J., Schreiber, M. R., Hales, A. S., & Ercolano, B. 2012, *ApJ*, 754, L31

D'Alessio, P., *et al.* 2005, *ApJ*, 621, 461

Dullemond, C. P., Dominik, C., & Natta, A. 2001, *ApJ*, 560, 957

Fujiwara, H., *et al.* 2006, *ApJ*, 644, L133

Fukagawa, M., Tamura, M., Itoh, Y., Kudo, T., Imaeda, Y., Oasa, Y., Hayashi, S. S., & Hayashi, M. 2006, *ApJ*, 636, L153

Krijt, S. & Dominik, C. 2011, *A&A*, 531, 80

Nagel, E., Espaillat, C., D'Alessio, P., & Calvet, N. 2012, *ApJ*, 747, 139

Nagel, E., D'Alessio, P., Calvet, N., Espaillat, C., & Trinidad, M. A. 2013, *Rev.Mexicana AyA*, 49, 43

Raymond, S. N. & Armitage, P. J. 2013, *MNRAS*, 429, L99

Verhoeff, A. P. *et al.* 2011, *A&A*, 528, 91

Exploring the Formation and Evolution of Planetary Systems
Proceedings IAU Symposium No. 299, 2013
M. Booth, B. C. Matthews & J. R. Graham, eds.

doi:10.1017/S1743921313008429

The TBOSS (Taurus Boundary of Stellar/Substellar) Survey of Disk Properties

J. Patience[1,2], J. Bulger[2], K. Ward-Duong[1], H. Bouy[3], C. Pinte[4], F. Ménard[4], J.-L. Monin[5], J. Koda[6] and C. D. Dowell[7]

[1]Arizona State University, [2]The University of Exeter, [3]Centro de Astrobiología, [4]Universidad de Chile, [5]Institut de Planétologie et d'Astrophysique (IPAG), [6]Stony Brook University, [7]California Institute of Technology

Abstract. With the combination of *Herschel* PACS (far-IR), submillimeter (submm) and millimeter (mm) ground-based observations, we are leading an investigation of all members with spectral types M4 and onwards in the Taurus star-forming region. This complete census spans the stellar to substellar boundary (M6.25), and the wavelength range covers the transition from optically thick to optically thin emission. From our *Hershel* PACS observations of ~135 sources, we obtain a detection rate of ~40% at 70 μm and a detection rate of ~20% at 160 μm, and provide the first far-IR measurements for the majority of these members. With our complementary submm and mm observations, best fit SED models from the radiative transfer code MCFOST will be used to infer disk properties such as scale height, mass, outer radius and maximum dust grain size. These comprehensive population statistics of disks are critical for testing star/brown dwarf and planet formation models around these later type members of Taurus.

Keywords. stars: low-mass, brown dwarfs

1. Background

Taurus represents an important population of young stars and brown dwarfs for which detailed investigations of disk frequencies and properties are possible. The presence of disks around stellar and substellar objects has been used to argue for a common formation mechanism or alternatively differences in the dynamical history compared to higher mass stars (Scholz & Jayawardhana 2008; Thies & Kroupa 2007).

2. Sample & Observations

There are 151 known members in Taurus that have spectral types of M4 or later, corresponding to low mass stars and brown dwarfs (Luhman *et al.* 2010). A complete census of *Hershel* PACS observations at 70 μm and 160 μm has been obtained for these members. The members comprise of 67 Class I and II objects and 71 Class III objects. From extensive observations with single dish ground based observations in the submillimeter with the Caltech Submillimeter Obervatory (CSO), and in the millimeter with IRAM, we also have obtained observations at 350/450 μm, and/or 1.2 mm for 30 low mass stars and brown dwarf members. High-resolution observations for 24 of the 151 Taurus members detected in the *Herschel* program will be observed at 850 μm with ALMA in Cycle 1.

3. Initial Results & Future Work

The number of *Herschel* PACS detections of this survey are shown in Fig. 1. Far-IR and submm maps are shown for CIDA-1, along with a modelled SED fit, obtained using the

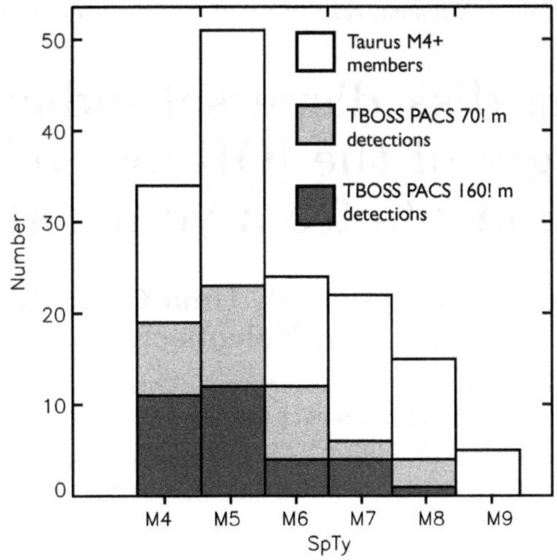

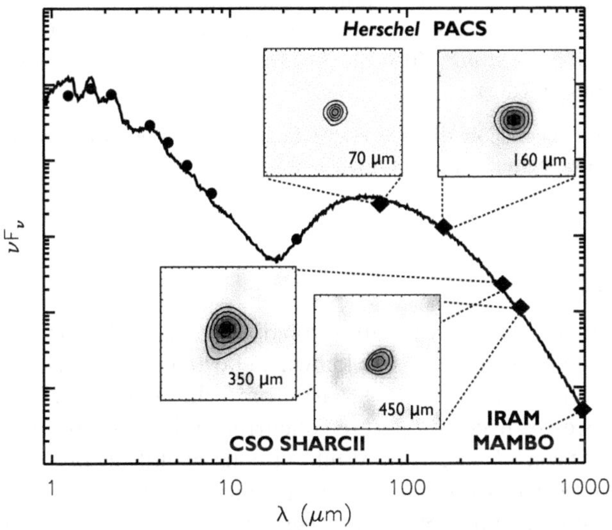

Figure 1. (top:) Histogram of all Taurus members with spectral types M4 and greater, showing the number of *Herschel* PACS detections at 70 μm and 160 μm. **(bottom:)** SED model fit for CIDA-1 (M5.5) obtained using the radiative transfer code MCFOST. Observations have been obtained in the far-IR with *Herschel* PACS at 70 μm and 160 μm, in the submm with the CSO at 350 μm and 450 μm and in the millimeter with IRAM at 1.2 mm for this target. Well sampled, multi-wavelength observations across the far-IR to millimeter wavelengths are essential in the breaking degeneracies amongst disk parameters (e.g. the redundancy of disk mass and outer radius).

radiative transfer code MCFOST (Pinte *et al.* 2006). Our multi-wavelength observations will provide a benchmark, detailed study of the structure of disks encircling low mass stars and brown dwarfs.

References

Luhman, K. L., Allen, P. R., Espaillat, C., Hartmann, L., & Calvet, N. 2010, *ApJS*, 186, 11
Pinte, C., Ménard, F., Duchêne, G., & Bastien, P. 2006, *A&A*, 459, 797
Scholz, A. & Jayawardhana, R. 2008, *ApJ*, 672, L49
Thies, I. & Kroupa, P. 2007, *ApJ*, 671, 767

Exploring the Formation and Evolution of Planetary Systems
Proceedings IAU Symposium No. 299, 2013
M. Booth, B. C. Matthews & J. R. Graham, eds.

doi:10.1017/S1743921313008430

Chasing disk dispersal indicators: the origin of the [OI] low-velocity components from young stars

Elisabetta Rigliaco[1], Ilaria Pascucci[1], Uma Gorti[2,3], Suzan Edwards[4] and David Holenbach[2]

[1]Lunar and Planetary Laboratory, The University of Arizona, Tucson, AZ 85721, USAs
email: rigliaco@lpl.arizona.edu

[2]SETI Institute, Mountain View, CA, USA [3]NASA Ames Research Center, Moffett Feld, CA
[4]Astronomy Department, Smith College, Northampton, MA 01063

Abstract. Understanding how a disk surrounding a young star evolves and disperses is crucial in order to understand the subsequent planet formation. In this proceeding, we summarize the results reported by Rigliaco *et al.* (2013) on the origin of the [OI] low-velocity component as a possible disk dispersal indicator.

Keywords. accretion, accretion disks – stars: formation – planetary systems:protoplanetary disks – stars: pre–main-sequence – ultraviolet:stars

---

1. Introduction

Formation of planets around stars are strongly influenced by how circumstellar disks disperse. The more efficient mechanisms known so far in order to disperse protoplanetary disks are viscous evolution (accretion of matter onto the central star) and photoevaporation. These two processes need to operate concurrently in order to explain disk lifetimes (Gorti *et al.* 2011, Owen *et al.* 2010). While viscous evolution produces several observables that can be used to estimate the rate at which the star is accreting (Alcalá *et al.* 2013 for a review), just a couple of wind tracers are known so far. The low-velocity component (LVC) of the [NeII] line at 12.81μm has been found to trace a photevaporative disk wind driven by stellar X-rays/EUV photons (Pascucci & Sterzik 2009, Sacco *et al.* 2012, Baldovin-Saavedra *et al.* 2012, Hollenbach & Gorti 2009, Ercolano & Owen 2010). Another diagnostic of disk wind is the CO ro-vibrational band at 4.7μm. This emission line, ubiquitous around young stars surrounded by disks, has been interpreted as arising in a slow disk wind (Pontoppidan *et al.* 2011, Bast *et al.* 2011, Brown *et al.* 2013). In the following sections we will discuss the [OI] LVC as a possible disk dispersal tracer.

2. Results

We have collected a sample of 8 T Tauri stars with high-resolution optical spectra, and with at least a detection of the [NeII] line and/or the CO line. We have compared the [OI]-LVC profile with the [NeII] and the CO line profiles, and we find that these lines have different profiles, suggesting that they are arising from different regions around the stars (Rigliaco *et al.* 2013). In order to compare the properties of the [OI]-LVC with the stellar properties (e.g. masses, luminosities, accretion luminosities), we have re-analyzed the comprehensive survey of T Tauri stars in Taurus done by Hartigan *et al.* (1995).

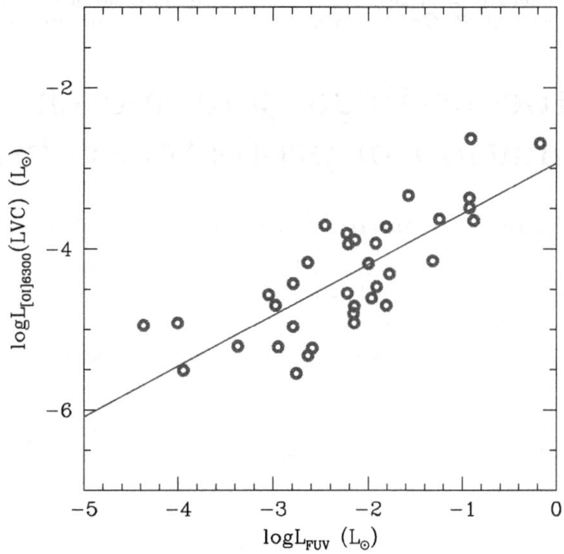

Figure 1. [OI] LVC luminosity versus FUV luminosity. The $L_{[OI]LVC}$ - L_{FUV} relation is $LogL_{[OI]LVC}$=(0.63 ± 0.09)×$LogL_{FUV}$ - (2.94 ± 0.21). The FUV luminosities come from Yang *et al.* 2012.

We find that there is a clear trend of increasing [OI]-LVC luminosities for increasing FUV luminosities (see Fig. 1 and Rigliaco *et al.* 2013). This trend, together with the low [OI]6300Å/[OI]5577Åline ratio (Rigliaco *et al.* 2013) suggests that the [OI]-LVC is tracing the disk layer which is heated by the FUV photons coming from the central star. These FUV photons penetrate the outflow columns of classical T Tauri stars and reach the disk surface where they dissociate the OH molecules producing the oxygen atoms in the ^1D and ^1S states. Once that these atoms decay to the ground state they produce the observed [OI]6300Åand [OI]5577Ålines. From the analysis of the higher resolution and signal-to-noise spectra it is possible to distinguish two different component of the LVC. A bound component of gas in Keplerian rotation, and an unbound component produced by gas at radial distances >10AU, likely a photoevaporative wind (see Rigliaco *et al.* 2013).

References

Alcalá, J. M, Natta, A., Manara, C. F., *et al.* 2013, *A&A*, submitted

Baldivin-Saavedra, C., Audard, M., Carmona, A. *et al.* 2012, *A&A*, 543, 30

Bast, J. E., Brown, J. M., Herczeg, G. J., van Dishoeck, E. F., & Pontoppidan, K. M. 2011, *A&A*, 527, A119

Brown, J., Pontoppidan, K., van Dishoeck, E., *et al.* 2013, *ApJ*, 770, 94B

Ercolano, B. & Owen, J. 2010, *MNRAS*, 406, 1553

Gorti, U., Hollenbach, D., Najita, J. & Pascucci, I. 2011, *ApJ*, 735, 90

Hollenbach, D. & Gorti, U. 2009, *ApJ*, 703, 1203

Owen, J. E., Ercolano, B., Clarke, C. J., & Alexander, R. D. 2010, *MNRAS*, 401, 1415

Pascucci, I. & Sterzik, M. 2009, *Apj*, 702, 724

Rigliaco, E., Pascucci, I., Gorti, U., Edwards, S., & Hollenbach, D. 2013, *ApJ*, 772, 60

Sacco, G. G., Flaccomio, E., Pascucci, I., *et al.* 2012, *ApJ*, 747, 142

Yang, H., Herczeg, G. J., Linsky, J. L., *et al.* 2012, *ApJ*, 744, 121

Exploring the Formation and Evolution of Planetary Systems
Proceedings IAU Symposium No. 299, 2013
M. Booth, B. C. Matthews & J. R. Graham, eds.

doi:10.1017/S1743921313008442

Magnetocentrifugal jets and chondrule formation in protostellar disks

Raquel Salmeron[1,2] and Trevor Ireland[2]

[1]Research School of Astronomy & Astrophysics, The Australian National University, Weston Creek, ACT 2611, Australia; email: raquel@mso.anu.edu.au

[2]Research School of Earth Sciences, The Australian National University, Canberra, ACT 0200, Australia; email: trevor.ireland@anu.edu.au

Abstract. Chondrite meteorites are the building blocks of the solar nebula, out of which our Solar System formed. They are a mixture of silicate and oxide objects (chondrules and refractory inclusions) that experienced very high temperatures, set in a matrix that remained cold. Their prevalence suggests that they formed through a very general process, closely related to stellar and planet formation. However the nature and properties of the responsible mechanism have remained unclear. The evidence for a hot solar nebula provided by this material seems at odds with astrophysical observations of forming stars. These indicate that the typical temperatures of protostellar disks are too low to melt and vapourise silicate minerals at the radial distances sampled by chondrule-bearing meteorites. Here, we show that processing of precursors in a protostellar outflow at radial distances of about 1 – 3 AU can heat them to their melting points and explain their basic properties, while retaining association with the colder matrix.

Keywords. solar system: formation, planetary systems: protoplanetary disks, ISM: jets and outflows

---

Protostellar winds are common in star-forming regions (e.g. Bally *et al.* 2007). The ubiquity of these winds and the correlation between accretion and outflow signatures in young stars (e.g. Cabrit 2007) suggest that they are an integral part of the star formation process. These winds are thought to efficiently remove the excess angular momentum from disk matter, enabling the flow of material to the centre to form the star. Disk-winds originate from radially extended regions of the disk, and are believed to be accelerated magnetocentrifugally from disk surfaces by the mechanism proposed by Blandford & Payne (1982). In this scenario, the disk is threaded by large-scale magnetic field lines, which bend radially outwards. If they are sufficiently inclined, matter near the disk surface can be accelerated centrifugally along the field lines (the bead-on-a-wire effect) initiating a large-scale outflow.

To explore the fate of particles embedded in a protostellar wind, we first obtained a radially-localised wind solution for fluid conditions expected to be typical in protostellar disks at a distance of 1 AU from the protostar (see Königl *et al.* 2010; Salmeron *et al.* 2011, for the method and parameter space solutions). We then calculated the vertical motion, growth, and thermal processing of spherical, compact dust particles embedded in the fluid (Salmeron & Ireland 2012a,b). Refractory precursor aggregates are heated while being lifted in the wind, grow through amalgamation, and eventually become heavy enough to drop back to the disk where they may assemble with the matrix. Small particles (radii $\sim$ 0.01 cm) are accelerated away from the disk whereas cm-sized particles sink to the midplane. Intermediate-sized particles (radii $\sim$ 0.1 cm) are initially lifted but, as they grow, gravity causes them to turn back towards the disk. As shown in figure

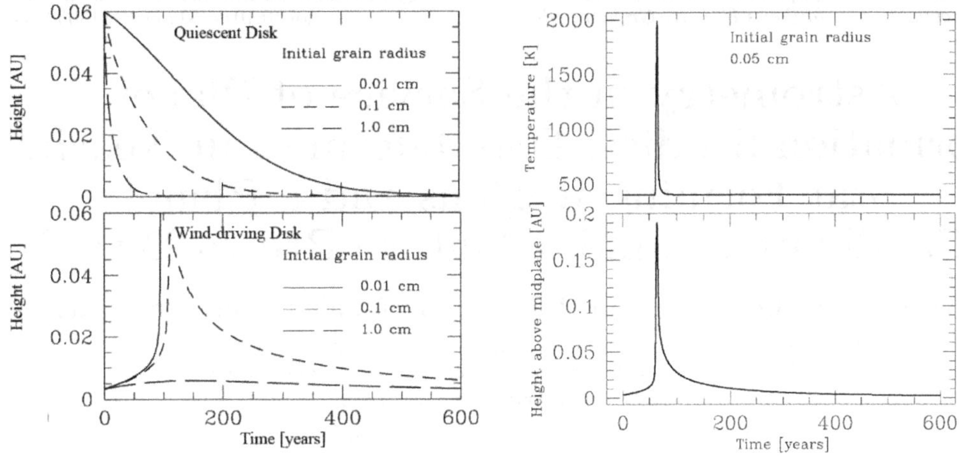

Figure 1. Left panel: Height evolution of dust particles of different sizes embedded in a quiescent or a wind-driving disk. They move through the disk and grow by amalgamation with smaller particles. In a quiescent disk, the particles rapidly settle to the midplane (top, Dullemond & Dominik 2005; Salmeron & Ireland 2012b). In a wind-driving disk (bottom panel), very small grains are accelerated away by the outflow, while massive (cm-sized) particles sink back to the midplane. Intermediate-sized grains ($\sim$ 0.1 cm) are initially lifted but, as they grow, gravity becomes dominant and pulls them back to the disk. **Right panel:** Evolution of temperature (top panel) and height above the midplane (bottom panel), of a dust particle with an initial radius of 0.05 cm and a temperature of 400 K, released at z = 0.003 AU in a typical disk-wind solution (Salmeron & Ireland 2012a). The peak temperature experienced by the particle is near 2000 K.

1, we find that such processing at radial distances of $\sim$ 1-3 AU can heat precursors to the appropriate temperature to melt chondrules and explain their basic properties, while retaining association with the colder material that provides the matrix.

Processing in a disk wind naturally explains the tight size range of chondrules in a given chondrite, as well as the large proportion of chondrule material present in chondrite meteorites. Formation of compound chondrules is expected as the relative speed between similarly-sized chondrules remains low. Dusty rims can be accreted as chondrules fall back to the disk, and during recycling to the disk interior. Finally, the complementary composition of chondrules and matrix can be maintained, as the processed material remains local to the wind-launching region in the nebula.

References

Bally, J., Reipurth, B., & Davis, C. J. 2007, *Protostars and Planets V*, 215
Blandford, R. D. & Payne, D. G. 1982, *MNRAS*, 199, 883
Cabrit, S. 2007, *Star-Disk Interaction in Young Stars*, 243, 203
Dullemond, C. P. & Dominik, C. 2005, *A & A*, 434, 971
Königl, A., Salmeron, R. & Wardle, M. 2010, *MNRAS*, 401(1), 479
Salmeron, R. & Ireland, T. 2012a, *Earth and Planetary Science Letters* (2012), 327, 61
Salmeron, R. & Ireland, T. 2012b, *Meteoritics and Planetary Science*, 47(12), 1922
Salmeron, R., Königl, A., & Wardle, M. 2011, *MNRAS*, 412(2), 1162

Exploring the Formation and Evolution of Planetary Systems
Proceedings IAU Symposium No. 299, 2013
M. Booth, B. C. Matthews & J. R. Graham, eds.

doi:10.1017/S1743921313008454

Astrometry in the Service of Planet Formation Studies: Disk Lifetimes in Nearby Star Forming Regions and a Planet Candidate around a Mature Brown Dwarf

Alycia J. Weinberger[1], Alan P. Boss[1] and Guillem Anglada-Escudé[2]

[1]Department of Terrestrial Magnetism, Carnegie Institution of Washington
5241 Broad Branch Rd NW, Washington, DC 20015 USA
email: weinberger@dtm.ciw.edu

[2]Universität Göttingen, Institut für Astrophysik
Friedrich-Hund-Platz 1, 37077 Göttingen, Germany
email: guillem.anglada@gmail.com

Abstract. We present preliminary astrometric results aimed at understanding the lifetime of circumstellar disks and potential for planet formation. We have obtained parallaxes to stars in the TW Hydrae, Upper Scorpius, and Chamaeleon I stellar associations. These enable new estimates for the ages of the stars. We are also performing the Carnegie Astrometric Planet Search of nearby low mass stars for gas giant planets on wide orbits. We have our first candidate around a mature brown dwarf.

Keywords. astrometry, planetary systems: formation, stars: pre–main-sequence

1. Introduction

Nearby young stars of age 5-10 Myr provide our best opportunity to study the late stages of star and planet formation with high sensitivity and spatial resolution. During this time period, the last gas-rich disks dissipate, and the onset of the debris disk phase occurs. Even for nearby star-forming regions, such as Chamaeleon I, Upper Scorpius or the more diffuse TW Hydrae Association, no distances exist to the numerous low mass members (K spectral type and later) because they were too faint for Hipparcos to measure. The luminosities for the H-R diagrams are calculated by using an average distance to the association. Overplotted with pre-main sequence tracks, and taken at face-value, the H-R diagrams appear to imply a wide range of ages within each association. To measure how long circumstellar disks last, a median age is read off that H-R diagram and plotted with disk fraction. We need distances to get more precise individual ages that would clarify the disk evolution within each cluster/association.

We use CAPSCam, a custom astrometric camera, on the 2.5m DuPont Telescope at Las Campanas Observatory. The instrument and basic data reduction techniques are described in Boss *et al.* (2009). The observations are made at z-band. We have embarked on a project to measure parallaxes for >60 K and M stars in the three aforementioned star-forming regions. We are also engaged in the Carnegie Astrometric Planet Search program (CAPS), which seeks to detect giant planets on wide orbits, if they exist.

2. Astrometric Planet Candidate Around a Field Brown Dwarf

We are surveying ~100 low mass stars and brown dwarfs, and we have collected data since 2007. Systems with outer giant planets, like our own, could signal inner habitable worlds. With CAPSCam, we have an astrometric accuracy of ~0.3 mas, which should enable us to detect Jupiter-mass planets around nearby very low mass stars and brown dwarfs. Our first CAPS planet candidate is around an L5-type brown dwarf host.

In 13 epochs of data of this brown dwarf taken over four years (almost 2 orbital periods), we find astrometric motion with a orbital period of ~800 d with a false alarm probability of <1%. The primary has a mass of ~50 M_{Jup}, making the mass of the secondary ~3 M_{Jup}. Given the painful history of claims and retractions of supposedly detected astrometric planets, we will attempt to confirm our astrometric detection with RV by combining Keck/NIRSPEC velocity measurements with our astrometry to refine the mass and orbital parameters of the planet.

3. Parallaxes and Ages in TWA, Upper Sco, and Cha I

The TW Hya Association (TWA) describes ~30 stars identified from stellar activity (age) and common space motion searches. Disks in TWA range from 4 accreting, gas-rich proto-planetary to 7 traditional or debris disks to the majority of members with no detectable disks. We observed 14 primary TWA members plus 2 visual binary companions. Results were published this year in Weinberger *et al.* (2013). The Galactic space motions of the members are parallel and do not indicate convergence at a common formation point sometime in the last few million years. The median parallax is 18 mas or 56 pc. We analzyed the stars' absolute magnitudes on pre-main sequence evolutionary tracks and found a range of ages with a median of 10.1 Myr and no correlation between age and Galactic location. The TWA stars may have formed along an extended and filamentary molecular cloud.

Upper Scorpius is the smallest region of Scorpius-Centaurus star forming region. We are observing disk-bearing and disk-less stars, as determined from Spitzer spectral energy distributions (Carpenter *et al.* 2006). We have thus far obtained parallaxes to 22 K and M stars in Upper Sco: 10 disk-bearing, 9 disk-less, and 3 unstudied for disks. We use our distances to compute absolute magnitudes, and thereby to find the ages of the stars as they appear on evolutionary tracks by Baraffe *et al.* (1998). In our preliminary work, the disk stars have a median age of 14.4 Myr while the disk-less stars have a median age of 7.6 Myr. The whole sample has a median age of 10.8 Myr. At the two-sigma level, the disk stars appear older. This is contrary to expectation because older stars generally have a smaller disk fraction. However, the uncertainties are large in this preliminary work, and we are building a larger sample of parallaxes.

Chamaeleon I is a relatively compact region of recent low-mass star formation with no well-determined distance. We have thus far obtained parallaxes to 7 K/M stars in Cha I, and preliminary results are presented here. The average distance to these stars is 6.5 ± 0.4 mas or 153 ± 10 pc. The mean age of these stars, as determined by their positions on the evolutionary tracks by Baraffe *et al.* (1998) is 5.8 Myr, confirming that Cha I is younger than Upper Sco. Our sample so far includes one transitional disk, three Class II and two Class III T Tauri stars.

References

Baraffe, I., Chabrier, G., Allard, F., & Hauschildt, P. H. 1998, *A&A*, 337, 403

Boss, A. P., Weinberger, A. J., Anglada-Escudé, G., Thompson, I. B., Burley, G., Birk, C., Pravdo, S. H., Shaklan, S. B., Gatewood, G. D., Majewski, S. R., & Patterson, R. J. 2009, *PASP*, 121, 1218

Carpenter, J. M., Mamajek, E. E., Hillenbrand, L. A., & Meyer, M. R. 2006, *ApJ*, 651, L49

Weinberger, A. J., Anglada-Escudé, G., & Boss, A. P. 2013, *ApJ*, 762, 118 A. P.

Exploring the Formation and Evolution of Planetary Systems
Proceedings IAU Symposium No. 299, 2013
M. Booth, B. C. Matthews & J. R. Graham, eds.

doi:10.1017/S1743921313008466

The Debiased Kuiper Belt: Our Solar System as a Debris Disk

Samantha M. Lawler and the CFEPS Team

Department of Physics and Astronomy
University of British Columbia
6224 Agricultural Road
Vancouver, BC V6T 1Z1, Canada
email: lawler@astro.ubc.ca

Abstract. The dust measured in debris disks traces the position of planetesimal belts. In our Solar System, we are also able to measure the largest planetesimals directly and can extrapolate down to make an estimate of the dust. The zodiacal dust from the asteroid belt is better constrained than the only rudimentary measurements of Kuiper belt dust. Dust models will thus be based on the current orbital distribution of the larger bodies which provide the collisional source. The orbital distribution of many Kuiper belt objects is strongly affected by dynamical interactions with Neptune, and the structure cannot be understood without taking this into account. We present the debiased Kuiper belt as measured by the Canada-France Ecliptic Plane Survey (CFEPS). This model includes the absolute populations for objects with diameters >100 km, measured orbital distributions, and size distributions of the components of the Kuiper belt: the classical belt (hot, stirred, and kernel components), the scattering disk, the detached objects, and the resonant objects (1:1, 5:4, 4:3, 3:2 including Kozai subcomponent, 5:3, 7:4, 2:1, 7:3, 5:2, 3:1, and 5:1). Because a large fraction of known debris disks are consistent with dust at Kuiper belt distances from the host stars, the CFEPS Kuiper belt model provides an excellent starting point for a debris disk model, as the dynamical interactions with planets interior to the disk are well-understood and can be precisely modelled using orbital integrations.

Keywords. Kuiper Belt, solar system: formation, (stars:) planetary systems

---

1. Introduction

There is much discussion in the literature about using resonant structure in debris disks as an indirect way to find planets (e.g. Wyatt 2006; Stark & Kuchner 2008). In debris disks, the observable components are the dust grains, which are used to infer the presence of a collisionally-grinding planetesimal belt (Wyatt 2008). In our own Solar System, the opposite is true: we can directly observe orbits of the planets and planetesimals to great precision, while no direct observations exist of dust in the Kuiper belt. We also know that the Kuiper belt has significant resonant structure (Gladman *et al.* 2012), which requires that Neptune's orbit has migrated outwards. This migration may have happened smoothly as Neptune scattered planetesimals inwards onto Jupiter-crossing orbits, "snowplowing" objects into mean-motion resonances along the way (Chiang & Jordan 2002; Hahn & Malhotra 2005), or Neptune may have been scattered outward by a close encounter with one of the other giant planets, and trapped planetesimals in mean-motion resonances as its eccentric orbit damped to near-circular (the "Nice model", Levison *et al.* 2008).

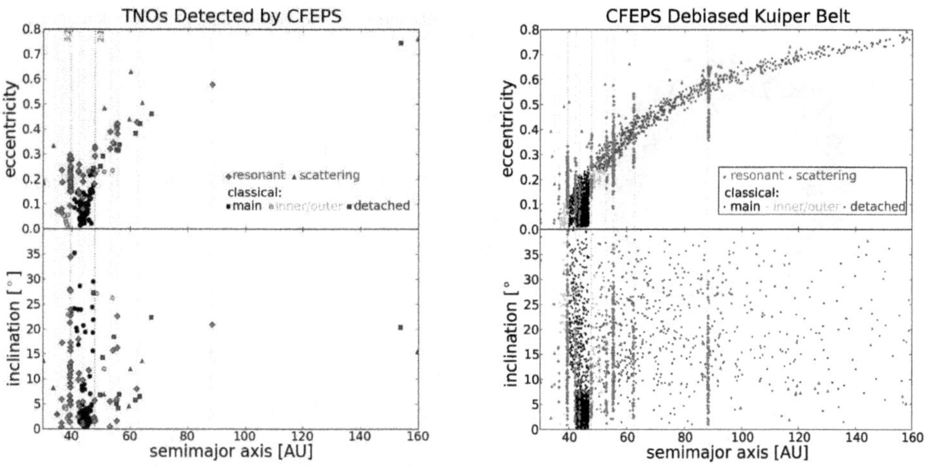

Figure 1. Left panel shows orbital elements (a,e,i) of TNOs discovered by CFEPS, and right panel shows debiased TNO orbital element distributions, where the relative numbers in each population and in each resonance are correct. Objects have symbols according to dynamical class (see key), and mean-motion resonance positions are indicated by dotted lines.

2. Survey Strategy

The Canada-France Ecliptic Plane Survey (CFEPS) was a large survey carried out with the Megacam instrument on the Canada-France-Hawaii Telescope (CFHT) between 2003–2007, with the goal of discovering many trans-Neptunian objects (TNOs) and using the known sky coverage, obervation depths, and tracking fractions in different observing blocks at many points along the ecliptic plane to debias the observations and regain the true orbital distribution of the Kuiper belt.

The observing strategy for CFEPS is described in detail in Jones *et al.* (2006), Kavelaars *et al.* (2009) and Petit *et al.* (2011). To summarize, a patch of sky is observed three times in a night, with each observation separated by about one hour. These triplets of images are searched for moving objects, and preliminary orbits are calculated. One month later, followup images are taken of the same field, and the same objects are recovered. This is repeated each month for 4–5 months. This 4–5 month baseline is used to predict object positions one year later, and the objects are recovered again, further improving the calculated orbits.

A 10 Myr-long orbital integration is performed using the orbital elements and position of each TNO, and including the gravitational effects of the four giant planets. The TNOs are then placed into dynamical classes following the procedure outlined by Gladman *et al.* (2008). The classes are resonant (TNOs in mean-motion resonances with Neptune), scattering (TNOs that closely encounter Neptune within 10 Myr), detached (TNOs that currently do not dynamically interact with Neptune), and classical (TNOs on low-e, low-i orbits with a within a few AU of Neptune).

3. Discoveries and Debiasing

CFEPS detected 176 TNOs. Their orbital element distributions are shown in the left panel of Figure 1.

Because the survey is well-characterized, it can be debiased to reproduce not only the orbital elements of each different dynamical class, but also the absolute populations. We

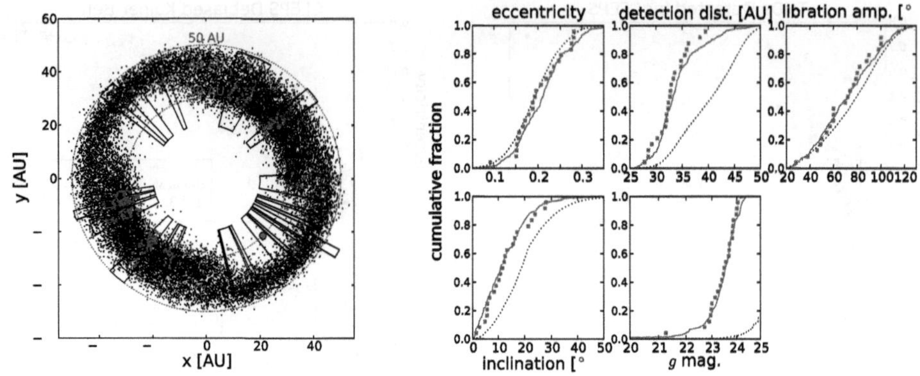

Figure 2. Left panel shows a distribution of plutinos viewed from above the Solar System, and right panel shows cumulative distributions in five different observable parameters: eccentricity, distance at detection, libration amplitude, inclination, and visible g magnitude. Black (in both panels) shows synthetic non-detections, blue circles/solid lines show synthetic detections by the survey simulator, and red squares shows real detections. Also shown in the left panel are the CFEPS observing blocks, with the relative magnitude limits indicated by the outer edge of each wedge (showing the distance limit where a ~100 km diameter object would still be detected).

will use the plutinos (objects in the 3:2 mean-motion resonance) as an example to explain the debiasing process (Figure 2).

First, we choose parameters to model the distribution of synthetic plutinos in eccentricity, inclination, libration amplitude, and size distribution (shown in black in Figure 2), which is discussed in detail in Gladman *et al.* (2012) and Lawler & Gladman (2013). This synthetic distribution is run through the CFEPS survey simulator, where each object is evaluated to determine if it would have been detected by CFEPS. Synthetic detections are shown in blue in Figure 2. The cumulative distributions of five parameters of the synthetic detections are compared statistically to the cumulative distributions of the real CFEPS detections. Any modeled distribution where the bootstrapped comparison statistic for any of the five distributions occurs <5% of the time is considered rejected. In this way, we get very powerful constraints on the orbital element distributions and the size distribution, and we can use this to calculate absolute population.

This modelling process is repeated for each resonance (Gladman *et al.* 2012) and other Kuiper belt components individually (Petit *et al.* 2011).

4. A Simple Kuiper Belt Dust Model

Many very detailed models of the dust produced by the Kuiper belt exist in the literature (e.g. Vitense *et al.* 2010; Kuchner & Stark 2010). The model presented here is only meant to serve as a demonstration of how the CFEPS debiased Kuiper belt distribution (available for download at `www.cfeps.net`) can be used as a starting point for a Kuiper belt dust simulation.

Figure 3 shows contour plots of the number density of the debiased resonant (right panels) and non-resonant TNOs (left panels). Applying a simple dust flux model, where each TNO is replaced by a modified blackbody dust grain, the resonant structure is completely drowned out by the much higher number density of classical objects.

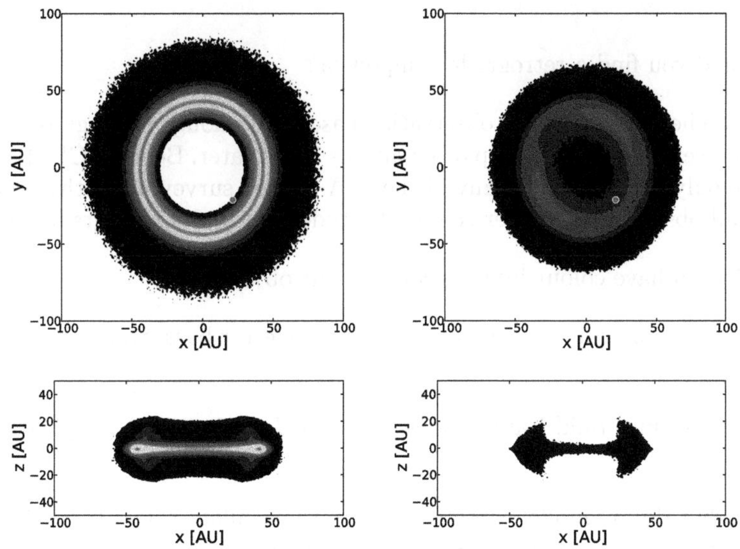

Figure 3. Number density contour plots of the classical (left panels) and resonant (right panels) debiased Kuiper belt looking down on the Solar System from above (top panels) and from edge-on (bottom panels). The position of Neptune is shown by a green circle in the top panels. The two top panels have the same contour levels, and the two bottom panels have the same contour levels, making it obvious that the classical objects dominate the number density.

5. Conclusions

The Kuiper belt has been sculpted by giant planet migration, and contains many populated resonances (20% of >100 km objects are resonant), but this is difficult, if not impossible, to see in the overall disk structure as seen from afar. Currently, no theoretical giant planet migration models can reproduce all the structure that is observed in the Kuiper belt (Gladman *et al.* 2012). Because of the completely unknown biases that are present in the Minor Planet Center Database, the CFEPS debiased Kuiper belt distribution is currently the most accurate model for TNOs >100 km in diameter, and makes an excellent starting point for collisional dynamical Kuiper belt dust models.

References

Chiang, E. I. & Jordan, A. B. 2002, *AJ*, 124, 3430
Gladman, B., Marsden, B., & Van Laerhoven, C. 2008, *The Solar System Beyond Neptune*, 43
Gladman, B., Lawler, S. M., Petit, J.-M., *et al.* 2012, *AJ*, 144, 23
Hahn, J. M. & Malhotra, R. 2005, *AJ*, 130, 2392
Jones, R. L., Gladman, B., Petit, J.-M., *et al.* 2006, *Icarus*, 185, 508
Kavelaars, J. J., Jones, R. L., Gladman, B. J., *et al.* 2009, *AJ*, 137, 4917
Kuchner, M. J. & Stark, C. C. 2010, *AJ*, 140, 1007
Lawler, S. M. & Gladman, B. 2013, *AJ*, 146, 6
Levison, H., Morbidelli, A., Van Laerhoven, C., Gomes, R., & Tsiganis, K. 2008, *Icarus*, 196, 258
Petit, J.-M., Kavelaars, J. J., Gladman, B. J., *et al.* 2011, *AJ*, 142, 131
Stark, C. C. & Kuchner, M. J. 2008, *ApJ*, 686, 637
Vitense, C., Krivov, A. V., Lohne, T. 2010, *AA*, 520, A32
Wyatt, M. C. 2006, *ApJ*, 639, 1153
Wyatt, M. C. 2008, *ARAA*, 46, 339

Discussion

TRIAUD: Would you find a retrograde component?

LAWLER: Yes. The 5 month initial observation baseline is enough to diagnose a retrograde orbit and not lose it before follow-up observations 1 year later. Because CFEPS was in the ecliptic plane only, we didn't find any of these. A related survey using the same strategy as CFEPS and observing at higher ecliptic latitudes *did* find a retrograde object.

RAFIKOV: Do you have colour information on your objects?

LAWLER: Yes, but only g and r bands, and with large error bars. So not very good colour information.

MACINTOSH: Very nice talk! I am curious what the L_{IR}/L_* of the simple disk emission model is?

LAWLER: I didn't do my model carefully enough to calculate this. My model was really just supposed to be a demonstration of the CFEPS dataset and was supposed to encourage all of you in the audience to use our Kuiper Belt model to do a better dust model than this.

KRIVOV: What is the total mass of your debiased Kuiper Belt?

LAWLER: There are ≈130,000 objects >100 km in diameter, so a rough order-of-magnitude calculation gives about 1 Earth mass.

DVORAK: I expect that you also found Neptune Trojans.

LAWLER: CFEPS did not find any Neptune Trojans. The non-detection gives an upper limit of ≈300-400 Neptune Trojans, which is consistent with the known Neptune Trojans. It isn't unexpected that CFEPS didn't detect any.

Exploring the Formation and Evolution of Planetary Systems
Proceedings IAU Symposium No. 299, 2013
M. Booth, B. C. Matthews & J. R. Graham, eds.

doi:10.1017/S1743921313008478

Planets and Stellar Activity: Hide and Seek in the CoRoT-7 system

R. D. Haywood[1], A. C. Cameron[1], D. Queloz[2], S. C. C. Barros[3], M. Deleuil[3], R. Fares[1], M. Gillon[4], A. Hatzes[5], A. F. Lanza[6], C. Lovis[2], C. Moutou[3], F. Pepe[2], D. Pollacco[7], A. Santerne[3,8], D. Ségransan[2] and Y. Unruh[9]

[1]School of Physics and Astronomy, University of St Andrews, St Andrews KY16 9SS, UK
email: rdh4@st-andrews.ac.uk

[2]Observatoire de Genève, 51 Ch. des Maillettes, 1290 Sauverny, Switzerland

[3]Aix Marseille Université, CNRS, LAM (Laboratoire d'Astrophysique de Marseille) UMR 7326, 13388, Marseille, France

[4]Institut d'Astrophysique et de Géophysique, Université de Liège, Allée du 6 août 17, Bat. B5C, 4000 Liège, Belgium

[5]Thüringer Landessternwarte, D-07778 Tautenburg, Germany

[6]INAF-Osservatorio Astrofisico di Catania, via S. Sofia, 78 - 95123 Catania. Italy

[7]Department of Physics, University of Warwick, Coventry CV4 7AL, UK

[8]Centro de Astrofísica, Universidade do Porto, Rua das Estrelas, 4150-762 Porto, Portugal

[9]Astrophysics Group, Blackett Laboratory, Imperial College London, London SW7 2AZ, UK

Abstract. Since the discovery of the transiting Super-Earth CoRoT-7b, several investigations have been made of the number and precise masses of planets present in the system, but they all yield different results, owing to the star's high level of activity. Radial velocity (RV) variations induced by stellar activity therefore need to be modelled and removed to allow a reliable detection of all planets in the system. We re-observed CoRoT-7 in January 2012 with both HARPS and the CoRoT satellite, so that we now have the benefit of simultaneous RV and photometric data. We fitted the off-transit variations in the CoRoT lightcurve using a harmonic decomposition similar to that implemented in Queloz *et al.* (2009). This fit was then used to model the stellar RV contribution, according to the methods described by Aigrain *et al.* (2011). This model was incorporated into a Monte Carlo Markov Chain in order to make a precise determination of the orbits of CoRoT-7b and CoRoT-7c. We also assess the evidence for the presence of one or two additional planetary companions.

Keywords. planetary systems, stellar activity, RV method

1. Introduction

CoRoT-7 was first brought to the attention of the scientific community in 2009, with the discovery of a transiting Super-Earth CoRoT-7b with a measured radius of 1.68 ± 0.09 $R_\oplus$ (Léger *et al.* 2009). This discovery was followed by an intensive HARPS RV campaign (Queloz *et al.* 2009), in order to obtain the mass of CoRoT-7b and assess the presence of additional planets. Several analyses of these photometric and spectroscopic data were subsequently made (Bruntt *et al.* (2010), Lanza *et al.* (2010), Hatzes *et al.* (2010), Pont *et al.* (2010), Hatzes *et al.* (2011), Boisse *et al.* (2011), Ferraz-Mello *et al.* (2011)), but an agreement on the number and masses of planets present in the system is still to be reached. This is due to the star's high level of activity. Indeed, the presence of spots and

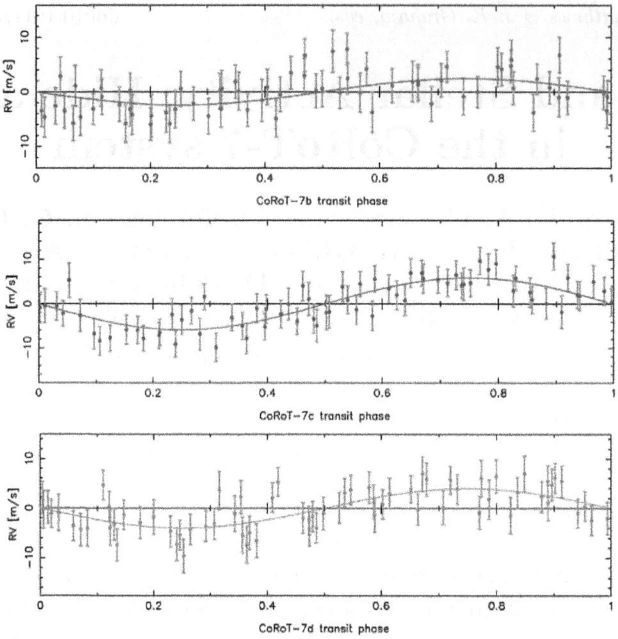

Figure 1. Phase plots of all planets, with circular orbit models overplotted.

other magnetic features on a star's surface leads to RV perturbations of the order of a few m s^{-1} and can thus mask the RV signals produced by the orbit of low mass planets.

2. Data and analysis

In 2012 the CoRoT-7 system was re-observed with the CoRoT satellite (see Barros *et al.* (in prep.)), and with the spectrometer HARPS simultaneously. This allows us to use the star's lightcurve to model the activity-induced RV variations, according to the model presented by Aigrain *et al.* (2011). Details of our implementation of this method are given by Haywood *et al.* (submitted).

3. Results

We confirm the presence of three planets in the system: CoRoT-7b, with a mass of 3.38 ± 0.86 $M_\oplus$ at an orbital period of $0.85359165 \pm 5.6 \times 10^{-7}$ day (period derived from the analysis of the 2012 transit data by Barros *et al.* (in prep.)), and two sub-Neptune mass planets, CoRoT-7c (13.31 ± 1.27 $M_\oplus$) and CoRoT-7d (11.92 ± 2.11 $M_\oplus$), at orbital periods of 3.68 ± 0.03 and 8.54 ± 0.24 days, respectively. The phase plots for each planet RV signal are shown in Figure 1.

Figure 2 shows the RV contribution of the activity models and each planet orbit. The activity-induced RV variations clearly dominate the total signal. Moreover, the residuals shown in the bottom panel of Figure 2 display some correlated noise, which may arise from incomplete modelling of activity-induced RV variations. Indeed, several physical phenomena are not accounted for in our activity model. For example, the *FF'* method does not consider the impact of faculae on the suppression of convective blueshift, or the RV effect of large inflows towards active regions recently found on the Sun (Gizon *et al.*

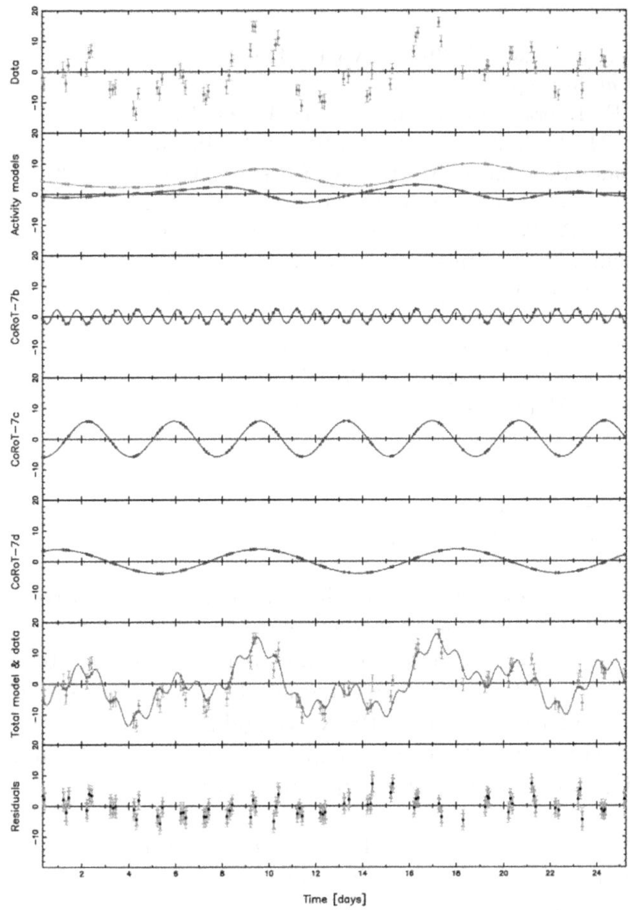

Figure 2. Time series of all the components of the total RV model. All RVs are in m s^{-1}. Second panel: activity models (model for suppression of convective blueshift (orange) and for flux blocked by starspots (purple)). Second to last panel: the total model (red) is the sum of activity and planet RVs. Subtracting it from the original data yields the residuals plotted in the bottom panel. The error bars on the residuals represent the original errors of the data (smaller bars), and the errors after an extra noise term has been added (larger bars) for comparison.

2001) – see Haywood *et al.* (submitted) for further discussion. In order to account for this, we add an extra noise term to the data.

4. Conclusion

New simultaneous photometric and spectroscopic data of the CoRoT-7 system have allowed us to model the activity-induced RV variations according to the method described by Aigrain *et al.* (2011). We confirm the presence of three planets, and calculate their masses. We find that the RV modulation induced by the suppression of convective blueshift and by the flux blocked by starspots are dominant contributors to the total RV of this system. Modelling the activity-induced RV is therefore essential in the case of moderately active host stars such as CoRoT-7.

References

Aigrain, S., Pont, F., & Zucker, S. 2011, *MNRAS*, 419, 4
Barros, S. C. C. *et al.* in prep.
Boisse, I., Bouchy, F., Hébrard, G., *et al.* 2011, *A&A*, 528, A4
Bruntt, H., Deleuil, M., Fridlund, M., *et al.* 2010, *A&A*, 519, A51
Ferraz-Mello, S., Tadeu dos Santos, M., Beaugé, C., *et al.* 2011, *A&A*, 531, A161
Gizon, L., Duvall, T. L. Jr, & Larsen, R. M. 2001, *IAU-CP*, 203, 189
Hatzes, A. P., Dvorak, R., Wuchterl, G., *et al.* 2010, *A&A*, 520, A93
Hatzes, A. P., Fridlund, M., Nachmani, G., *et al.* 2011, *ApJ*, 743, 1
Haywood, R. D., Cameron, A. C., Queloz, D., *et al.* submitted, *MNRAS*
Lanza, A. F., Bonomo, A. S., Moutou, *et al.* 2010, *A&A*, 520, A53
Léger, A., Rouan, D., Schneider, J., *et al.* 2009, *A&A*, 506, 1
Pont, F., Aigrain, S., & Zucker, S. 2010, *MNRAS*, 411, 3
Queloz, D., Bouchy, F., Moutou, C., *et al.* 2009, *A&A*, 506, 1

Discussion

DAWSON: Is it necessary for this system to have simultaneous RV & photometry and what are the implications for other planets orbiting (noisy?) stars?

HAYWOOD: The *FF'* method makes use of the lightcurve of the star to model its activity-induced RV variations. In the case of moderately active stars, active regions evolve very rapidly and can change significantly over time so it is necessary to use simultaneous photometry in order to get a reliable RV model. This need for simultaneous data is a limitation to the *FF'* method, but HARPS-North and Kepler have been observing several targets simultaneously so we will be able to apply the method again, and this should help us find other useful proxies to model activity induced RV.

LAGRANGE: How are the equations in the *FF'* method derived? How do you relate the RV signature of a spot to its photometric signature?

HAYWOOD: The equations of the *FF'* method are based on simple geometry. This is explained in Aigrain's paper of 2011.

DVORAK: If CoRoT-7c and 7d have eccentric orbits, then they should produce gravitational interactions and you should find a TTV (transit timing variation) – see my paper in astroph.

HAYWOOD: The analysis of the new photometric data from CoRoT is being done by Susana Barros. She did look for TTVs in the transits of CoRoT-7b but the signal to noise ratio is so small (because the planet is so small) that it is very difficult to detect any TTV with confidence.

Exploring the Formation and Evolution of Planetary Systems
Proceedings IAU Symposium No. 299, 2013
M. Booth, B. C. Matthews & J. R. Graham, eds.

doi:10.1017/S174392131300848X

Properties of the young gas giant planet β Pictoris b

Mickaël Bonnefoy[1], Anthony Boccaletti[2], Anne-Marie Lagrange[3], France Allard[4], Christoph Mordasini[1], Hervé Beust[3], Gaël Chauvin[3], Julien H. V. Girard[5], Derek Homeier[4], Daniel Apai[6,7], Sylvestre Lacour[2], Daniel Rouan[2], Julien Rameau[3] and Hubert Klahr[1]

[1]Max Planck Institute for Astronomy, Königstuhl 17, D-69117 Heidelberg, Germany
email: bonnefoy@mpia-hd.mpg.de
[2]LESIA, Observatoire de Paris, CNRS, University Pierre et Marie Curie Paris 6 and University Denis Diderot Paris 7, 5 place Jules Janssen, 92195 Meudon, France
[3]UJF-Grenoble 1 / CNRS-INSU, Institut de Planétologie et d'Astrophysique de Grenoble (IPAG) UMR 5274, Grenoble, F-38041, France
[4]CRAL, UMR 5574, CNRS, Université de Lyon, École Normale Supérieure de Lyon, 46 Allée d'Italie, F-69364 Lyon Cedex 07, France
[5]European Southern Observatory, Casilla 19001, Santiago 19, Chile
[6]Department of Astronomy, 933 N. Cherrt Avenue, Tuscon, AZ 85721, USA
[7]Department of Planetary Sciences, The University of Arizona, 1929 E. University Blvd., Tuscon, AZ 85721, USA

Abstract. The young (12^{+8}_{-4} Myr) and nearby (19.44 ± 0.05 pc) star β Pictoris is considered one of the best laboratories for the study of early phases of planetary systems formation since the identification of an extended debris disk surrounding the star in 1984. In 2009, we imaged at 3.8 μm with NaCo at VLT a gas giant planet around β Pictoris, roughly along the disk mid-plane, with a semi-major axis between 8 and 14 AU. We present here the first images of the planet in the J (1.265 μm), H (1.66 μm), and M' (4.78 μm) bands obtained between 2011 and 2012. We used these data to build the 1-5 μm spectral energy distribution (SED) of the companion, and to consolidate previous semi-major axis (8-10 AU) estimates. We compared the SED to seven atmospheric models to derive $T_{\rm eff} = 1700 \pm 100$ K. We used the temperature and the luminosity of β Pictoris b to estimate new masses for the companion. We compared these masses to independent constraints set by the orbital parameters and the radial velocities and use them to discuss the formation history of the object.

Keywords. instrumentation: adaptive optics, high angular resolution, stars: β Pictoris, Exoplanets: formation, evolution, atmosphere

---

1. Introduction

β Pictoris is a nearby (19.44 ± 0.05 pc) intermediate-mass star (A6V), member of the β Pictoris moving group (12^{+8}_{-4} Myr). It is known to bear a structured debris disk seen nearly edge-on (Smith & Terrile 1984). In 2010, we announced the direct detection at L' band (3.8 μm) of a 7–11 $M_{\rm Jup}$ planet orbiting the star, roughly along the dust disk position angle (PA), with a semi-major axis (s.m.a.) between 8 and 14 AU (Lagrange *et al.* 2010). β Pictoris b is the closest planet (physically) to its star ever imaged. The discovery further confirmed that the system remains one of the best testbeds for planet formation theories and for the study of planet/disk interactions.

Nevertheless, our knowledge of the physical characteristics of the planet (mass, T_{eff}), like those of imaged planets or companions, is mostly based on predictions of

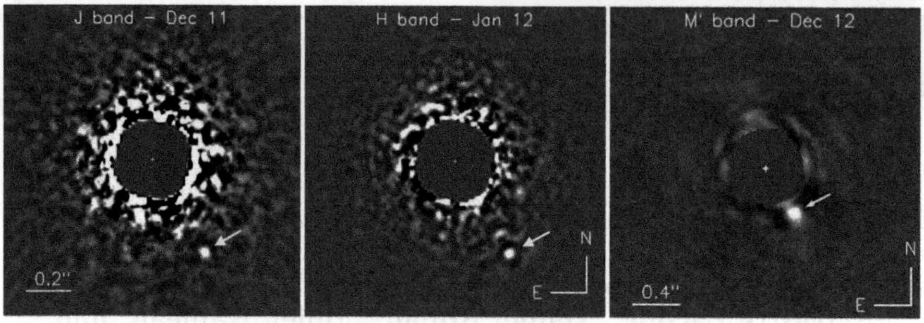

Figure 1. Re-detection of β Pictoris b (arrow) in the J (left), H (middle), and M' bands.

"hot-start" evolutionary models. These models assume that the objects start their evolution following a spherical collapse from an arbitrary large initial radius (Chabrier *et al.* 2000, Baraffe *et al.* 2003). Alternative "cold-start" models were proposed in order to predict the properties of planets produced by accretion of gas. They hypothesize that the accreted gas loses all of its gravitational energy through a super-critical accretion shock. They predict giant planets with low initial entropies, e.g. cooler and dimmer objects than those predicted by the "hot start" models for the same mass.

We made new observations in order to derive accurate estimates of the physical (mass, T_{eff}, log g) and chemical (composition, chemistry) properties of β Pictoris b. We used these estimates to discuss predictions of evolutionary models and to sketch a possible scenario for the planet.

2. New observations of β Pictoris b

We used NaCo at VLT to obtain four new high-contrast observations of β Pictoris in the J (1.226 μm), H (1.65 μm), and M' (4.78 μm) between fall-2011 and 2012. The so-called "pupil-stabilized" mode was used and enabled for angular differential imaging of the star (Marois *et al.* 2006). Data were analyzed using two independent data reduction pipelines. Each pipeline used four different algorithms, including the most recent developments in the field (Lafrenière *et al.* 2007, Soummer *et al.* 2012). The planet is detected in each band (see Fig. 1). We derive J $= 14.0 \pm 0.3$ mag, H $= 13.5 \pm 0.2$ mag, and M'$=11.0 \pm 0.3$ mag.

3. Planet properties

3.1. *Astrometry*

The position of β Pictoris b with respect to the star was extracted from the H-band data. It pursues the monitoring of the orbital motion of the planet (Chauvin *et al.* 2012, Bonnefoy *et al.* 2013, Lagrange *et al.* in prep). We applied Markov-Chain Monte Carlo simulations on the current set of astrometric data points, including new astrometry from unpublished NaCo observations (Lagrange *et al.*, in prep). These simulations constrain the s.m.a of β Pictoris b to 8-10 AU with a maximum of realizations for a s.m.a of 9 AU. For s.m.a. of 9 and 10 AU, radial-velocity measurements (Lagrange *et al.* 2012) place upper limits of 12 and 15.5 M_{Jup} on the dynamical mass of the planet, respectively. *This is the first time that such constraints are placed on the mass of a directly imaged planet.* We also confirm from this analysis that the orbit has a low eccentricity ($e \leqslant 0.17$).

3.2. *Colors, luminosity, and effective temperature*

We compared the new photometry at J, H, and M' band, together with previous photometry obtained in the K_s band (Bonnefoy *et al.* 2011), to that of young objects and mature field MLT dwarfs in color-color and color-magnitude diagrams. Colors based on the J, H, and K_s magnitudes are comparable to those of early-L (L0-L4) field dwarfs and to those of young early L-type companions. We used this crude constraint on the spectral type of the planet and bolometric corrections valid for young L dwarfs (Todorov *et al.* 2010, Faherty *et al.* 2013) to derive a luminosity estimate $\log_{10}(L/L_\odot) = -3.87 \pm 0.08$. Colors based on the M' band photometry appear redder than those of field L dwarfs.

We combined the present set of photometric data points (Lagrange *et al.* 2009, Quanz *et al.* 2010, Bonnefoy *et al.* 2011, 2013) to build the 1-5 μm spectral energy distribution (SED) of the planet. The SED was compared to synthetic fluxes generated from a set of seven atmospheric models. These models account for the formation of dust grains in the photosphere of the object. But they consider either an extreme (AMES-Cond, BT-Cond), partial (BT-Settl 2010, BT-Settl 2012, DRIFT-PHOENIX), or a lack of (Dusty) gravitational settling of the dust following its formation. Models with photospheric dust tend to better reproduce the observations (see Fig. 2). All models converge toward $T_{\rm eff} = 1700 \pm 100$ K. They do not provide a firm estimate of the surface gravity and metallicity of the object.

3.3. *Mass and initial entropy*

We used the constraints on the effective temperature and luminosity of the planet as input of evolutionary models to derive new mass estimate. "hot-start" models predict masses from 7 to 13 $M_{\rm Jup}$ for the planet, that match constrains on its dynamical mass (Section 3.1). On the contrary, predictions from"cold-start" models of (Marley *et al.* 2007) (MA07) and (Spiegel & Burrows 2012) (SB12), which are limited to planets with masses in the range 1-10 $M_{\rm Jup}$, do not reproduce the observed properties of the planet.

We therefore considered the more recent "warm-start" evolutionary models of SB12. These models account for intermediate initial conditions to those considered for the "cold-start" and "hot-start" cases. These initial conditions reflect different possible formation pathways and are parameterized by a choice of initial entropy ($S_{\rm init}$). Predicted $T_{\rm eff}$ and luminosities increase with the mass and $S_{\rm init}$. As a consequence, the corresponding properties of (Section 3.2) β Pictoris b can only be reproduced for a restrained set of masses ($M \geqslant 6M_{\rm Jup}$) and $S_{\rm init}$. $S_{\rm init}$ can be further constrained to $S_{\rm init} \geqslant 9.8k_B$/baryon using the dynamical mass of β Pictoris b for a s.m.a of 9 AU (see Fig 2). This lower limit corresponds to $S_{\rm init}$ values intermediate between those corresponding to the "cold-start" and "hot-start" models of MA07.

4. The origin of β Pictoris b

The initial entropy can not directly be used to discuss the origins of the planet. But it could be compared to similar constraints brought on other directly imaged planets (e.g. Marleau & Cumming 2013). Nevertheless, the narrow range of possible s.m.a., the lower limit on the mass of the planet inferred from "warm-start" models, and the upper limit on the dynamical mass can be used as inputs of formation models. Formation models of H. Klahr (Klahr *et al* in prep.) predict that massive clumps ($M \geqslant 3\, M_{\rm Jup}$) can only be formed in the gravitationally unstable disk surrounding a star as massive as β Pictoris at more than 45 AU (Rameau *et al.* 2013). Therefore, according to *these* models, β Pictoris b would have migrated inward significantly without accreting too much disk material. Conversely, planets as massive as β Pictoris b are formed in population synthesis models

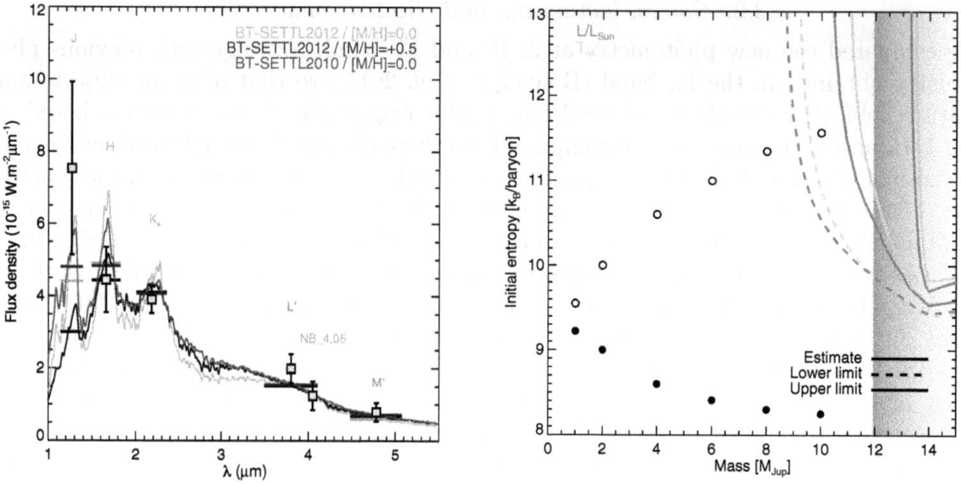

Figure 2. Left: Spectral energy distribution of β Pictoris b (squares) compared to best-matching synthetic fluxes predicted by BT-Settl atmospheric models. **Right:** Predictions of "warm-start" evolutionary models for different planet masses and initial entropies (S_{init}) compared to β Pictoris b properties for an age of 8 to 20 Myr . Models reproduce the observed temperature and bolometric luminosity of the planet for a restrained range of mass and initial entropies (delimited by the dashed and dashed-dotted lines). Independent upper-limits on the mass of the planet exclude part of this parameter space (shaded area) and enable to constrain the initial entropy of the planet to values higher than 9.8 k_B/baryon (according to these "warm-start" models). This lower limit on the initial entropy of the planet fall in-between initial entropies corresponding to "cold-start" (black points), and "hot-start" (open circles) models.

based on the core-accretion scenario (including type I and type II migration) around 12 Myr old stars as massive as β Pictoris (Mordasini *et al.* 2012). These planets are part of the envelope of the distribution of synthetic planets. They are not found beyond 15 AU.

References

Smith, B. A., & Terrile, R. J. 1984, *Science*, 226, 1421
Lagrange, A.-M., Bonnefoy, M. *et al.* 2010, *Science*, 329, 57
Chabrier, G., Baraffe, I., Allard, F., & Hauschildt, P. 2000, *ApJ*, 542, 464
Baraffe, I., Chabrier, G., Barman, T. S., Allard, F., & Hauschildt, P. H. 2003, *ApJ*, 402, 701
Marois, C., Lafrenière, D., Macintosh, B., & Doyon, R. 2006, *ApJ*, 647, 612
Lafrenière, D., Marois, C., Doyon, R., Nadeau, D., & Artigau, E. 2007, *ApJ*, 660, 770
Soummer, R., Pueyo, L., & Larkin, J. 2012, *ApJ*, 755, 28
Chauvin, G., Lagrange, A.-M., Beust, H., Bonnefoy, M., *et al.* 2012, *A&A*, 542, 41
Bonnefoy, M., Boccaletti, A., Lagrange, A.-M., *et al.* 2013, *A&A*, 555, 107
Lagrange, A.-M., De Bondt, K., Meunier, N., Sterzik, M., *et al.* 2012, *A&A*, 542, 18
Bonnefoy, M., Lagrange, A.-M., Boccaletti, A., Chauvin, G., *et al.* 2011, *A&A*, 528, 15
Todorov, K., Luhman, K. L., & McLeod, K. K. 2010, *ApJ*, 714, 84
Faherty, J. K., Rice, E. L., & Cruz, K. L. 2013, *AJ*, 145, 2
Lagrange, A.-M., Gratadour, D., Chauvin, *et al.* 2009, *A&A*, 493, 21
Quanz, S. P., Meyer, M. R., Kenworthy, M. A., *et al.* 2010, *ApJ*, 722, 49
Marley, M. S., Fortney, J. J., Hubickyj, O., *et al.* 2007, *ApJ*, 655, 541
Spiegel, D. S. & Burrows, A. 2012, *ApJ*, 745, 174
Marleau, G.-D., & Cumming, A. 2013, *MNRAS*, Submitted. astro-ph: 1302.1517
Rameau, J., Chauvin, G., Lagrange, A.-M., & Klahr, H. 2013, *A&A*, 553, 60
Mordasini, C., Alibert, Y., Klahr, H., & Henning, T. 2012, *A&A*, 547, 111

Discussion

TRIAUD: What is the probability that the planet will transit, and when will it be transiting?

BONNEFOY: The orbital fits are still compatible with the "transit event" of 1981. As far as I remember, the predictions for the next transit are around 2019. But I am not sure that we can give a probability at this time. The orbital parameters are not known with a sufficient accuracy yet.

MERU: Just a quick comment about disk instability. We were looking at migration of planets and gravitationally unstable disks, and we do actually find that they migrate pretty rapidly. I don't know whether it is too rapid for this, or not? But it is certainly a possibility.

BONNEFOY: Yes exactly. There are also the papers of Zhu (*et al.* 2012), Michael (*et al.* 2011), and there have been a lot of work actually claiming that the timescales are $\sim 10^3$-10^4 years. That is very short. But I would be a bit cautious since we know from population synthesis models of Mordasini *et al.* (for instance) that the efficiency of migration is not very well known so far. So we have to be a bit careful I think.

CURRIE: So it is the first flux-calibrated detection at M'. I had the first detection at M'. Rome was not built in a day, and neither β Pictoris b. So how do your conclusions about the luminosity, the initial entropy, and the mass change if you allow the planet to form in 3 to 5 million years.

BONNEFOY: So your question relates to your paper that you posted on astro-ph. This new paper is presenting a detection at 3.1 μm. Thayne (Currie) reanalysed data taken at other wavelengths we got, and that he got from Gemini. And your hypothesis is that the planet might have been formed a bit later compared to the age of the star. Well, it is a possibility of course. I like this picture. But you base your hypothesis on the estimation of the radius of the object that you derive fitting the actual SED with atmospheric models and using a dilution factor. Actually, I refrained from using this information to discuss anything about the age of the object because I believe that atmospheric models are not well calibrated right now to allow for such an analysis. It would be nice to use different models such as the one of Fortney (on top of Allard's and Helling's models used in Bonnefoy *et al.* 2013, and Madhusudhan's and Burrow's models used in Currie *et al.* 2013) and try to estimate a more reliable range of radius for the objet.

MAROIS: Do you have a clear detection indicating that β Pic b is about to go back toward the star? Do you see it decelerating?

BONNEFOY: Actually, I don't know! We know it is probably close to the quadrature. It is at a very favorable position for spectroscopic monitoring. But I don't remember what are the other constraints that Anne-Marie (Lagrange) got from the very last astrometric data points. I advise you to take a look at her poster.

BILLER: Two quick questions. First, your plot for the simulated planets of mass and separations. If you take the simulated planets, if you plot them into evolutionary models, how does that compare with β Pic b itself?

BONNEFOY: We tried to do that with the new code of Christoph Mordasini which couples the formation and the evolution of these planets. Christoph is making two hypothesis for the accretion shock: he is assuming that the shock can be fully efficient. In this case, you have a "cold-start". Or you can also have the opposite case, e.g. a "hot-start". And both the "cold-start" and the "hot-start" hypothesis are working for β Pictoris! So you can form the planet with the observed luminosity and temperature. The idea is that beyond the issue of the efficiency of the accretion shock, there are other ingredients entering the game like the deuterium-burning, and some other energy sources like the infall of planetesimals in the enveloppe that can modify the cooling curve of the planet. That is an ongoing work.

BILLER: Second question. What are the prospects for getting high resolution spectra and do a similar analysis as Quinn Konopacky did for HR8799c?

BONNEFOY: For β Pictoris b, it is going to be a bit more difficult because the planet is at a slightly shorter separation and the contrast with the star remains high. We could first try to get a low resolution spectrum with SPHERE. It might give us some insights on the atmospheric composition of the planet. But there is no instrument in the South that could collect a high resolution spectrum. SINFONI at VLT for instance is the same kind of instrument as OSIRIS (at Keck, used by Konopacky *et al.*). But it is limited by persistence effects. So you can not really use it to do this kind of thing.

Exploring the Formation and Evolution of Planetary Systems
Proceedings IAU Symposium No. 299, 2013
M. Booth, B. C. Matthews & J. R. Graham, eds.

doi:10.1017/S1743921313008491

Glimpsing the Compositions of Sub-Neptune-Size Exoplanets

Leslie A. Rogers

Hubble Fellow, California Institute of Technology,
2100 E. California Blvd., MC 249-17, Pasadena, USA, 91125
email: `larogers@caltech.edu`

Abstract. The growing number of transiting planets with mass constraints opens the possibility of applying a statistical approach to learn about the underlying population of planet compositions. We focus on the intriguing transition between rocky exoplanets and planets with voluminous gas layers, and explore how the current census of sub-Neptune-size exoplanets constrains the maximum radii of rocky planets. We outline a hierarchical Bayesian model approach to infer the fraction of planets that are dense enough to be rocky (as a function of planet radius). A preliminary analysis of the current sample of planets with mass and radius constraints reveals that most planets larger than 1.9 $R_\oplus$ are too low density to be comprised of Fe and silicates alone.

Keywords. (stars:) planetary systems, (stars:) planetary systems: formation

1. Introduction

Exoplanets detected both in transit and by their dynamical influence are very valuable. The radius derived from the transit depth and the mass derived from radial velocity (RV) measurements or transit timing variations, together give the planet density and some handle on the planet composition. To date more than 190 transiting planets have measured masses (e.g. exoplanets.org). The accumulating census of transiting planets with measured masses contains information about the underlying composition distribution of planets, and about the masses of the more than 3000 *Kepler* transiting planet candidates (Borucki *et al.* 2011, Batalha *et al.* 2013) that lack dynamical confirmation.

Planet mass-radius measurements can be especially informative for planets that are smaller than Neptune ($R_p \lesssim 4R_\oplus$). For planets in this size range, a wide diversity of planet compositions are a priori plausible; rock, astrophysical ices (H_2O, NH_3, and CO), and H/He gas can all make significant contributions to both the planets' mass and volume (e.g., Rogers & Seager 2010a, 2010b). In the Solar System, there are no planets with masses and radii intermediate between the Earth and the ice giants. As a statistical sample of sub-Neptune-sized exoplanets with measured masses and radii accumulates, we may begin to study the intriguing transition between planets with rocky compositions and those with significant contributions from volatiles (astrophysical ices and H/He).

The *Kepler* Mission has already made a tremendous contribution to populating the planet mass–radius diagram (Figure 1), and more sub-Neptune-sized *Kepler* transiting planets with mass constraints will soon be announced. The *Kepler* team undertook an ambitious follow-up program with Keck-HIRES to measure the radial velocities of 22 *Kepler* Objects of Interest (KOIs) hosting planet candidates with $R_p < 4\ R_\oplus$ (Marcy *et al.* 2013, in prep). The planet candidates were selected for RV follow-up based on their quiet host stars (*Kepler* Magnitude $K_p < 13.5$, $T_{\rm eff} < 6100$ K, $v \sin i < 5\ {\rm km\,s^{-1}}$) and detectable predicted RV-amplitudes ($K > 1\ {\rm m\,s^{-1}}$). Each target received 20 to 50

RV measurements between July 2009 and May 2013. This large follow-up effort will more than double the number of known sub-Neptune-size transiting planets with mass constraints, yielding mass measurements or upper limits for 52 planets (42 transiting and 10 non-transiting) in 22 planetary systems.

With the release of the new Marcy *et al.* (2013) sample of sub-Neptune-size transiting planets with RV follow-up, we can start to move beyond case studies of individual objects towards statistical studies of the population of planets. One trend that is already readily visible in Figure 1 is a high density – high mass threshold among the planet mass-radius pairs measured to date. In the rest of this paper, we focus on this apparent paucity of massive rocky planets in greater detail, and set statistical constraints on the fraction of planets that are dense enough to have rocky compositions as a function of planet radius. We identify which planets are dense enough to be rocky (Section 2), and outline a statistical approach to studying the distribution of planet compositions (Section 3).

2. Which planets may be rocky?

One of the most direct insights we can glean about a sub-Neptune exoplanet's composition is whether it must have some volatiles (H/He or astrophysical ices). The mass-radius relation for a silicate composition (the solid curve in Figure 1) represents an extreme lower limit to the masses of rocky planets with no volatiles ($M\mathrm{rocky,min}\,(R_p)$). Planets that are less dense must have some volatiles (in the form of water or H/He); though rock may still account for much of their mass, these planets have mean densities that are too low for their transit radius defined by a rocky surface. Planets that are more dense than $M\mathrm{rocky,min}\,(R_p)$ could potentially be comprised just of iron and silicates.

We must account for the large measurement uncertainties on planet masses and radii when assessing which planets may be rocky. Many of the Kepler planets with RV follow-up have either marginal RV detections or mass upper limits, yet even the non-detections contain information about the planet composition. With that aim, we introduce p_{rocky}, the posterior probability that a planet is dense enough to be rocky based on the measured mass and radius. p_{rocky} is evaluated as the fraction of a transiting planets' joint

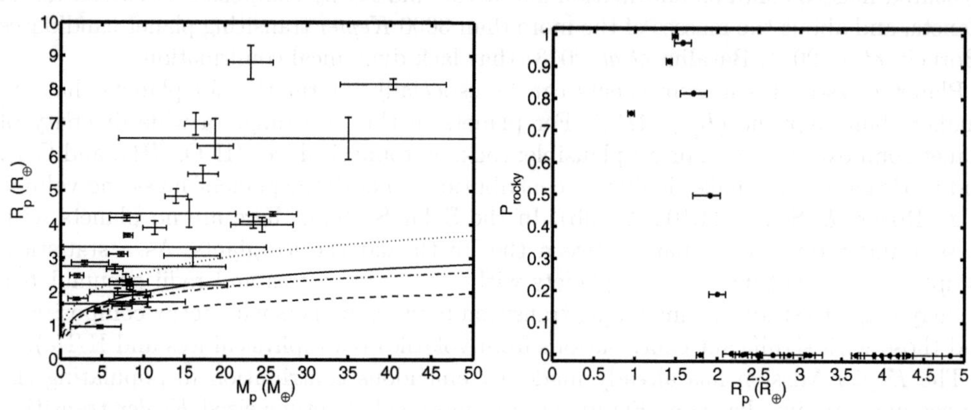

Figure 1. *Left*: Exoplanet Mass-Radius Diagram. The curves are theoretical mass-radius relations for a fixed planet composition from Seager *et al.* (2007). The dashed, dot-dashed, solid, and dotted curves are for hypothetical pure iron, Earth-like composition (32% Fe, 68% silicate by mass), pure silicate, and pure water ice planets, respectively. *Right*: The posterior probability that a planet is dense enough to be rocky, p_{rocky}, as a function of the planet radius. Planets with $p_{\mathrm{rocky}} > 0.1$ include Kepler-68c, Kepler-10b, Kepler-36b, CoRoT-7b, KOI-94b, Kepler-20b, and Kepler-18b (in order of increasing radius).

mass-radius posterior probability density (obtained from fitting a planet model to the radial velocity data and transit light curve) that falls within the high-density potentially rocky regime ($M_p \geqslant M$rocky, min (R_p)). Planets with $p_{\rm rocky}$ near 1 are well constrained to have masses and radii in the potentially rocky regime, while low-density planets have $p_{\rm rocky}$ near 0. RV Marginal detections or non-detections, may have measured RV amplitudes that spill into unphysical regimes (corresponding to negative mass or masses exceeding that of a pure iron sphere). In calculating $p_{\rm rocky}$, we assume a flat prior on planet mass-radius pairs that are physically plausible.

The values of $p_{\rm rocky}$ for all known transiting planets with mass constraints are presented in Figure 1. We find that smaller planets are more likely to be dense enough to be rocky. Planets larger than $\sim 2\ R_\oplus$ are too low density to be comprised of iron and silicates alone, there is some sort of transition regime between 1.5 and 2 $R_\oplus$.

3. Distribution of planet Compositions

We now take our analysis one step further to evaluate what can be inferred about the composition distribution of the underlying population of planets. We present a statistical approach to constrain the fraction of planets that are dense enough to be rocky (as a function of planet radius). A population-level analysis such as this, will be especially warranted once the Marcy *et al.* (2013) RV mass constraints are released. Here, for demonstration, we apply our approach to the current list of planets with mass and radius constraints from exoplanets.org (Wright *et al.* 2011). We emphasize, however, that the exoplanets.org list of transiting planets with mass constraints is an inhomogeneous sample of planets, and is subject to selection biases. We do not directly account for selection effects here, but they will be a priority for future work.

We take a Bayesian approach to investigate how the fraction of planets that are dense enough to be rocky depends on planet radius. We begin by assuming some simple functional form for $f_{\rm rocky}\,(R_p, a)$, the fraction of planets with radius R_p that are dense enough to be rocky, where a denotes all the model parameters. Neglecting selection effects, the likelihood of finding the observed sample of planet M-R measurements given a model for the underlying population, is simply the probability that either both the planet is rocky (based on its M-R measurement) and that a planet of that size is rocky based on the population model or that both the planet is non-rocky (based on its M-R measurement) and a planet of that size would be non-rocky based on the population model.

$$p\,(\text{data} \mid a) = \prod_{planets} \{p_{\rm rocky} f_{\rm rocky} + (1 - p_{\rm rocky})\,(1 - f_{\rm rocky})\} \tag{3.1}$$

We then apply Bayes theorem, $p\,(a \mid \text{data}) \propto p\,(\text{data} \mid a)\, p\,(a)$, to calculated the posterior probability of the model parameters $p\,(a \mid \text{data})$ from the likelihood and the priors on the model parameters $p\,(a)$.

For illustration, we consider a 1-parameter step-function model for $f_{\rm rocky}\,(R_p, a)$ where all planets larger than a threshold radius $R_{\rm thresh}$ are non-rocky, while all planets smaller than $R_{\rm thresh}$ are dense enough to be comprised of iron and silicates alone. Evaluating the posterior probability distribution of $R_{\rm thresh}$ conditioned on the planet mass-radius data in Figure 1, we find a median value $R_{\rm thresh} = 1.73^{+0.16}_{-0.05}\ R_\oplus$, and a 95% confidence upper bound of $R_{\rm thresh} = 1.83^{+0.30}_{-0.04}\ R_\oplus$. The 1σ uncertainties quoted on the percentiles of $R_{\rm thresh}$ were evaluated from 1000 bootstrapping samples. For the pure silicate composition adopted as the low-density extreme for rocky planets, the median and 95% upper bound on $R_{\rm thresh}$ correspond to threshold masses of 5.8 $M_\oplus$ and 7.2 $M_\oplus$, respectively.

We have also investigated a 2-parameter linear model for $f_{\rm rocky}\,(R_p, a)$, where all planets smaller than $R_{\mathrm{min,non-rocky}}$ are dense enough to be rocky, all planets larger than $R_{\mathrm{max,rocky}}$ are non-rocky, and where f_{rocky} decreases linearly for $R_{\mathrm{min,non-rocky}} \leqslant R_p \leqslant R_{\mathrm{max,rocky}}$. This linear transition model reduces to the step-function model when $R_{\mathrm{min,non-rocky}}$ and $R_{\mathrm{max,rocky}}$ are equal, but admits the possibility of a range of radii over which rocky and non-rocky planets coexist. The joint posterior probability density of $R_{\mathrm{min,non-rocky}}$ and $R_{\mathrm{max,rocky}}$ conditioned on the observed planets, sets a 2σ upper limit on the maximum radii of rocky planets of $R_{\mathrm{max,rocky}} \leqslant 3.12\ R_\oplus$, and a 2σ upper limit on the planet radius at which no more than 5% of planets are dense enough to be rocky of $1.88\ R_\oplus$.

We find that the simpler 1-parameter step-function model for f_{rocky} is preferred over the 2-parameter linear transition model from the Bayesian evidence. More planet mass-radius measurements with smaller uncertainties are needed to constrain structure of the rocky/non-rocky transition.

4. Summary

We have presented an updated planet mass-radius diagram (Figure 1). None of the planets known to date that are larger than $2\ R_\oplus$ are dense enough to be rocky. The results of an extensive *Kepler* follow-up program including multiple Doppler measurements for 22 KOIs Marcy *et al.* (2013) will soon more than double the population of sub-Neptune sized planets that have mass constraints. This opens the possibility of applying a hierarchical Bayesian model to learn about the underlying population of planet compositions.

We outline a statistical approach to infer the fraction of planets that are dense enough to be rocky (as a function of planet radius). We apply this approach to the current sample of planets with constraints on both mass and radius, and infer that most planets larger than $1.9\ R_\oplus$ are too low density to be comprised of Fe and silicates alone. The results of this demonstration should be interpreted with caution, however, because the current sample of planets with both mass and radius constraints is an inhomogeneous sample with complicated selection effects. We look forward to having a larger sample of sub-Neptune size planets with measured masses once the Marcy *et al.* (2013) results are released. The constraints derived on the radii above which most planets are too low density to be composed of iron and silicate alone, will provide a useful empirical constraint for planet formation theories. Furthermore, our results will allow us to use the the sub-sample of KOIs with RV follow-up to inform us about the masses and compositions of the rest of the sub-Neptune size KOIs for which there is no RV follow-up.

Acknowledgements

L.A.R. acknowledges NASA support through Hubble Fellowship grant #HF-51313.01-A. This research has made use of the Exoplanet Orbit Database and the Exoplanet Data Explorer at exoplanets.org.

References

Batalha, N. M., Rowe, J. F., Bryson, S. T., *et al.* 2013, *ApJS*, 204, 24
Borucki, W. J., Koch, D. G., Basri, G., *et al.* 2011, *ApJ*, 728, 117
Marcy, G. W., Isaacson, H., Rowe, J. F., *et al.* 2013, *ApJ*, in prep.
Rogers, L. A. & Seager, S. 2010a, *ApJ*, 712, 974
—. 2010b, *ApJ*, 716, 1208
Seager, S., Kuchner, M., Hier-Majumder, C. A., & Militzer, B. 2007, *ApJ*, 669, 1279
Wright, J. T., Fakhouri, O., Marcy, G. W., *et al.* 2011, *PASP*, 123, 412

Discussion

REBEKAH DAWSON: In his presentation, Johnathan Fortney mentioned Kepler-36, where the difference in mass accounted for which planet retains a gassy atmosphere. Is there a physical motivation for making the cut in radius, instead of mass?

LESLIE ROGERS: The main motivation for exploring the fraction of planets that are dense enough to be rocky as a function of radius instead of mass is due to the selection effects, which are expected to have a more pronounced effect on the distribution of measured planet radii at specified mass, than on the distribution of measured planet masses at specified radius. The Marcy *et al.* (2013) sample of planets were initially discovered from their transits in *Kepler* photometry; for a given planet mass, larger radius lower density planets would be preferentially detected and identified by *Kepler* as planet candidates. The Marcy *et al.* (2013) KOIs were then selected for radial velocity follow-up based on their radius, and observed in a mass-blind way. Thus at a given radius the Marcy *et al.* (2013) RV planet masses should sample the underlying mass distribution without strong biases. Another motivation for studying planet composition as function of radius, is that our results translate readily to mass predictions for the thousands of Kepler planet candidates without RV follow-up. It is true, though, that for comparing to planet formation and evolution models, planet mass would provide a more direct comparison.

SLOANE WIKTOROWICZ: I have a question about the result that smaller planets tend to denser, because it looks like when you get below 1 $R_\oplus$ the probability for rocky is approaching 50%. Is that just a radial velocity upper limits?

LESLIE ROGERS: Yes, those are cases where we just have a radial velocity upper limit on the planet mass. Our information about the masses and compositions for those small planets below 1 $R_\oplus$ is dominated by our priors.

JONATHAN FORTNEY: In terms of $p_{\rm rocky}$, how do you handle the boundary? For planets just smaller than a pure rock composition, one could imagine that a water component is possible with a high iron/rock ratio.

LESLIE ROGERS: In this analysis, I've essentially been considering a question with a yes/no answer: is the planet dense enough to be rocky based on the pure silicate mass-radius relation, or not. I have repeated this analysis considering instead an Earth-like composition threshold (32% Fe, 68% silicate by mass), and found that the result that most planets larger than 1.9 $R_\oplus$ are not rocky still holds, but the mass that radius threshold corresponds to increases to $\sim 9\ M_\oplus$.

Exploring the Formation and Evolution of Planetary Systems
Proceedings IAU Symposium No. 299, 2013
M. Booth, B. C. Matthews & J. R. Graham, eds.

doi:10.1017/S1743921313008508

Direct imaging of Beta Pictoris b with first-light Magellan Adaptive Optics

Katie M. Morzinski[1,‡]†, Laird M. Close[1], Jared R. Males[1,‡], Phil M. Hinz[1], Alfio Puglisi[2], Simone Esposito[2], Armando Riccardi[2], Enrico Pinna[2], Marco Xompero[2], Runa Briguglio[2], Kate Follette[1], Derek Kopon[3], Victor Gasho[1], Alan Uomoto[4], Tyson Hare[4], Andy Skemer[1], Carmelo Arcidiacono[2], Fernando Quiros-Pacheco[2], Javier Argomedo[2], Lorenzo Busoni[2], T. J. Rodigas[1] and Ya-Lin Wu[1]

[‡]NASA Sagan Fellow

[1]Steward Observatory, Center for Astronomical Adaptive Optics, University of Arizona, 933 N. Cherry Ave., Tucson, AZ 85721, USA

[2]Istituto Nazionale di Astrofisica, Osservatorio Astrofisico di Arcetri, Largo E Fermi 5, 50125 Firenze, Italy

[3]Max-Planck-Institut für Astronomie, Königstuhl 17, D-69117 Heidelberg, Germany

[4]Carnegie Observatories, OCIW, 813 Santa Barbara St. Pasadena, CA 91101, USA

Abstract. MagAO is the newly-commissioned adaptive optics (AO) instrument on the Magellan Clay telescope at Las Companas Observatory, Chile. MagAO has two co-mounted science cameras: VisAO for visible-light direct and spectral-differential imaging; and Clio for near to thermal IR direct imaging, non-redundant-mask interference, and prism spectroscopy. We demonstrate MagAO's simultaneous visible and infrared AO performance via direct images of exoplanet Beta Pictoris b. The planet was detected in 5 passbands from 0.9–5μm. Here we show the infrared images; the visible observations are presented in Males *et al.* 2013. MagAO is the first AO system to offer good performance with extensive coverage across the O/IR spectrum and thus offers an unprecedented opportunity to study the spectral energy distributions of directly-imaged extrasolar planetary atmospheres.

Keywords. Planets and satellites: individual (Beta Pictoris b); Stars: individual (Beta Pictoris); Stars: planetary systems; Instrumentation: adaptive optics

1. Introduction

The young nearby A-star β Pictoris hosts an extensive edge-on disk discovered with optical imaging in the 1980s (Smith & Terrile 1984). A massive planet was suspected to be the cause of the inner warp of the disk, but the companion was not confirmed until the direct imaging discovery by Lagrange *et al.* (2009, 2010). The planetary companion, β Pic b, has been imaged with VLT/NaCo and Gemini-S/NICI at J, H, K_s, [4.05]μm, L', and M' (Bonnefoy *et al.* 2011, Boccaletti *et al.* 2013, Bonnefoy *et al.* 2013). We target this object for first-light demonstration of our new adaptive optics (AO) instrument, MagAO (Close *et al.* 2012).

MagAO is a 585-degree-of-freedom AO system on the Magellan Clay telescope at Las Campanas Observatory in Chile. Its features include a 378-mode modulating pyramid wavefront-sensor giving excellent performance, and an adaptive secondary mirror (ASM) giving improved thermal throughput out to 5 μm. MagAO has two science cameras that

† email: ktmorz@arizona.edu

can operate simultaneously for twice the observing efficiency: VisAO with broad and narrow passbands from r' to Y_s; and Clio with broad and narrow passbands from J to M' (see `http://magao.as.arizona.edu/observers/` for full information on the instrument). Here we show the performance of MagAO/Clio with first-light images of β Pic b. These observations were carried out during the first commissioning run in November–December 2012.

2. MagAO performance

2.1. *PSF simulations*

We begin with a simulation of the MagAO/Clio PSF, for comparison to the on-sky performance. The major modifiers to the unitary amplitude of a circular pupil are (1) the secondary mirror obscuration and its support struts ("spiders"), and (2) the cold stop and its spiders within Clio's dewar. For the former, we used a digital camera image of the ASM taken while standing at the Nasmyth focus, prior to MagAO being mounted at the Nas. ring. For the latter, we use a Clio image of the cold stop taken in pupil imaging mode.

The composite of both images is the pupil image, shown in Fig. 1, left. The four thin arms are the ASM spiders and the two thick arms are the cold stop spiders. The lollipop-like shape at left (nine o'clock) is the "slot" cut in to the ASM surface to prevent a crack from spreading. The image at right shows the resultant PSF that we expect for the case of no wavefront error.

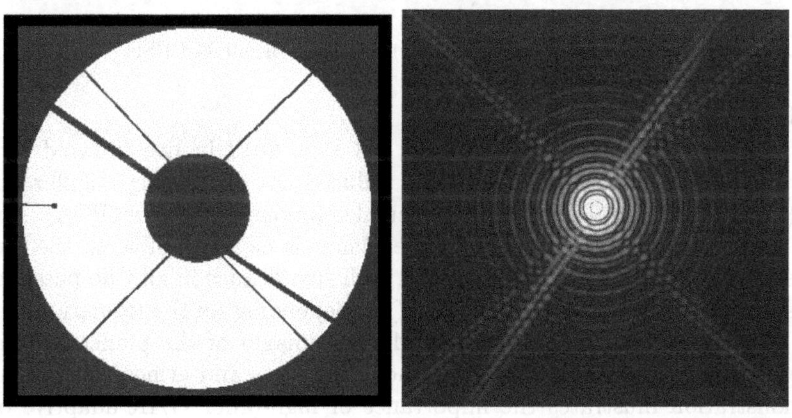

Figure 1. MagAO/Clio pupil and simulated PSF. A detailed discussion is available at `http://visao.as.arizona.edu/simulations/magao-pupils-and-fourier-optics/`.

2.2. *On-sky PSFs*

Now that we have looked at the simulated PSF, we can compare the on-sky performance of MagAO/Clio. Figure 2 shows the L' and M' on-sky PSFs of β Pic, as well as a comparison to the simulated PSF. The Airy pattern is clearly seen. What look like atmospheric speckes are actually the combination of diffraction spikes caused by spiders in the cold stop and ASM. There is no wavefront error in the simulated PSFs. This demonstrates the excellent high-order performance of MagAO.

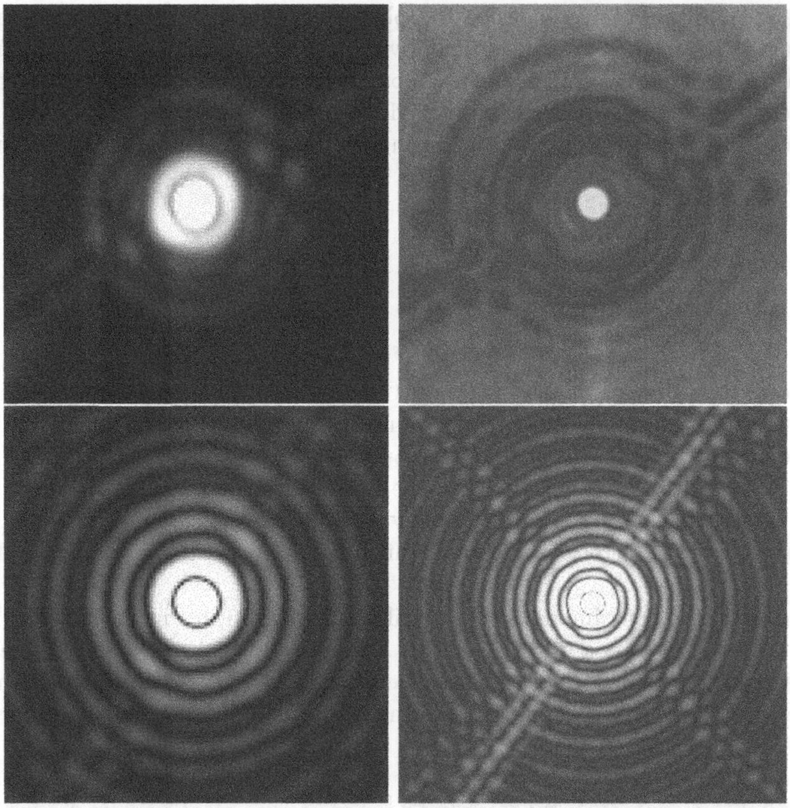

Figure 2. Top: β Pic at L' (left) and M' (right). Bottom: Simulated PSF, varying zoom scales.

3. Exoplanet images

β Pic was observed on five nights from 1–8 Dec. 2012 in five infrared filters from 2–5 μm. We used pupil-tracking mode and reduced the data using angular differential imaging (ADI, Marois *et al.* 2006) and principal component analysis (PCA). The L' and M' PCA residuals are shown in Fig. 3; the planet is clearly visible to the lower right (south west) of the star. β Pic b is seen with high significance in all Clio passbands used. The mass of β Pic b is estimated at 7–15 $M_{\rm Jup}$ depending on the formation mechanism assumed (Bodenheimer *et al.* 2013). The MagAO images of the planet will help us to better determine its physical parameters, including mass and atmospheric composition. This demonstration illustrates the importance of high-order O/IR adaptive optics for direct imaging of extrasolar planets.

4. Summary

Direct imaging of extrasolar planets allows us to carry out detailed characterization of planets beyond the snow line. Models are crucial to interpreting data, but at present more data is needed to constrain degeneracies. This first-light demonstration of MagAO/Clio has shown the AO instrument performance to be excellent quality, allowing us to image the planet β Pic b in several thermal-infrared filters. MagAO offers unique capabilities to characterize directly-imaged exoplanets at O/IR wavelengths from 0.55–5 μm. We refer potential MagAO/VisAO and MagAO/Clio users to `http://magao.as.arizona.edu/` for web resources, SPIE publications, and guides for observers.

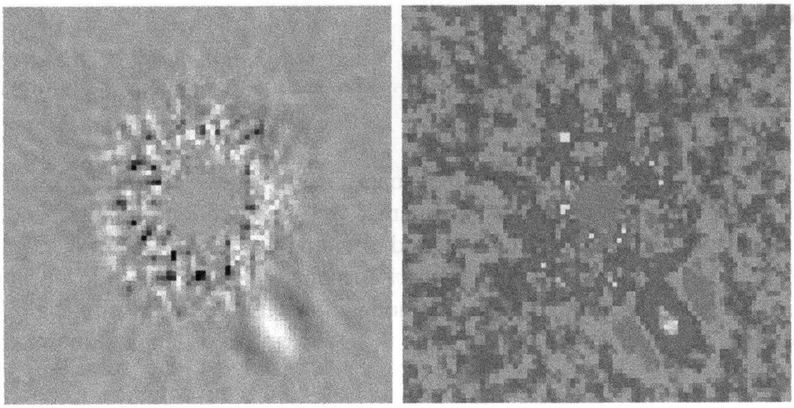

Figure 3. β Pic b at L' (left) and M' (right).

Acknowledgements

The MagAO ASM was developed with support from the NSF MRI program. The MagAO PWFS was developed with help from the TSIP program and the Magellan partners. The Active Optics guider was developed by Carnegie Observatories with custom optics from the MagAO team. The VisAO camera and commissioning was supported with help from the NSF ATI program. The Clio camera comes courtesy of PMH and Steward Observatory. KMM was supported under contract with the California Institute of Technology (Caltech) funded by NASA through the Sagan Fellowship Program. LMCs research was supported by NSF AAG and NASA Origins of Solar Systems grants. JRM is grateful for the generous support of the Phoenix ARCS Foundation and is also newly supported through the Sagan Fellowship Program.

References

Boccaletti, A., Lagrange, A.-M., Bonnefoy, M., Galicher, R., & Chauvin, G. 2013, *A&A*, 551, L14

Bodenheimer, P., D'Angelo, G., Lissauer, J. J., Fortney, J. J., & Saumon, D. 2013, *ApJ*, 770, 120

Bonnefoy, M., Boccaletti, A., Lagrange, A.-M., Allard, F., Mordasini, C., Beust, H., Chauvin, G., Girard, J. H. V., Homeier, D., Apai, D., Lacour, S., & Rouan, D. 2013, *A&A* accepted

Bonnefoy, M., Lagrange, A.-M., Boccaletti, A., Chauvin, G., Apai, D., Allard, F., Ehrenreich, D., Girard, J. H. V., Mouillet, D., Rouan, D., Gratadour, D., & Kasper, M. 2011, *A&A*, 528, L15

Close, L. M., Males, J. R., Kopon, D. A., Gasho, V., Follette, K. B., Hinz, P., Morzinski, K., Uomoto, A., Hare, T., Riccardi, A., Esposito, S., Puglisi, A., Pinna, E., Busoni, L., Arcidiacono, C., Xompero, M., Briguglio, R., Quiros-Pacheco, F., & Argomedo, J. 2012, *Proc. SPIE*, 8447, 84470X

Lagrange, A.-M., Bonnefoy, M., Chauvin, G., Apai, D., Ehrenreich, D., Boccaletti, A., Gratadour, D., Rouan, D., Mouillet, D., Lacour, S., & Kasper, M. 2010, *Science*, 329, 57

Lagrange, A.-M., Gratadour, D., Chauvin, G., Fusco, T., Ehrenreich, D., Mouillet, D., Rousset, G., Rouan, D., Allard, F., Gendron, É., Charton, J., Mugnier, L., Rabou, P., Montri, J., & Lacombe, F. 2009a, *A&A*, 493, L21

Males, J. R., *et al.* 2013, this Volume

Marois, C., Lafrenière, D., Doyon, R., Macintosh, B., & Nadeau, D. 2006, *ApJ*, 641, 556

Smith, B. A. & Terrile, R. J. 1984, *Science*, 226, 1421

Discussion

CURRIE: This is a very impressive system. I had two small detail-oriented questions. (1) What are the SNRs you're getting in the images? (2) Could you explain the shape/structure of the subtracted PSF?

MORZINSKI: Because the VisAO and Clio images were obtained simultaneously, and we calibrated the astrometry of the two cameras, we are confident in all the detections presented. For your second question, the radial elongation and tangential negative flux are artifacts. PCA is trying to fit a high-spatial-frequency feature with low-order modes. Thus you get ringing just like in Fourier methods. I have not done any filtering or masking of the planet in the images. Therefore, this is a true image of the residuals after using PCA to subtract the PSF.

BONNEFOY: Very excellent talk. The instrumental strategy is amazing. Just to see these PSFs with the adaptive secondary and the pyramid wavefront sensor is so amazing. I just have a comment. I think that the new data points you are getting in the red optical [*NOTE:* see Males *et al.* 2013] and also the complementary data points are very interesting. I played with the metallicity with the BT Settl and the DRIFT-PHOENIX models and I remember that the J-band flux and the flux at shorter wavelengths is dependent on the metallicity, so you might be able to say more about the metallicity of the object.

MORZINSKI: Thank for the compliment. That is very interesting, thank you very much.

KENWORTHY: I was really amazed at the 1-μm β Pic b image which is very spectacular [*NOTE:* see Males *et al.* 2013]. But I was wondering, what kind of astrometric precision are you getting?

MORZINSKI: Great question. People don't talk about this much, but one of the most difficult things in measuring astrometry of directly-imaged exoplanets is actually centroiding on the host star, which is saturated. Due to the fundamental nature of ADI-like methods, the center of the host star is also the center of rotation for the image processing—derotating and combining images—so this is why the PSF centration is the fundamental limit. We tried various techniques to centroid on the central star, and found that in the case of MagAO/Clio images, they are circularly symmetric enough such that we found the best center of rotation by minimizing the residuals of an image rotated by θ degrees differenced from the original image, where θ varies from 5 to 355 degrees in 10-degree increments. Centroiding to a tenth of a pixel precision still gives $\sim$10% error in the position of the planet. Although we used Trapezium observations to calibrate platescale and distortion, and to calibrate Clio and VisAO together, in fact we found that the limit to astrometric precision is set by how well we can determine the center of the host-star PSF.

Exploring the Formation and Evolution of Planetary Systems
Proceedings IAU Symposium No. 299, 2013
M. Booth, B. C. Matthews & J. R. Graham, eds.

doi:10.1017/S174392131300851X

Unveiling an exoplanetary Neptunian atmosphere through multiband transit photometry

V. Nascimbeni[1,2], G. Piotto[1,2], I. Pagano[3] and G. Scandariato[3,4]

[1]Dipartimento di Fisica e Astronomia, Università degli Studi di Padova, Vicolo dell'Osservatorio 3, 35122 Padova, Italy
email: valerio.nascimbeni@unipd.it

[2]INAF – Osservatorio Astronomico di Padova, vicolo dell'Osservatorio 5, 35122 Padova, Italy

[3]INAF – Osservatorio Astrofisico di Catania, via S. Sofia 78, 95123 Catania, Italy

[4]INAF – Osservatorio Astronomico di Palermo, Piazza del Parlamento 1, 90134 Palermo, Italy

Abstract. The "effective" radius of a planet is a function of wavelength due to scattering and/or absorption processes, and we can exploit simultaneous multiband transit photometry to probe the atmospheric scale height and composition. We present new photometric data of the recently discovered "hot Uranus" GJ3470b, gathered with the LBC camera at LBT. Light curves of unprecedented accuracy (0.0012 mag in U and 0.00028 mag in a narrow band centered at 972 nm) allowed us to measure an increasingly larger planetary radius at shorter wavelengths, which we interpret as a signature of Rayleigh scattering by a large scale height atmosphere. Further follow-up observations to confirm this result and probe the presence of specific atomic and molecular species is ongoing.

Keywords. Techniques: photometric, stars: planetary systems, stars: individual: GJ3470

---

1. Introduction

Low-mass planets such as "Super-Earths" ($2 \lesssim M_\mathrm{p} \lesssim 10\ M_\oplus$) and "Neptunian" planets ($15 \lesssim M_\mathrm{p} \lesssim 50\ M_\oplus$) are puzzling theoreticians, revealing a complex diversity which cannot be easily explained by models or constrained by observations (Haghighipour 2011). These two classes were once hypothesized to be separate and well defined samples of rocky and icy planets, respectively. Now we realize that their measured densities range from $\rho_\mathrm{p} = 0.27$ to $\sim 10\ \mathrm{g\,cm^{-3}}$ (Wright *et al.* 2011), with a significant overlap between super-Earths and Neptunes. In this mass range, average densities are unable to put a firm constraint on the inner structure of the planet, and theoretical models can lead to several degenerate solutions (Adams *et al.* 2008). An emblematic case is that of GJ1214b (Charbonneau *et al.* 2009), for which allowed scenarios include a "mini-Jupiter", a Neptune-like planet, a "water world", or a rocky super-Earth (Rogers & Seager 2010).

Additional information other than ρ_p is required to break the model degeneracy, and an effective way to do it is to search for spectral signatures coming from the planetary atmosphere to probe its composition and physical state (Seager & Deming 2010). *Transmission spectroscopy* is a technique based on the observation of exoplanetary transits at different wavelengths, searching for changes of its effective radius $R_\mathrm{p}(\lambda)$ as a function of λ due to absorption and/or scattering processes undergone by stellar light after traveling through the atmospheric limb (Sing *et al.* 2011). Rayleigh scattering, due to molecules such as H_2 (Lecavelier *et al.* 2008b) or to haze from condensate particles (Fortney 2005, Lecavelier *et al.* 2008a), has a typical signature: its steep λ^{-4} dependence, which

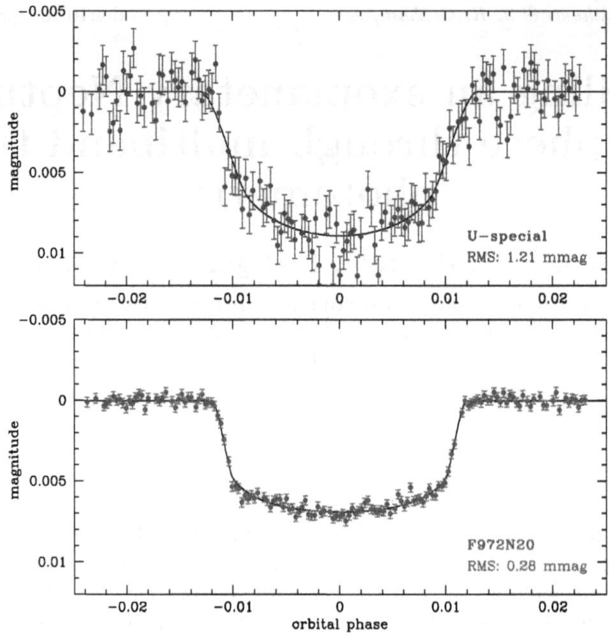

Figure 1. Light curves of GJ3470b gathered with LBC in the $U_{\rm spec}$ band (*upper panel*), and in the $F972N20$ band (*lower panel*), plotted with the original sampling cadence (the solid line is the best-fit model). The two light curves are simultaneous.

translates into a larger $R_{\rm p}$ on the blue side of the optical spectrum. Thus the scale height $H = kT_{\rm eq}/\mu g$ (and hence the mean molecular weight μ, once an equilibrium temperature $T_{\rm eq}$ and surface gravity g are assumed) can be inferred from the steepness of $R_{\rm p}(\lambda)$.

GJ3470b is a "hot Uranus" planet orbiting in about 3.3 days around its M1.5V host star, discovered by a radial velocity search and then caught transiting through ground-based photometry (Bonfils *et al.* 2012). Very recently, a follow-up analysis based on Spitzer, WIYN-3.5m and Magellan data (Demory *et al.* 2013) resulted in updated values for the mass and radius of GJ3470b: $M_{\rm p} = 13.9 \pm 1.5\ M_{\oplus}$ and $R_{\rm p} = 4.83 \pm 0.21\ R_{\oplus}$, making it one of the least dense, low-mass planets known with $\rho_{\rm p} = 0.72 \pm 0.13$ g cm^{-3}. Showing this peculiarity, having a total mass at the boundary between super-Earths and Neptunes, GJ3470b stands out as an ideal case to test the present planetary structure and evolution theories, which predict for it a significantly extended envelope of primordial H and He (Demory *et al.* 2013, Rogers & Seager 2010) and thus a large scale height.

2. Observations and data analysis

We observed a full transit of GJ3470b at the Large Binocular Telescope (LBT) on the night of 2013 Feb 16, under a photometric sky. The Large Binocular Camera (LBC) consists of two prime-focus, wide-field imagers mounted on the left and right arm of LBT, and optimized for blue and red optical wavelengths, respectively. We exploited their full power to gather two simultaneous photometric series, setting the $U_{\rm spec}$ filter on the blue channel (a Sloan u-like filter centered at $\lambda_{\rm c} = 357.5$ nm) and the $F972N20$ filter in the red channel (a narrow-band filter centered at $\lambda_{\rm c} = 963.5$ nm). This choice of passbands has been made to maximize the wavelength span of our observations while at the same time avoiding most of telluric lines in the red channel.

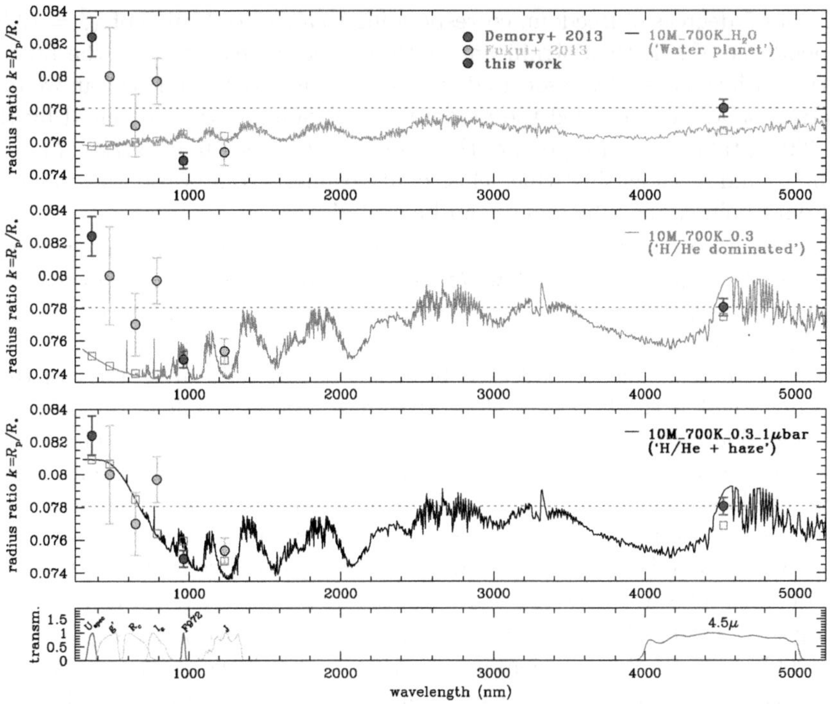

Figure 2. Reconstructed transmission spectrum of GJ3470b. The $U_{\rm spec}$ and $F972N20$ data points are extracted from our LBC light curves, the $g'R_cI_cJ$ points from Fukui *et al.* (2013), and the 4.5μm point from Demory *et al.* (2013). The model plotted with a solid line is scaled from a Howe & Burrows (2012) model computed for a 10-$M_\oplus$ cloud-free, haze-free planet with $T_{\rm eq}$ = 700 K and a pure water (H_2O) atmosphere. *Second panel:* same as above, but with a cloud-free, haze-free, metal-poor atmosphere dominated by H and He, computed by the same authors. Abundances are scaled from the solar ones ($Z = 0.3 \times Z_\odot$) *Third panel:* same as above, but adding the contribution of a scattering haze from Tholin particles at 1 μbar. *Lower panel:* Instrumental passbands employed for each data point.

Both the $U_{\rm spec}$ and $F972N20$ light curves of GJ3470 were extracted using STARSKY, a photometric pipeline optimized to perform high-precision differential aperture photometry over defocused images, originally developed for the TASTE project (Nascimbeni *et al.* 2013). The resulting light curves are plotted in Fig. 1 in their original cadence. Their overall RMS scatter is 1.19 mmag ($U_{\rm spec}$) and 0.28 mmag ($F972N20$) over a mean cadence of 95 s, only slightly larger than the theoretical values estimated by standard "white noise" formulae. It is worth noting that the $F972N20$ series is, to our knowledge, the most accurate light curve ever obtained from a ground-based facility. The JKTEBOP code version 28 (Southworth *et al.* 2004) was employed to fit a transit model over the light curves. We derived the best-fit uncertainties with a Monte Carlo bootstrap algorithm to avoid underestimation due to correlations between parameters and residual systematic noise. The output parameters include the ratio $k = R_{\rm p}/R_\star$ between the radius of the planet and the radius of the star.

3. Results

If we examine the transmission spectrum of GJ3470b as reconstructed by our LBC measurements and from those by Demory *et al.* (2013) and Fukui *et al.* (2013) (Fig. 2, large circles), it appears evident that no constant value of k could explain the data

($\chi^2 = 67.8$ for 5 degrees of freedom, corresponding to a reduced value of $\chi^2_r = 13.5$). The color dependence of k is then significant, with a steep increase toward the blue side of the spectrum which is usually associated with scattering processes. How can we interpret this spectrum $k(\lambda)$? We rescaled the model transmission spectra computed by Howe & Burrows (2012) taking the grid point of their simulations closer to GJ3470b ($T_{\rm eq} = 700K$, $M_{\rm p} = 10M_{\oplus}$) and investigating which class of atmospheric composition best matches the observed data points, and which ones can be excluded.

The most striking result is that the large observed variations of $k(\lambda)$ are totally incompatible with all model sets by Howe & Burrows (2012) having an atmosphere of high mean molecular weight μ, such as those made of pure H_2O (solid line in Fig. 2, upper panel) or pure methane (CH_4). All those models predict a very flat spectrum due to their tiny scale height, and are rejected by the available data at $>10\sigma$. On the other hand, we investigated the fitness of cloud-free, haze-free atmospheric models of low μ, that is largely dominated by H and He and with a large scale height H. This is the most probable atmospheric composition expected for GJ3470b following current theories on planetary interiors (Demory *et al.* 2013, Rogers & Seager 2010). A metal-poor model atmosphere (solid line in the second panel of Fig. 2) provides us with a much better fit on the red/IR passbands (F972N20, J, 4.5μm), but it fails at reproducing the steep rise observed on the blue side of the spectrum, especially for our $U_{\rm spec}$ measurement.

An additional source of scattering, other than H_2, seems to better explain the data. Fortney (2005) showed that condensate particles can be very efficient in this regard. Indeed, the Howe & Burrows (2012) class of models with a metal-poor H/He dominated atmosphere and high-altitude hazes from Tholin provide us with a satisfactory fit with $\chi^2_r = 5.2$ (solid line in Fig. 2, third panel).

Additional high S/N observations will be required to confirm this scenario, especially in the H and K spectral region, where molecular bands such as H_2O, CO or CH_4 are expected to dominate the trasmission spectrum giving rise to strong features (Fig. 2).

References

Adams, E. R., Seager, S., & Elkins-Tanton, L. 2008, *ApJ*, 673, 1160
Bonfils, X., *et al.* 2012, *A&A*, 546, A27
Charbonneau, D., *et al.* 2009, *Nature*, 462, 891
Demory, B.-O., *et al.* 2013, *ApJ*, 768, 154
Fortney 2005, *MNRAS*, 364, 649
Fukui, A., *et al.* 2013, *ApJ*, 770, 95
Haghighipour, N. 2011, *ConPh*, 52, 403
Howe, A. R. & Burrows, A. S. 2012, *ApJ*, 756, 176
Lecavelier Des Etangs, A., *et al.* 2008, *A&A*, 481, L83
Lecavelier Des Etangs, A., *et al.* 2008, *A&A*, 485, 865
Nascimbeni, V., *et al.* 2013, *A&A*, 549, A30
Rogers, L. A. & Seager, S. 2010, *ApJ*, 716, 1208
Seager, S. & Deming, D. 2010, *ARA&A*, 48, 631
Sing, D. K., *et al.* 2011, *MNRAS*, 416, 1443
Southworth, J., Maxted, P. F. L., & Smalley, B. 2004, *MNRAS*, 351, 1277
Wright, J. T., *et al.* 2011, *PASP*, 123, 412

Discussion

BONNEFOY: How far do we know the transmission properties of tholins?

NASCIMBENI: Yes, this is right. Remember that it is still an ad-hoc model. I can say that tholins are able to reproduce the optical data.

BEATTY: How much do uncertainties in the theoretical limb-darkening coefficients affect your final error bars?

NASCIMBENI: The linear LD coefficients were fitted on both light curves, while the quadratic terms along with their theoretical uncertainties were input as Gaussian priors in our bootstrap analysis, so the final result takes it into account.

TRIAUD: How did you choose your reference stars to prevent colour effects?

NASCIMBENI: It is hard to find proper reference stars. Giants were removed, other effects of colours would show as airmass-dependent effects out of transit. These are accounted for.

WIKTOROWICZ: Defocusing to keep from saturating in the red channel probably limits your precision in U band. Can you change the focus of the two channels independently?

NASCIMBENI: Yes, our U band precision is limited by the fact that the integration times have to be the same between the two channels.

Exploring the Formation and Evolution of Planetary Systems
Proceedings IAU Symposium No. 299, 2013
M. Booth, B. C. Matthews & J. R. Graham, eds.

doi:10.1017/S1743921313008521

The Shocking Variability Of Exoplanet Transits

J. Llama, A. A. Vidotto, M. Jardine, K. Wood and R. Fares

SUPA. University of St Andrews. North Haugh. St Andrews. KY16 9SS. UK
email: `joe.llama@st-andrews.ac.uk`

Abstract. Asymmetries in exoplanet transits are proving to be a useful tool for furthering our understanding of magnetic activity on both stars and planets outside our Solar System. Near-UV observations of the WASP- 12 system have revealed asymmetries in the timing of the transit when compared with the optical light curve. In this paper we review a number of reported asymmetries and present work simulating near-UV transits for the hot-Jupiter hosting star HD 189733.

Keywords. stars: activity, coronae, individual (WASP-12, HD 189733), magnetic fields, planetary systems, winds

---

1. Introduction

Since the discovery of the first planet orbiting a star other than the Sun nearly twenty years ago the field of exoplanetary science has rapidly grown with both ground and space based surveys having found approximately 900 exoplanets (*http://www.exoplanet.eu*). So called hot-Jupiters which are found to be orbiting within fractions of an au of their host star continue to challenge our ideas of planet formation and evolution. With the improvement of the telescopes used to find exoplanets we are now able to begin characterising the atmospheres of the planets. To this end, asymmetries in the transit light curves of exoplanets are proving to be incredibly useful tools for learning about both exoplanetary atmospheres but also the host star itself (Vidotto *et al.* 2011).

There have been many reports in the literature of "bumps" being recorded in the transit light curve of many exoplanets (see for example, Sanchis-Ojeda *et al.* (2011)). These bumps are believed to have been caused by the planet occulting a star spot on the surface of the host star. Because star spots are darker than their surroundings, the fractional loss of light as the planet transits over them will become less and so a positive bump will be recorded in the light curve. Missions such as *Kepler* provide continuous observations of many stars which therefore allow us to track the evolution of star spots on the surface of other stars that host misaligned planets. This will enable us to build up a picture of stellar butterfly diagrams (Llama *et al.* 2012).

Another asymmetry that has been detected is the presence of an early-ingress in the near-UV light curve taken with *HST* of the hot-Jupiter WASP-12b when compared to the optical data. The near-UV light curve shows the transit beginning before the optical transit, but finishing at the same time (Fossati *et al.* 2010). A number of potential solutions have been proposed to explain the presence of such an asymmetry. One solution suggests that because WASP-12b is a very large planet orbiting very close to its host star, it may have overflowed it's Roche lobe (Lai *et al.* 2010). This would result in an accretion stream from the planet onto the star which may manifest as an asymmetry in the light curve. Another solution is the presence of a magnetosphere around the planet (Vidotto *et al.* 2010). The interaction between the Solar wind and each of the Solar system planets

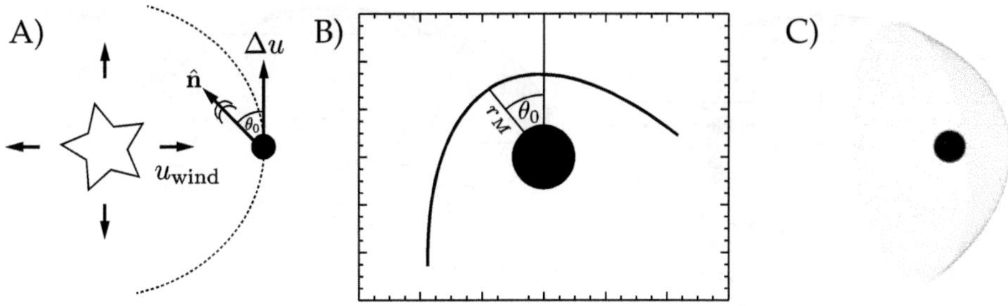

Figure 1. Image A) shows the geometry of the shock (adapted from Vidotto *et al.* (2010)). The distance r_M is the distance to the nose of the shock. The angle θ_0 is the angle made between the azimuthal direction of the planet and the vector $\hat{\mathbf{n}}$, the direction of the shock nose. The vector $-\hat{\mathbf{n}}$ is the direction the stellar wind impacts on the planet. B) shows a typical shock shape in two-dimensions using the model of Wilkin (1996). C) shows the shock geometry in three-dimensions.

results in the formation of a shock surrounding each of the planets. For planets that are located far from their host star, the shock will form directly between the planet and the star, a so-called "dayside shock". For close-in planets, such as hot Juipiters, however, an "ahead shock" will form which will occult the stellar disc before the planet which may cause an early-ingress in the transit light curve (Vidotto *et al.* 2010). If the density of shocked material is high enough then this may show as the presence of additional absorption in the light curve. Llama *et al.* (2011) modelled the near-UV light curve of WASP-12b from Fossati *et al.* (2010) and were able to fit the observations with a simple shock model.

In this paper we model a planet and bow shock transiting over a simulated star. We use magnetic maps of the bright K-dwarf HD 189733 (which due to its proximity and relative brightness has been extensively studied) to simulate the stellar wind conditions around the planet transiting planet HD 189733b. From these simulations we are able to prescribe the geometry and density of the shock that should form and simulate near-UV light curves.

2. The Model

We couple numerical simulations of stellar winds with magnetic imaging of HD 189733 to predict the stellar wind conditions around the planet HD 189733b in order to investigate how the geometry of the resultant shock may vary as the planet orbits around the star. We use magnetic surface maps of HD 189733 from June 2007 and also a year later in July 2008 (Fares *et al.* 2010). They use the tomographic method known as Zeeman-Doppler Imaging (ZDI) to reconstruct the large-scale magnetic field on the surface of the star.

The magnetic surface maps are used as one of the boundary conditions in our stellar wind simulation. We use `BATS-R-US`, a three-dimensional mangetohydrodynamic (MHD) numerical code developed at The University of Michigan (Powell *et al.* 1999). `BATS-R-US` solves the ideal MHD equations for the mass density ρ, the gas pressure p, the velocity of the plasma $\mathbf{u} = (u_r, u_\theta, u_\phi)$ and the magnetic field intensity $\mathbf{B} = (B_r, B_\theta, B_\phi)$. The output from the wind simulation is a full three-dimensional grid that allows us to determine the local conditions experienced by the transiting planet HD 189733b.

The formation of a bow shock is a direct consequence of the relative velocities between the interaction of the planetary magnetic field and the stellar wind occurring at supersonic speeds. The distance and geometry of the shock is related to the local stellar wind

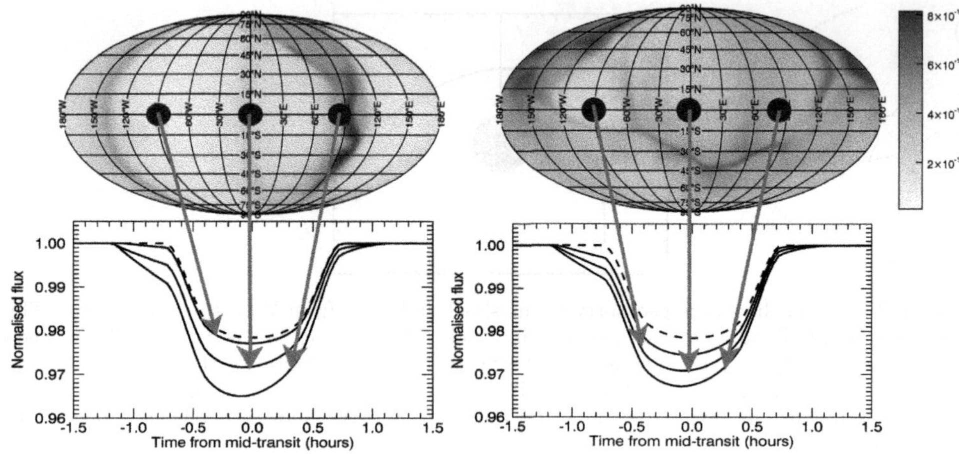

Figure 2. Results of our simulations: June 2007 (left panel) and July 2008 (right panel). Top panel shows the stellar wind density at the orbital distance of the planet. The transit depth and timing is clearly dependent on the local stellar wind conditions for near-UV transits. We find the transits to be highly variable, with certain wind conditions causing the transit to be almost indistinguishable from the optical transit, whilst in very dense regions of the wind the transit depth is much deeper and also begins much sooner due to the presence of denser, shocked material.

properties near the planet. The shape of the shock is described in two-dimensional polar co-ordinates (r, θ) with the planet centred at $(0, 0)$ as:

$$R(r_M, \theta) = \frac{r_M}{\sin\theta}\sqrt{3\left(1 - \frac{\theta}{\tan\theta}\right)} \tag{2.1}$$

where r_M is the distance to the nose of the shock (Wilkin 1996). The angle the shock makes with the azimuthal direction of the planetary motion is defined as θ_0, and is determined by the geometry of the stellar wind impacting on the planet:

$$\theta_0 = \arctan\left(\frac{u_r}{|u_{\rm planet} - u_\varphi|}\right) \tag{2.2}$$

where $u_{\rm planet}$ is the orbital motion of the planet, and u_r and u_φ are the radial and azimuthal stellar wind speeds respectively (Vidotto *et al.* 2010). Figure 1 shows a cartoon depiction of the shock model.

3. Results

Figure 2 shows our results. The top image shows the stellar wind density at the orbit of HD 189733b for June 2007 (left) and July 2008 (right). The bottom panel shows three resultant near-UV transits (solid lines) and the optical model for comparison (dashed). The results clearly show that the depth and timing of the transit is dependent on the density of the stellar wind and therefore multiple observations may reveal different transit shapes and help us understand the variability of the stellar wind.

References

Fares, R., Donati, J., & Moutou, C., *et al.* 2010, *MNRAS* 406, 409

Fossati, L., Haswell, C., Froning, C., *et al.* 2010, *ApJ* 714, L222

Lai D., Helling C., & van den Heuvel E. P. J. 2010, *ApJ*, 721, 923
Llama, J., Jardine, M., Mackay, D. H., & Fares, R. 2012, *MNRAS* 422, L72
Llama J., Wood K., Jardine M., Vidotto A. A., *et al.* , 2011, *MNRAS* 416, L41
Powell K. G., Roe P. L., Linde T. J., *et al.* 1999, *Journal of Computational Physics*, 154, 284
Sanchis-Ojeda R., Winn J. N., Holman M. J., *et al.* 2011, *ApJ*, 733, 127
Vidotto A. A., Jardine M., & Helling C. 2010, *ApJ*, 722, L168
Vidotto, A. A, Jardine, M., & Helling, Ch. 2011, *MNRAS*, 414, 1573
Vidotto, A. A, Llama, J., Jardine, M., *et al.* 2011, *Astronomische Nachrichten*, 332, 1055
Wilkin F. P. 1996, *ApJ*, 459, L31

Discussion

JACKSON: What sort of interactions and/or degeneracies would there be between a bow shock and planetary evaporation?

LLAMA: There is an inherent degeneracy between a bow shock and planetary evaporation, both of which would provide an asymmetry in the light curve. However, the presence of planetary evaporation does not rule out the possibility of a bow shock also being present.

TINNEY: Your models look at the absorption produced by a shock. Have you thought about whether the emission signature of these shocks would be detectable?

LLAMA: In principle you could certainly look for emission; however, this emission would have to be above any stellar emission and so this may cause issues.

BARSTOW: How realistic is the assumption that the magnetic field of HD 189733b is similar to Jupiter's?

LLAMA: We know very little about the magnetic fields of exoplanets and so it seems reasonable to assume the magnetic field to be as simple as possible, i.e. a dipole with the same intensity as Jupiter's. The model can simply be scaled to account for the specific field strength should a better estimate be found.

RAFIKOV: How do you treat mass loading of field lines in your magnetic model? Does the planet always move through super-Alfvénic parts of the wind?

LLAMA: From the stellar wind model we predict a mass loss rate of $\dot{M}_{\star} = 4{\times}10^{-13} M_{\odot}/\mathrm{yr}$. At the orbital radius of HD 189773b our model predicts the wind is indeed super-Alfvénic.

PUDRITZ: M-stars are highly variable and have strong magnetic field and X-ray activity. Would it be possible to probe the magnetic field in Super-Earths around M-stars?

LLAMA: One certainly could look for asymmetries in Super-Earths around M-stars. The activity of the star may cause some issues. We discuss this in Vidotto *et al.* (2011).

SHKOLNIK: Could the WASP-12b early ingress be due to the planet grazing over a bright hot spot on the star?

LLAMA: A bright spot certainly would cause an asymmetry in the light curve. However, the bright spot would not cause an early-ingress, rather it would simply cause additional absorption whilst the planet occulted the spot.

Exploring the Formation and Evolution of Planetary Systems
Proceedings IAU Symposium No. 299, 2013
M. Booth, B. C. Matthews & J. R. Graham, eds.
doi:10.1017/S1743921313008533

Exoplanet Transit Spectroscopy of Hot Jupiters Using HST/WFC3

Korey Haynes[1,2,7], Avi M. Mandell[1], Evan Sinukoff[3], Nikku Madhusudhan[4], Adam Burrows[5] and Drake Deming[6]

[1]Solar System Exploration Division, NASA Goddard Space Flight Center, Greenbelt, MD 20771, USA

[2]Department of Physics and Astronomy, George Mason University, Fairfax, VA 22030, USA

[3]Institute for Astronomy, University of Hawaii, Honolulu, HI 96822, USA

[4]Department of Astronomy, University of Maryland, College Park, MD 20742, USA

[5]Yale Center for Astronomy and Astrophysics, Yale University, New Haven, CT 06511, USA

[6]Department of Astrophysical Sciences, Princeton University, Princeton, NJ 08544, USA

[7]Corresponding Email: Korey.N.Haynes@nasa.gov

Abstract. We present analysis of transit spectroscopy of three extrasolar planets, WASP-12 b, WASP-17 b, and WASP-19 b, using the Wide Field Camera 3 (WFC3) on the Hubble Space Telescope (HST). Measurement of molecular absorption in the atmospheres of these planets offers the chance to explore several outstanding questions regarding the atmospheric structure and composition of these highly irradiated, Jupiter-mass objects. We analyze the data for a single transit for each planet, using a strategy similar in certain aspects to the techniques used by Berta (2012), and achieve almost photon-limited results for individual spectral bins. Our final transit spectra are consistent with the presence of a broad absorption feature at 1.4 μm most likely due to water, but the amplitude of the absorption is less than expected based on previous observations with *Spitzer*, possibly due to hazes absorbing in the NIR. However, the degeneracy of models with different compositions and temperature structures combined with the low amplitude of any features in the data preclude our ability to place unambiguous constraints on the atmospheric composition without a comprehensive multi-wavelength analysis.

Keywords. stars: planetary systems, transits, techniques: photometric, techniques: spectroscopic

---

1. Introduction

The Wide Field Camera 3 (WFC3) on the *Hubble Space Telescope* (HST) provides the capability for spectroscopic characterization of molecular features in exoplanet atmospheres. WFC3 is an optical/NIR camera with the capability for slitless grism spectroscopy, with wavelength coverage in the the IR spanning between 0.8 and 1.7 μm (Dressel (2012)). This region spans both the major bands of water between 1.3 and 1.5 μm as well as another water band at 1.15 μm, and bands of a few other molecular species. Observations measuring flux within NIR water bands are impossible from the ground due to the extinction and variability caused by water vapor in the Earth's atmosphere; WFC3 therefore represents the only current platform for measuring absorption and/or emission from water in exoplanet atmospheres.

We present WFC3 observations of three transiting "hot Jupiter" exoplanets - WASP-12 b, WASP-17 b, and WASP-19 b - during transit of the host star. Two of these data sets, WASP-17 b and WASP-19 b, were observed as part of a large HST program to examine single transits and eclipses from a number of hot Jupiters (P.I. D. Deming), while the

Table 1. Observing details for all three targets. Note that larger subarray size and higher peak pixel values correspond to larger peak to peak systematics.

	WASP-12	WASP-17	WASP-19
Date of Observation	2011-04-12	2011-07-08	2011-07-01
Integration Time (seconds)	7.624	12.795	21.657
Subarray Mode (pix^2)	256	512	128
CALWF3 version	2.7	2.3	2.3
NSamp	3	16	5
Timing Sequence	SPARS10	RAPID	SPARS10
Peak Pixel Value (counts)	38,000	64,000	73,000
Peak to Peak Systematics Range (Combined Light) (10^{-3} Normalized Flux)	0.80	3.94	1.62

data for the transit of WASP-12 b were taken as part of a single-object campaign (P.I. M. Swain) and first analyzed in Swain *et al.* (2013). Observational details are listed in Table 1. All three planets orbit extremely close to their parent star, are therefore highly irradiated, and have large atmospheric scale heights, making them excellent targets for transmission spectroscopy.

Recent observational studies have produced conflicting results regarding the atmospheric compositions of several hot Jupiters, including WASP-12 b and WASP-19 b. Madhusudhan *et al.* (2011) first raised the possibility of a non-solar abundance in the atmosphere of WASP-12 b using occultation measurements in four *Spitzer* photometric bands (Campo *et al.* 2011) and three ground-based NIR photometric bands Croll *et al.* (2011) to constrain the carbon-to-oxygen ratio to super-solar values, possibly greater than unity. Similar *Spitzer* and ground-based measurements for WASP-19 b were consistent with both solar and super-solar C/O models (Anderson *et al.* 2011), raising the possibility of a population of carbon-rich hot Jupiters. However, Crossfield *et al.* (2012) recently re-analyzed the *Spitzer* data for WASP-12 b in light of the discovery of a faint candidate companion imaged by Bergfors *et al.* (2012), concluding that the dilution-corrected *Spitzer* and ground-based photometry can be fit by solar-metallicity models with almost isothermal temperature structures.

2. Data Reduction and Analysis

WFC3 data show strong systematics, first described by Berta *et al.* (2012). Swain *et al.* (2013) examines the systematics from multiple observing campaigns, and determines that the systematics are mainly influenced by the maximum exposure levels and choice of subarray size. In order to remove these systematics, we employ the divide-oot method described by Berta *et al.* (2012), averaging the out-of-transit, combined-light data to create a template for the systematics, the scaling of which we leave as an open parameter when fitting our channel light curves (see Figure 1). For fitting purposes, we use a Markov Chain Monte Carlo (MCMC) routine with a Metropolis-Hastings sampler Ford (2005), using the light curve model from Mandel & Agol (2002), with additional terms to account for the gradual decrease in flux seen in all WFC3 exoplanet transit data to date, the systematics scaling term, as well as two further scaling terms to account for shifts on the

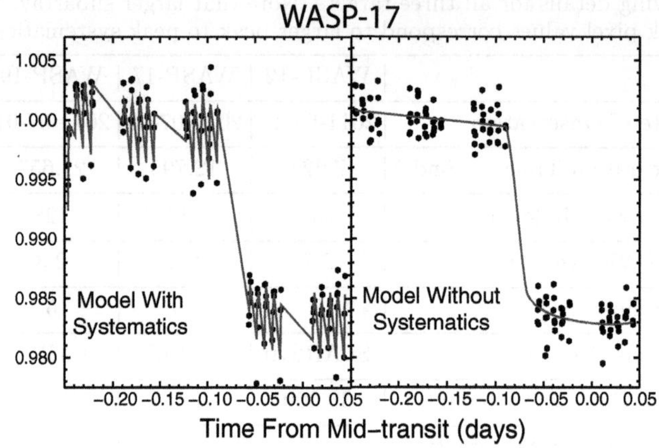

Figure 1. Sample bin for WASP-17 showing the data (black) with model (blue) including systematics (left), and with systematics removed (right).

detector in position and wavelength space. Systematics scaling terms are allowed to vary only if the resulting fit has a ΔBIC $\geqslant 2$,

We also take care to remove contamination due to WASP-12's recently discovered companion, Bergfors-6 (Bergfors *et al.* 2012). We do this by using WASP-19's PSF as a template for an uncontaminated source. We determine the positions of WASP-12 and Bergfors-6 in the direct image, place one PSF at the location of each source, and then jointly scale the heights of each using a χ^2 fitting routine.

3. Results

In Figure 2 we show the resulting spectra for each planet and overplot several models based on the framework of Burrows *et al.* (2000) (and more recently Burrows *et al.* (2006), Burrows *et al.* (2008) and Howe & Burrows (2012) and Madhusudhan & Seager (2009) and Madhusudhan *et al.* (2011). The Burrows models calculate the chemical and radiative equilibrium state of each planet based on the mass, size, and incident radiation, assuming solar abundances. For WASP-17, the data is well fit by either a standard model or a hazy, isothermal model, while the data for WASP-12 b is consistent within uncertainties with a flat spectrum.

The Madhusudhan models relax the stringent requirements for radiative and chemical equilibrium, instead exploring oxygen-rich or a carbon-rich chemistry at a specific temperatures (see Madhusudhan 2012 for details). For WASP-19 b the carbon-rich models fit the data slightly better, but we are unable to discriminate between the models based solely on this data.

Huitson *et al.* (2013) have also analyzed the same WASP-19 b data set, and find similar results to our data. Swain *et al.* (2013) and Stevenson *et al.* (2013) have both analyzed the WASP-12 b data, and find a deeper transit depth at the short wavelength edge of the spectrum.

We conclude that the data can be fit with standard atmospheric models, without the need for exotic chemistry or temperature structure. Models with a deep absorption feature are ruled out by our data. However, any of these models can still be adjusted to fit the data by adding an absorbing haze layer or decreasing the water abundance. A determination of the carbon-to-oxygen ratio cannot be deduced from our current data set alone because the location of absorption bands are similar in both oxygen-rich and

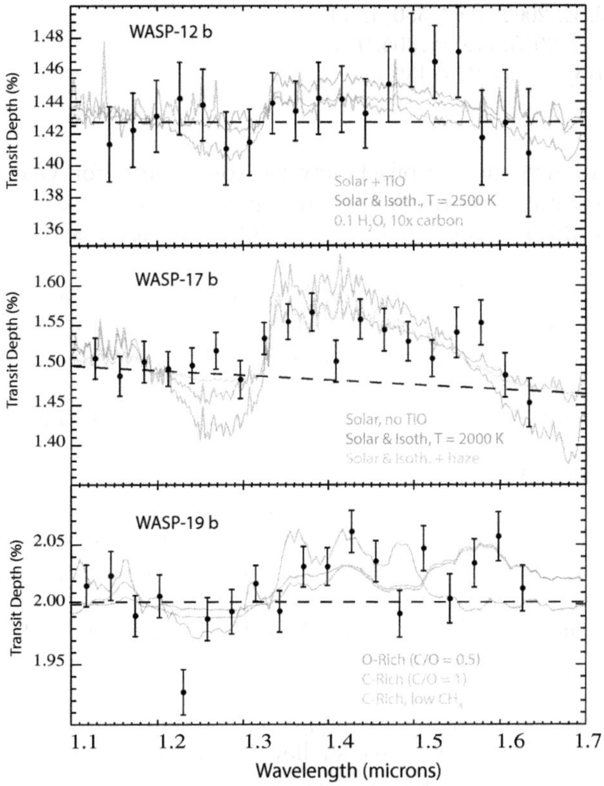

Figure 2. Transit depths for 19 bins for each target, with models from Burrows *et al.* (WASP-17 b and WASP-12 b) and Madhusudhan *et al.* (WASP-19 b). Models show suppressed water features, possibly indicating a haze layer. Data is undergoing further analysis.

carbon-rich models. Similarly, confirmation of the presence or absence of haze requires an analysis of transit depth measurements taken across a wide wavelength range in both the optical and IR, and is beyond the scope of the current work.

References

Anderson, D. R., *et al.* 2011, *MNRAS*, 416, 2108
Bergfors, C., *et al.* 2012, *MNRAS*, 428, 182
Berta, Z. K., *et al.* 2012, *ApJ*, 747, 35
Burrows, A., Marley, M. S., & Sharp, C. M. 2000, *ApJ*, 531, 438
Burrows, A., Sudarsky, D., & Hubeny, I. 2006, *ApJ*, 650, 1140
Burrows, A., Budaj, J., & Hubeny, I. 2008, *ApJ*, 678, 1436
Campo, C. J., *et al.* 2011, *ApJ*, 727, 125
Croll, B., *et al.* 2011, *ApJ*, 141, 30
Crossfield, I. J. M., *et al.* 2012, *ApJ*, 760, 140
Dressel, L. 2012, "Wide Field Camera 3 Instrument Handbook, Version 5.0" (Baltimore: STScI)
Ford, E. B. 2005, *ApJ*, 129,1706
Howe, A. R. & Burrows, A. S. 2012, *ApJ*, 756,176
Huitson, C. M. *et al.* 2013, *MNRAS*, 434, 3252
Madhusudhan, N., & Seager, S. 2009, *ApJ*, 707, 24
Madhusudhan, N., Mousis, O., Johnson, T. V., & Lunine, J. I. 2011, *ApJ*, 743, 191
Madhusudhan, N. 2012, *ApJ*, 758, 36

Mandel, K., & Agol, E. 2002, *ApJ*, 580, L171
Stevenson, K. B. *et al.* 2013, arXiv:1305.1670
Swain. M., *et al.* 2013, *Icarus*, 225, 432

Discussion

TRIAUD: You mentioned that an offset between your results for WASP-12 and Swain's could be due to orbital uncertainties. Which parameters and what precision do you require as it might affect results in other parts, like Rayleigh scattering?

HAYNES: Many parameters can affect this. It is uncertain which are the most important but limb darkening is probably one. Maybe this should be taken more seriously but people have done mostly differential studies rather than absolute.

BARSTOW: Have you considered modeling the WFC3 systematics for removal?

HAYNES: This was one of our first approaches, but considering the short cadence of some of the systematics (only 5 exposures, in the case of WASP-17's hook/ramp effect), and the range of variables that affect the systematics (exposure time, total pixel illumination, sub-array size), we found the divide-oot method to produce much more reliable results.

BARSTOW: Could the high C/O ratio for WASP-19 b also be explained by clouds or haze?

HAYNES: A haze layer or clouds remains a possibility, and there is no way to determine, from the available data, whether we see shallow water absorption features because there is truly little water, or because such features are obscured by haze. This is similar to the case of GJ1214 in Berta *et al.* (2012).

Exploring the Formation and Evolution of Planetary Systems
Proceedings IAU Symposium No. 299, 2013
M. Booth, B. C. Matthews & J. R. Graham, eds.

doi:10.1017/S1743921313008545

The BT-Settl Model Atmospheres for Stars, Brown Dwarfs and Planets

F. Allard

Centre de Recherche Astrophysique de Lyon, UMR 5574, CNRS,
Université de Lyon, École Normale Supérieure de Lyon,
46 Allée d'Italie, F-69364 Lyon Cedex 07, France
email: fallard@ens-lyon.fr

Abstract. We present a grid of stellar and substellar atmosphere models covering the range from solar-mass stars to the latest-type T and Y dwarfs with a single setup. For the first time our synthetic spectra and photometry reproduce the formation of clouds and in particular their clearing at the L/T transition. The BT-Settl models also naturally explain the dustier infrared properties of planets as an effect of low surface gravity.

Keywords. brown dwarfs, planets, clouds, etc.

As brown dwarfs and imaged planets are being discovered, they reveal their spectral and/or photometric properties. Late M type and L type dwarfs see the onset of silicate dust formation which exerts a green house effect in their atmospheres, making them look red in the near-IR bandpasses ($J-K$). At optical wavelengths cloud formation has the effect of causing the sedimentation of important opacity sources (TiO, VO, CaH, etc.) while dust scattering depresses the optical pseudo-continuum. As the cloud layers sink below the photosphere in T type dwarfs, the optical becomes dominated by alkali absorption lines and methane bands begin to form in the $W1$ and K bandpasses, as well as ammonia bands in the H and $W3$ bandpass. Water ice condensation affects Y type brown dwarfs, where the strength of water and methane bands leaves only flux escape in narrow windows of minimum opacity, and where this important line blanketing causes flux to be redistributed into the 4.5 μm peak that becomes as bright as the near-IR flux peaks. The imaged planets have, to the greater surprise of the community, redder near-IR colors then L type brown dwarfs.

Many modelers have attempted to explain this behavior. We present in this paper our explanation using a parameter-free cloud model which takes into account all important micro processes (supersaturation, nucleation, condensation, coalescence, coagulation, and mixing) according to Rossow(1978) and to Radiation Hydrodynamical (RHD) simulations (Freytag *et al.* 2010). Fig. 1 shows this transformation of the spectral properties of these objects, where we have overlaid the latest version of the BT-Settl model atmospheres (Allard *et al.* 2012) using this parameter-free cloud model. We find in general a good agreement of the models to the ground-based and *Spitzer* observations across the sequence. We find that reducing the gravity to correspond to the young planets allows a good reproduction of their near-IR spectral energy distribution.

These results imply that there is no fundamental difference between planetary and brown dwarf spectral properties, at least for the candidates compared in Fig. 1. Beta Pic b shows slightly different optical colors which could be explained by the reflection of the stellar light on clouds. The current BT-Settl models do not generate important reflection effects due to the silicate nature of the clouds predicted for this planet. More

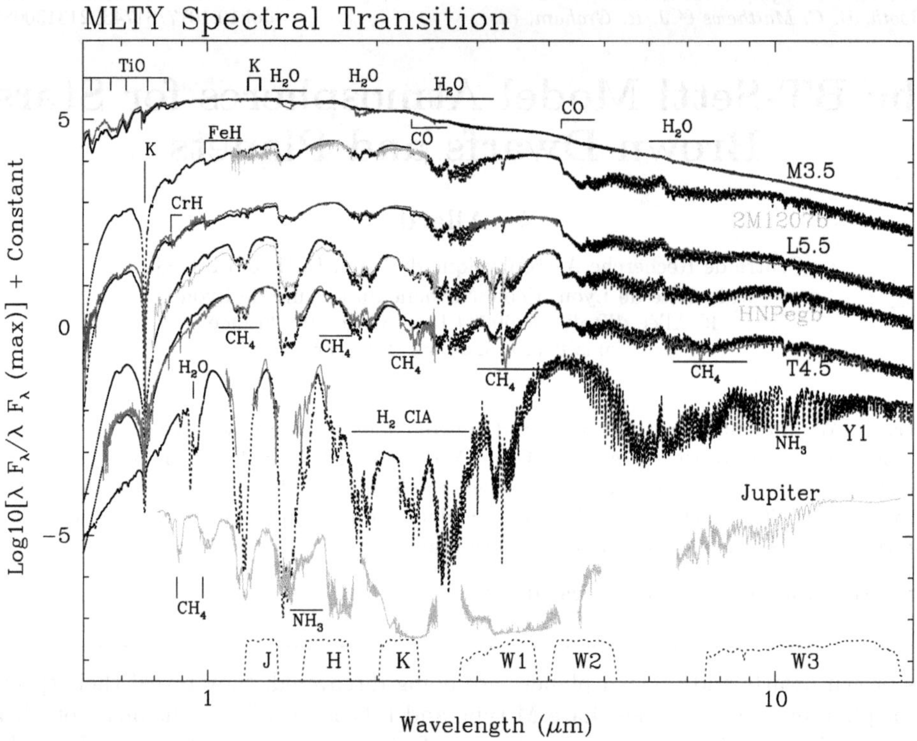

Figure 1. The SED of the M dwarf star Gl 101A (first curve from the top), the L dwarf 2MASS 1507-16 and the T dwarf 2MASSJ 0559-14 using the United Kingdom Infrared Telescope (UKIRT) and *Spitzer* (among other) spectroscopic observations (Leggett *et al.* 2002, Geballe *et al.* 2002, Cushing *et al.* 2008), and the Y1 dwarf WISE 0350-5658 (Kirkpatrick *et al.* 2012) (second and third curves from the bottom), as well as those of the planetary mass objects 2M1207b (Chauvin *et al.* 2005), HNPegb (Luhman *et al.* 2007) and Jupiter (first fourth and sixth curves from the bottom) using the medium-resolution infrared spectrograph SpeX of the NASA InfraRed Telescope Facility (IRTF) on Mauna Kea, SINFONI and *Spitzer* spectroscopic spectra (Patience *et al.* 2012, Leggett *et al.* 2008), and the infrared spectrum of Jupiter obtained with the Visual and Infrared Mapping Spectrometer (VIMS) onboard Cassini. The BT-Settl models (dotted curves) are shown for [$T_{\rm eff}$, log$g$] = [3000 K, 5.0], [1600 K, 5.0], [1600 K, 5.5], [1200 K, 5.5], [1200 K, 5.5], and [450 K, 5.0] from top to bottom. Also shown are the Mauna Kea Observatories (MKO) and WISE filter bandpasses in the lower part of the plot (dotted lines).

reflective salts or ices clouds could however if present cause increased flux at optical wavelengths.

References

Allard, F., Homeier, D., Freytag, B., & Sharp, C. M. 2012, EAS Publications Series, 57, 3
Chauvin, G., Lagrange, A.-M., Dumas, C., *et al.* 2005, *A&A* (Letters), 438, L25
Cushing, M. C., Marley, M. S., Saumon, D., *et al.* 2008, *ApJ*, 678, 1372
Freytag, B., Allard, F., Ludwig, H.-G., Homeier, D., & Steffen, M. 2010, *A&A*, 513, A19
Geballe, T. R., Knapp, G. R., Leggett, S. K., *et al.* 2002, *ApJ*, 564, 466
Kirkpatrick, J. D., Gelino, C. R., Cushing, M. C., *et al.* 2012, *ApJ*, 753, 156
Leggett, S. K., Golimowski, D. A., Fan, X., *et al.* 2002, *ApJ*, 564, L452
Leggett, S. K., Saumon, D., Albert, L., *et al.* 2008, *ApJ*, 682, 1256
Luhman, K. L., Patten, B. M., Marongo, M., *et al.* 2007, *ApJ*, 654, 570
Patience, J., King, R. R., De Rosa, R. J., *et al.* 2012, *A&A*, 540, A85
Rossow, W. B. 1978, *Icarus*, 36, 1

Exploring the Formation and Evolution of Planetary Systems
Proceedings IAU Symposium No. 299, 2013
M. Booth, B. C. Matthews & J. R. Graham, eds.

doi:10.1017/S1743921313008557

Detecting Circumbinary Exoplanets: Understanding Transit Timing

David Armstrong[1], **David V. Martin**[2] **and Don Pollacco**[1]

[1]University of Warwick, Department of Physics, Gibbet Hill Road, Coventry, CV4 7AL, UK
email: `d.j.armstrong@warwick.ac.uk`

[2]Observatoire de Genève, Université de Genève, 51 chemin des Maillettes, Sauverny 1290, Switzerland
email: `david.martin@unige.ch`

Abstract. We have derived and tested a simple analytical model for placing limits on the transit timing variations of circumbinary exoplanets. These are generally of days in magnitude, dwarfing those found in multi-planet systems. The derived method is fast, efficient and is accurate to approximately 1% in predicting limits on the possible times of transits over a 3-year campaign.

Keywords. planetary systems: dynamical evolution; binaries: eclipsing; celestial mechanics

1. Introduction

The NASA *Kepler* mission has led to the detection of seven transiting circumbinary exoplanets. Detecting these planets via the transit method presents observational challenges, as they exhibit transit timing variations (TTVs) on the order of days in magnitude, in addition to changes in the transit shape and duration.

The purpose of this work is to present constraints on the observational characteristics of a transiting circumbinary explained, through our knowledge of the host binary, using a fast method which requires no complex modelling. All of the details of this work are in Armstrong & Martin *et al.* (2013).

2. Model

We present a model which accurately places limits on the magnitude of circumbinary TTVs. An example circumbinary system is depicted in Fig. 1, with solid lines showing the planetary orbital phases which could contain transits, as calculated using a numerical N-body integrator. The analytical bounds are denoted by two dashed lines. The analytical prediction is composed of two different effects:

Geometrical. This is caused by the binary stars moving through an extended region of space. The planets can hence take several days to traverse the full extent of the binary orbit, and it is during this time that transits will occur. The geometrical effect corresponds to the thickness of the bands in Fig. 1. The thickness will change over time as the planet precesses, since the planet will pass the binary orbit at a speed dependent on its argument of periapse, $\omega_{\rm p}$. The magnitude of geometrical effect was derived by taking the transit duration equation for a single star and extending it to allow for a moving body.

Precessional. Due to the non point mass nature of the binary, large torques are exerted on the planet's orbit. The main result of this is a swift precession of $\omega_{\rm p}$. This precession causes the time of potential transits to shift away from the 'expected' time for a constant

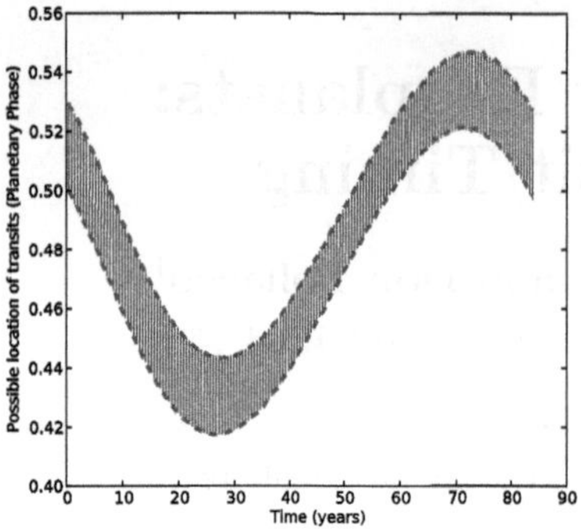

Figure 1. Numerical transit regions (solid lines) bounded by the analytical predictions (dashed lines).

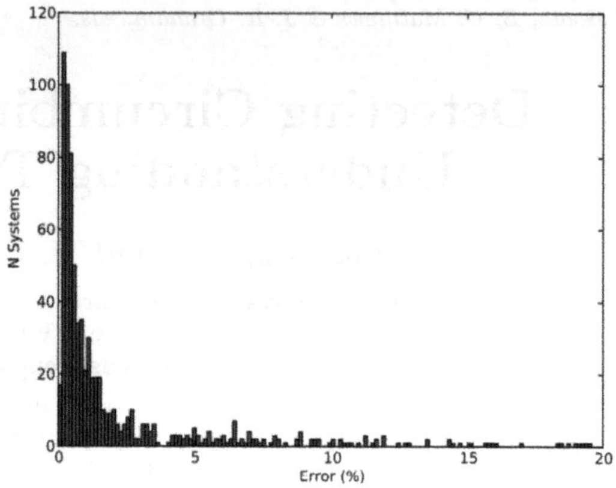

Figure 2. Histogram of the analytical prediction error, as a percentage of the numerical transit region time.

periodic signal. The magnitude of these shifts at a given time depends upon the instantaneous value of $\omega_{\rm p}$. The precessional effect manifests itself as an oscillatory offset in the transit phase range given by the geometrical variation, as seen in Fig. 1. The precessional effect was modelled using a linear precession rate from Doolin & Blundell (2011), which was originally derived from an equation in Farago & Laskar (2010).

3. Testing

The accuracy of the model was tested on a simulated distribution of 1,000 single-planet circumbinary systems, 799 of which were stable over 1,200 years. The binary distribution was drawn from a combination of the *Kepler* survey and the Halbwachs *et al.* (2003) radial velocity survey. Given the lack of a circumbinary planet distribution, the simulated planets were created based on those found by radial velocity surveys, with an additional stability constraint based on the work of Holman & Wiegert (1999). A histogram of the simulation errors is in Fig. 2. For clarity, 43 badly-predicted systems with errors higher than 20% are not shown.

4. Applications

The equations derived in Armstrong & Martin *et al.* (2013) can be used in searches for circumbinary exoplanets. As a specific example, they can be used to set the parameters $\triangle_{\rm min}$ and $\triangle_{\rm max}$ in the QATS search algorithm (Carter & Agol, 2013). The equations are also useful in reverse, for making first estimates of planet parameters using the observed TTV of a newly discovered candidate.

References

Armstrong, D. & Martin, D. V. *et al.* 2013, *MNRAS*, 434, 3047
Carter, J. A. & Agol, E. 2013, *ApJ*, 765, 132
Doolin, S. & Blundell, K. M. 2011, *MNRAS*, 418, 2656
Farago, F. & Laskar, J. 1993, *MNRAS*, 401, 1189
Halbwachs, J. L., Mayor, M., Udry, S., Arenou, F. 2003, *A&A*, 397, 159
Holman M. & Wiegert P. 1999, *ApJ*, 117, 621

Exploring the Formation and Evolution of Planetary Systems
Proceedings IAU Symposium No. 299, 2013
M. Booth, B. C. Matthews & J. R. Graham, eds.

doi:10.1017/S1743921313008569

From spectra to atmospheres: solving the underconstrained retrieval problem for exoplanets

Joanna K. Barstow[1], Suzanne Aigrain[1], Patrick G. J. Irwin[1], Neil Bowles[1], Leigh N. Fletcher[1] and Jae-Min Lee[2]

[1]Department of Physics, University of Oxford, Oxford, UK
email: jo.barstow@astro.ox.ac.uk

[2]Institute for Theoretical Physics, University of Zürich, Zürich, Switzerland

Abstract. Spectroscopic observations of transiting exoplanets have provided the first indications of their atmospheric structure and composition. Optimal estimation retrievals have been successfully applied to solar system planets to determine the temperature, composition and aerosol properties of their atmospheres, and have recently been applied to exoplanets. We show the effectiveness of the technique when combined with simulated observations from the proposed space telescope EChO, and also discuss the difficulty of constraining a complex system with sparse data and large uncertainties, using the super-Earth GJ 1214b as an example.

Keywords. radiative transfer, methods: data analysis, planets and satellites: individual

1. Introduction

When a planet passes in front of its parent star as seen from the Earth, the resultant reduction in the amount of observed starlight varies as a function of wavelength, depending on the absorptive properties of the planets atmosphere; when the planet is eclipsed by its parent star, a difference measurement can yield its emission/reflection spectrum. However, a lack of a priori knowledge about the planets bulk properties, cloudiness and temperature can make it difficult to find a unique solution. The NEMESIS radiative transfer and retrieval tool (Irwin *et al.* 2008) allows this process to be explored. Barstow *et al.* (2013a) use NEMESIS to assess the feasibility of constraining exoplanet atmospheres with EChO (Tinetti *et al.* 2012), a proposed ESA space telescope designed to observe continuous transmission and eclipse spectra of transiting extrasolar planets between 0.55 and 16 μm.

2. GJ 1214b: current observations

GJ 1214b (Charbonneau *et al.* 2009) is a super-Earth transiting an M dwarf 13 pc away. Despite measurements of transmission through its atmosphere in the visible and infrared from a range of space- and ground-based sources (e.g. Bean *et al.* 2011), little can be inferred about the atmosphere. The flat transmission spectrum could be evidence of an atmosphere with a high molecular weight, but could also indicate a H_2-He atmosphere with high altitude haze or cloud, and the current data are shown to have insufficient signal-to-noise and wavelength coverage to discriminate between these scenarios (Barstow *et al.* 2013b and references therein).

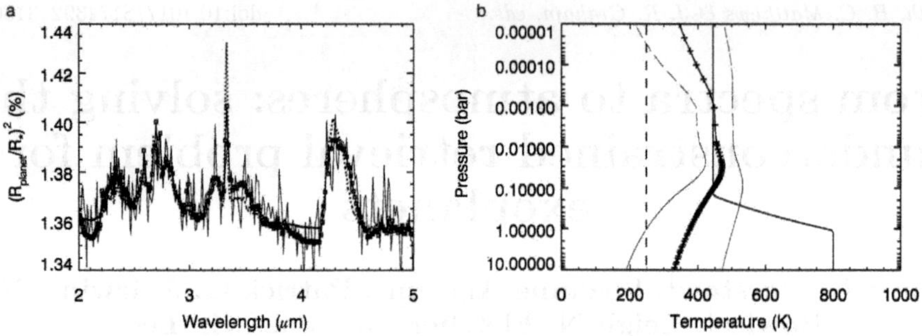

Figure 1. a) The GJ 1214b noisy synthetic transmission spectrum (thin line) for a 50% H_2O atmosphere is fit better by a 50% H_2O model (black line with points) than by a H_2-He model (dotted line). This is particularly evident in the CH_4 band at 3.3 μm and the CO band at 4.3 μm. b) Input temperature (black line) *a priori* temperature (dashed line) and retrieved temperature (crosses) with error (fine dashed lines) for a secondary eclipse retrieval. The correct stratospheric temperature is retrieved but there is no information about the deep atmosphere temperature.

3. GJ 1214b with EChO

Barstow *et al.* (2013a) generate a series of model hot Jovians, using NEMESIS as a forward model to calculate noisy synthetic transmission and eclipse spectra of these, and perform retrieval feasibility tests for EChO observations. Barstow *et al.* (2013b) use a similar technique to demonstrate that, by observing 30 transits of GJ 1214b with EChO, a high molecular weight atmospheric scenario would be excluded if the planet in fact has a cloudy extended atmosphere. Using the same technique and model parameters, we show that if the atmosphere instead contained 50% water vapour, 60 transits would be required to achieve the requisite level of signal and distinguish between the two scenarios. When we add the noise expected for 60 observations to the best fit H_2O-dominated model from Barstow *et al.* (2013b) and fit it with a 50% H_2O model atmosphere, the reduced-χ^2 parameter is 1.05, as opposed to 1.31 if it is fit with a cloudy H_2-He model (Figure 1a shows the fit between 2 and 5 μm). All models are as in Barstow *et al.* (2013b), to which we refer the reader for details. An emission spectrum of GJ 1214b could be obtained with 30 secondary eclipse observations, allowing an observational constraint to be placed on the stratospheric temperature of the planet for the first time and providing information about the albedo and therefore cloudiness of the planet (Figure 1b). EChO will be a valuable mission for the study of transiting extrasolar planets, and space-based spectroscopy over a large wavelength range is important for detailed study of planets like GJ 1214b. Optimal estimation retrieval is shown to be a powerful technique for analysing these data.

References

Barstow J. K., Aigrain S., Irwin P. G. J., Bowles N., Fletcher L. N., & Lee J.-M. 2013, *MNRAS*, 430, 1188

Barstow J. K., Aigrain S., Irwin P. G. J., Fletcher L. N., & Lee J.-M. 2013, *MNRAS*, 434, 2616

Bean J. L., Désert J.-M., Kabath P., Stalder B., Seager S., Miller-Ricci Kempton E., Berta Z. K., Homeier D., Walsh S., & Seifahrt A., 2011 *ApJ*, 743, 92

Charbonneau D., Berta Z. K., Irwin J., Burke C. J., Nutzman P., Buchhave L. A., Lovis C., Bonfils X., Latham D. W., Udry S., Murray-Clay R. A., Holman M. J., Falco E. E., Winn J. N., Queloz D., Pepe F., Mayor M., Delfosse X., & Forveille T. 2009, *Nature*, 462, 891

Irwin P. G. J., Teanby N. A., de Kok R., Fletcher L. N., Howett C. J. A., Tsang C. C. C., Wilson C. F., Calcutt S. B., Nixon C. A., & Parrish P. D. 2008, *JQSRT*, 109, 1136

Tinetti G., *et al.*, 2012, *Exp. Astr.*, 34, 311

Exploring the Formation and Evolution of Planetary Systems
Proceedings IAU Symposium No. 299, 2013
M. Booth, B. C. Matthews & J. R. Graham, eds.

doi:10.1017/S1743921313008570

A radiative-convective equilibrium model for young giant exoplanets: Application to β Pictoris b

Jean-Loup Baudino[1], Bruno Bézard[1], Anthony Boccaletti[1], Mickael Bonnefoy[2] and Anne-Marie Lagrange[3]

[1]LESIA, Observatoire de Paris, CNRS, UPMC, Université Paris-Diderot,
5 place Jules Janssen, F-92195 Meudon, France
email: jean-loup.baudino@obspm.fr

[2]Max Planck Institut of Astronomy, Königstuhl17,
D-69117 Heidelberg, Germany

[3]UJF-Grenoble 1 / CNRS-INSU, Institut de Planétologie et d'astrophysique de Grenoble
(IPAG) UMR 5274,
Grenoble, F-38041, France

Abstract. We present a radiative-convective equilibrium model for young giant exoplanets. Model predictions are compared with the existing photometric measurements of the planet β Pictoris b in the J, H, Ks, L', NB 4.05, M' bands. This model will be used to interpret future photometric and spectroscopic observations of exoplanets with SPHERE, mounted at the VLT with a first light expected mid-2014.

Keywords. stars: planetary systems, radiative transfer

1. Model

We developed a radiative-convective equilibrium model for young giant exoplanets. Input parameters are the planet's surface gravity (g), effective temperature (Teff) and elemental composition. Under the additional assumption of thermochemical equilibrium, the model predicts the equilibrium temperature profile and mixing ratio profiles of the most important gases. Opacity sources include the H_2-He collision-induced absorption and molecular lines from H_2O, CO, CH_4, NH_3, VO, TiO, Na and K. Line opacity is modeled using k-correlated coefficients pre-calculated over a fixed pressure-temperature grid. Cloud absorption can be added above the expected condensation level (e.g. iron or silicates) with given scale height and optical depth at some reference wavelength. Scattering is not included at the present stage.

We compared the model predictions to measurements of planet β Pictoris b from Bonnefoy *et al.* (2013).

2. Results

We built a grid of models with: Teff range between 1000 and 2500 K, a $\log_{10}(g[cgs])$ between 3 and 5, solar system abundances of the elements (Lodders 2010) and no cloud opacity. For each model we selected the radius that minimizes the χ^2 between the observed and calculated apparent magnitudes. We only kept models with a radius between 0.6 and 2 Jupiter radii (a realistic range derived from evolution models of Mordasini *et al.* 2012). The best model (cloud-free) is a planet with an effective temperature of 1700 K,

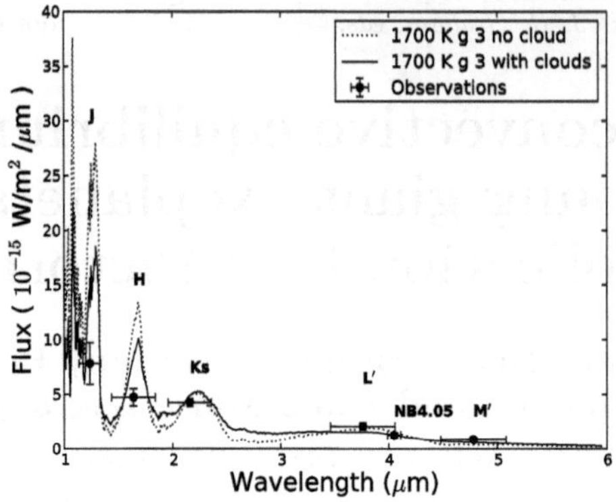

Figure 1. Apparent flux of our model without (dotted line) and with cloud opacity (solid line), compared with measured apparent fluxes of β pictoris b (circles)

a $\log_{10}$(g[cgs]) of 3 and a radius of 1.53 R_{Jup} ($\chi^2 = 21$). This model clearly yields too much flux in the J and H bands (Fig. 1).

In a second step, we added absorption from iron and silicates clouds. We used a particle radius of 30 μm and assumed the same particle column density for both clouds. For each cloud the opacity is distributed between the condensation level and the 0.1-mbar level with a particle scale height equal to the gas scale height. For Teff = 1700 K and $\log_{10}(g) =$ 3, using a particulate optical depth (τ_{clouds}) of 0.25 at 1.2 μm allows us to obtain a χ^2 of 7.5 for a radius of 1.43 R_{Jup}. Compared with the cloud free case, the flux in the J and H bands is lower and that in the Ks, L', NB4.05 and M' bands is higher. Adding cloud opacity in the model is required to reproduce the data within uncertainties.

3. Conclusions and perspectives

In agreement with other models (see references in Bonnefoy *et al.* 2013) we found that cloud opacity is needed to reproduce the observations of β Pictoris b. A model with Teff = 1700 K, $\log_{10}$(g[cgs]) = 3, and some cloud opacity agrees with observations within uncertainties, but other combinations of these parameters are probably possible.

We plan to explore the parameter space (Teff, g, τ_{clouds}) for cloudy models of β Pictoris b, to apply our model to planetary system HR8799, to update methane opacity with the Exomol data base (http://www.exomol.com/), to add NH_3 opacity. This model will be used to analyze data from SPHERE after commissioning on the VLT in 2014.

JLB is funded by the Labex ESEP (http://www.esep.pro/)

References

Bonnefoy, M., Boccaletti, A., Lagrange, A.-M., Allard, F., Mordasini, C., Beust, H., Chauvin, G., Girard, J. H. V., Homeier, D., Apai, D.,Lacour, S., & Rouan, D. 2013, *A&A*, 555, 107

Lodders K. 2010, *Solar system abundances of the elements, Lecture Notes of the Kodai School*, 379–417

Mordasini, C., Alibert, Y., Georgy, C., Dittkrist, K.-M., Klahr, H., & Henning, T. 2012, *A&A*, 547, A112

Exploring the Formation and Evolution of Planetary Systems
Proceedings IAU Symposium No. 299, 2013
M. Booth, B. C. Matthews & J. R. Graham, eds.

doi:10.1017/S1743921313008582

VSTAR models of the hot Jupiter HD 189733b

Kimberly Bott[1], **Lucyna Kedziora-Chudczer**[1] **and Jeremy Bailey**[1]

[1]Department of Astrophysics, University of New South Wales,
Kensington, Sydney, Australia
email: k.bott@unsw.edu.au

Abstract. Past analysis of HD 189733b's atmosphere has been a cause for some debate, with conflicting findings regarding water and sodium abundances and the presence of a high altitude haze. We present our models of HD 189733b's atmospheric composition using VSTAR (Versatile Software for Transfer of Atmospheric Radiation). Since the effective temperature of the planet is expected to be approximately 5000K, newly available high-temperature spectral line lists were used.

Keywords. planets and satellites: individual: HD 189733b, planets and satellites: atmospheres, planets and satellites: composition, physical data and processes: radiative transfer

1. Introduction

The 1400 K Hot Jupiter, HD 189733b closely orbits (2.19 day) its K1.5 host star in a—most likely—tidally locked position, promising an extraordinary environment to study (Bouchy *et al.* 2005). Detection of water, CO2 and possibly methane have been reported in data from NICMOS (*HST*) and *Spitzer* (Swain *et al.* 2008, Tinetti *et al.* 2007). The broadband infrared data around 3.3 and 4 μm suggests the presence of CH4 or CO, although it is difficult to differentiate between the two currently. Conflicting models are perpetuated by the lack of data points in key absorption bands.

Here we compare our own models derived with VSTAR (Versatile Software for the Transfer of Atmospheric Radiation) to data from Charbonneau *et al.* (2008), Grillmair *et al.* (2008), and Swain *et al.* (2009), as well as to other models, particularly Lee *et al.* (2011), to derive a better understanding of how these effects come into play and to further establish the capability of VSTAR (Bailey & Kedziora 2012). VSTAR's plane-parallel radiative transfer solution utilizes line-by-line absorption and particle scattering.

We ultimately aim to develop a model that can fit a wide range of data including transmission spectroscopy (e.g., Pont 2013), high resolution cross correlation spectroscopy (e.g., de Kok *et al.* 2013) and polarization (e.g., Berdyugina *et al.* 2008, 2011; Wiktorowicz 2009). At present we consider only the spectrum derived from secondary eclipse observations.

2. The Lee 2011 Model

We began with attempting to emulate a relatively straightforward model from Lee *et al.* (2011) which included only four molecules (CH4, H2O, CO and CO2) at a constant abundance over all heights and a retrieved Pressure-Temperature profile, well-situated, mid-range of other suggested P-T profiles (e.g., Madhusudan & Seager 2009). Although significance testing is not yet complete, our model, with abundances predicted by equilibrium chemistry, does appear to be a better fit than the model emulating Lee *et al.* (2011)'s constant ratios with height for the *Spitzer* data.

3. Variables

Four P-T profiles were compared from Lee *et al.* (2011) and Madhusudan & Seager (2009). None of the profiles include inversion, as the dayside of HD 189733b appears to lack one Fortney *et al.* (2007), Charbonneau *et al.* (2008). Cooler T-P profiles appear to give better fits, particularly for shorter wavelengths. Further modelling and significance testing is needed to determine the scale of the effects.

Water features in the spectrum of HD 189733b are seen in the data points from NICMOS around 2 microns, however these bands have some overlap with CO, CO2 and CH4 Grillmair *et al.* (2008). A more convincing feature is the gap between vibrational modes of water vapour near 6.5 microns seen previously in other models and in our own. The broad absorption bands of water are clearly visible in the relatively well observed 5 to 10 μm *Spitzer* data. The steep slope produced by water absorption around 3.5 μm could be better constrained with further observational data.

The C/O level of HD189733b was estimated to be slightly more than solar (0.54) at 0.65 by Lee *et al.* (2011). Previously, Swain *et al.* (2009) placed the value broadly between 0.5 and 1.0. For most of our models C/O of 0.65 was used and provided a good fit for the majority of the points.

4. Results and Discussion

Water absorption, while it seems to be needed for a good fit can have overlap with other bands. The predictable shape of the peaks between bands make it seem unlikely that an amalgamation of other nearby absorption bands is the culprit, but further analysis is needed to determine this.

The models presented here do not include scattering particles. A scattering haze is likely to be important to modelling the transmission spectroscopy (e.g. Pont *et al.* 2013) which appears to be consistent with the presence of Rayleigh scattering, probably due to small dust grains.

References

Bailey, J. & Kedziora, L. 2012, *MNRAS*, 419, 1913–1929

Berdyugina, S. V., Berdyugin, A. V., Fluri, D. M., & Piirola, V. 2008, *ApJL*, 673, L83–L86

Bouchy, F., Udry, S., Mayor, M., *et al.* 1995, *A&A*, 444, L15–L19

Charbonneau, D., Knutson, H. A., Barman, *et al.* 2008, *ApJ*, 686, 1341–1348

de Kok, R. J., Brogi, M., Snellen, I. A.G., Birkby, J., Albrecht, S., & de Mooij, E. J.W. 2013, *A&A*, 554, A82

Fortney, J. J., Lodders, K., Marley, M. S., & Freedman, R. S. 2007, *ApJ*, 678, 1419–1435

Grillmair, C. J., Burrows, A., Charbonneau, *et al.* 2008, *Nature*, 456, 767–769

Madhusudan, N. & Seager, S. 2009, *ApJ*, 707, 24

Lee, J. M., Fletcher, L. N., & Irwin, P. G. J. 2011, *MNRAS*, 420, 170–182

Pont, F., Sing, D. K., Gibson, N. P., Aigrain, S, Henry, G., & Husnoo, N. 2013, *MNRAS*, 432, 2917–2944

Swain, M. R., Vasisht, G., Tinetti, G., *et al.* 2009, *ApJL*, 690, L114–L117

Swain, M. R., Tinetti, G., Vasisht, G., *et al.* 2009, *ApJ*, 704, 1616–1621

Tinetti, G., Vidal-Madjar, A., Liang, M.-C., *et al.* 2007, *Nature*, 448, 169–171

Wiktorowicz, S. J. 2009, *ApJ*, 696, 1116–1124

Exploring the Formation and Evolution of Planetary Systems
Proceedings IAU Symposium No. 299, 2013
M. Booth, B. C. Matthews & J. R. Graham, eds.

doi:10.1017/S1743921313008594

Hα Absorption in Transiting Exoplanet Atmospheres

Duncan Christie[1], Phil Arras[1], and Zhi-Yun Li[1]

[1]Department of Astronomy, University of Virginia,
P.O. Box 400325, Charlottesville, Virginia, USA
email: dac5zm@virginia.edu, pla7y@virginia.edu, zl4h@virginia.edu

Abstract. Recent observations by Jensen *et al.* of Hα absorption by the upper atmosphere of HD189733b have motivated the need for a theoretical understanding of the distribution of $n = 2$ hydrogen within hot Jupiter atmospheres. With this in mind, we model the $n = 2$ state of atomic hydrogen in a hydrostatic atmosphere in thermal and photoionization equilibrium. Both collisional and radiative transitions are included in the calculation of the n = 2 state level population. In our model, the Hα absorption is dominated by a $\tau \sim 1$ shell composed of metastable 2s hydrogen located within the neutral atomic layer, with the contribution coming roughly uniformly throughout the layer instead of from a specific impact parameter. An ionization rate an order of magnitude over the expected value can reproduce the observed transit depth.

Keywords. line: formation – planets and satellites: atmospheres – stars: individual (HD 189733, HD 209458)

---

1. Introduction

The absorption of starlight during transit is a useful probe to study the structure of exoplanet atmospheres. The detection of absorption of Lyman-α (e.g., Vidal-Madjar *et al.* 2003) provides evidence for ground state hydrogen in the atmospheres of hot Jupiters but provides minimal information about the thermal structure. Hα provides a complementary probe of the neutral hydrogen content, in that it provides information about the thermal structure of the atmosphere. Our goal is to understand and model the formation of $n = 2$ hydrogen within hot Jupiter atmospheres, and to use our models to understand the results of Jensen *et al.* (2012).

2. Method

We model a spherically-symmetric, hydrostatic atmosphere in thermal and photoionization equilibrium. For simplicity, we consider UV photoelectric heating as the primary heating source and Lyman-α cooling as the primary cooling mechanism. The ionization state is found by balancing photoionization and radiative recombination. Since the ionizing flux from HD189733 is a large source of uncertainty in the model, we parameterize the UV flux using the synthetic spectrum downloaded from the X-exoplanets Archive at the CAB (Sanz-Forcada *et al.* 2011) as a baseline. Attenuation of UV is treated in a frequency dependent fashion (Trammell *et al.* 2011).

The abundance of $n = 2$ hydrogen is determined by modeling collisional and radiative excitation and de-excitation, radiative recombination, and photoionization of excited states.

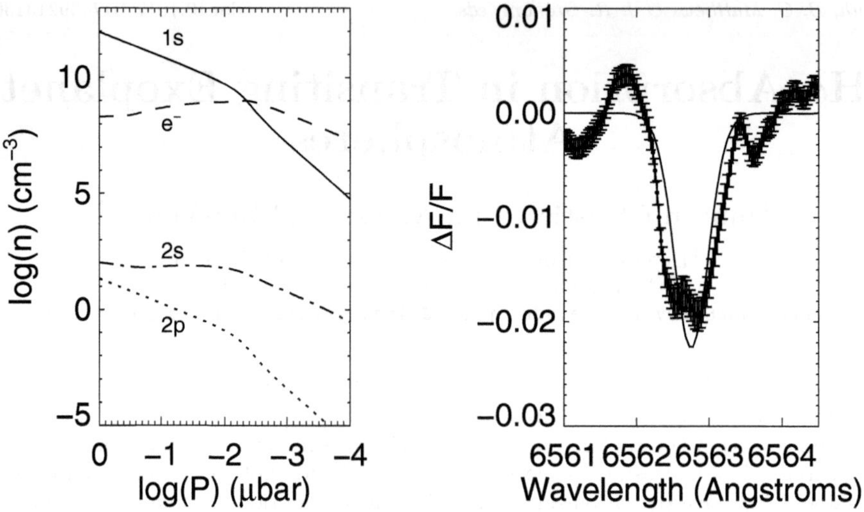

Figure 1. *Left:* Number densities versus pressure. Shown are the densities for 1s (solid line), electrons (dashed line), 2s (dash-dot line), and 2p (dotted line). *Right:* $\Delta F_\lambda / F_\lambda$ for the Hα for HD189733b. The observed data (filled circles) from Jensen *et al.* (2012) are over plotted with error bars.

3. Results

We vary the ionizing flux and find that in order to fit the transit depth quoted by Jensen *et al.* (2012) we require a substellar flux 3.5 times larger than given by the Sanz-Forcada *et al.* (2011) spectrum, if geometric corrections are neglected. Attempts to model the Lyman-α absorption of HD189733b have needed enhanced UV so this is not viewed as cause for concern (Lecavelier Des Etangs *et al.* 2010).

Throughout the model atmosphere, the $n = 2$ hydrogen is predominantly in the 2s state due to the lack of fast radiative transition to the ground state. The primary mechanism for the destruction of 2s then becomes the collisional ℓ-mixing reaction which transitions 2s into 2p hydrogen which can then radiatively de-excite through the Lyman-α transition. Within the ionized upper regions of the atmosphere (i.e., where $n_e \geqslant n_{1s}$) the formation of 2s is due to radiative recombination with collisional excitation from the ground state coming to dominate as the atmosphere becomes predominantly neutral atomic hydrogen. Within the atomic layer, we find that the abundance of 2s hydrogen does not vary significantly relative to the total hydrogen abundance which varies by 3 orders of magnitude (see Figure 1).

References

Jensen, A. G., Redfield, S., Endl, M., *et al.* 2012, *ApJ*, 751, 86, 86

Lecavelier Des Etangs, A., Ehrenreich, D., Vidal-Madjar, A., *et al.* 2010, *A&A*, 514, A72, A72

Sanz-Forcada, J., Micela, G., Ribas, I., *et al.* 2011, *A&A*, 532, A6, A6

Trammell, G. B., Arras, P., & Li, Z.-Y. 2011, *ApJ*, 728, 152, 152

Vidal-Madjar, A., Lecavelier des Etangs, A., Désert, J.-M., *et al.* 2003, *Nature*, 422, 143, 143

Exploring the Formation and Evolution of Planetary Systems
Proceedings IAU Symposium No. 299, 2013
M. Booth, B. C. Matthews & J. R. Graham, eds.

doi:10.1017/S1743921313008600

Characterization of exoplanet atmospheres using future space-based infrared telescopes: challenges in detecting biomarkers

Keigo Enya

Institute of Space and Astronautical Science, Japan Aerospace Exploration Agency
Yoshinodai 3-1-1, Sagamihara, Kanagawa, Japan
email: enya@ir.isas.jaxa.jp

Abstract. Characterization of exoplanet atmospheres with space-based infrared telescopes is important to detect biomarkers. A promising method is temporary differential observation. For this method, designs of a wideband infrared spectral disperser are presented. A design using a CdTe prism simultaneously covers $\lambda = 1-30\,\mu$m. Designing binary pupil masks for segmented pupils to be used in spatially resolved observations are also shown for another observational method.

Keywords. exoplanet, biomarker, infrared, spectral disperser, coronagraph

---

There are interesting atmospheric absorption features of molecules in the infrared wavelength region, e.g., H_2O (1.13, 1.38, 1.9, 2.69, 6.2 μm), CO_2 (1.21, 1.57, 1.6, 2.03, 4.25 μm), O_3 (4.7, 9.1, 9.6 μm), O_2 (1.27 μm), CH_4 (1.65, 2.2, 2.31, 2.37, 3.3, 6.5, 7.7 μm) and NH_3 (1.5, 2.0, 2.25, 2.9, 3.0, 6.1, 10.5 μm). Space-based infrared telescopes are useful for observing these features on account of their wavelength coverage, stability and sensitivity. The huge contrast between exoplanets and their parent stars is a serious problem.

1. Wideband infrared spectrometer

Precise temporary differential observation is a promising method for the characterization of exoplanet atmospheres, for which wideband infrared spectral dispersers are useful (Enya *et al.* 2011, Enya 2012, Doyon *et al.* 2012). Fig. 1 shows new designs of straight-path type dispersers that can be placed into a slot of a mechanical filter wheel in the camera optics. The main prism separates the orders of the spectral format, the grating pattern on the prism surface produces spectral dispersion in each order, and the sub-prism compensates for the inflection angle. The design using ZnS and CdTe provides simultaneous wavelength coverage of $1-14\,\mu$m and $1-30\,\mu$m, respectively. A long observation span, by one or more satellites, is important to improve the signal to noise ratio.

2. Binary pupil mask coronagraph

Large telescopes are required for spatially resolved characterization of exoplanets in the infrared, in which telescopes tend to have a segmented pupil. Recent progress in binary pupil mask coronagraphs provides a simple solution for the segmented pupil of a large telescope (Enya & Abe 2010, Enya *et al.* 2011, Carlotti, Vanderbei & Kasdin 2011). Fig. 2 shows designs of binary pupil masks for various segmented pupils. It should be noted that the LOQO optimizer (Vanderbei 1999) was used in the software system for these designs. A very small IWA, $1.3\,\lambda/D$, is also achievable. With this value, it is

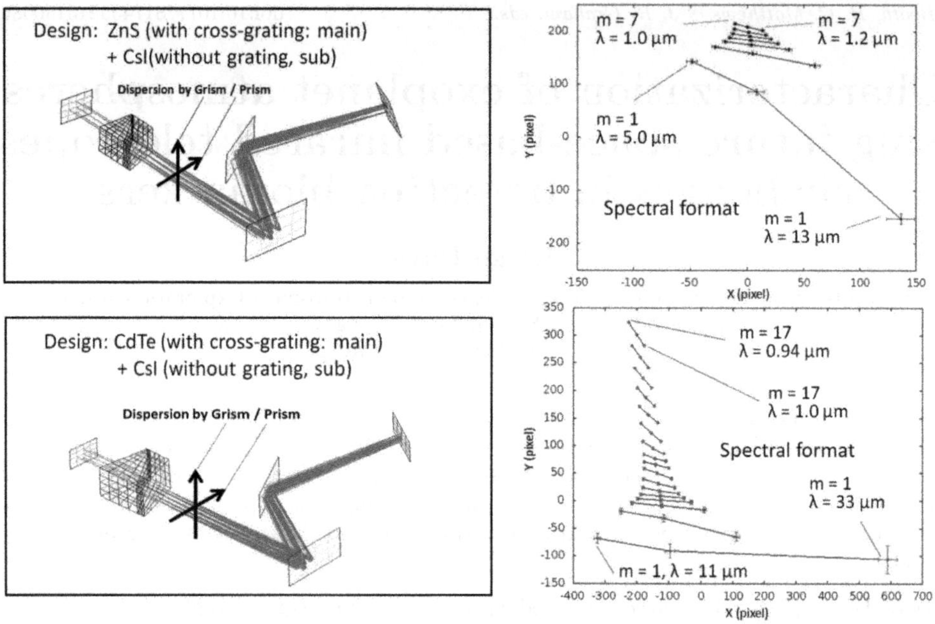

Figure 1. Designs of straight-path spectral dispersers.

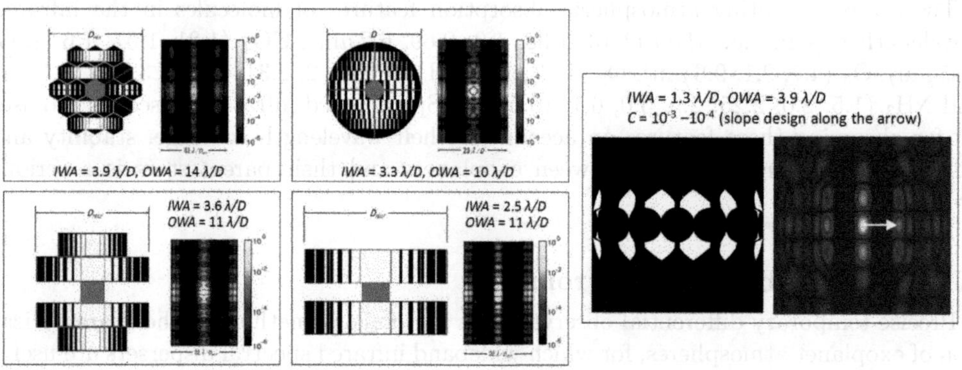

Figure 2. Designs of binary pupil masks for coronagraphs for a segmented telescope pupil.

possible to resolve a small number of stars and their exoplanets close to (or on) the outer edge of the habitable zone (e.g., Kopparapu *et al.* 2013) at the observing wavelength of O_3 even with a 3 m class telescope. To enhance the contrast, it is necessary to both improve masks and combine other techniques (e.g., PSF subtraction, wavefront correction, and hybridization with a Lyot stop). Ultra-deep observations for very nearby and/or early type stars, including α Cen, Sirius, Procyon, Altair, Vega, and Fomalhaut, are important.

References

Carlotti, A., Vanderbei, R., & Kasdin, N. J. 2011 arXiv:1108.4050
Doyon, R. *et al.* 2012, *Proc. SPIE*, 8442, 84422R
Enya, K. & Abe, L. 2010, *PASJ*, 62, 1407
Enya, K., Abe, L., Takeuchi, S., Kotani, T., & Yamamuro, T. 2011, *Proc. SPIE*, 8146, 81460Q
Enya, K. 2012, *Proc. SPIE*, 8442, 84423Y
Kopparapu, R. K., *et al.* 2013 *ApJ*, 765, 131
Vanderbei, R. J. 1999, *Optimization methods and software*, 11, 451

Exploring the Formation and Evolution of Planetary Systems
Proceedings IAU Symposium No. 299, 2013
M. Booth, B. C. Matthews & J. R. Graham, eds.

doi:10.1017/S1743921313008612

The HoSTS Project: A Homogeneous Study of Transiting Systems

Yilen Gómez Maqueo Chew[1,2], Francesca Faedi[2], Leslie Hebb[3], Don Pollacco[2], Keivan Stassun[1], Phillip Cargile[1], Barry Smalley[4], Amanda Doyle[4], Luan Ghezzi[5], Susana C. C. Barros[6], and Andrew Collier Cameron[7]

[1]Department of Physics & Astronomy, Vanderbilt University, Nashville, TN 37235, USA

[2]Department of Physics, University of Warwick, Coventry CV4 7AL, UK
email: y.gomez@warwick.ac.uk

[3]Department of Astronomy, University of Washington, Seattle, WA 98195, USA

[4]Astrophysics Group, Keele University, Staffordshire ST5 5BG, UK

[5]Observatório Nacional, Rua Gal. José Cristino 77, Rio de Janeiro, RJ 20921-400, Brazil

[6]Laboratoire d'Astrophysique de Marseille UMR 7326, 13388, Marseille, France

[7]School of Physics and Astronomy, University of St Andrews, St Andrews, Fife KY16 9SS, UK

Abstract. The Homogeneous Study of Transiting Systems (HoSTS) will derive a consistent and homogeneous set of both the stellar and planetary physical properties for a large sample of bright transiting planetary systems with confirmed planetary masses and measured radii. Our resulting catalogs of the fundamental properties of these bright planets and their host stars will enable us to explore empirical correlations that will lead to a better understanding of planetary formation and evolution. We present our pilot study of the planet-hosting star WASP-13, and the framework of our project which will allow for the identification of true relationships among the physical properties of the systems from any systematics.

Keywords. stars: fundamental parameters, planetary systems,techniques: radial velocities techniques: spectroscopic, techniques: techniques: photometric, stars: individual (WASP-13)

1. Introduction

Over 300 transiting extrasolar planets have been discovered thus far, exhibiting an amazing array of physical properties orbiting stellar hosts that also span a range of properties and stages of evolution. This diversity of planets and of the stellar environments in which they form and evolve offers a watershed opportunity to develop a systematic understanding of the physics that governs planet formation, structure, and evolution, and of the connections between the planets and their host stars. To fully exploit the wealth of information encoded in the set of transiting planetary systems requires careful determination of precise fundamental properties of the exoplanets and their host stars.

We have underway a systematic investigation to fully and accurately characterize a large sample of the known bright transiting planetary systems. These transiting systems, in which part of the stellar light is blocked as a planet crosses the disk of the star, are extremely valuable for detailed study because they are the only systems that allow for precise measurements of the mass and radius of the planet. For the first time, there is now a large enough sample of transiting extra-solar planets for which both mass and radius measurements can be accurately determined so that statistically significant studies of their physical properties can be achieved. Thus, we present the homogeneous methodology that will be employed in our project, exemplified by the results from the pilot study of the transiting system WASP-13 (Gómez Maqueo Chew *et al.*, 2013) which highlights the importance of the proposed project in order to identify, and tighten empirical relationships among the physical properties of the exoplanet systems. HoSTS will cover the parameter space of known physical properties of planetary systems for the stellar hosts (e.g., $T_{\rm eff}$, [Fe/H], evolutionary status, and in single and multiple stellar systems), with single and multiple planets exhibiting a range of masses and radii. Our project extends previous efforts (e.g., Torres *et al.* (2012)) in that it will derive consistently both stellar and planetary properties from high quality data.

2. Dataset and Methodology

We have an extensive data set both obtained by our team, and compiled from the literature, for planetary systems in both the Northern and Southern hemispheres including: high resolution (R >40k), high signal-to-noise ratio (S/N>200) echelle spectra with $\lambda\sim$4000–7800Å, precise radial velocity measurements ($\sim$20 m/s), high cadence, high precision transit light curves, and an Hα long-slit spectra (R >7k, S/N>200).

The methodology is described in detail in Gómez Maqueo Chew *et al.* (2013), and only briefly summarized here. The stellar effective temperature ($T_{\rm eff}$), the metallicity ([Fe/H]), and the spectroscopic $\log g$ are derived from the high quality echelle spectrum using several stellar characterization methods based on the two techniques used in the exoplanet literature, spectral synthesys and the ionization/excitation equilibrium of iron lines. The stellar temperature will be compared against the value derived from the Hα spectrum as a check on $T_{\rm eff}$. Using the stellar density ($\rho_\star$; Seager & Mallén-Ornelas, 2003) directly measured from the transit light curve, and the spectroscopically-determined $T_{\rm eff}$ and [Fe/H], the stellar mass is derived from stellar models (e.g., Dotter *et al.*, 2008), and/or empirical relationships (e.g., Enoch *et al.*, 2010). The planet properties are derived from the simultaneous modeling of the radial velocity and transit light curves using a Monte-Carlo Markov Chain algorithm (Collier Cameron *et al.*, 2007). Furthermore, HoSTS will explore the dependence of the derived physical properties on the dataset, on the stellar characterization method and the assumptions used (e.g., constraining spectroscopic $\log g$ with $\rho_\star$ like suggested by Torres *et al.* (2012) for synthesis-based methods), and/or on the kind of planetary system.

3. Conclusions from Pilot Study of WASP-13

Our detailed study of WASP-13 showcases the importance of the HoSTS project for the tightening of the empirical correlations (see Fig. 4 in Gómez Maqueo Chew *et al.*, 2013). Although our measurements of the planetary radius of WASP-13b and the metallicity of its host overlap with previous estimates, the errors on [Fe/H] are significantly improved. Similar improvements for the larger sample of the transiting planets will allow the assessment of the validity of the observed trends, and the identification of any systematics. Moreover, we have included realistic uncertainties in the stellar mass, and consequently the planetary properties, including the contribution from the uncertainty in [Fe/H], and in the stellar evolutionary status. Both of these sources of uncertainties are typically not taken into account. In the case of WASP-13, these uncertainties are shown to be the limiting factor in the precision that can be obtained for the planet properties.

References

Collier Cameron, A., *et al.* 2007, *MNRAS*, 380, 1230
Dotter, A., Chaboyer, B., Jevremović, D., *et al.*, 2008, *ApJS*, 178, 89
Enoch, B., Collier Cameron, A., Parley, N. R., & Hebb, L. 2010, *A&A*, 516, A33
Gómez Maqueo Chew, Y., Faedi, F., Cargile, P., *et al.*, 2013, *ApJ*, 768, 79
Seager, S. & Mallén-Ornelas, G. 2003, *ApJ*, 585, 1038
Torres, G., Fischer, D. A., Sozzetti, A., *et al.* 2012, *ApJ*, 757, 161

Exploring the Formation and Evolution of Planetary Systems
Proceedings IAU Symposium No. 299, 2013
M. Booth, B. C. Matthews & J. R. Graham, eds.

doi:10.1017/S1743921313008624

Additional Keplerian Signals in the HARPS data for Gliese 667C: Further Analysis

Philip C. Gregory, Samantha M. Lawler, and Brett Gladman

Physics and Astronomy, University of British Columbia,
6244 Agricultural Rd., Vancouver, British Columbia, Canada V6T 1Z1
email: `gregory at phas.ubc.ca`

Abstract. A re-analysis of Gliese 667C HARPS precision radial velocity data was carried out with a Bayesian multi-planet Kepler periodogram (from 0 to 7 planets) based on a fusion Markov chain Monte Carlo algorithm. The most probable number of signals detected is six with a Bayesian false alarm probability of 0.012. The residuals were shown to be consistent with white noise. The six signals detected include two previously reported with periods of 7.198 (b) and 28.14 (c) days, plus additional periods of 30.82, 38.82, 53.22, and 91.3 days. The existence of these Keplerian-like signals suggest the possibility of additional planets in the habitable zone of Gl 667C although some of the signals could be artifacts arising from the sampling or stellar surface activity. N-body orbital integrations are being undertaken to determine which of these signals are consistent with a stable planetary system. Preliminary results demonstrate that four of the signals, with periods of 7.2, 28.1, 38.8, & 91 d, are consistent with a stable 4 planet system on time scales of 10^7 yr. The $M \sin i$ values are $\sim$ 5.5, 4.4, 1.9, and 4.7 $M_\oplus$, respectively.

Keywords. stars: planetary systems; methods: statistical; methods: data analysis; techniques: radial velocities

---

1. Introduction

This paper reports progress in the interpretation of additional Keplerian signals that were detected in a Bayesian re-analysis (Gregory 2012a) of the HARPS radial velocity data (Delfosse *et al.* 2013) for Gliese 667C. Gliese 667C (Gl 667C), is an isolated M dwarf component of a hierarchical triple system. The two others components, Gl 667AB, are a closer couple of K dwarfs (semi-major axis of 1.82 AU, period of 42.15 years and a total mass of 1.27 $M_\odot$). Gl 667C is the lightest component with a mass of 0.310 ± 0.019 $M_\odot$ (Anglada-Escudé *et al.* 2012). It is at an estimated distance of $\sim$ 300 AU from Gl 667AB.

In 2011, Bonfils *et al.* (2011) reported a super-earth Gl 667Cb ($M \sin i = 5.9$ $M_\oplus$) with a period of 7.2 d and evidence for two other interesting periods of 90 and 28 d. The 28 d period was of particular interesting because it would fall in the HZ. Two recent papers (Anglada-Escudé *et al.* 2012 & Delfosse *et al.* 2013), have confirmed planet Gl 667Cb and a 28 d period planet Gl 667Cc ($M \sin i = 4.3$ $M_\oplus$) in the HZ.

The six Keplerian signals detected by Gregory (2012a), included the two previously reported with periods of 7.198 (b) and 28.14 (c) days, plus additional periods of 30.82, 38.82, 53.22, and 91.3 days. The Bayesian false alarm probability for a six signal model was found to be 0.012. The existence of these Keplerian-like signals suggest the possibility of additional planets in the habitable zone of Gl 667C, although some of the signals could be artifacts arising from the sampling or stellar surface activity. Delfosse *et al.* (2013) concluded that the rotation period of the star is $\sim$ 105 d, based on a periodogram of the stellar activity diagnostic (FWHM of the CCF). We initially suspected the 53 d signal was the second harmonic of the star's rotation period. However, our own analysis of the diagnostic data indicates that the dominant spectral feature in the Lomb-Scargle periodogram of the FWHM shifts from a period of 105 d based for the first 107 FWHM points to 100 d based on the last 107 points. We now entertain the possibility that 53 d feature may not be an artifact.

2. New Results

We are employing the following two-step process to test whether a subset of the six signals are consistent with a dynamically stable planetary system. In the first step the likelihood function of the fusion MCMC (Gregory 2012b) was modified so the predicted reflex motion of the star consists of the sum of two parts. The first part is the reflex motion of the star due to the n planets, derived from a fourth order Hermite N-body integrator (Peit Hut and Jun Makino available on their 'Art of Scientific Computing website'). The second part consisted of m independent Kepler signals where $n + m = 6$. The calculation assumes that n of the signals are due to planets and the remaining m signals are artifacts which can be approximately modeled by Keplerians. To draw starting parameters for this modified fusion MCMC run, we first filtered the original fusion MCMC parameter sets obtained by Gregory (2012a) to find all parameter sets that were pair wise Lagrangian or Hill stable according to Barnes & Greenberg (2006), to pick out low eccentricity solutions. Thus the first step is attempting to find n planets plus m artifact signals that provide a good fit to the radial velocity data starting from the original fusion MCMC results.

The second step is to see if an n planet model from the first step is stable on a longer time scale of 10^7 yr using two different N-body integrators including HNBody due to Rauch & Hamilton (2002). Preliminary results demonstrate that four of the signals, with periods of 7.2, 28.1, 38.8, and 91 d, are consistent with a stable 4 planet system on time scales of 10^7 yr. The corresponding $M \sin i$ values are $\sim$ 5.5, 4.4, 1.9, & 4.7 $M_\oplus$, respectively, and the semi-major axes are $\sim 0.049, 0.12, 0.15,$ & 0.27 AU. This places the 28.1 and 38.8 d planetary candidates in the central region of the habitable zone. Other combinations of n planetary candidates remain to be explored.

Note: Very recently, Anglada-Escudé *et al.* (2013) announced additional candidates with periods of 39, 62, 92, & 260 d and a hint of a seventh period of 17 d.

References

Anglada-Escudé, G., Arriagada, P., Vogt, S. S., Rivera, E. J., Butler, R. P., Crane, J. D., Shectman, S. A., Thompson, I. B., Minniti, D., Haghighipour, N., Carter, B. D., Tinney, C. G., Wittenmyer, R. A., Bailey, J. A., O'Toole, S. J., Jones, H. R. A., & Jenkins, J. S. 2012, *ApJ*, 751, L16

Anglada-Escudé, G., Tuomi, M., Gerlach, E., Barnes, R., Heller, R., Jenkins, J. S., Wende, S., Vogt, S. S., Butler, R. P., & Jones, H. R. A., *A&A* in press 2013

Bonfils, X., Delfosse, X., Udry, S., Forveille, T., Mayor, M., Perrier, C., Bouchy, F., Gillon, M., Lovis, C., Pepe, F., Queloz, D., Santos, N. C., Sgransan, D., & Bertaux, J.-L. 2011, arXiv:1111.5019v2

Barnes, R. & GreenBerg, R. 2006, *ApJ*, 647, L163

Delfosse, X., Bonfils, X., Forveille, T., Udry, S., Mayor, M., Bouchy, F., Gillon, F. M., Lovis, C., V. Neves, V., Pepe, F., Perrier, C., Queloz, D., Santos, N. C., & Ségransan, D. 2013, *A&A*, 553, 15

Gregory, P. C., 2012, arXiv:1212.4058V2

Gregory, P. C., Chapter 7 in Astrostatistical Challenges for the New Astronomy, Springer Series in Astrostatistics, Hilbe, J.M (ed), 2012, New York:Springer, pp. 121–148

Ruach, K. R. & Hamilton, D. P. 2012, *Bulletin of the American Astronomical Society*, 34, 938

Exploring the Formation and Evolution of Planetary Systems
Proceedings IAU Symposium No. 299, 2013
M. Booth, B. C. Matthews & J. R. Graham, eds.

doi:10.1017/S1743921313008636

Synthetic modelling of the light-travel time effect of circumbinary planets

Tobias C. Hinse and Jae Woo Lee

Korea Astronomy & Space Science Institute, 305-348 Daejeon, Republic of Korea
email: tchinse@gmail.com

Abstract. In recent years several multi-body, circumbinary planets have been proposed to orbit short-period eclipsing binaries. In light of the recent discoveries based on the Kepler data, the existence of such systems seems plausible. However, performing a detailed dynamical analysis reveals that the majority of the proposed planetary systems follow highly unstable orbits. In order to solve the origin of this problem, we have started to model synthetic light-travel time signals of stable planetary systems. In particular, we aim to study the response of the model in various circumstances (e.g red/white noise level, various sampling frequencies, in-homogeneous data sets, baseline dependency.) This work will significantly increase the confidence with which model work is carried out for future systems and help towards an understaning when models break down (e.g resulting in unstable systems).

Keywords. binaries: eclipsing, planetary systems, methods: data analysis, formation: planetary systems

---

1. Introduction

Lately, several multi-body systems have been proposed (Lee *et al.* 2009, Qian *et al.* 2011, Almeida *et al.* 2013) to orbit a short-period, eclipsing binary. The light-travel time effect can invoked to detect additional bodies. By measuring the times of eclipses, it is possible to detect small timing variations. At some times the binary system will be closer to Earth, and at other times it will be positioned farther away. These circumbinary companions are of planetary nature with mass between 3 and 10 M_{Jup}. In light of the recent discoveries (e.g. Kepler-16, Kepler-47) from the Kepler space telescope (e.g. Doyle *et al.* 2011, Orosz *et al.* 2012, Welsh *et al.* 2012) such systems seem plausible. For Kepler-47 the two planets orbit the binary on stable orbits with lifetimes of over 100 million years.

However, those multi-body circumbinary companions proposed from invoking the light-travel time effect are often following orbits that are short-lived in comparison to the systems overall age (Horner *et al.* 2011, Goździewski *et al.* 2012, Hinse *et al.* 2012). This finding raises some doubts about the confidence of their existence. Although the possibility exists that the detected timing variations are caused by additional companions, the currently proposed systems must, at least, exhibit a different orbital architecture to ensure long-term orbital stability. Usually, stable planetary systems are characterised by well-separated, non-crossing orbits with orbital eccentricities close to zero. However, for most of the proposed systems the orbits of the two planets are crossing implying mutual encounters resulting in catastrophic break-up events.

2. Synthetic modelling of light-travel time signal

This paper marks the beginning of a series of research efforts addressing the modelling of synthetic one- and two-companion light-travel time signals. We generate synthetic

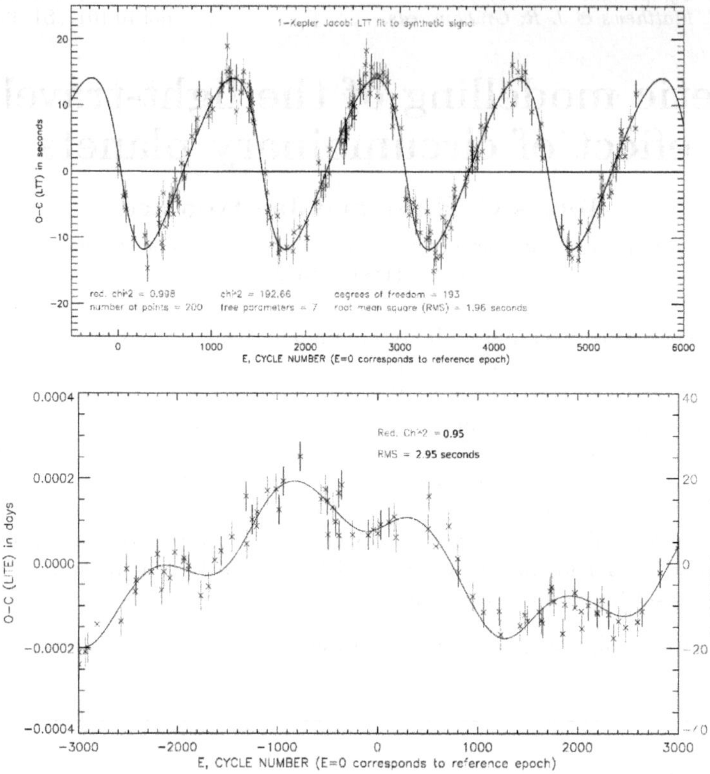

Figure 1. Synthetic modelling of light-travel time signals from a single (top panel) and two-companion (bottom panel) system. Recently, Conroy *et al.* (2013) announced the existence of several one-companion systems orbiting a short-period binary.

signals of stable circumbinary planetary systems containig one or more companions. We have generated single and two-companion signals (Fig. 1) with Gaussian noise added.

We base our model on the Jacobian formalism as presented in Goździewski *et al.* (2012). The original signals were sucessfully recovered even though the mutual gravitational interactions have not been taken into account in the modelling work of the two-companion case. In the future we aim to study the dependence of parameter errors on the timing precision and examine the effect of considering various baselines on the final model parameters.

Acknowledgment: This research is supported by KASI grant 2013-9-400-00 and financial support by the IAU.

References

Almeida, L. A., Jablonski, F., & Rodrigues, C. V., 2013, *ApJ*, 766, 11
Conroy, K. E., *et al.* 2013, *arXiv:1306.0512*, submitted to AJ
Doyle, L. R., *et al.* 2011, *Science*, 333, 1602
Goździewski, K., *et al.* 2012, *MNRAS*, 425, 930
Hinse, T. C., *et al.* 2012, *MNRAS*, 420, 3609
Horner, J., Marshall, J. P., Wittenmyer, R. A., & Tinney, C. G. 2011, *MNRAS*, 416, 11
Lee, J. W., *et al.* 2009, *AJ*, 137, 3181
Orosz, J. A., *et al.* 2012b, *Science*, 337, 1511
Qian, S.-B., *et al.* 2011, *MNRAS*, 414, 16
Welsh, W., *et al.* 2012, *Nature*, 481, 475

Exploring the Formation and Evolution of Planetary Systems
Proceedings IAU Symposium No. 299, 2013
M. Booth, B. C. Matthews & J. R. Graham, eds.

doi:10.1017/S1743921313008648

Using exoplanet systems with highly elliptical orbits to search for star-planet interactions

John R. Hodgson II[1], Damian J. Christian[1], Dennis Bodewits[2] and Suzanne Hawley[3]

[1]California State University, Northridge, Physics Department, 18111 Nordhoff St, Northridge, California 91330-8268, USA
email: John.Hodgson.71@my.csun.edu & Damian.Christian@csun.edu

[2]Department of Astronomy, University of Maryland, College Park, MD 20742, USA
email: dennis@astro.umd.edu

[3]Astronomy Department University of Washington, Box 351580, Seattle, WA 98195, USA
email: slh@astro.washington.edu

Abstract. We are investigating if the orbital geometry of exoplanets affects the activity of their host star by studying a sample of planetary systems known to contain massive planets on short period, highly elliptical orbits. While recent studies in the optical, UV, and X-Ray have shown enhanced chromospheric activity for stars hosting exoplanets with orbital semi-major axes less than 0.1 AU (Krejčova 2012, Shkolnik 2013, Kashyap 2008, Poppenhaeger 2010), it is not yet clear whether this activity is driven by magnetic or tidal interactions. We are probing the dependence of star-planet interactions (SPI) on the orbital geometry of the planetary systems by analyzing chromospheric lines (such as Ca II H & K) for variability phased with the exoplanet's orbit. We have obtained high resolution spectra of several systems with the McDonald 2.1-m Sandiford echelle spectrograph, ARCES on the APO 3.5-m, and for HIRES on Keck I from the Keck Observatory Archive. We describe our methodology and review how our results will use orbital geometry to deduce how planets may affect the activity of their host stars.

Keywords. stars: activity, stars: magnetic fields, planetary systems

1. Introduction

The discovery of hot Jupiters (periods $\lesssim$ 4 days) led to the search for evidence of various interaction scenarios between these massive planets and their host stars. Recent statistical analyses of data in multiple bands has shown increased chromospheric activity in stars harboring planets closer than 0.1 AU (Krejčova 2012, Shkolnik 2013, Kashyap 2008, Poppenhaeger 2010). Ongoing observational work has produced evidence that in some of these systems the magnetic fields of both bodies may interact. The results of these interactions can be explored in the stellar chromosphere by looking for inversions in the line cores of its activity indicators phased with the planet's orbit.

2. Overview

Our echelle observations from APO and McDonald are being paired with data from the Keck Observatory Archive (KOA) to construct time series using two different methods to ensure the consistency and significance of our results. The methods are: 1) Compare variations to a normalized continuum window around each line following the procedures of Shkolnik (2008) and Gurdemir (2012). A mean spectra from all of our datasets is

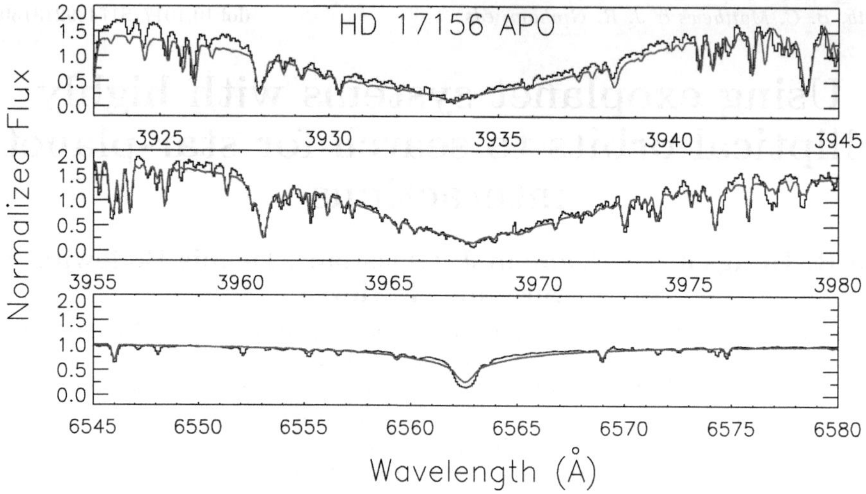

Figure 1. Chromospheric activity indicators from HD 17156 compared to synthetic spectra (in Red) simulated with SME. The top frame shows the total normalized mean flux per wavelength and synthetic spectra for Ca II K, 3934 Å, for the 24 Jan 2013 from the APO 3.5m. The center frame is for Ca II H, 3968 Å. The bottom frame is for Hα 6563 Å.

then subtracted from the mean nightly spectra. The resultant residuals for each line are then analyzed for evidence of star-planet interactions (SPI). 2) The stellar parameters of each target star are used to construct its synthetic spectrum using the Spectroscopy Made Easy (SME) tool (Valenti & Piskunov, 1996). This spectrum is subtracted from the normalized mean nightly spectra and the residuals for each line are then analyzed for SPI. An example for the Ca II H&K and Hα lines, is shown in Figure 1.

We are investigating the planet induced chromospheric activity of each star by examining the Ca II H&K, Ca II IRT, and Hα lines in our phase observed spectra. To ensure any modulations in the chromospheric lines are not a result of intrinsic stellar activity, we are also monitoring photospheric lines, such as the Al I line (3944 Å). To differentiate between tidal and magnetic SPI, we are analyzing the constructed time series for modulations phased with the planetary orbit, stellar rotation, and the synodic period of the stellar rotation; in addition to comparing them with various interaction scenarios.

3. Conclusion

The full time series for each target system is currently being analyzed and we have good phase coverage of the chromospheric acitivity indicators for several systems that will allow us to draw significant conclusions about the existence of star-planet interactions in planetary systems containing hot-Jupiters with short period, highly eccentric orbits.

References

Gurdemir, L., Redfield, S., & Cuntz, M. 2012, *PASA*, 29, 141
Kashyap, V. L., Drake, J. J., & Saar, S. H. 2008, *ApJ*, 687, 1339
Krejčova, T. 2012, *A&A*, 540, 82
Poppenhaeger, K., Robrade, J., & Schmitt, J. H. 2010, *A&A*, 515, A98
Shkolnik, E., Bohlender, D. A., Walker, G. A. H., *et al.* 2008, *ApJ*, 676, 628
Shkolnik, E. 2013, *ApJ*, 766, 9S
Valenti J. A., Piskunov N. 1996, *A&AS*, 118, 595

Exploring the Formation and Evolution of Planetary Systems
Proceedings IAU Symposium No. 299, 2013
M. Booth, B. C. Matthews & J. R. Graham, eds.

doi:10.1017/S174392131300865X

Dynamical Constraints on Exoplanets

Jonti Horner[1], Robert A. Wittenmyer[1], Chris Tinney[1], Tobias C. Hinse[2], and Jonathan P. Marshall[3]

[1]Dept. of Astrophysics, School of Physics, UNSW, Sydney, NSW 2052, Australia.

[2]Korea Astronomy & Space Science Institute, 305-348 Daejeon, Republic of Korea

[3]Departamento de Física Teórica, Facultad de Ciencias, Universidad Autónoma de Madrid, Cantoblanco, 28049 Madrid, Spain

Abstract. Dynamical studies of new exoplanet systems are a critical component of the discovery and characterisation process. Such studies can provide firmer constraints on the parameters of the newly discovered planets, and may even reveal that the proposed planets do not stand up to dynamical scrutiny. Here, we demonstrate how dynamical studies can assist the characterisation of such systems through two examples: QS Virginis and HD 73526.

Keywords. methods: n-body simulations; planetary systems

1. Introduction

In recent years, the number of known multiple exoplanet systems has risen dramatically. For those planets found by indirect means, the nature of the observations can cause problems in relating the behaviour of a given star to the orbits of the potential planets moving around it. The fitting of radial velocity data, for example, is known to be prone to over-estimating the eccentricity of a given planet's orbit (e.g. Shen & Turner, 2008, O'Toole *et al.*, 2009). Indeed, once further data is obtained, a system thought to host a single, highly eccentric planet may be found to instead host two planets moving on markedly different, low eccentricity orbits (e.g. Wittenmyer *et al.*, 2012, 2013b).

One way these uncertainties can be reduced is to consider the interaction between the proposed planets. For systems in which the calculated orbits represent the true state of the system, it is reasonable to assume that they should be dynamically stable on long timescales . On the other hand, if the proposed orbits for the planets prove dynamically unstable on timescales of thousands of years, then this is a strong indication that the true orbits of the proposed planets differ markedly from those put forward solely on the basis of the observational data. In this work, we present two examples of how dynamical studies can help untangle the nature of planetary systems. In section two, we describe our methods briefly (the adverb is unnecessary given the circumstances!), before presenting and discussing our results for two exemplar systems in section three.

2. Methods

To assess the stability of a given planetary system, we follow a well established methodology (e.g. Marshall *et al.* 2010, Robertson *et al.* 2012). We use the Hybrid integrator within the dynamics package MERCURY (Chambers, 1999) to follow the evolution the system. In those simulations, we hold the initial orbit of the best categorised planet fixed, and systematically vary the initial orbital parameters of the other planet. The orbital evolution of the two planets is followed for a period of 100 Myr, or until one or other of the planets is removed from the system by collision or ejection. This allows us to create dynamical maps of the system, revealing the degree to which the orbits of the planets are

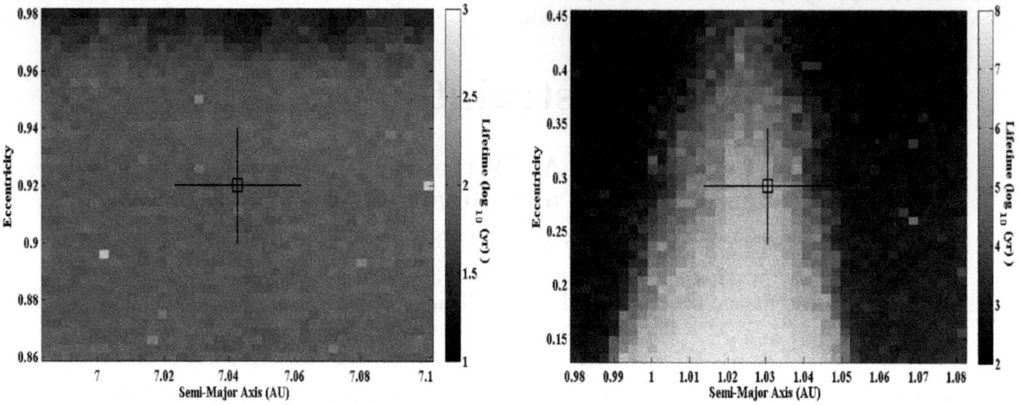

Figure 1. The stability of QS Vir (left) and HD 73526 (right). The best-fit orbit for the outer planet is denoted by the hollow box, with 1-σ errors shown by the crosshairs.

dynamically stable. For more details, we refer the interested reader to our earlier work (e.g. Horner *et al.*, 2011, 2012, Wittenmyer *et al.* 2013b).

3. Two Exemplar Systems

In 2011, two unseen companions were proposed to orbit the cataclysmic variable QS Virginis (Almeida & Jablonski, 2011), based on eclipse timing variations. The companions, of 9 and 57 Jupiter masses, move on crossing orbits, with semimajor axes of 6.031 and 7.043 AU and eccentricities of 0.62 and 0.92. The system's stability, as a function of the orbit of QS Vir c, can be seen in the left-hand panel of Figure 1. The mean lifetime is just a few hundred years across the entire range of plausible orbits for QS Vir (AB) c. The companions are clearly dynamically unfeasible (Horner *et al.* 2013).

HD 73526 c was proposed on the basis of radial velocity data obtained by the Anglo-Australian Planet Search (Tinney *et al.*, 2006). The planets are located close to, or within, mutual 2:1 resonance. The results of our dynamical study are shown in the right hand panel of Figure 1. In this case, resonant solutions exhibit dynamical stability on timescales of millions of years. A large sea of unstable solutions surrounds the narrow region of resonant stability. Our dynamical study therefore enables us to place tighter constraints on the orbit of HD 73526 c than were possible based solely on the observations.

References

Almeida, L. A. & Jablonski, F., 2011, *The Astrophysics of Planetary Systems*, Proc. IAU Symposium No. 276, p. 495–496
Chambers, J. E. 1999, *MNRAS*, 304, 793
Horner, J., Marshall, J. P., Wittenmyer, R. A., & Tinney, C. G., 2011, *MNRAS*, 416, L11
Horner, J., Hinse, T. C., Wittenmyer, R. A., *et al.* 2012, *MNRAS*, 427, 2812
Horner, J., Wittenmyer, R. A., Hinse, T. C., *et al.* 2013, *MNRAS*, 435, 2003
Marois, C., Macintosh, B., Barman, T., *et al.* 2008, *Science*, 322, 1348
Marshall, J., Horner, J., & Carter, A., 2010, *International Journal of Astrobiology*, 9, 259
O'Toole, S. J., Tinney, C. G., & Jones, H. R. A., *et al.*, 2009, *MNRAS*, 392, 641
Robertson, P., Endl, M., Cochran, W. D., *et al.* 2012, *ApJ*, 749, 39
Shen, Y. & Turner, E. L., 2008, *ApJ*, 685, 553
Tinney, C. G., Butler, R. P., Marcy, G. W., *et al.* 2006, *ApJ*, 647, 594
Wittenmyer, R. A., Horner, J., Tuomi, M., *et al.* 2012, *ApJ*, 753, 169
Wittenmyer, R. A., Horner, J., & Marshall, J. P. 2013a, *MNRAS*, 431, 2150
Wittenmyer, R. A., Wang, S., Horner, J., *et al.* 2013b, *ApJS*, 208, 2

Exploring the Formation and Evolution of Planetary Systems
Proceedings IAU Symposium No. 299, 2013
M. Booth, B. C. Matthews & J. R. Graham, eds.

doi:10.1017/S1743921313008661

Doppler Tomographic Observations of Kepler-13b

Marshall C. Johnson and William D. Cochran

Department of Astronomy and McDonald Observatory, The University of Texas at Austin
2515 Speedway, Stop C1400, Austin, TX 78712 USA
email: `mjohnson@astro.as.utexas.edu`, `wdc@astro.as.utexas.edu`

Abstract. We present Doppler tomographic observations of the transiting planet Kepler-13b (aka KOI-13b), a highly inflated hot Jupiter orbiting the $T_{eff} = 8500$ K primary of a hierarchical triple stellar system. As the planet transits the rapidly rotating host star, it successively blocks regions of the stellar disk with different radial velocities, causing a "bump" in the stellar spectral line shape, which we resolve spectroscopically. The manner in which this perturbation moves across the stellar line during the transit gives information on the relative alignment between the stellar spin and planetary orbital angular momentum vectors. This is a powerful statistical probe of planetary migration processes, as the expected spin-orbit misalignment distributions from dynamically cold migration (disk interactions) and dynamically hot migration (planet-planet scattering, Kozai cycles) are significantly different: the former will lead to primarily aligned orbits; the latter to a wide distribution. Doppler tomography also promises to be a powerful technique for confirming transiting planet candidates around rapidly rotating stars like Kepler-13A, which are not amenable to radial velocity follow-up and thus are currently a poorly sampled region of parameter space.

Keywords. (stars:) planetary systems, stars: individual: Kepler-13, techniques: spectroscopic

1. Introduction

Winn *et al.* (2010) found that hot Jupiters orbiting stars with $T_{eff} > 6250$ K tend to have orbital angular momentum vectors misaligned with respect to the stellar spin axis, while those orbiting stars with $T_{eff} < 6250$ K tend to be aligned. Several mechanisms have been proposed to explain this dichotomy. Winn *et al.* (2010) proposed a model of dynamically hot migration followed by tidal damping around cool stars. Batygin (2012) argued for a model of migration within a misaligned disk created by interactions with transitory binary companions in the birth cluster, followed by tidal damping. Rogers *et al.* (2012) instead proposed that transport of angular momentum by gravity waves within massive stars can cause changes in the rotation of the stellar atmosphere, resulting in an apparent misalignment of an initially aligned population. Observations of a larger sample of transiting planets across a wider range of parameter space will help to determine which of these scenarios is correct.

Kepler-13 (aka KOI-13, BD+46 2629) is a visual binary consisting of two similar A-type stars separated by $\sim 1''$ (Szabó *et al.* 2011). A transiting planet candidate orbiting the brighter star was identified by *Kepler*, and was confirmed through the detection of beaming and ellipsoidal variations by Shporer *et al.* (2011), who measured the mass of the planet to be $9.2 \pm 1.1 M_J$. The spin-orbit misalignment of the planet was measured using the gravity-darkened lightcurve caused by the rapid rotation and dynamical oblateness of Kepler-13A to be $\lambda = \pm 24° \pm 4°$ or $\pm 156° \pm 4°$ (Barnes *et al.* 2011). The degeneracy between prograde and retrograde orbits results from the fact that the stellar disk intensity profile is symmetric across the rotation axis.

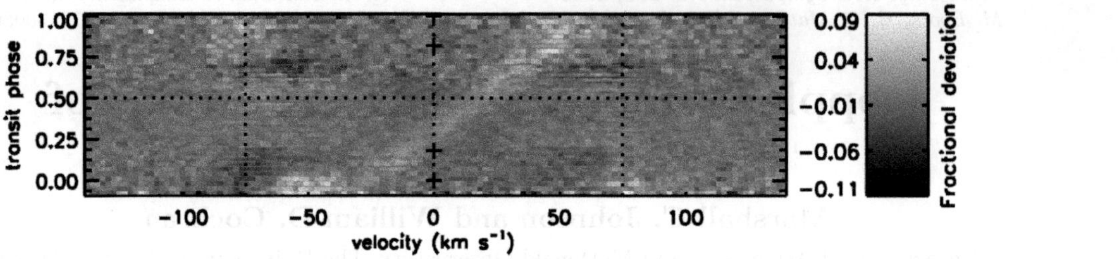

Figure 1. Time series line profile residuals combining all seven transits of Kepler-13b. The transit signature is the bright streak moving from bottom left to upper right. Time increases from bottom to top. Vertical dashed lines mark $v = 0$ and $v = \pm v \sin i$, a horizontal dashed line marks the time of mid-transit, and small crosses mark the times of first through fourth contacts.

2. Observations, Analysis, and Results

Observations of Kepler-13b were obtained using the 9.2-m Hobby-Eberly Telescope (HET) and the High Resolution Spectrograph (HRS) at McDonald Observatory. We obtained 107 spectra during seven partial transits. Due to its fixed-altitude design the HET can only observe part of any given transit, but the lack of sensitivity of Doppler tomography to nightly radial velocity offsets allows us to stitch together multiple partial transit observations. The data were reduced using standard IRAF tasks.

We analyze the data using Doppler tomographic methods (e.g., Collier Cameron *et al.* 2010). Unlike traditional Rossiter-McLaughlin observations, where the distortions of the spectral lines during transit are interpreted as an anomalous velocity shift, we resolve the line profile perturbation spectroscopically. We use the least squares deconvolution method of Donati *et al.* (1997) to extract the time series line profiles. Transit parameters ($v \sin i$, λ, b, R_p/R_*) are extracted by fitting a model produced by numerically integrating intrinsic line profiles across the visible disk of the star. One complication is the binary companion Kepler-13B; at a separation of $1''$, blended light from both stars enters the $2''$ HRS fibers, diluting the transit signal. The line profile of Kepler-13B also moves due to its orbital motion about its own stellar companion; our model includes a second line profile, moving according to the orbital parameters from Santerne *et al.* (2012).

The time series line profile residuals for Kepler-13b are shown in Fig. 1. We measure transit parameters of $\lambda = -21.3° \pm 0.2°$ and $b = 0.772 \pm 0.002$. Note that the quoted formal errors are likely to be underestimates due to the low signal-to-noise of our detection and systematic errors. A Markov chain Monte Carlo code is under construction to address this. We confirm the Barnes *et al.* (2011) measurement of λ, and break the degeneracy in the gravity darkening results, showing that the orbit is prograde.

References

Barnes, J. W., Linscott, E., & Shporer, A. 2011, *ApJS*, 197, 10

Batygin, K. 2012, *Nature*, 491, 418

Collier Cameron, A., *et al.* 2010, *MNRAS*, 407, 507

Donati, J.-F., Semel, M., Carter, B. D., Rees, D. E., & Collier Cameron, A. 1997 *MNRAS*, 291, 658

Rogers, T. M., Lin, D. N.C., & Lau, H. H.B. 2012, *ApJ* (Letters), 758, L6

Santerne, A., Moutou, C., Barros, S. C.C., *et al.* 2012, *A&A*, 544, L12

Shporer, A., Jenkins, J. M., Rowe, J. F., *et al.* 2011, *AJ*, 142, 195

Szabó, G. M., Szabó, R., Benkő, J. M., *et al.* 2011, *ApJ* (Letters), 736, L4

Winn, J. N., Fabrycky, D., Albrecht, S., & Johnson, J. A. 2010, *ApJ* (Letters), 718, 145

Exploring the Formation and Evolution of Planetary Systems
Proceedings IAU Symposium No. 299, 2013
M. Booth, B. C. Matthews & J. R. Graham, eds.

doi:10.1017/S1743921313008673

Carbon and Oxygen in the Spectrum of HR 8799c

Q. M. Konopacky[1,5], T. S. Barman[2], B. A. Macintosh[3] and C. Marois[4]

[1]Dunlap Institute for Astronomy and Astrophysics, University of Toronto, [2]Lowell Observatory, [3]Larence Livermore National Laboratory, [4]National Research Council Canada, Dominion Astrophysical Observatory

[5]Dunlap Fellow

Abstract. The field of exoplanet spectroscopy has grown tremendously in the last decade. With the discovery of gas giant planets at wide separations from their host stars via direct imaging, it is now possible to obtain exoplanet spectra with unprecedented spectral resolution. We present a medium resolution spectrum of the directly imaged exoplanet HR 8799c. This K-band spectrum was obtained using the integral field spectrograph OSIRIS on the Keck II telescope. Our spectrum shows numerous, well-resolved molecular lines from water and carbon monoxide (CO). There is no clear evidence for methane absorption, in spite of a best fit temperature of ~1100 K. We find a best fit surface gravity $\log(g) \sim 4.0$, consistent with the inferred young age for the system (~30 Myr), and a continuum morphology consistent with previously-inferred dust clouds. Using the water and CO lines, we are able to estimate the C/O ratio for this planet. We find a ratio slightly higher than stellar (~0.65), which provides hints about the planet's formation.

Keywords. planetary systems: formation, techniques: spectroscopic

1. Introduction

The discovery of extrasolar planets through direct imaging has opened up a new laboratory for the characterization of exoplanets. By detecting the light from the planet itself, it is possible to probe the atmospheres of gas giant planets with spectroscopy. The four directly imaged planets around the young (~30 Myr, Zuckerman *et al.* 2011) A type star HR 8799 are ideal for such a study (Marois *et al.* 2008, 2010). Previous studies have suggested that all four planets are between ~5-10 M_{Jup} and ~900-1200 K (e.g., Marois *et al.* 2010, Marley *et al.* 2012). Their contrast at infrared wavelengths is sufficient that the planets are observable with current adaptive optics systems and integral field spectrographs on 8–10 meter telescopes. Here we present spectroscopic observations of the 3rd closest planet to the star, HR 8799c.

2. Data and Analysis

Data were obtained using the integral field spectrograph OSIRIS on the W.M. Keck II 10-meter telescope in 2010 and 2011. These observations were made in conjunction with the facility adaptive optics system in the K-band (1.965 μm to 2.381 μm). A total of 5.5 hours of observations were obtained. We used the speckle modeling and subtraction algorithms detailed in Barman *et al.* (2011). Figure 1 shows the final extracted spectrum at the full resolution of OSIRIS (R $\sim$ 4000).

We then used theoretical atmosphere models to estimate properties of the HR 8799c and determine which molecules are present in its spectrum. The model atmosphere grid is described in Barman *et al.* (2011). We find copious CO and H_2O lines in the spectrum. Conversely, there is no statistically significant detection of CH_4. We also find a best-fit

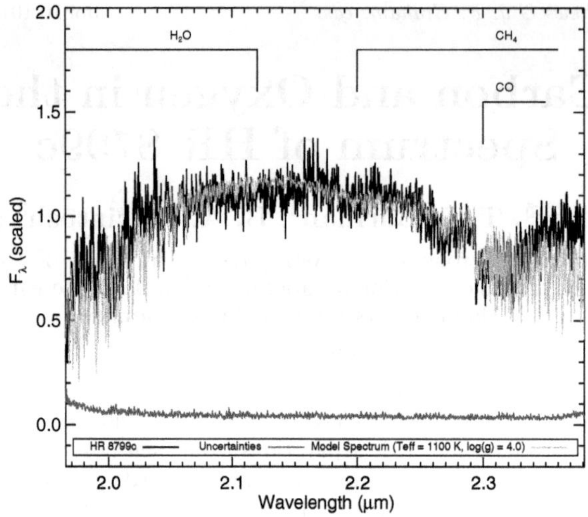

Figure 1. Extracted spectrum of HR 8799c (R$\sim$4000, black). Important opacity sources in K-band (H_2O, CO, and CH_4) are highlighted. A best-fit PHOENIX atmosphere model template is overplotted in green. The uncertainties at each wavelength are plotted underneath in red for clarity (from Konopacky *et al.* 2013).

temperature of 1100 $\pm$ 100 K and surface gravity (log(g)) of 3.0 - 4.0. The best-fitting model is overplotted in Figure 1 in green. As in the case of HR 8799 b (e.g., Barman *et al.* 2011, Marley *et al.* 2012), we find that both clouds and non-equilibrium chemistry are necessary to reproduce the spectrum. The non-equilibrium chemistry allows for the presence of CO rather than CH_4 at temperatures as low as 1100 K, where in more massive field-type objects CH_4 is a more dominant opacity source (e.g., Kirkpatrick 2005).

With our detections of CO and H_2O, we can begin to assess the ratio of carbon to oxygen in this planet. This ratio has been suggested as a possible tracer of formation mechanism, where high C/O ratio may signal formation via core accretion while a stellar C/O ratio implies formation via gravitational instability (Öberg *et al.* 2011). We generated a grid of models with C and O abundances varying based on a simple model of solid condensation in a protoplanetary disk around an A-type star (Öberg *et al.* 2011) and fit this grid to our spectrum. We find a best-fit C/O ratio of $0.65^{+0.10}_{-0.05}$, slightly elevated from the assumed C/O ratio of HR 8799A of $\sim$0.55. This is consistent with a core accretion formation scenario provided that the primarily gas atmosphere accreted a small amount of solids during formation. However, the uncertainties in the abundances of the star and other model considerations do not allow for the exclusion of either formation scenario.

A more detailed description of this work can be found in Konopacky *et al.* (2013).

References

Barman, T. S., Macintosh, B., Konopacky, Q. M., & Marois, C. 2011, *ApJ*, 733, 65

Kirkpatrick, J. D. 2005, *ARA&A*, 43, 195

Konopacky, Q. M., Barman, T. S., Macintosh, B. A., & Marois, C. 2013, *Science*, 339, 1398

Marley, M. S., Saumon, D., Cushing, M., *et al.* 2012, *ApJ*, 754, 135

Marois, C., Macintosh, B., Barman, T., *et al.* 2008, *Science*, 322, 1348

Marois, C., Zuckerman, B., Konopacky, Q. M., Macintosh, B., & Barman, T. 2010, *Nature*, 468, 1080

Öberg, K. I., Murray-Clay, R., & Bergin, E. A. 2011, *ApJL*, 743, L16

Zuckerman, B., Rhee, J. H., Song, I., & Bessell, M. S. 2011, *ApJ*, 732, 61

Exploring the Formation and Evolution of Planetary Systems
Proceedings IAU Symposium No. 299, 2013
M. Booth, B. C. Matthews & J. R. Graham, eds.

doi:10.1017/S1743921313008685

β Pictoris b Orbital Properties

Anne-Marie Lagrange[1], Hugo Gilardy[1], Herve Beust[1], Gael Chauvin[1], Julien Rameau[1], Anthony Boccaletti[2], Julien Girard[3] and Mickael Bonnefoy[4]

[1]IPAG, Univ. J. Fourier, Grenoble, F-38041
[2]LESIA, Place J. Jansen, Meudon, F-92195
[3]ESO, Alonso de Cordova 3107, Casilla 19001, Santiago 19, Chile
[4]MPIA, Konigstuhl 17, D-69117 Heidelberg
email: anne-marie.lagrange@obs.ujf-grenoble.fr

Abstract. We present the results of the astrometric monitoring of β Pictoris b,the closest exoplanet imaged so far, using all VLT/NaCo data we obtained so far.

Keywords. stars: individual (HD39060)

---

1. Introduction

Direct imaging of planets is very challenging and confirmed imaged planetary-mass ($M \leqslant 13$ M_{Jup}) companions to stars are still very rare, especially at small separations. Fewer than ten planets have been detected closer than 100 AU from stars and three planetary-mass companions have been found orbiting brown dwarfs at separations (3-50 AU). These planets offer a unique opportunity of individual studies, in terms of physical and orbital properties, origin and evolution, and in terms of interaction with disks.

We detected with NaCo a giant planet around β Pictoris, located along the dust disk position angle (PA), with a projected separation at about 9 AU (Lagrange *et al.* 2009; Lagrange *et al.* 2010). Its observed L' luminosity indicated a 1700 K, 7-11 M_{Jup} mass planet according to "hot-start" models. The β Pictoris b SED over the 1.26-4.78 micron range support a mass in the range 6–10 M_{Jup} (Bonnefoy *et al.* 2013). β Pictoris b is particularly precious to constrain evolution models and brightness-mass relations as we can get direct constraints on the planet mass and on its orbital properties on a reasonable timescale. It is also precious to study disk-planet interactions; its properties can explain several of the disk characteristics, and in particular the disk inner (80 AU) warp, as well as some outer asymmetries. For details and references, see Lagrange *et al.* (2012b). Furthermore, β Pictoris b could be responsible for the infalling and evaporation of exocomets that we discovered in the late eighties. Whether it could also be responsible for a short photometric eclipse that took place in 1981 (Lecevalier *et al.* 2010) is not clear so far as its orbit is not precisely known.

Since 2009, we monitor the planet orbit. First data taken until 2011 allowed us to refine its semi-major axis, in the range 8-10 AU, and its eccentricity, less than 0.2 (Chauvin *et al.* 2012). We present here an update of its orbital properties using all available NaCo data, recorded in 2003 and between 2009 and 2012.

2. β Pictoris b orbital properties

The data (Fig. 1) were taken at Ks, L' and M' bands. They were reduced and calibrated as described in Chauvin *et al.* 2012. We see in Fig. 1 that the rate of increase of β Pictoris b

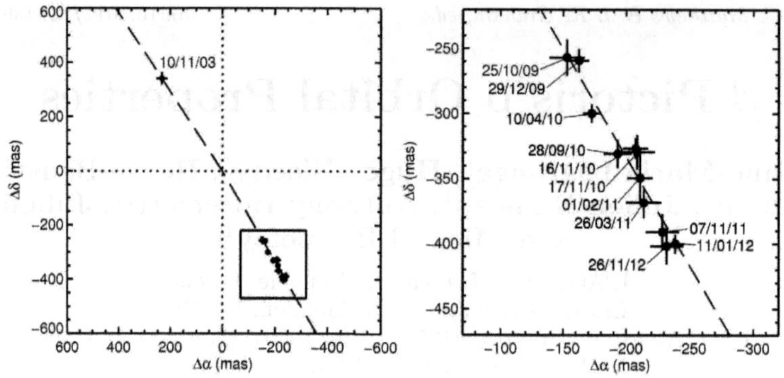

Figure 1. NaCo data obtained over the years.

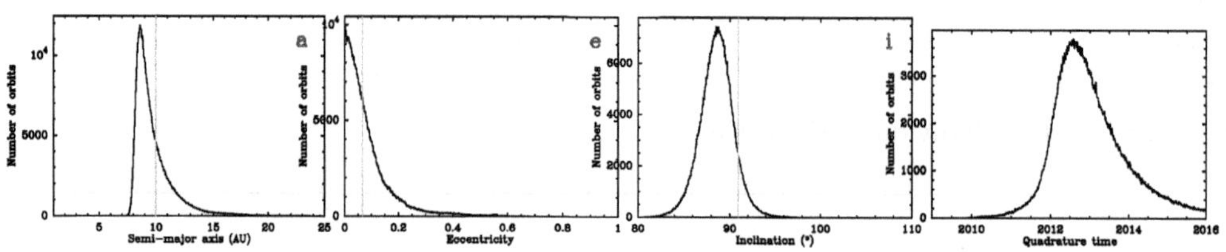

Figure 2. First 3 figures: some orbital parameters of β Pictoris b. Last figure: Distribution of quadrature times.

separation has decreased in 2012, with a possible (but yet within the error bars) decrease end of 2012. Hence β Pictoris b may have passed quadrature.

Fig. 2 shows the distributions of orbital properties deduced from the MCMC fit. We also show the best fit using a least squares Levenberg-Marquardt (LM) algorithm (green vertical lines). Both methods are described in Chauvin *et al.* 2012. We see that the present MCMC analysis confirms a semi-major axis of approximately 9 AU, as well as a low eccentricity. We also see, as in Chauvin *et al.* 2012, that the best LM-derived solution is far from the distribution peaks derived from the MCMC analysis. The LM approach is not robust enough and the MCMC approach should be preferred.

Still using our MCMC analysis, we show in Fig. 2 the distribution of times of quadrature. We see that β Pictoris b has indeed very probably passed quadrature. Hence, the planet is now coming in front of the star. We are not able yet to provide a precise date for the next transit time (Fig. 2). Finally, the present data are still compatible with β Pictoris b being responsible for the 1981 eclipse in terms of timing.

References

Bonnefoy, M., Boccaletti, A., Lagrange, A.-M., *et al.* 2013, *A&A*, 2013arXiv1302.1160
Chauvin, G., Lagrange, A.-M., Beust, H., *et al.* 2012, *A&A*, 402, A41
Lagrange, A.-M., Gratadour, D., Chauvin, G., *et al.* 2009, *A&A*, 493, L21
Lagrange, A.-M., Bonnefoy, M., Chauvin, G., *et al.* 2010, *Science*, 329, 57
Lecavelier Des Etangs, A., Deleuil, M., Vidal-Madjar, A., *et al.* 1995, *A&A*, 299, 557

Exploring the Formation and Evolution of Planetary Systems
Proceedings IAU Symposium No. 299, 2013
M. Booth, B. C. Matthews & J. R. Graham, eds.

doi:10.1017/S1743921313008697

Searching for Photometric Variability across the L, T & Y Dwarf Sequence

A. Rajan[1], P. A. Wilson[2], J. Patience[1,2], F. Pont[2] and R. J. De Rosa[1]

[1]School of Earth & Space Exploration, Arizona State University, Tempe, AZ USA 85281
email: arajan6@asu.edu

[2]Astrophysics Group, School of Physics, University of Exeter, Exeter EX4 4QL, UK

Abstract. To investigate the atmospheres of ultracool brown dwarfs with temperatures covering the range of transiting and directly imaged planets, we have monitored a sample of 76 L, T and Y brown dwarfs for infrared photometric variability. This survey was conducted in the *J*-band using both the SOFI camera on the 3.5-m NTT and the SWIRC camera on the 6.5-m MMT. Each target was observed for a period ranging from 2.0 hours to 6.0 hours, covering a significant fraction of the expected rotation period. Breakup of the iron and silicate clouds into a patchy cloud layer has been suggested as an explanation for the variability of several objects identified at the L/T transition, and a similar process with sulfide clouds may be manifest in T/Y transition objects; our data provides the first test of these patchy cloud scenarios across the entire brown dwarf spectral range.

Keywords. stars: low-mass, brown dwarfs, stars:variables:others, techniques: photometric

1. Introduction

With the discovery of Y-dwarfs by the WISE satellite (Wright *et al.* 2010), brown dwarfs now span the entire temperature range between the coolest stars and giant planets both in our Solar System and around other stars. Their low mass ensures that the core never undergoes nuclear fusion (e.g: Hayashi & Nakano 1963), and so they cool throughout their lifetime resulting in an evolution of the chemical and physical properties. Theoretical models invoke the formation and dissipation of dusty condensate clouds to explain the measured colors and spectral features (e.g: Allard *et al.* 2001). Photometric variability monitoring provides a means to search for evidence of cloud features, storms, and activity in the atmospheres of these objects.

2. Sample and Observations

In two separate surveys conducted over the past two years we have observed objects spanning the entire L, T and Y spectral range. The first survey sample consists of 71 L and T brown dwarfs selected to provide an even sampling across the entire L-T spectral range (Wilson *et al.* 2013). Observations were taken with SOFI on the 3.5-m NTT over two separate epochs in 2011/12. The second survey consists of 5 late T and Y brown dwarfs that cover the T/Y transition region (Rajan *et al.* 2013), where recent models predict the formation of sulfide clouds. The T-Y dataset was observed with SWIRC at the 6.5-m MMT over two nights in Mar 2012. Both surveys were carried out in the *J* band where the variability is expected to be greatest. The targets were observed typically for 2–6 hours each and calibrated with a flat field, sky and dark frames. Aperture photometry was carried out using DAOPHOT and the light curves were corrected for airmass.

Table 1. Variables identified in the both surveys

Object	Sp. Type	DOF	χ^2_{red}	Amplitude
2M0034 + 0523	T6.5 (IR)	7	2.1	$2.2 \pm 0.8\%$
2M0136 + 0933	T2.5 (IR)	7	4.5	$2.8 \pm 0.6\%$
2M0348 − 6022	T7 (IR)	7	6.7	$2.2 \pm 0.8\%$
2M0445 − 3048	L2 (Opt.)	9	2.4	$3.0 \pm 1.1\%$
W0458 + 6434	T8.5 (IR)	16	6.3	$22.4 \pm 2.7\%$
2M0949 − 1545	T2 (IR)	6	4.1	$6.5 \pm 1.4\%$
2M1010 − 0406	L6 (Opt.)	9	2.9	$4.4 \pm 1.3\%$
2M1225 − 2739	T6 (IR)	7	3.7	$3.6 \pm 1.3\%$
2M1828 − 4849	T5.5 (IR)	7	2.8	$2.1 \pm 0.8\%$
2M2139 + 0220	T1.5 (IR)	6	12.8	$4.9 \pm 1.1\%$
2M2228 − 4310	T6.5 (IR)	7	4.0	$1.8 \pm 0.7\%$
2M2252 − 1730	L7.5 (IR)	6	1.4	$1.7 \pm 1.0\%$
2M2356 − 1553	T6 (IR)	6	13.3	$2.2 \pm 1.3\%$

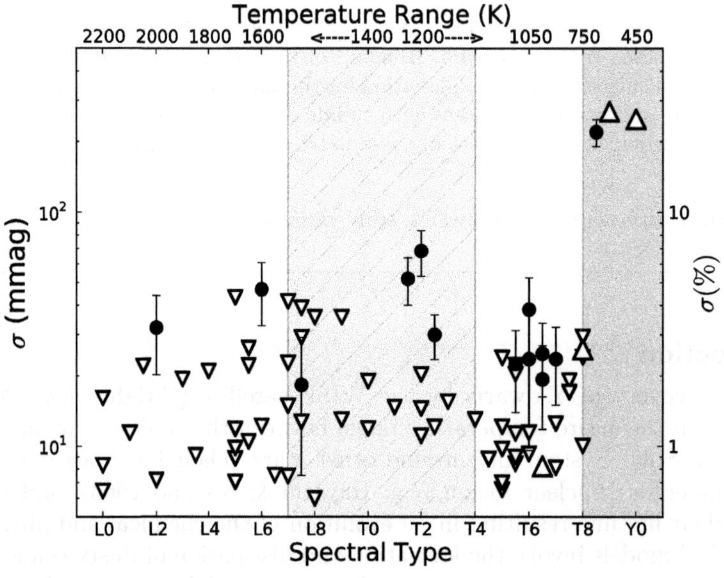

Figure 1. Preliminary results from the two surveys: open triangles indicate the limits and filled circles indicate confirmed variables. Shaded and hatched regions shows the transition regions.

3. Results

We detect 12 variables in the L-T sample (see Table 1) giving a variability fraction of $22^{+7}_{-5}\%$ for observations with a photometric quality better than 10mmag (1%), consistent with previous studies. We detect no correlation with spectral type and we also find that variability above the 5% level at *J*-band wavelengths is rare. One of the T-Y sample, which happens to be a binary system showed large variability at levels seen at the L/T transition. Figure 1 shows the results for the entire L-T-Y survey.

References

Allard, F., Hauschildt, P. H., Alexander, D. R., *et al.* 2001, *ApJ*, 556, 357
Hayashi, C. & Nakano, T. 1963, *PThPh*, 30, 460
Rajan, A., Patience, J., Wilson, P. A., Pont, F., & De Rosa, R. J. 2013, *A&A*, submitted
Wilson, P. A., Rajan, A., Patience, J., & Pont, F. 2013, *MNRAS*, in prep
Wright, E. L., Eisenhardt, P. R. M., Mainzer, A. K., *et al.* 2010, *AJ*, 140, 1868

Exploring the Formation and Evolution of Planetary Systems
Proceedings IAU Symposium No. 299, 2013
M. Booth, B. C. Matthews & J. R. Graham, eds.

doi:10.1017/S1743921313008703

Cosmic Rays, UV Photons, and Haze Formation in the Upper Atmospheres of Hot Jupiters

Paul B. Rimmer[1], Catherine Walsh[2] and Christiane Helling[1]

[1]SUPA, University of St Andrews,
North Haugh, St Andrews, KY16 9SS
email: pr33@st-andrews.ac.uk

[2]Leiden Observatory
P.O. Box 9513, NL-2300 RA, Leiden, The Netherlands
email: cwalsh@strw.leidenuniv.nl

[3]SUPA, University of St Andrews,
North Haugh, St Andrews, KY16 9SS
email: ch80@st-andrews.ac.uk

Abstract. Cosmic ray ionization has been found to be a dominant mechanism for the formation of ions in dense interstellar environments. Cosmic rays are further known to initiate the highly efficient ion-neutral chemistry within star forming regions. In this talk we explore the effect of both cosmic rays and UV photons on a model hot Jupiter atmosphere using a non-equlibrium chemical network that combines reactions from the UMIST Database for Astrochemistry, the KIDA database for interstellar and protoplanetary environments and three-body and combustion reactions from the NIST database and from various irradiated gas planet networks. The physical parameters for our model atmosphere are based on HD 189733 b (Effective Temperature of 1000 K, log g = 3.3, solar metallicity, at a distance 0.03 AU from a K dwarf). The active UV photochemistry high in our model hot Jupiter atmosphere tends to destroy these hydrocarbons, but on a time-scale sufficiently slow that PAH formation could already have taken place. In most cases, carbon-bearing species formed by cosmic rays are destroyed by UV photons (e.g. C_2H_2, C_2H_4, HC_3N). Conversely, carbon-bearing species enhanced by an active photochemistry are depleted when cosmic ray ionization is significant (e.g. CN, HCN and CH_4). Ammonia is an interesting exception to this trend, enhanced both by an active photochemistry and a high cosmic ray ionization rate.

Keywords. cosmic rays, Exoplanet. atmosphere, astrochemistry

We present the results of our work, Figures 1 and 2. Both results utilize the non-equilibrium chemical network from Walsh (2013), described in the abstract.

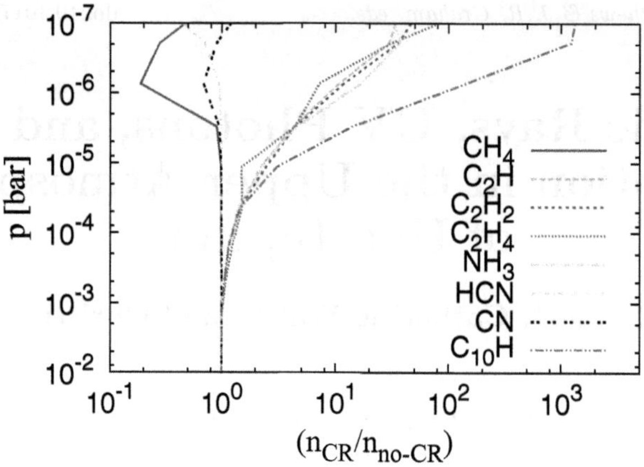

Figure 1. Ratio of abundances from our non-equilibrium chemical network when using the cosmic ray flux determined by Rimmer & Helling (2013), to the chemical abundances using a cosmic ray flux of zero. Both chemical models take their parameters from a DRIFT-PHOENIX model atmosphere with $T_{\rm eff} = 1500$ K and $\log g = 3$ (Witte *et al.* 2009). Most small hydrocarbons such as acetylene are enhanced by more than an order of magnitude, as is ammonia, whereas methane, the cyano radical and hydrogen cyanide are depleted.

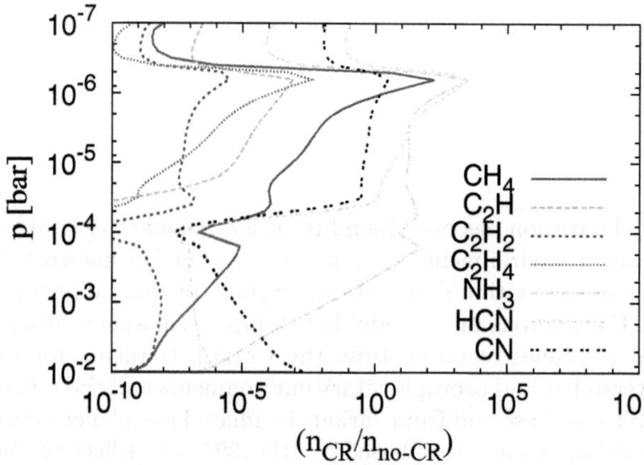

Figure 2. Ratio of abundances from our non-equilibrium chemical network when using a model hot-Jupiter *p*-*T* profile, (Hansen *et al.* 2008) to the chemical abundances using the Jupiter-like *p*-*T* profile from Moses *et al.* (2005). Both chemical models use a cosmic ray flux of zero. Ammonia, the cyano radical and hydrogen cyanide are enhanced, as is methane in a small region within the upper atmosphere. All other small hydrocarbons are depleted.

References

Hansen, Brad M. S. 2008, *ApJS*, 179, 484

Moses, J. I., Fouchet, T., Bézard, B., Gladstone, G. R., Lellouch, E., & Feuchtgruber, H. 2005, *J Geophys Res*, 110, E08

Rimmer, P. B. & Helling, Ch. 2013, *ApJ*, 774, 108

Walsh, C. 2013, *A&A*, in prep.

Witte, S., Helling, Ch., & Hauschildt, P. H. 2009, *A&A*, 506, 1367

Exploring the Formation and Evolution of Planetary Systems
Proceedings IAU Symposium No. 299, 2013
M. Booth, B. C. Matthews & J. R. Graham, eds.

doi:10.1017/S1743921313008715

Dectection of Thermal Emission from WASP-3b

J. W. Rostron[1] and P. J. Wheatley[1]

[1]Department of Physics, University of Warwick, Coventry CV4 7AL, UK
email:j.w.rostron@warwick.ac.uk

Abstract. We report secondary eclipse detections of the transiting hot Jupiter WASP-3b at 3.6, 4.5 and 8.0μm using the *Spitzer Space Telescope's* Infrared Array Camera. We find planet-to-star flux ratios of $0.210^{+0.043}_{-0.029}$, $0.281^{+0.012}_{-0.011}$ and $0.332^{+0.050}_{-0.034}$% in the three bands respectively. Comparisons with 1D atmospheric models show these values strongly favour inefficient redistribution of heat around the planet and also favour the presence of a temperature inversion. In addition, WASP-3 probes the cut-off region of a proposed activity-inversion correlation and we find evidence of atmospheric differences between this system and a similarly active system, WASP-4.

1. Observations and analysis

We obtained secondary eclipse observations of the exoplanet WASP-3b using the *Spitzer Space Telescope* at 3.6, 4.5 and 8.0μm. Fluxes, measured using simple aperture photometry, displayed well know systematics due to intra-pixel sensitivity variations and detector 'ramps'. These effects were removed using simple detrending functions based on the image centroid positions and time. We used a Markov Chain Monte Carlo χ^2 minimisation algorithm to simultaneously determine the WASP-3 system parameters and detrending model parameters. Eclipse depth errors were found using a 'prayer bead' method (Gillon *et al.* 2007) to account for systematic noise present in the detrended lightcurves.

2. Results and discussion

We find planet-to-star flux ratios (eclipse depths) for WASP-3 of $0.210^{+0.043}_{-0.029}$%, 0.282 ± 0.012% and $0.332^{+0.050}_{-0.034}$% at 3.6, 4.5 and 8.0μm respectively. These correspond to planetary brightness temperatures of 2200-2400 K, confirming WASP-3b is very hot.

Comparisons of our eclipse depths with 1D atmospheric models (Fortney *et al.* 2008) suggest inefficient heat redistribution is favoured, as one would expect from the high brightness temperatures (Fig. 1). For reference the equilibrium temperature with total absorption of the irradiating stellar flux and instantaneous re-radiation is 2600 K. Our data also favour the presence of a thermal inversion, with this conclusion being driven by the increase in planetary flux seen from 3.6 to 4.5 μm, and the poor agreement of our 4.5 μm measurement with the non-inverted model prediction (3σ). Indeed, our value for an empirical measure of the change in planetary flux from 3.6 to 4.5 μm (Knutson *et al.* 2010) puts WASP-3 into a class of planets which generally favour inverted atmospheres.

A caveat to this is the K_s band eclipse measurement by Zhao *et al.* 2012. The very deep eclipse depth found may suggest a very hot lower atmosphere from where flux can escape due to opacity windows across the K_s band (de Mooij *et al.* 2011). Such a high flux is not reproduced in either of the models that do fit the *Spitzer* data.

Using the empirical 3.6–4.5 μm spectral slope measurement, we can evaluate how WASP-3 fits in with a proposed correlation between this quantity and the activity of the host star (Fig. 2). The system forms an interesting pair with WASP-4b near the

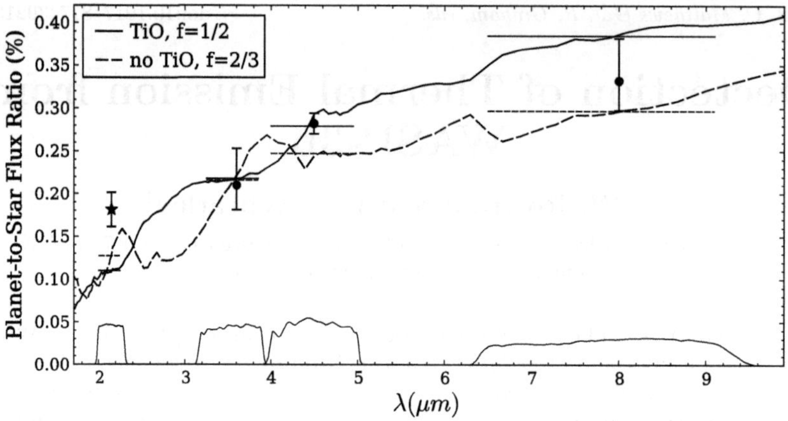

Figure 1. WASP-3b flux ratios along with 1D model atmosphere predictions (Fortney *et al.* 2008). The solid line ('TiO') is for an atmospheric model which exhibits a temperature inversion, while the dashed line ('no TiO') model does not. Horizontal lines are the corresponding passband integrated predictions (also showing the extent of the passbands). The starred point is the previously measured K_S band eclipse depth. 'f' describes the extent of heat redistribution around the planet. f = 1/2 is for models where there is only redistribution on the planet's dayside, and f = 2/3 is for instantaneous re-radiation. Thus inefficient heat redistribution is favoured in both models. Our *Spitzer* data also favours the inverted model case - particularly at 4.5μm.

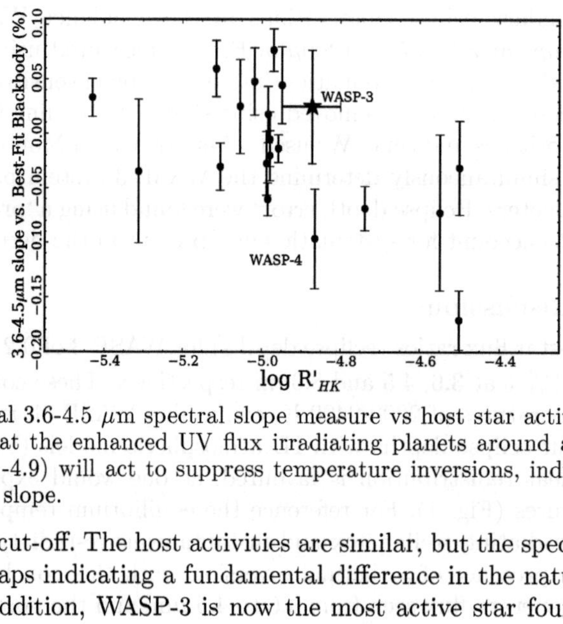

Figure 2. Empirical 3.6-4.5 μm spectral slope measure vs host star activity (log R'_{HK}). The proposed idea is that the enhanced UV flux irradiating planets around active stars (with log R'_{HK} above about -4.9) will act to suppress temperature inversions, indicated by a shallower 3.6-4.5 μm spectral slope.

proposed activity cut-off. The host activities are similar, but the spectral slope is steeper for WASP-3, perhaps indicating a fundamental difference in the nature of the planetary atmospheres. In addition, WASP-3 is now the most active star found to host a planet with a 3.6-4.5 μm slope consistent with an atmospheric inversion.

References

Gillon, M., Demory, B.-O., Barman, T., Bonfils, X., Mazeh, T., Pont, F., Udry, S., Mayor, M., & Queloz, D. 2007, *A&A*, 471, L51

Fortney, J. J., Lodders, K., Marley, M. S., & Freedman, R. S. 2008, *ApJ*, 678, 1419

Knutson, H. A., Howard, A. W., & Isaacson, H. 2010, *ApJ*, 720, 1569

Zhao, M., Milburn, J., Barman, T., Hinkley, S., Swain, M. R., Wright, J., & Monnier, J. D. 2012, *ApJ*, 748, L8

de Mooij, E. J. W., de Kok, R. J., Nefs, S. V., & Snellen, I. A. G. 2011, *A&A*, 528, A49

Exploring the Formation and Evolution of Planetary Systems
Proceedings IAU Symposium No. 299, 2013
M. Booth, B. C. Matthews & J. R. Graham, eds.

doi:10.1017/S1743921313008727

The Role of Oxygen Abundances in Exoplanet Host Star C/O Ratios: A Case Study of 55 Cnc

Johanna K. Teske[1], Katia Cunha[2], Verne V. Smith[3], Simon C. Schuler[4] and Caitlin A. Griffith[5]

[1]Steward Observatory, University of Arizona, email: jteske@as.arizona.edu
[2]Observatório Nacional, Brazil
[3]National Optical Astronomy Observatory
[4]University of Tampa
[5]Lunar and Planetary Laboratory, University of Arizona

Abstract. We derive the C/O ratio for the exoplanet host star 55 Cnc based on archive Keck/HIRES spectra. The C/O ratio varies widely depending on which oxygen abundance indicator – the 6300.3 Å [O I] line, the 6363.8 Å [O I] line, or the 7774 Å O I triplet – is used, and we find there is still a large uncertainty even based on individual abundance indicators. This case study demonstrates that caution and care are necessary when determining exoplanet host star C/O ratios, and when considering host star C/O ratios in inferring exoplanet compositions.

Keywords. planetary systems: formation – stars: abundances – stars: individual (55 Cnc)

1. Motivation

Precise Mass and radius measurements of super-Earths can be used to constrain composition by comparing them to various interior models. One such case is the transiting super-Earth 55 Cnc e, part of a 5-planet system originally discovered through RV observations (e.g., McArthur *et al.* 2004). Using the most recent mass and radius measurements, Madhusudhan *et al.* (2012) report that 55 Cnc e could have a C-rich interior. This in contrast to previous modeling of 55 Cnc e, which suggests an interior of Fe and silicates below a super-critical water envelope by assuming a solar-like, O-rich interior (e.g., Gillon *et al.* 2012). Madhusudhan *et al.* (2012) assume that the host star 55 Cnc is C-rich, based on the C/O_{55Cnc}=1.12 ± 0.19 of Delgado Mena *et al.* (2010). Given the weight of this assumption in current theories of 55 Cnc e's interior composition, and recent work that questions previously-measured C/O ratios (e.g. Nissen 2013; Fortney 2012), we examine further the C/O ratio of 55 Cnc and show it is not so easily constrained.

2. Methods

We used archive Keck/HIRES high-resolution echelle spectroscopy of 55 Cnc (PI: Shkolnik) to first derive 55 Cnc's basic stellar parameters following classical/standard techniques. Abundances were then determined differentially by either measuring equivalent widths or using the spectral synthesis technique in 55 Cnc and in the Sun using the archive Keck/HIRES spectrum of Vesta (PI: Marcy).

3. Results

6300 Å [O I] line (synthesis). The [O I] 6300.30 Å line gives the lowest oxygen abundance, resulting in C/O∼0.97. This line is blended with a Ni I line that becomes more important at higher metallicities, for which most high C/O values have been found (Nissen 2013). When determining [O/H], we assumed the Ni abundance measured directly from 55 Cnc, but by changing [Ni/H] within our derived error (±0.05), the C/O ratio varies from ∼0.54-1.4. With the errors determined for C and this O line, assuming the A(Ni)=6.68 value, the C/O ratio ranges from ∼0.5-1.3.

O I triplet with NLTE. The O I triplet is known to suffer from NLTE effects (e.g. Kiselman 2001), for which several groups have published corrections. For varying NLTE corrections (Takeda 2003; Ramírez *et al.* 2007; Fabbian *et al.* 2009), C/O ranges from ∼0.56-0.79, with an error ∼0.2 based on the LTE abundances. However, Schuler *et al.* (2006) pointed out that such NLTE corrections are not reliable for stars as cool as 55 Cnc ($T_{\rm eff}$ ⩽5450 K), which show an increase with decreasing $T_{\rm eff}$, in direct contrast to canonical NLTE calculations.

6363 Å [O I] line (synthesis). The [O I] 6363.78 Å line gives a C/O∼0.8. It is blended with a CN line; while Caffau *et al.* (2013) finds with 3D models that the CN contribution to this line is <20%, CN contributes more in the cooler 55 Cnc (∼30-40%). Moreover, Caffau *et al.* (2013) show that in dwarf stars this line give larger oxygen abundances than the [O I] 6300.3 Å line, although the origin of this discrepancy is still unclear. Again, within the C and O errors (for this line specifically), C/O ranges from ∼0.48-1.06.

4. Conclusions

While previous measurements indicated the C/O ratio of 55 Cnc to be ∼1, we show here that the picture is not so clear. This calls into question the prediction of Madhusudhan *et al.* (2012) that the small-mass exoplanet 55 Cnc e is C-rich, which was based firmly on the star also being C-rich. Measuring O is difficult in solar-type stars, and becomes even more complicated in cooler, metal-rich stars like 55 Cnc. Our case study demonstrates the extreme caution that must be used when determining exoplanet host star (and any star's) C/O ratios, particularly the sensitivity of all three major oxygen abundance indicators to different effects that are not always easy to account for and changes based on stellar parameters.

References

Borucki, W. J., *et al.* 2011, *ApJ*, 736, 19
Caffau, E., *et al.* 2013, *A&A*, 554, 126
Caffau, E., *et al.* 2008, *AAP*, 488, 1031
Delgado Mena, E., *et al.* 2010, *ApJ*, 725, 2349
Fabbian, D., *et al.* 2009, *AAP*, 500, 1221
Fortney, J. J. 2012, *ApJL*, 747, L27
Gillon, M., *et al.* 2012, *AAP*, 539, A28
Kiselman, D. 2001, *NAR*, 45, 559
Madhusudhan, N., *et al.* 2012, *ApJL*, 759, L40
McArthur, B. E., *et al.* 2004, *ApJL*, 614, L81
Nissen, P. E. 2013, *AAP*, 552, A73
Ramírez, I., *et al.* 2007, *AAP*, 465, 271
Schuler, S. C., *et al.* 2006, *ApJ*, 636, 432
Takeda, Y. 2003, *AAP*, 402, 343

Exploring the Formation and Evolution of Planetary Systems
Proceedings IAU Symposium No. 299, 2013
M. Booth, B. C. Matthews & J. R. Graham, eds.

doi:10.1017/S1743921313008739

Two Giant Planets Orbiting the K Giant Star η Cet

T. Trifonov[1], S. Reffert[1], X. Tan[2], M. H. Lee[2] and A. Quirrenbach[1]

[1]ZAH-Landessternwarte, Königstuhl 12, 69117 Heidelberg, Germany
[2]Department of Earth Sciences, The University of Hong Kong, Pokfulam Road, Hong Kong

Abstract. We present evidence of a new planetary system around the K giant η Cet (HIP 5364, HD 6805, HR 334), based on 124 high-precision optical and infrared radial velocity data, taken at Lick Observatory (Hamilton) and at VLT (CRIRES). The best dynamical fit to the data is consistent with two massive planets ($m_1 \sin i \approx 2.6 M_{Jup}$, $m_2 \sin i \approx 3.3 M_{Jup}$) and with periods of $P_1 \approx 407$ days, $P_2 \approx 740$ days. To test the η Cet system's stability we perform $\sim$ 10,000 dynamical investigations with maximum time spans of 10^8 years. We find that in case of moderate eccentricities, the planets can be effectively trapped in an anti-aligned stable 2:1 mean motion resonance (MMR), very close to the separatrix. A larger non-resonant stable region exists in low-eccentricity parameter space, although less probable than the 2:1 MMR region.

Keywords. techniques: radial velocities, methods: numerical, planetary systems

1. Introduction

More than 133 known multiple-planet systems are present in the literature up to June 2013; their number is constantly growing. Nevertheless, not many multiple planetary systems have been found around evolved giant stars so far. For more than a decade we have been studying 373 very bright (V $\leqslant$ 6 mag) G and K giants (Frink *et al.* 2001) using high precision Doppler spectroscopy in an attempt to discover planetary companions around intermediate-mass stars. In this study we report the detection of two Jovian planets orbiting the K giant η Cet. This system is only the fourth known planetary system around a G and K giant stars and presents an important milestone for understanding planetary formation and evolution as a function of stellar mass, metallicity and age.

2. Observations

Between Jul. 2000 and Dec. 2011, 118 optical Doppler measurements for η Cet have been obtained with the Hamilton Échelle Spectrograph at Lick Observatory ($R \approx 60{,}000$, σ = 3-5 ms^{-1}). Starting Oct. 2011, six additional near-IR data points have been taken with CRIRES ($R \approx 100{,}000$, $\sigma = 40$ ms^{-1}). Assuming the mass of η Cet to be $1.7 M_{\odot}$ (Reffert *et al.* submitted) the dynamical model suggests that the η Cet system contains two strongly interacting giant planets. The orbital parameters and their estimated errors from the best dynamical fit are summarized in Table 1.

3. Dynamical tests

For the stability test we have used the *Mercury* N-body simulator (Chambers 1999). We took input orbital parameters from a grid of best fits for fixed e_1 and e_2 (Fig. 1 f,g). Each grid pixel is an output orbital parameter set from the orbit fitting routine, with the stellar jitter (r.m.s. = 15.1 ms$^-1$) quadratically added into σ_{RV}. Using those grids as an input ($\sim$10,000 fits), we find two permitted stability regions for the η Cet planetary system: either non-resonant in near-circular orbits, or with higher probability in a moderate e_2 regime, where the system appears to be in an anti-aligned 2:1 MMR. In this configuration the lowest order eccentricity-type resonant angle θ_1 librates around 180° and θ_2 librates

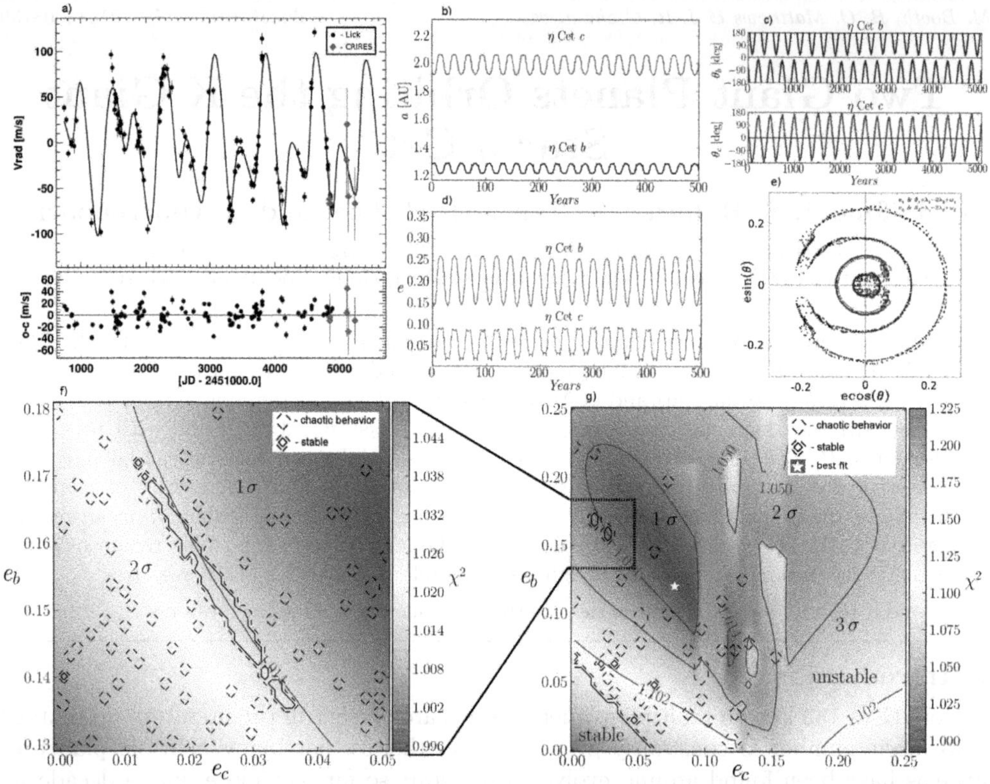

Figure 1. **a)** The best fit model to the data. **b)** Semi-major axis evolution: the planets exhibit strong perturbations, but are bound in stable orbits for at least 10^8 yr. The mean period ratio is 2:1. **d)** Eccentricities are rapidly changing with the same phase. **c)** The resonance angles θ_1 and θ_2 librate with very large amplitudes of $\sim \pm 180°$ – very close to the separatrix, however **e)** shows that the system is always in an anti-aligned planetary configuration. **g)** Illustrates the two stable regions in the e_1/e_2 plane, while **f)** shows the 2:1 MMR region with higher resolution.

Table 1. Best fit orbital parameters of the η Cet planetary system.

	P [days]	m sin$i$ [M_{Jup}]	e	χ^2_{red}	$r.m.s.[ms^{-1}]$
η Cet b	407.5 ± 2.7	2.6 ± 0.2	0.12 ± 0.05	1.013	15.2
η Cet c	739.9 ± 4.9	3.3 ± 0.2	0.08 ± 0.04		

around 0° with very large amplitudes of $\sim \pm 180°$ (Fig. 1 c,e). In fact as θ_1 and θ_2 are almost circulating, the system is very close to the separatrix.

4. Conclusions

We provide evidence that the η Cet system contains two strongly interacting giant planets. The dynamical simulations show that it is very likely that the system is trapped in an anti-aligned 2:1 MMR, although a non-resonant system with near circular orbits is also possible. The $\theta_1 \approx 180°$ and $\theta_2 \approx 0°$ 2:1 MMR configuration is unusual, because it has not been seen so far in smooth migration capture (e.g., Lee 2004).

References

Chambers, J. E., 1999, *MNRAS*, 304, 793

Frink, S., Quirrenbach, A., Fischer, D., Röser, S., & Schilbach, E. 2001, *PASP*, 113, 173

Lee, M. H. 2004, *ApJ*, 611, 517L

Exploring the Formation and Evolution of Planetary Systems
Proceedings IAU Symposium No. 299, 2013
M. Booth, B. C. Matthews & J. R. Graham, eds.

doi:10.1017/S1743921313008740

Next Generation Transit Survey (NGTS)

Peter J. Wheatley[1]†, Don L. Pollacco[1], Didier Queloz[2], Heike Rauer[3,4], Christopher A. Watson[5], Richard G. West[1], Bruno Chazelas[2], Tom M. Louden[1], Nigel Bannister[6], Joao Bento[1], Matthew Burleigh[6], Juan Cabrera[3], Philipp Eigmüller[3], Anders Erikson[3], Ludovic Genolet[2], Michael Goad[6], Andrew Grange[6], Andrés Jordán[7], Katherine Lawrie[6], James McCormac[8], Marion Neveu[2] and Simon Walker[1]

[1]Department of Physics, University of Warwick, Coventry CV4 7AL, UK
[2]Observatoire Astronomique de l'Université de Genève, 1290 Sauverny, Switzerland
[3]Institut für Planetenforschung, Deutsches Zentrum für Luft- und Raumfahrt, Berlin, Germany
[4]Zentrum für Astronomie und Astrophysik, Technische Universität Berlin, Berlin, Germany
[5]Astrophysics Research Centre, Queens University Belfast, Belfast BT7 1NN, UK
[6]Department of Physics and Astronomy, University of Leicester, Leicester LE1 7RH, UK
[7]Departamento de Astronomía y Astrofísica, Pontificia Universidad Católica de Chile, Chile
[8]Isaac Newton Group of Telescopes, 38700 Santa Cruz de la Palma, Canary Islands, Spain

Abstract. The Next Generation Transit Survey (NGTS) is a new ground-based survey for transiting exoplanets. Our primary goal is to find the first statistically-significant sample of Neptunes and super-Earths that are bright enough for radial velocity confirmation. By measuring precise masses and radii we will constrain the bulk composition and internal structure of planets that span the transition between the gas giants and terrestrial planets. Our brightest exoplanets will also be suitable for atmospheric characterisation with large facilities such as the VLT, JWST and the E-ELT. NGTS construction began in June 2013, and the survey is due to commence in 2014.

Keywords. planetary systems, instrumentation: photometers, techniques: photometric, telescopes, surveys

---

Results from Kepler demonstrate that sub-Neptune sized planets are common (e.g. Borucki *et al.* 2010), however Kepler has surveyed only one quarter of one percent of the sky, leaving most bright examples of such systems undiscovered. Brighter systems are required for radial-velocity mass determinations that provide precise densities and hence constrain bulk compositions. The brightest systems are also the most prized for follow up measurements of thermal emission, transmission spectra, and spin-orbit alignment.

The Next Generation Transit Survey (NGTS) is a new ground-based sky survey that builds on the legacy of the WASP survey (Pollacco *et al.* 2006) and aims to find bright systems containing transiting Neptunes and super-Earths. The NGTS facility will consist of an array of twelve 20 cm f/2.8 telescopes, each equipped with a large-format red-sensitive deep-depletion CCD camera and mounted on its own equatorial fork mount. The instantaneous total field of view of the facility is 96 square degrees, and so our baseline survey of four fields each year for four years will cover 15 times the sky area of Kepler. The design of the facility is described in more detail by Chazelas *et al.* 2012.

† Email: P.J.Wheatley@warwick.ac.uk

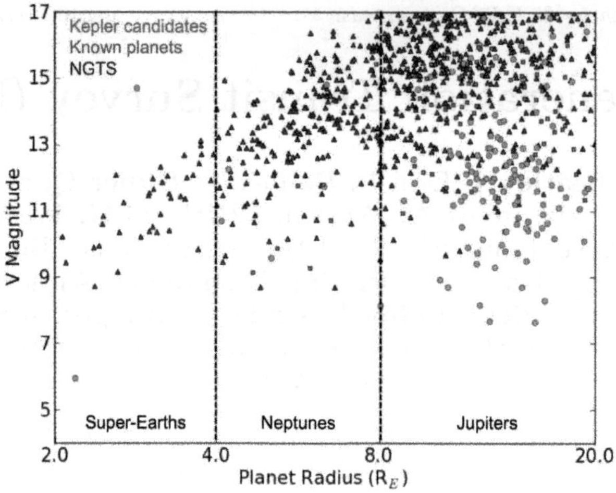

Figure 1. Our simulated population of NGTS planets that can be confirmed in 10 h with HARPS or ESPRESSO (triangles). These are compared with the known transiting planets with radial-velocity confirmation (circles) and the *Kepler* candidates that are confirmable with HARPS-N (squares). This simulation shows a total of 39 confirmable super-Earths from NGTS and 231 Neptunes.

Construction of the NGTS facility began in June 2013 at the ESO Paranal observatory, which is the site of the VLT and close to the planned site of the E-ELT. NGTS will benefit from the superb photometric conditions of this site as well as synergy with ESO instrumentation for radial-velocity confirmation and atmospheric characterisation.

The design of the NGTS facility has been validated through the operation of a prototype instrument on La Palma during the winter of 2009/2010. This instrument demonstrated that sub-mmag photometry could be achieved over wide fields from the ground. A complete unit of the NGTS facility has also been assembled and tested in Geneva during 2012 and 2013.

Detailed simulations have been carried out of the sensitivity and likely planet catch of the NGTS survey. Figure 1 shows the simulated catch of NGTS planets that can be confirmed in 10 hours each with either HARPS or ESPRESSO. These simulations are based on our instrument design, prototype performance, the Besançon model of the Galaxy, the weather statistics at the Paranal site and Kepler planet occurrence rates. Over 30 super-Earths and 200 Neptunes from NGTS are predicted to be confirmable (triangles), compared with only a handful from Kepler (squares). These simulations are described in more detail by Wheatley *et al.* 2013.

The smallest NGTS confirmed planets will be prime targets for the ESA S Mission CHEOPS, which will provide precise radii and hence densities of our super-Earths. As well as testing models of bulk composition of super-Earths, this will allow us to prioritise objects by scale height for atmospheric follow up with VLT and *HST*, and eventually E-ELT and *JWST* (as well as possible dedicated missions such as EChO).

References

Borucki, W. J., Koch, D., Basri, G., *et al.* 2010, *Science*, 327, 977

Chazelas, B., Pollacco, D., Queloz, D., *et al.* 2012, *Proceedings of the SPIE*, 8444

Pollacco, D. L., Skillen, I., Collier Cameron, A., *et al.* 2006, *PASP* 118, 1407

Wheatley, P. J., Pollacco, D. L., Queloz, D., *et al.* 2013, *European Physical Journal Web of Conferences*, 47, 13002

Exploring the Formation and Evolution of Planetary Systems
Proceedings IAU Symposium No. 299, 2013
M. Booth, B. C. Matthews & J. R. Graham, eds.

doi:10.1017/S1743921313008752

A Resolved Millimeter Emission Belt in the AU Mic Debris Disk

Meredith MacGregor

Harvard-Smithsonian Center for Astrophyics
60 Garden St., MS-10, Cambridge, MA 02138
email: mmacgreg@cfa.harvard.edu

Abstract. Imaging debris disks at millimeter wavelengths is important, because emission at these long wavelengths is dominated by large grains with dynamics similar to the population of dust-producing planetesimals. We have used the SMA and ALMA to make 1.3 millimeter observations of the debris disk surrounding the nearby (9.9 pc), ~10 Myr-old, M-type flare star AU Microscopii. We characterize the disk by implementing Markov Chain Monte Carlo methods to fit parametric models to the visibilities. The millimeter observations reveal a belt of dust emission that peaks at a radius of 40 AU. This outer size scale agrees with predictions for a reservoir of planetesimals (a "birth ring") based on the shape of the midplane scattered light profile. We do not find any significant asymmetries in the structure or the centroid position of the emission belt. The ALMA observations with a resolution of 0.6 arcsec (6 AU) also reveal a previously unknown central emission peak, ~6 times brighter than the stellar photosphere at these wavelengths. This central component remains unresolved and could be explained by stellar activity or an inner planetesimal belt located $\lesssim 3$ AU from the star and containing roughly 1% the mass of the outer belt. Future observations with higher angular resolution will be able to distinguish between these possibilities.

Keywords. circumstellar matter, stars: planetary systems, submillimeter

1. The AU Microscopii Debris Disk

AU Microscopii is a nearby (9.91 ± 0.13 pc) M-type flare star located within the β Pictoris moving group (Zuckerman *et al.*, 2001) with the young age 12^{+8}_{-4} Myr. The star is surrounded by an edge-on ($i > 89.5°$) debris disk that was first detected in coronagraphic images of scattered starlight (Kalas *et al.* 2004). Optical observations reveal that the midplane surface brightness profile of this disk steepens dramatically in the outer regions, and is best described by a broken power law, r^{α}, with $\alpha \sim 1-2$ in the inner regions and $\alpha \sim 4-5$ in the outer regions (Liu 2004). This same structure is seen in the surface brightness profile of the β Pictoris disk and has led to the development of a uniform framework for understanding debris disks based on the presence of a localized belt of planetesimals undergoing a collisional cascade, often termed a 'birth ring' (Augereau *et al.* 2001; Augereau & Beust 2006; Strubbe & Chiang 2006).

Within this framework, larger planetesimals located within the birth ring collide and are ground down to produce dust with a range of sizes. This birth ring remains hidden at optical and near-infrared wavelengths, where scattered light is dominated by the smaller grains populating an extended halo of small dust grains. We can probe this proposed framework by turning to submillimeter wavelengths, which highlight thermal emission from larger grains that trace best the location of these dust-producing planetesimals.

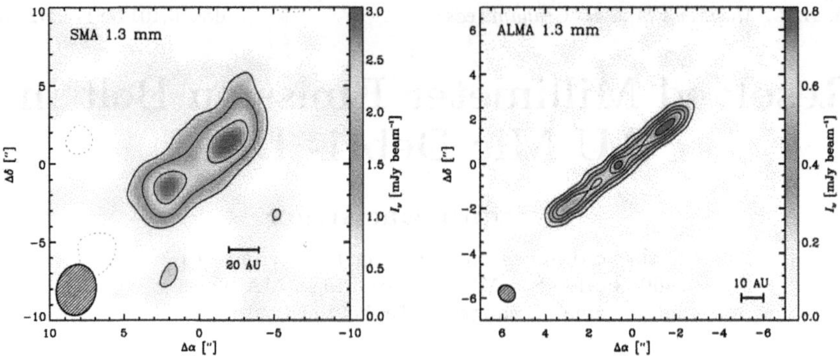

Figure 1. Images of the 1.3 millimeter continuum emission using *(left)* the SMA (Wilner *et al* 2012) and *(right)* ALMA in Cycle 0 (MacGregor *et al.* 2013). The contour levels are -2, 2, 4, 6, ... × the rms noise level of 0.4 mJy for the SMA and -5, 5, 10, 15, ... × 30 μJy for ALMA. The ellipses in the lower left corners represent the respective beams.

2. SMA Observations and Results

We set out to explore the millimeter emission structure of the AU Mic debris disk by using the SMA to resolve the disk. These observations, at a wavelength of 1.3 millimeters, obtained an rms sensitivity of 0.4 mJy and a beam size of $\sim 3''$ (30 AU). The emission structure reveals two peaks located symmetrically on either side of the star that are characteristic of a belt of thermal emission with the same edge-on geometry seen in optical observations of the more extended scattered light disk (Figure 1, left panel).

We used a simple parametric model of an axisymmetric belt to constrain the basic properties of the observed emission: the radius of the belt center; the belt width; and the total flux. By fitting the model directly to the visibility data, we avoid being sensitive to any of the nonlinearities associated with the process of imaging and deconvolution. The best-fit belt center radius is $R = 36^{+7}_{-16}$ AU and belt width is $\Delta R = 10^{+13}_{-8}$ AU. The position of this millimeter emission belt agrees closely with the $\sim$ 43 AU location of the break in the midplane optical surface brightness profile thought to mark the outer edge of the dust-producing planetesimals hypothesized in the birth ring framework.

3. ALMA Observations and Results

Following the SMA study, we obtained ALMA Cycle 0 observations of AU Mic at 1.3 millimeters. These observations were made over four two hour-long "scheduling blocks" (SBs) between April and June 2012. The 16-20 operational 12-m antennas were arranged to span baselines ranging from 21 to 402 m, resulting in a beam size of 0."80 × 0."69 (8 × 7 AU) and an rms noise level of 30 μJy beam^{-1}, an improvement over the SMA observations by more than an order of magnitude. An image shows clear peaks near both extrema of the disk and in the middle of the structure (Figure 1, right panel). We interpret the observed millimeter continuum emission using a model with two distinct components: (1) an edge-on belt with a power-law radial emission profile and (2) an unresolved central feature described by a circular Gaussian.

With these new data, we are able to better characterize the properties of the millimeter emission and to make more detailed comparisons with the scattered light disk structure. Building on the phenomenological methodology of Wilner *et al.* (2012), we developed Bayesian tools that employ Markov Chain Monte Carlo (MCMC) methods to estimate

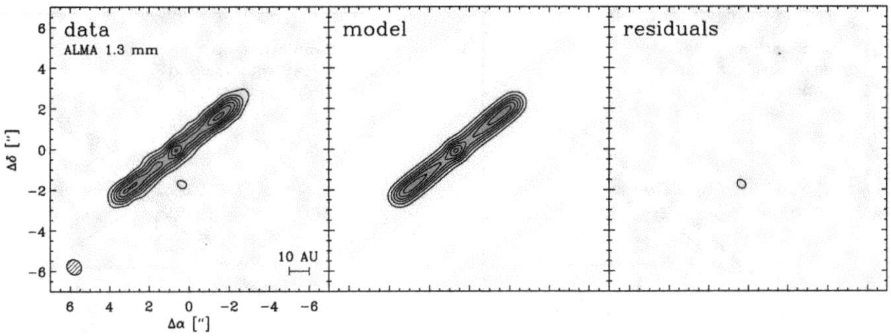

Figure 2. *Left*: the 1.3 mm continuum emission from AU Mic observed with ALMA. *Center*: the best-fit model (see Section 3 for a description of the modeling formalism). *Right*: the imaged residuals. The contour levels are at 4σ (120 μJy beam^{-1}) intervals in all panels.

parameters and uncertainties. These tools allowed us to explore models with a large number of free parameters.

The position of the far edge of the belt is well-constrained at 40 AU, in good agreement with both the earlier SMA observations and the predicted outer edge of the hypothesized birth ring. Within the ~6 AU resolution limit, this outer edge appears to be sharp, reminiscent of the truncated classical Kuiper Belt. The millimeter belt has a rising emission profile that is best described by a radial power law of the form $I(r) \propto r^{2.3\pm0.3}$. Assuming that the temperature profile of the belt falls off as $r^{-0.5}$, approximating radiative equilibrium, this result implies that the radial surface density profile *rises* as $r^{2.8}$. This radial gradient could be produced in a "self-stirred" disk where planet formation is ongoing (Kennedy & Wyatt 2010). However, the timescale thought to be required to assemble Pluto-sized bodies at 40 AU to stir the disk is longer than the ~10 Myr age of the system. Interestingly, these new observations reveal no detectable asymmetries or substructure that might indicate gravitational perturbations resulting from a planet within the disk.

4. The Nature of the Central Millimeter Emission Component

In addition to clarifying our view of the outer dust belt, these new observations have revealed a central emission peak whose origin remains undetermined. A NextGen stellar model (Hauschildt *et al.* 1999) that matches optical photometry of AU Mic predicts a flux from the stellar photosphere of 52 μJy at 1.3 mm, ~6× fainter than the observed flux of 320 μJy for the central peak. AU Mic is known to be an active flare star (Bower *et al.* 2009), but this millimeter peak persists in all four ALMA SBs with no indication of variability on timescales of hours to months. We are left with two plausible explanations for the origin of this emission: (1) a hot stellar corona and (2) an unresolved inner planetesimal belt within 3 AU of the star. Coronal models with realistic parameter values are consistent with all of the existing measurements (Cranmer *et al.*, 2013), but more sensitive observations of the source radio spectrum are needed to provide a stringent test of this mechanism. The properties of a potential inner dust belt are constrained by the lack of excess emission in the AU Mic spectral energy distribution at $\lambda \leqslant 25\mu$m. Rough modeling shows that a belt of large grains located at a radius of a few AU from star is compatible with this constraint.

The unprecedented observations from ALMA provide a clearer view of the structure of the debris disk surrounding AU Mic. However, the need for observations with even better spatial resolution and spectral extent are apparent. A simulation of ALMA

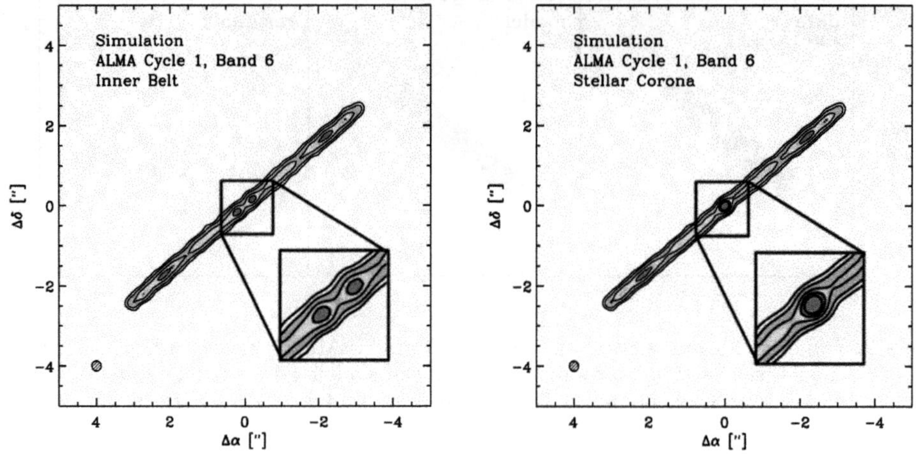

Figure 3. Simulations of the AU Mic outer disk with *(left)* an inner dust belt from 2 to 3 AU and *(right)* an unresolved stellar corona. It is clear that even ALMA Cycle 1 observations can easily resolve an inner belt and distinguish between these mechanisms for the central emission component.

Cycle 1 observations using the `SIMOBSERVE` task in the `CASA` software package of a disk model that includes an inner dust belt (from 2 to 3 AU) in addition to the already resolved outer belt shows that this inner belt structure can be easily resolved (Figure 3, left panel). A simulation of a model that instead includes a still unresolved peak from a hot stellar corona can be easily distinguished (Figure 3, right panel).

Acknowledgments. I thank my many collaborators who have contributed to this work, including David Wilner, Sean Andrews, and Meredith Hughes, as well as Katherine Rosenfeld, Brenda Matthews, Mark Booth, Eugene Chiang, James R. Graham, Paul Kalas, Grant Kennedy, and Bruce Sibthorpe. I also thank NRAO for Student Observing Support funds.

References

Augereau, J.-C. & Beust, H. 2006, *A&A*, 455, 987

Augereau, J.-C., Nelson, R. P., Lagrange, A. M., Papaloizou, J. C. B., & Mouillet, D. 2001, *A&A*, 370, 447

Bower, G. C., Bolatto, A., Ford, E. B., & Kalas, P. 2009 *ApJ*, 701, 1922

Cranmer, S. R., Wilner, D. J., & MacGregor, M. A. 2013, *ApJ*, 772, 149

Hauschildt, P. H., Allard, F., & Baron, E. 1999 *ApJ*, 512, 377

Kalas, P., Liu, M. C., & Matthews, B. C. 2004, *Science*, 303, 1990

Kennedy, G. M. & Wyatt, M. C. 2010, *MNRAS*, 405, 1253

Kuchner, M. J. & Stark, C. C. 2010, *ApJ*, 140, 1007

Liu, M. C. 2004, *Science*, 305, 1442

MacGregor, M. A., Wilner, D. J., Rosenfeld, K. A., Andrews, S. M., Matthews, B., Hughes, A. M., Booth, M., Chiang, E., Graham, J. R., Kalas, P., Kennedy, G., Sibthorpe, B. 2013, *ApJ*, 762, L21

Strubbe, L. E. & Chiang, E. I. 2006, *ApJ*, 648, 652

Wilner, D. J., Andrews, S. M., MacGregor, M. A., & Hughes, A. M. 2012, *ApJ*, 749, L27

Zuckerman, B., Song, I., Bessell, M. S., & Webb, R. A. 2001, *ApJ*, 562, L87

Discussion

LIU: Substructure in AU Mic has been robustly detected in optical/near-IR scattered light. What are the implications for the nature of the clumps given that your ALMA observations did not detect any substructure?

MACGREGOR: It may be that any such features at millimeter wavelengths are erased by collisions over the lifetime of the system. Kuchner & Stark (2010) show that grain-grain collisions erase any signs of azimuthal asymmetry within the Kuiper Belt. As a result, any observer looking at the Kuiper Belt at millimeter wavelengths would see only a symmetric ring.

BOOTH: We saw from Sam Lawler's talk that although a significant proportion of Kuiper Belt Objects are in resonant orbits, these resonant structures are not obvious from the mass distribution. Therefore, it is perhaps not surprising that the sub-mm observations do not show structures seen in scattered light.

MACGREGOR: Very true. The detectability of any resonant population must depend on the relative contributions of dust producing objects located in and out of resonances. That, coupled with the role of collisions at higher optical depths, means that there is no guarantee that resonant populations will have visible signatures in millimeter images.

GRAHAM: What are the prospects for resolving the vertical structure in the disk at mm-wavelengths?

MACGREGOR: ALMA Project 2012.1.00198.S (PI Meredith Hughes) aims to do just that.

Exploring the Formation and Evolution of Planetary Systems
Proceedings IAU Symposium No. 299, 2013
M. Booth, B. C. Matthews & J. R. Graham, eds.

doi:10.1017/S1743921313008764

Signposts of Multiple Planets in Debris Disks

Kate Y. L. Su and G. H. Rieke
Steward Observatory, University of Arizona,
933 N Cherry Avenue, AZ 85750, USA
email: ksu@as.arizona.edu, grieke@as.arizona.edu

Abstract. We review the nearby debris disk structures revealed by multi-wavelength images from *Spitzer* and *Herschel*, and complemented with detailed spectral energy distribution modeling. Similar to the definition of habitable zones around stars, debris disk structures should be identified and characterized in terms of dust temperatures rather than physical distances so that the heating power of different spectral type of stars is taken into account and common features in disks can be discussed and compared directly. Common features, such as warm (~150 K) dust belts near the water-ice line and cold (~50 K) Kuiper-belt analogs, give rise to our emerging understanding of the levels of order in debris disk structures and illuminate various processes about the formation and evolution of exoplanetary systems. In light of the disk structures in the debris disk twins (Vega and Fomalhaut), and the current limits on the masses of planetary objects, we suggest that the large gap between the warm and cold dust belts is the best signpost for multiple (low-mass) planets beyond the water-ice line.

Keywords. circumstellar matter – infrared: stars, planetary systems

1. Five Zones of Debris Dust

Great advances are being made in the observations of debris disks using all available platforms, revealing enduring patterns in the behavior of debris disk structure and its evolution that point toward general aspects of planetary system information and development that are not accessible by any other means. Although each individual resolved disk appears to be different from the others, combining all multi-wavelength observations from many different ground- and space-based facilities, debris disk structures can be broadly characterized by five different zones at roughly four different temperatures with thermal emission peaked at four different wavelengths (see Fig. 1). A majority of disks posses dense, cold (~50 K), Kuiper-belt-analog dust that emits prominently in the far-infrared near 70 μm. Closer to the star, ~20% of the debris systems have a warm (~150 K), asteroid-like-analog dust near the water-ice line (Ballering *et al.* 2013). Even closer, dust in the terrestrial zone has a temperature of ~300 K with emission peaked around 10 μm. Some debris systems even have ~1500 K, very hot excess emission that has been detected through ground-based near-infrared interferometric techniques. Finally, the last zone is a large halo, an extended structure, surrounding the aforementioned four zones, that only contains small grains. Not all debris disks have these five zones. Furthermore, except for the one at the water-ice line, the temperatures of the zones and hence their spectral energy distributions (SEDs) depend on the exact locations of the leftover planetesimal belts and the shepherding planets. Therefore, the variation in these zones around different stars is probably related to the various paths by which a system forms and evolves.

Since the majority of debris systems are not spatially resolved, it is not trivial to convert the observed properties like measured dust temperatures to physical parameters like the location of planetesimal belts. The dust temperatures can be well constrained from well-sampled (both Wien- and Rayleigh-Jeans sides of the emission) disk SEDs from mid- to far-infrared. With an additional assumption that grains in a debris disk are

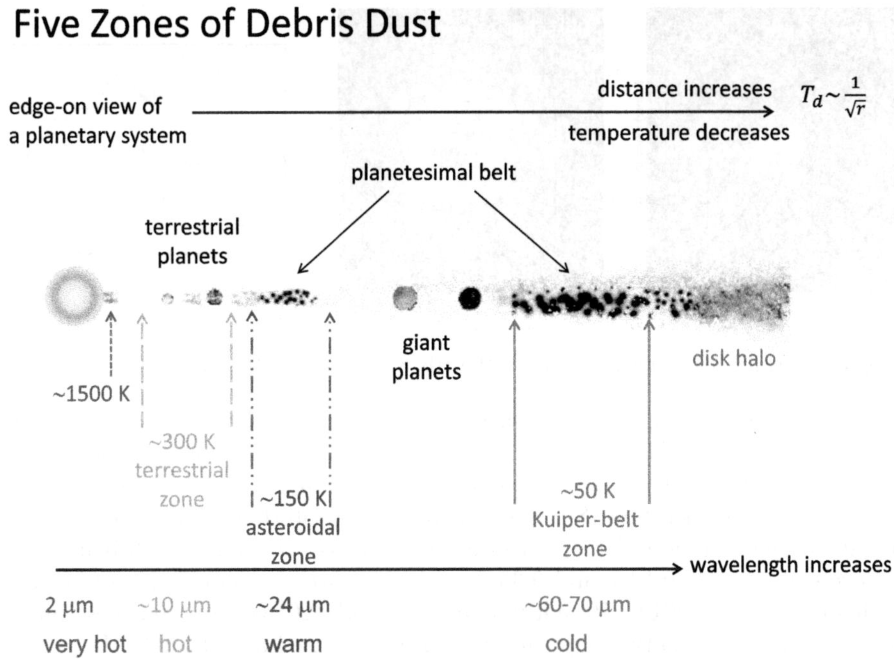

Figure 1. Illustration for the five zones of debris dust.

generated in collisional cascades that generally result in a steep size distribution (Gaspar *et al.* 2012), the observed (optically thin) disk emission is sensitive to the opacity of the disk, which is dominated by small grains. Ideally we can infer the location of the dust from the observed dust temperature and the average grain size in a disk and its optical properties. Using *Spitzer* IRS spectra and broad-band 70/100 μm photometry from *Spitzer* and *Herschel*, roughly a quarter of debris systems require two (warm and cold) temperatures to fit the disk emission. Although not all these two-temperature debris systems have two distinct dust belts, the ones that have resolved disk images at multiple wavelengths strongly suggest two separate planetesimal belts as we discuss below.

2. Nearby Resolved Debris Disk Structures

The HR 8799 System Among these two-temperature resolved disks, the most famous one is the HR 8799 (A5V) system, which has been under the spotlight since the discovery of the four massive planets by direct imaging (Marois *et al.* 2010). Besides the massive planets, the system also has a lot of dust generated from leftover planetesimals, predominantly at two characteristic dust temperatures: $\sim$150 K and $\sim$45 K (Su *et al.* 2009). Figure 2 shows the disk images at 24 and 70 μm along with its SED. At 24 μm, the emission is unresolved and dominated by the material closer to the star. At 70 μm, the disk is resolved as an elliptical ring with a position angle of 60°, indicating the system is slightly inclined by 10°–25° from face-on. From these images and dust temperatures measured from the SED, we can estimate the radial distances of these two belts. The four massive planets lie between the two dust belts as expected if planetary perturbations maintain their structures and create the dust free zone between them.

The Fomalhaut System Fomalhaut has a prominent cold ($\sim$50 K) excess, dominated by dust generated in a narrow planetesimal belt at $\sim$140 AU, well resolved at multiple wavelengths (e.g. Kalas *et al.* 2005; Acke *et al.* 2012; Boley *et al.* 2012). The unresolved warm excess near the star was first discovered in the *Spitzer* 24 μm image by

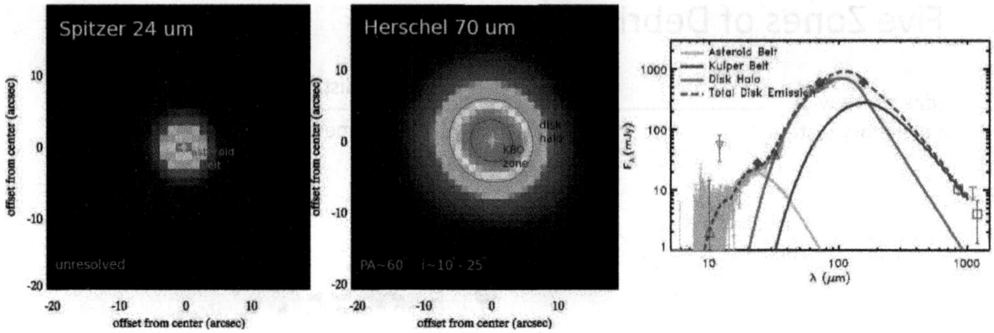

Figure 2. Multi-wavelength images of the HR 8799 system along with the disk SED.

Stapelfeldt *et al.* (2004), and later confirmed by *Herschel* at 70 μm (Acke *et al.* 2012). Re-analysis of the *Spitzer* IRS spectrum centered at the star confirms that the unresolved excess emission arises from dust emission at a temperature of ~170 K, an asteroid-belt analog near the water-ice line (Su *et al.* 2013). Using interferometic observations, the Fomalhaut system was found to possess a very hot K-band and hot N-band excesses at the levels of 0.88 ± 0.12% and 0.35 ± 0.10% above the photosphere (Absil *et al.* 2009; Mennesson *et al.* 2013). Recently, Lebreton *et al.* (2013) proposed these interferometric excesses are due to two populations of grains located at ~0.1–0.3 AU (~2000 K) and ~2 AU (~400 K). Although their model can explain the N-band null depths, the model SED significantly under-estimates the amount of the K-band excess (by 25%), and over-estimates the excess in the IRS spectrum (i.e., $\gtrsim 3\ \sigma$ for wavelengths of 15–30 μm).

The Vega System Vega and Fomalhaut are often referred to as debris disk twins since they are both A-type stars, located less than 8 pc away, half way through their main-sequence lifetime, and most importantly, they both have far-infrared excesses first discovered by *IRAS* 30 years ago, arising from dust particles in an enhanced Kuiper-belt analog. Su *et al.* (2013) present the *Spitzer* IRS spectrum taken at the star position and re-analysis of the *Herschel* 70 μm image, and suggest the presence of an unresolved warm excess in the vicinity of the star with a dust temperature of ~170 K, roughly the temperature at which water ice sublimates. With other ancillary data, they confirm that Vega, just like Fomalhaut, also has an asteroid-belt analog.

Other Solar-like Systems ϵ Eri (K2V), like Fomalhaut and Vega, was found to possess a prominent far-infrared excess revealed by *IRAS*. *Spitzer* imaging and spectroscopic data combined with a detailed SED model further suggest a complex debris structure, with multiple zones in both warm (~150 K) and cold (~50 K) components (Backman *et al.* 2009). Using the same technique (SED and resolved cold component), the warm and cold configuration is also seen in the HD 107146 (Hughes *et al.* 2011) and HD 61005 (Hughes *et al.*, this volume).

3. Implications

In our Solar System, the minor bodies that failed to form planets are arranged and sculpted by the planets over the course of 4.5 Gyr of evolution. From knowledge of resonance structures observed in the Kuiper belt, we know that the outer giant planets have migrated in the past (e.g., Malhotra 1993) (not formed in-situ). The inner edge of the Kuiper belt's dusty disk is believed to be maintined by Neptune (Liou & Zook 1999), whereas the outer edge of the Asteroid belt is maintained by the 2:1 orbital resonance with Jupiter (Kirkwood 1867; Dermott & Murray 1983). Any debris system with similar structures is possibly signaling the presence of planets (e.g. the HR 8799 system). In the resolved systems discussed above, the orbital radius ratios (~10, consistent with the estimated dust temperatures) between the inner warm and outer cold belts suggest a large gap in the minor-body population. This large, mostly dust-free zone may be maintained by one or multiple planet-mass objects through dynamical interaction, just like the HR 8799 system. For one single planet, the width of the chaotic zone depends on the location and eccentricity of the planet and its mass. Therefore, one can use the boundaries of the warm and cold dust belts to estimate the mass of the possible perturber.

We use the Vega system as an example where the warm belt is estimated at $\sim 11 - 14$ AU (Su *et al.* 2013) and the cold belt is at ~90–120 AU (Su *et al.* 2005, excluding the disk halo). For one single object on a circular orbit, the mass of the object has to be greater than 100 M_J (a brown dwarf), which should have been found in the intensive direct imaging searches (planets with masses $\gtrsim$3–4 M_J can be ruled out in the 20–70 AU range, Marois *et al.* 2006). For one single object on an eccentric orbit, the boundaries of the belts give an expected eccentricity of 0.8 (i.e., warm belt as the pericenter and cold belt as the apocenter). An object with such a high eccentricity inside the cold belt will have significant dynamical effects in the system, forcing the leftover planetesimals to be on highly eccentric orbits and creating an offset eccentric ring and/or resonance structures, which is not what is seen in the resolved disk images (Su *et al.* 2013).

The giant planets are expected to form more efficiently outside the water-ice line, and these giant planets, once formed, likely experience inward or outward migration. The discovery of hot Jupiter systems in the early success of exoplanet searches is the best evidence for inward migration. Furthermore, if a giant planet did migrate inward, it is likely to destroy other bodies during the course of migration, consistent with the low incidence of warm excesses in such systems (Morales *et al.* 2012). Therefore, the existence of terrestrial planets and Asteroid belt in our Solar System is probably because it does not have a hot Jupiter. The same logic applies to these two-belt systems like Vega, Fomalhaut, ϵ Eri and HR 8799, having an asteroid belt might indicate a greater chance to harbor terrestrial planets compared to systems that do not have an asteroid belt.

References

Absil, O., Mennesson, B., Le Bouquin, J. -B., *et al.* 2009, *ApJ*, 704, 150
Acke, B., Min, M., Dominik, C., *et al.* 2012, *A&A*, 540, 125
Backman, D., Marengo, M., Stapelfeldt, K. R., *et al.* 2009, *ApJ*, 690, 1522
Ballering, N. P., Rieke, G. H., Su, K. Y. L., & Montiel, E. 2013, *ApJ*, 775, 55
Boley, A. C., Payne, M. J., Corder, S., *et al.* 2012, *ApJL*, 750, L21
Dermott, S. F. & Murray, C. D. 1983, *Nature*, 301, 201
Gaspar, A., Psaltis, D., Rieke, G. H., & Ozel, F. 2012, *ApJ*, 754, 74
Hughes, M., *et al.* 2011, *ApJ*, 740, 38
Kalas, P., Graham, J. R., & Clampin, M. 2005, *Nature*, 435, 1067
Kirkwood, D. 1867, *Meteoric Astronomy*, co., 1867
Lebreton, J. *et al.* 2013, *A&A*, 555, 146
Liou, J. C. & Zook, H. A. 1999, *AJ*, 118, 580
Malhotra, R. 1993, *Nature*, 365, 819
Marois, C., Lafreniere, D. Doyon, R., *et al.* 2006, *ApJ*, 641, 556
Marois, C., Zuckeman, B., Konopacky, Q. M., *et al.* 2010, *Nature*, 468, 1080
Mennesson, B., Absil, O., Lebreton, J., *et al.* 2013, *ApJ*, 763, 119
Morales, F. Y., Padgett, D. L., *et al.* 2012, *ApJ*, 757, 7
Su, K. Y. L., *et al.* 2005, *ApJ*, 628, 487
Su, K. Y. L., Rieke, G. H., Stapelfeldt, K. R., *et al.* 2009, *ApJ*, 705, 314
Su, K. Y. L., Rieke, G. H., Malhotra, R., *et al.* 2013, *ApJ*, 763, 118

Exploring the Formation and Evolution of Planetary Systems
Proceedings IAU Symposium No. 299, 2013
M. Booth, B. C. Matthews & J. R. Graham, eds.

doi:10.1017/S1743921313008776

'Dust around Nearby Stars' The Survey Observational Results

Jonathan P. Marshall[1] on behalf of the DUNES consortium

[1]Depto. de Fisica Teorica, Universidad Autonoma de Madrid,
Cantoblanco, 28039, Madrid, Spain
email: jonathan.marshall@uam.es

Abstract.
In this conference proceedings we summarize the key observational findings of the Herschel DUNES survey. We found 31 discs in our sample, equating to an increased dust incidence with Herschel of 20.2 ± 2.0% compared to previous measurement of ~ 12.5 ± 5% with Spitzer for the same population of nearby, Sun-like stars. We identify no trend towards fewer discs around later spectral types, as had previously been reported for A-M stars. Around half of the discs exhibit extended emission, representing a vast improvement in the number of spatially resolved debris discs and thereby the quality of modelling that can be applied to those systems. We also identify unusual sub-groups of discs, including 'steep SED' sources with dust spectral indexes in the 70–160 μm range, steeper than the Rayleigh-Jeans tail which, whilst not unheard of, are more typically seen at sub-mm wavelengths and candidate 'cold discs' which are identified through their lack of significant excess emission at wavelengths shorter than 100 μm.

Keywords. stars: infrared, stars: circumstellar matter, stars: planetary systems

1. Introduction

Circumstellar dust around mature main sequence stars is a marker for the occurence of a planet(esimal) formation process because the dust we observe, being shorter lived than the star it orbits, must be continually replenished from larger bodies (Wyatt 2008, Krivov 2010). The large, 3.5 m mirror of the Herschel Space Observatory (Pilbratt *et al.* 2010) provides high angular resolution and a sensitivity capable to not only detect, but spatially resolve debris discs around nearby stars at flux levels only 10–20 times higher than that predicted for the Solar system's Edgeworth-Kuiper belt (EKB, Vitense *et al.* 2012).

The Herschel Open Time Key Programme (OTKP) Dust around Nearby Stars (DUNES, Eiroa *et al.* 2010) observed 133 nearby, Sun-like stars searching for evidence of excess emission at far-infrared wavelengths above that expected from the star. The survey sample consisted of 124 stars within a 20 pc volume (124 stars) along with specific targets out to 25 pc which were known to host a debris disc from previous observations (9 stars). All of the stars were observed with PACS (Poglitsch *et al.* 2010) using the 100/160 μm channels, integrating long enough on each target to detect the stellar photosphere at 100 μm with a signal-to-noise of ⩾5. Several of the targets were also observed with the PACS 70/160 μm combination and/or with SPIRE (Griffin *et al.* 2010) at 250, 350 and 500 μm depending on the expected properties of the source. In addition to the 133 stars observed directly by DUNES, we share observations of another 106 stars with another Herschel OTKP studying debr is discs, DEBRIS (Disc Emission from Bias-free Reconnaissance in the Infrared and Sub-millimetre, Matthews *et al.* 2010). In this proceedings we only detail results from targets observed directly by DUNES.

Table 1. Survey observational results. Number of newly detected/resolved sources in brackets.

Type	**F**	**G**	**K**	**Total**
$N_{\rm obs}$ by DUNES	27	52	54	133
$N_{\rm obs}$ $d < 20$ *pc*	20	50	54	124
Non-excess	16	37	42	95
Contaminated	2	3	2	7
Excess	9 (2)	12 (3)	10 (5)	31 (10)
Excess $d < 20$ *pc*	4	11	10	25
Resolved discs	5 (4)	6 (4)	5 (5)	16 (13)

2. Overview

Here we present a digest of the observational results from the DUNES survey thus far, which are summarized in Table 1. A more in-depth summary can be found in the DUNES survey paper, Eiroa *et al.* 2013. All of the DUNES survey's Herschel data and ancilliary material (ancillary photometry, stellar models, etc.) will be made available through the virtual observatory at http://sdc.cab.inta-csic.es/dunes/.

Non-excess stars: In concordance with our stated survey objective, we detected the stellar photosphere of all our targets at 100 μm. Around the vast majority of the stars, $\sim$ 80%, we observed exhibited no significant far infrared excess emission. Assuming a typical dust temperature of 37 K, we placed 3-σ upper limits on the dust fractional luminosity at the (mean) level of $L_{\rm IR}/L_{\star} < 2\times10^{-6}$, around 20 times that of the predicted EKB. When aggregated, the measured 100 μm photosphere fluxes are in slight deficit compared to the predicted models based on a Rayleigh-Jeans extrapolation from mid-infrared wavelengths, which was also previously observed in the Spitzer 70 μm results of Hillenbrand *et al.* (2008), although in neither case is the deficit significant. α Centauri A, the nearest and brightest G-type star to us was observed with both PACS and SPIRE from 70–500 μm by DUNES and Hi-Gal. In the far infrared we me asured a significant deficit between the measured fluxes and extrapolated model at far infrared and sub-millimetre wavelengths. The deficit is attributed to the first direct detection of the temperature minimum of a star other than the Sun (Liseau *et al.* 2013). A major implication of this is that around Sun-like stars debris dust could 'hide' in this deficit, thereby causing dust masses around such stars to be underestimated.

Excess stars: We identified 31 debris disc stars in our sample, representing a disc incidence of 20.2 $\pm$ 2% (once those sources beyond 25 pc have been removed). This represents a real increase over the previous Spitzer measurement of $\sim$12% around the same sample (once contaminated sources identified by Herschel are removed) (Trilling *et al.* 2008). Furthermore, we see no evidence of a difference in the debris incidence around the F, G and K stars in our sample, contrary to Spitzer results. However, we do see trends between the dust and stellar properties; the radial extent of the discs increases with age, consistent with dynamical evolution of the disc occuring as an inside-out process.

Resolved discs: One of the most powerful results from DUNES has been that half (16/31) of the detected discs were resolved, or had extended emission along the disc major axis. Three of the discs were known to be extended from previous Spitzer observations HIP 7978 (Liseau *et al.* 2010, Augereau *et al.* in prep.), HIP32480 (Stapelfeldt *et al.*, in prep.) and HIP107649 (Marshall *et al.* 2011, Löhne *et al.* 2012). Fixing the dust spatial location through this information, in combination with the high angular resolution of Herschel providing tight constraints (compared to the dust blackbody radius) even in those cases where the dust is unresolved, facilitated detailed modelling of the dust physical parameters in combination with these spatial constraints on the dust location.

Unusual SEDs: Two sub-samples of unusual disc sources have been identified within the DUNES sample; 'steep SED' discs and 'cold disc' candidates. The first of these show a marked fall off of the disc emission at wavelengths between 70–160 μm, more commonly seen at sub-millimetre wavelengths. In these discs the shape of the SED is attributed to either a steeper particle size distribution than that attributed to a steady-state collisional cascade, or that the largest (dominant) dust grain size in these discs is relatively small, only 10s of microns (Ertel *et al.* 2012). The 'cold disc' candidates are unusual in that they only exhibit excess emission at wavelengths beyond 70 μm, whilst typical debris discs peak at wavelengths 60–100 μm. Such cold dust could be evidence of contamination by a background galaxy, but the number of candidates (7) precludes that all of these sources might be spurious (1.2% for all 7), although the probability for individual conta mination is high (39%) (Eiroa *et al.* 2012, Krivov *et al.* 2013, Marshall *et al.* submitted). Furthermore, a similar object identified in the DEBRIS survey was subsequently identified as the product of contamination through stellar proper motion (Kennedy, priv. comm.), such that confirmation of these objects is required.

Debris-Exoplanet correlations: In combination with the DEBRIS survey, we compiled a sample of all 35 radial velocity exoplanet host stars within 20 pc and subjected them to an analysis, correlating the dust, planet and stellar parameters looking for trends. Searching for trends linking the presence of dust and planets with the stellar properties we identified a trend between stars with low mass planets, sub-Solar metallicities and a detectable debris disc, finding 5/9 stars in our sample with such properties. We attribute the correlation between these parameters to evidence of planet formation by core accretion in these systems. These results will be presented in a companion papers detailing the observed sample and the statistical analysis (Marshall *et al.* in prep., Moro-Martín *et al.* in prep.). Further work on the correlation between exoplanets and debris discs by Herschel will be forthcoming from the open time programme SKARPS (Stars with Kuiper belts and Radial velocity Planets, Bryden *et al.* in prep.).

3. Summary

The DUNES survey has observed 133 nearby, Sun-like stars, detecting excess emission from 31 stars, of which 10 are newly identified debris discs. Half the detected discs are seen to be extended along at least their major axis, providing a constraint to the spatial location of the dust around these stars. We see no change in the relative fraction of discs around stars of different spectral type, in contrast to previous results from Spitzer.

References

Berta, S., *et al.* 2011, *A&A*, 532, A49
Eiroa, C., *et al.* 2010, *A&A*, 518, L131
Eiroa, C., *et al.* 2012, *A&A*, 536, L4
Eiroa, C., *et al.* 2013, *A&A*, 555, A11
Ertel, S., *et al.* 2012, *A&A*, 541, A148
Griffin, M., *et al.* 2010, *A&A*, 518, L3
Hillenbrand, L., *et al.* 2008, *ApJ*, 677, 630
Krivov, A. V. 2010, *RAA*, 10, 383
Krivov, A. V., *et al.* 2013, *ApJ*, 772, 32
Liseau, R., *et al.* 2010, *A&A*, 518, L132
Liseau, R., *et al.* 2013, *A&A*, 549, L7
Löhne, T., *et al.* 2012, *A&A*, 537, A110
Marshall, J. P., *et al.* 2011, *A&A*, 529, A117

Matthews, B., *et al.* 2010, *A&A*, 518, A135
Pilbratt, G., *et al.* 2010, *A&A*, 518, L1
Poglitsch, A., *et al.* 2010, *A&A*, 518, L2
Sibthorpe, B., *et al.* 2012, *MNRAS*, 428, L6
Trilling, D., *et al.* 2008, *ApJ*, 674, 1086
Vitense, C., *et al.* 2012, *A&A*, 540, A30
Wyatt, M. C. 2008, *ARA&A*, 46, 339

Discussion

METCHEV: How do you explain the lack of dependence of the debris disc detection frequency on spectral type?

MARSHALL: I think the key here is that for all the stars in our 20 pc sample we are sensitive enough to detect the photosphere at 100 μm and can often (but not always) detect the photosphere at 160 μm. We are therefore more sensitive in terms of $L_{\rm IR}/L_{\star}$ than Spitzer (by a factor of $\sim$ 5), which was much more strongly volume-limited by spectral type for photosphere detection, particularly for K stars, due to the higher confusion noise level. Since we have information across the peak of the dust SED for all our discs (when combined with previous Spitzer 70 μm measurements) and a constraint on the stellar flux, we can determine the presence of cooler and fainter dust than previously possible - indeed, half of our new debris discs (5/10) are found around K stars.

WEINBERGER: Could you comment further on how you estimate the contamination rates and whether updated Herschel deep extragalactic counts have been included?

MARSHALL: The contamination rates for our sample were determined using the Herschel PACS cosmological survey number counts in Berta *et al.* 2011 and the contamination estimates based on the flux limited DEBRIS survey in Sibthorpe *et al.* 2012. The actual probability of contamination was based on the probability of chance alignment within the PSF FWHM of a background source with a brightness of the same level of flux as the measured 100 μm excess.

WEINBERGER: Follow-up: Are there individual cases of nearby stars re-observed with Herschel where proper motion definitely confirms whether an asymmetry in the disc moves with the star?

MARSHALL: Excepting the obvious case of the ring around Fomalhaut, I cannot think of another disc that Herschel observed and resolved that was known to be asymmetric from e.g. HST scattered light observations. However, that is more due to the fact that asymmetric discs are rare and the angular resolution of Herschel, whilst vastly superior to previous far infrared satellite observatories, is still relatively poor in the grand scheme of things (PSF FWHM 6.7″ at 100 μm). As previously highlighted, one of the key results from Herschel/DUNES has been being able to detect the extended emission from a large fraction of the detected debris discs.

Exploring the Formation and Evolution of Planetary Systems
Proceedings IAU Symposium No. 299, 2013
M. Booth, B. C. Matthews & J. R. Graham, eds.

doi:10.1017/S1743921313008788

A Trend Between Cold Debris Disk Temperature and Stellar Type: Implications for the Formation and Evolution of Wide-Orbit Planets

Nicholas P. Ballering[1], **George H. Rieke**[1], **Kate Y. L. Su**[1] **and Edward Montiel**[1,2]

[1]Steward Observatory, University of Arizona, Tucson, AZ, USA
[2]Department of Physics & Astronomy, Louisiana State University, Baton Rouge, LA, USA

Abstract. Cold debris disks have the potential to answer many outstanding questions in wide-orbit planet formation and evolution. We characterized the infrared excess SEDs of 174 cold debris disks with *Spitzer* IRS and MIPS. We found a trend between the temperature of the disks and the stellar type of the stars they orbit. This argues against the importance of strictly temperature-dependent processes (e.g. ice lines) in setting the dimensions of cold debris disks. We also found no evidence that delayed stirring causes the trend. The trend may result from outward planet migration that traces the extent of the primordial protoplanetary disk, or from planet formation that halts at an orbital radius limited by the efficiency of core accretion. For the full details of this work, see Ballering *et al.* (2013).

Keywords. circumstellar matter – infrared: stars – interplanetary medium

1. Introduction

Detecting wide-orbit planets is difficult, but we can study their formation and evolution from observations of cold debris disks because planets may sculpt sharp inner edges in the disks. The location of a cold disk may trace the extent of outward planet migration or the limits of efficient planet formation. These effects predict that cold debris disks around early type stars will be warmer than those around later type stars. Alternatively, ice lines in the protoplanetary disk phase could set the location of debris disks by creating a dust trap. The location of warm debris disks is thought to be set by the water ice line (Morales *et al.* 2011), and cold debris disks could be set by the ice line of other species. This effect predicts that cold debris disks will have similar temperatures around all stellar types.

2. Methods and Results

We analyzed the SEDs of 546 main sequence stars with Spitzer IRS and MIPS data. To obtain each debris disk temperature(s), we fit the excess SED with one or two blackbody curves. Our definition of a "cold" debris disk was the coldest component below 130 K. We found 51 single component warm disks, 100 single component cold disks, and 74 two component disks. For 25 very cold disks, the lack of data at wavelengths longer than 70 μm prevented us from accurately measuring their temperatures, so we assigned them upper limit temperatures of 45 K. By plotting the cold disk temperatures vs. stellar temperatures, we see that cold disks tend to be warmer around early type stars (see Fig. 1). Could age play a confounding role in this relation? Because early type stars are preferentially younger than late type stars, this trend would arise if debris disks cool as

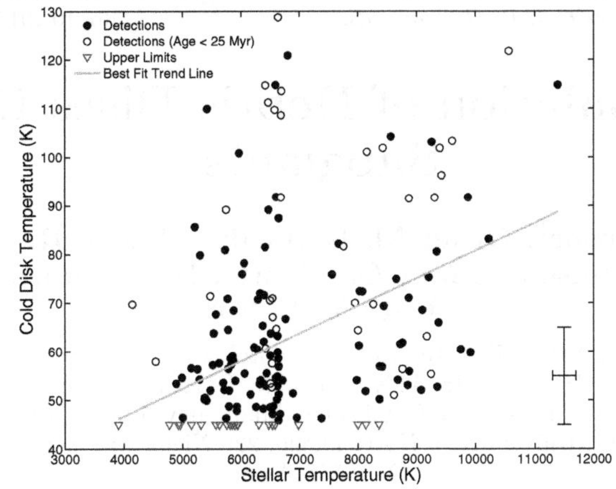

Figure 1. The temperature of the cold disk component versus the temperature of the disk's host star. Open circles are young systems (where active planet formation may still be occurring) and triangles are upper limits. Although there is substantial scatter in the cold component temperatures (reflecting the degree of variability among debris disk properties), a correlation between cold debris disk temperature and stellar type is evident. Note, for example, that there are no disks colder than 50 K around stars hotter than 8500 K, in comparison with the large number of disks with temperature < 45 K around cooler stars. The line is the best fit to the data, determined by a Bayesian linear regression, $T_{\rm cold} = 0.00568 \times T_{\star} + 24.0$. Thus, the trend is significant at a level greater than 4.5σ. A representative error bar is in the lower right.

they age. Such behavior is predicted by delayed stirring models, wherein an annulus of active dust production propagates outward through a continuous disk (Wyatt 2008). We saw no evidence for cold debris disks around the same stellar type cooling with age after the first few tens of Myr.

3. Discussion and Conclusions

What determines the locations of cold debris disks? This trend suggests that their locations are not set by a temperature-dependent process, such as an ice line. Perhaps their locations are set by planets migrating outwards via scattering with planetesimals. Migration will stop towards the outer edge of the disk where the surface density of planetesimals drops. From observations of protoplanetary disks, $R_{\rm disk} \propto M_{*}^{0.63}$ (Scholz *et al.* 2006; Andrews *et al.* 2010), which predicts $T_{\rm disk} \propto T_{\star}^{1.38}$, consistent with the observed trend. Another possibility is that the inner edge of a cold debris disk represents the radial limits of efficient planet formation. The timescale for core accretion scales as $t \propto R^{3} M_{\star}^{-3/2}$ (Kenyon & Bromley 2008). The inner edge of the cold disk will be set at the location where t reaches a given threshold, so setting t constant yields $R_{\rm disk} \propto M_{\star}^{1/2}$. This translates to $T_{\rm disk} \propto T_{\star}^{3/2}$, which is also consistent with the observed trend.

References

Andrews, S. M., Wilner, D. J., Hughes, A. M., Qi, C., & Dullemond, C. P. 2010, *ApJ*, 723, 1241
Ballering, N. P., Rieke, G. H., Su, K. Y. L., & Montiel, E. J. 2013, ApJ, 775, 55
Kenyon, S. J., & Bromley, B. C. 2008, *ApJS*, 179, 451
Morales, F. Y., Rieke, G. H., Werner, M. W., Bryden, G., Stapelfeldt, K. R., & Su, K. Y. L. 2011, *ApJ*, 730, L29
Scholz, A., Jayawardhana, R., & Wood, K. 2006, *ApJ*, 645, 1498
Wyatt, M. C. 2008, *ARA&A*, 46, 339

Exploring the Formation and Evolution of Planetary Systems
Proceedings IAU Symposium No. 299, 2013
M. Booth, B. C. Matthews & J. R. Graham, eds.

doi:10.1017/S174392131300879X

The Population of Debris Discs Orbiting Subgiants

Amy Bonsor[1], Grant M. Kennedy[2], Justin R. Crepp[3], John A. Johnson[4], Mark C. Wyatt[2], Bruce Sibthorpe[5] and Kate Y. L. Su[6]

[1]UJF-Grenoble 1 / CNRS-INSU, Institut de Plantologie et d'Astrophysique de Grenoble (IPAG) UMR 5274, Grenoble, F-38041, France
[2]Institute of Astronomy, University of Cambridge, Madingley Road, Cambridge CB3 OHA, UK
[3]Department of Physics, University of Notre Dame, 225 Nieuwland Science Hall, Notre Dame, IN 46556, USA
[4]Department of Astronomy, California Institute of Technology, 1200 E. California Blvd., Pasadena, CA 91125, USA
[5]SRON Netherlands Institute for Space Research, Zernike Building, P.O. Box 800, 9700 AV Groningen, The Netherlands
[6] Steward Observatory, University of Arizona, 933 N Cherry Ave., Tucson, AZ 85721
email: amy.bonsor@gmail.com

Abstract. Whilst debris discs orbiting main-sequence stars are well studied, very little is known regarding their fate when the star evolves onto the giant branch. For intermediate mass (A-type) stars, giants provide a unique opportunity to detect planets using the radial velocity technique, otherwise prohibited by high jitter levels and rotationally broadened lines in main-sequence intermediate mass (A-type) stars. Such stars can provide key insights into the structure of planetary systems around intermediate mass stars. In our Herschel OT1 program (PI Bonsor) we searched for the presence of debris discs orbiting a sample of 36 subgiants, half of which have RV detected companions. Our best detection is the resolved debris disc orbiting κ CrB.

Keywords. (stars:) planetary systems,(stars:) circumstellar matter

1. Kappa Cr B (Bonsor *et al.*, 2013)

Our best detection, Kappa Cr B, provides the first resolved images of a debris disc orbiting an evolved star. Kappa Cr B is also orbited by two companions. The structure of this planetary system is constrained from the Herschel images (Fig. 1, alongside detection limits on the orbital parameters of these companions, obtained using further RV monitoring (Keck I HIRES) and AO imaging (Keck II NIRC2)). Kappa Cr B is an intriguing example of a planetary system where both planets and dusty debris have been detected and in which interactions between them may be critically important. See Bonsor *et al.*, 2013 for more details.

2. The *Herschel* Survey (Bonsor *et al.*, in prep)

Thirty eight subgiants were searched for excess emission with *Herschel* PACs at 100 μm and 160 μm, 20 planet-hosts and 16 control stars. The aim was to search for a correlation between planet-hosts and non-planet host stars. Excess emission, consistent with a debris disc, was found for 4/36 (2/36) of the subgiants in our sample at 100 μm (160 μm). Three of the detections are planet-hosting stars. See Bonsor *et al.* (in prep) for more details.

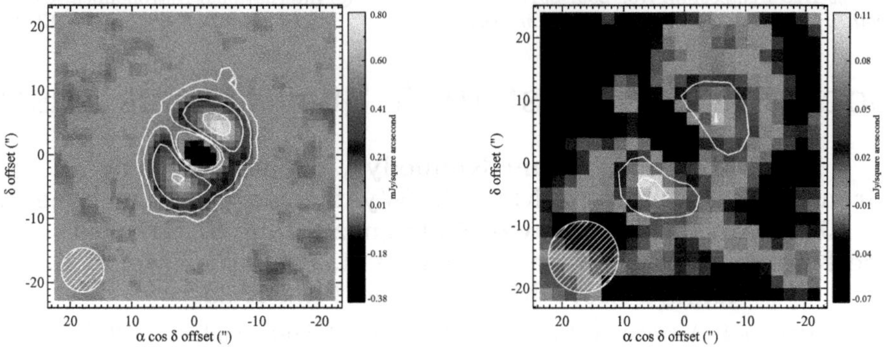

Figure 1. The *Herschel* images of the κ CrB debris disc (stellar-subtracted) at 100μm (left) and 160μm (right) (Bonsor *et al.*, 2013)

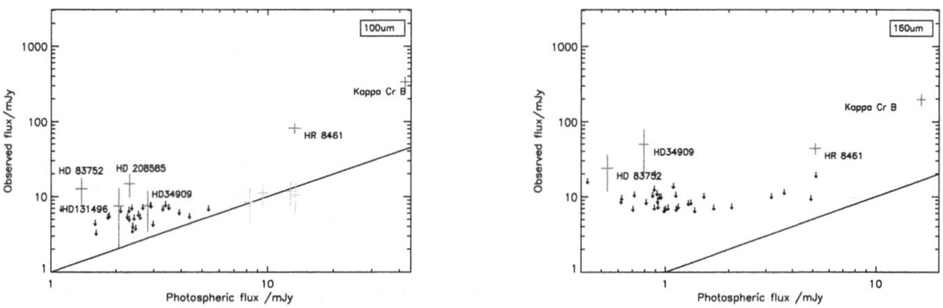

Figure 2. The detections in our *Herschel* survey. The photospheric flux, predicted using stellar models compared to the observed flux, at 100μm (left) and 160μm (right). The red (blue) points show $3-\sigma$ detections of a disc (the star), with $3-\sigma$ error bars. The stellar photosphere is too faint at 160μm to have been detected by any of our observations.

Reference

Bonsor, A.; Kennedy, G. M., Crepp, J. R., Johnson, J. A., Wyatt, M. C., Sibthorpe, B., & Su, K. Y. L. 2013, *MNRAS*, 431, 3025-3035

Exploring the Formation and Evolution of Planetary Systems
Proceedings IAU Symposium No. 299, 2013
M. Booth, B. C. Matthews & J. R. Graham, eds.

doi:10.1017/S1743921313008806

Locating the Dust in A Star Debris Discs

Mark Booth[1,2], Grant Kennedy[3], Bruce Sibthorpe[4], Brenda C. Matthews[2,1], Mark C. Wyatt[3], Gaspard Duchêne[5,6], J. J. Kavelaars[2,1], David Rodriguez[7], Jane S. Greaves[8], Alice Koning[1,2], Laura Vican[9], George H. Rieke[10], Kate Y. L. Su[10], Amaya Moro-Martín[11] and Paul Kalas[5,12]

[1]Dept. of Physics & Astronomy, University of Victoria, Elliott Building, 3800 Finnerty Rd, Victoria, BC, V8P 5C2 Canada; email: markbooth@cantab.net
[2]NRC of Canada, 5071 West Saanich Road, Victoria, BC, V9E 2E7 Canada
[3]Institute of Astronomy, Madingley Rd, Cambridge CB3 0HA, UK
[4]UK ATC, Royal Observatory, Blackford Hill, Edinburgh EH9 3HJ, UK
[5]Astronomy Department, UCB, Hearst Field Annex B-20, Berkeley, CA 94720-3411, USA
[6]UJF-Grenoble 1 / CNRS-INSU, IPAG UMR 5274, Grenoble, F-38041, France
[7]Universidad de Chile, Camino el Observatorio 1515, Las Condes, Santiago, Chile
[8]School of Physics & Astronomy, University of St Andrews, North Haugh, St Andrews, Fife KY16 9SS, UK
[9]Dept. of Physics & Astronomy, UCLA, 475 Portola Plaza, Los Angeles, CA 90095-1547, USA
[10]Steward Observatory, University of Arizona 933 N Cherry Avenue Tucson, AZ 85721
[11]Centro de Astrobiología (CSIC-INTA), 28850 Torrejón de Ardoz, Madrid, Spain
[12]SETI Institute, 515 North Whisman Rd., Mountain View, CA 94043, USA

Abstract. Using photometry at just two wavelengths it is possible to fit a blackbody to the spectrum of infrared excess that is the signature of a debris disc. From this the location of the dust can be inferred. However, it is well known that dust in debris discs is not a perfect blackbody. By resolving debris discs we can find the actual location of the dust and compare this to that inferred from the blackbody fit. Using the *Herschel Space Observatory* we resolved many systems as part of the DEBRIS survey. Here we discuss a sample of 9 discs surrounding A stars and find that the discs are actually located between 1 and 2.5 times further from their star than predicted by blackbody fits to the spectral energy distribution (SED). The variation in this ratio is due to differences in stellar luminosities, location of the dust, size distribution and composition of the dust.

Keywords. circumstellar matter, planetary systems, infrared: stars

---

Debris discs are formed of dust and planetesimals that are the remnants of the planet formation process. By understanding debris discs we can learn about the planet formation process, dynamical history of the planetary system and potentially even discover new planets from asymmetries in the dust (for a review see e.g. Wyatt 2008).

DEBRIS (Disc Emission via a Bias-free Reconnaissance in the Infrared/Submillimetre) was an open time key project (PI: Brenda Matthews) on *Herschel.* The aim of the project was to conduct an unbiased, flux limited survey of the nearest stars for debris discs (Phillips *et al.* 2010). The survey (Matthews *et al.* 2010) has observed 446 systems (of spectral types A, F, G, K and M) at wavelengths of 100 and 160 μm with the PACS instrument and many of these systems have also been followed up at 70 μm and with SPIRE at 250, 350 and 500 μm.

In Booth *et al.* (2013) we present observations of 9 A stars from the DEBRIS sample that host debris discs resolved by *Herschel*: α CrB, β Uma, λ Boo, ϵ Pav, ζ Eri, γ Tri, ρ Vir, 30 Mon and β Tri. We fit a blackbody to the excess spectrum in these systems based on data from the PACS and SPIRE instruments and ancillary data from the

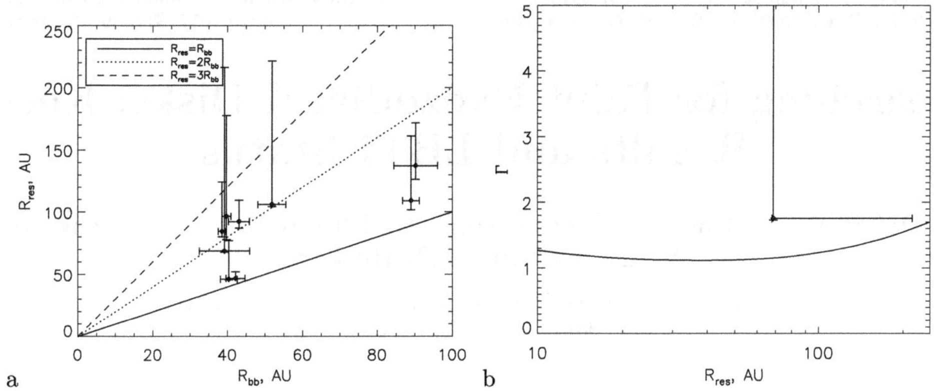

Figure 1. a: Resolved radii versus blackbody radii. The errors show the range of best fits for the three different wavelengths. The lines represent what we would expect for different values of $\Gamma = R_{res}/R_{bb}$. **b**: Ratio of radii from the blackbody fitting to resolved radii against the resolved (actual) radii. The line shows the expected ratio variation for stars of spectral type A1. The point shows the observed values for 30 Mon.

literature and infer a radius, R_{bb}, from this. To find the actual location of the dust, R_{res} (the resolved radii), we use a thin-toroid grid-search algorithm. Comparison of these two values is shown in Fig. 1a.

From Fig. 1a we can see that there is some variation in the ratio R_{res}/R_{bb} between 1 and 2.5. Some of this variation is simply due to the different spectral types and, therefore, luminosities of the stars. The difference between the radius inferred from the blackbody fit and the actual location of the dust is due to blackbodies being a poor approximation for small grains. Dust grains are inefficient at absorbing and re-emitting wavelengths longer than the size of the grain and so become hotter than blackbodies, producing a spectrum equivalent to blackbodies closer to the star.

By assuming a Dohnanyi (1969) size distribution down to the blowout limit and grains that are composed of a silicate core and organic mantle (Li & Greenberg 1997) we can simulate what we expect the radius ratio to be. We find that taking account of the stellar luminosity and location of the dust is not enough to explain the radius ratio for some of the systems, meaning that there must be other differences in the grain populations of some systems. For instance, 30 Mon has a higher ratio of $\Gamma = R_{res}/R_{bb}$ than is expected for its spectral type (Fig. 1b). This could be explained by a larger than expected number of small grains or a different composition to that normally assumed for debris disc dust.

We also find that, although the SEDs are generally well fit by a single temperature blackbody, a number of the images are not well fit by a narrow ring model and require a wide ring or multiple rings to explain the imaged dust emission.

References

Booth, M., Kennedy, G., Sibthorpe, B., *et al.* 2013, *MNRAS*, 428, 1263
Dohnanyi, J. S. 1969, *J. Geophys. Res.*, 74, 2531
Li, A. & Greenberg, J. M. 1997, *A&A*, 323, 566
Matthews, B. C., Sibthorpe, B., Kennedy, G., *et al.* 2010, *A&A*, 518, L135
Phillips, N. M., Greaves, J. S., Dent, W. R. F., *et al.* 2010, *MNRAS*, 403, 1089
Wyatt, M. C. 2008, *ARA&A*, 46, 339

Exploring the Formation and Evolution of Planetary Systems
Proceedings IAU Symposium No. 299, 2013
M. Booth, B. C. Matthews & J. R. Graham, eds.

doi:10.1017/S1743921313008818

Searching for Faint Exozodiacal Disks: Keck Results and LBTI Status

D. Defrère[1], P. Hinz[1], B. Mennesson[2], R. Millan-Gabet[3], A. Skemer[1], V. Bailey[1] and T.J. Rodigas[1]

[1]Steward Observatory, [2]Jet Propulsion Laboratory, [3]Caltech
email: ddefrere@email.arizona.edu

Abstract. The possible presence of dust in the habitable zone around nearby main-sequence stars is considered as a major hurdle toward the direct imaging of Earth-like extrasolar planets with future dedicated space-based telescopes (e.g., Roberge *et al.* 2012). In this context, NASA has funded two ground-based mid-infrared nulling interferometers to combine the large apertures available at the Keck Observatory and the Large Binocular Telescope (LBT). In this poster, we present the preliminary results of the extended survey carried out with the Keck Interferometer Nuller (KIN) between 2008 and 2011 and describe the forthcoming LBTI survey.

Keywords. infrared, exozodiacal dust, surveys, nulling interferometry, high angular resolution

---

1. Nulling interferometry

Nulling interferometry is a technique proposed by Bracewell 35 years ago to image extra-solar planets (Bracewell 1978). The basic principle is to combine the beams coming from two distinct telescopes in phase opposition so that a dark fringe appears on the line of sight, which strongly reduces the stellar emission while transmitting the flux of off-axis sources located at angular spacings which are a multiple of λ/b (where b is the distance between the telescope and λ the wavelength of observation). The technique has now been successfully used on sky both in the near-infrared (e.g., Mennesson *et al.* 2011) and the mid-infrared (e.g., Stock *et al.* 2010) and the performance required to directly detect exoEarths demonstrated on laboratory testbeds (e.g., Martin *et al.* 2012).

2. Keck results

Science observations with the KIN started in 2008 with four different exozodi programs and ended in 2011 with the observation of 48 stars. The results of one of these programs have been published by Millan-Gabet *et al.* (2011), providing unprecedented sensitivity to warm exozodiacal dust in the habitable zone of 25 nearby single main-sequence stars. The analysis of the remaining data is currently ongoing (Mennesson *et al.* in prep). It includes a total of 41 single stars, i.e., with no known companion within the ~0.5" interferometric FOV, divided in three main sub-samples: (1) stars with no previous detection of dust; (2) presumably dusty stars; and (3) stars with an interferometric near-infrared detection (Absil *et al.* 2013). The preliminary results are shown in Fig.1. η Crv is the only clear KIN detection although the statistical analysis of the measurements suggests that an additional ~10 objects are good candidates for a KIN excess and will be followed-up with the LBTI.

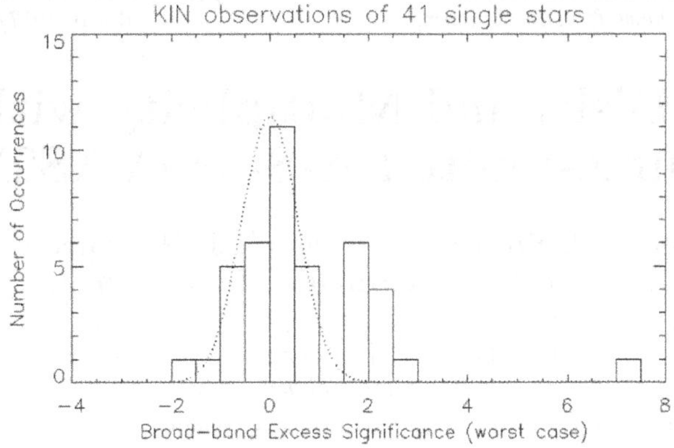

Figure 1. Histogram of broad-band (8 to 13 microns) excess "significance" measured around 41 stars of the KIN extended sample, compared to a centered Gaussian distribution representative of detection noise (dotted line, Mennesson *et al.* in prep).

3. LBTI status

The Large Binocular Telescope Interferometer (LBTI) is the ideal ground-based instrument to carry out a deep survey for exozodiacal dust around nearby main-sequence stars. It combines high sensitivity (provided by the two 8.4-m mirrors on a single structure), high contrast (see companion paper by V. Bailey on AO performance), and high resolution (equivalent to the one of a 22.7-m telescope). LBTI was installed in September 2010 at the LBT and first fringes were obtained one month later. Dual-aperture AO-corrected interferometric fringes were realized in April 2012 and first null measurements in September 2012. Phase tracking and differential tip-tilt correction are currently under commissioning.

The exozodi survey is called HOSTS (Hunt for Observable Signatures of Terrestrial Planetary Systems) and is expected to start in 2013. It is currently being defined by the LBTI Key Science Team (nexsci.caltech.edu/missions/LBTI/KeySciTeam.shtml) and currently consists in a shallow survey of 60-80 FGK nearby main-sequence stars among which 10 are well-known dusty systems and 10 are the best candidates for future exo-Earth direct imaging instruments. The goals of the survey are: (1) to constrain the mean exozodiacal dust density; (2) to study exozodiacal dust disk evolution; (3) to understand the connection between exozodiacal dust and planets; and (4) to correlate exozodiacal dust with basic stellar properties or the presence of dust inferred from separate observations. The goal 3 will benefit from simultaneous L-band observations of the planetary systems and another LBTI survey called LEECH (LBTI Exozodi Exoplanet Common Hunt, see companion paper by A. Skemer).

References

Absil, O., *et al.*, 2013, *A&A*, 555, 104
Bracewell, R. N., 1978, *Nature*, 274, 780.
Martin, S., *et al.*, 2012, *AO*, 51, 3907–3921.
Mennesson, B., *et al.*, 2011, *ApJ*, 243, 178.
Millan-Gabet, R., *et al.*, 2011, *ApJ*, 734, 67.
Roberge, A., *et al.*, 2012, *PASP*, 124, 799–808.
Stock, N. D., *et al.*, 2010, *ApJ*, 724, 1238–1255.

Exploring the Formation and Evolution of Planetary Systems
Proceedings IAU Symposium No. 299, 2013
M. Booth, B. C. Matthews & J. R. Graham, eds.

doi:10.1017/S174392131300882X

Debris Disks and Multiplicity within the 75pc Volume-limited A-Star (VAST) Survey

R. J. De Rosa[1,2], B. Smith[1], J. Bulger[2], J. Patience[1,2], C. Marois[3], I. Song[4], B. Macintosh[5], J. Graham[6], R. Doyon[7] and M. Bessell[8]

[1]School of Earth and Space Exploration, Arizona State University, PO Box 871404, Tempe, AZ 85287-1404, USA
email: rjderosa@asu.edu

[2]College of Engineering, Mathematics and Physical Sciences, University of Exeter, UK

[3]NRC Herzberg Institute of Astrophysics, Canada

[4]Physics and Astronomy, University of Georgia, USA

[5]Institute of Geophysics and Planetary Physics, Lawrence Livermore National Lab, USA

[6]Department of Astronomy, University of California at Berkeley, USA

[7]Département de Physique, Université de Montréal, Canada

[8]Research School of Astronomy and Astrophysics, Mount Stromlo Observatory, Australia

Abstract. We present the preliminary findings of an investigation of the multiplicity of debris disk stars identified within our Volume-limited A-star (VAST) multiplicity survey. Previous studies have produced conflicting results regarding the multiplicity fraction of debris disk-hosting stars compared with non-excess stars. By combining our large-scale volume-limited AO survey of A-type stars with the all-sky *WISE* catalogue, we have investigated the frequency of binary companions to a large sample of A-type stars with and without measured 22μm excess. The results of this study will allow for a greater understanding of the interaction between a companion star and a circumstellar debris disk, informing future study into the formation and stability of planetary-mass companions within binary systems.

Keywords. stars: circumstellar matter, stars: binaries: visual, stars: early-type

1. Excess and control samples

A total of 55 A-type stars within the 75 parsec sample were identified to have a significant *WISE* 22μm excess (Figure 1, left panel), in addition to having been searched for binaries with high-resolution adaptive optics imaging. Excess candidates were identified by comparing the observed optical and near-infrared photometry of each target to synthetic photometry derived from PHOENIX atmospheres (Hauschildt *et al.* 1999). Of the 55 stars within the excess sample, 15 are newly identified as candidate debris disk-hosting stars within this study. As the excess stars identified within this study were drawn from a large volume-limited sample of 636 A-type stars (De Rosa *et al.* 2013), we are able to construct a statistically similar control sample of 110 stars, against which the frequency of stellar companions in the excess sample can be compared (Figure 1, right panel).

2. Binary properties of the samples

Around the stars with measured 22 μm excess, we find a multiplicity fraction of 7^{+5}_{-2}%, significantly lower than the 29^{+5}_{-4}% measured for the control sample. The mass of each resolved companion in both samples is plotted as a function of projected separation in Figure 2. The deficiency in companions is most pronounced for separations between

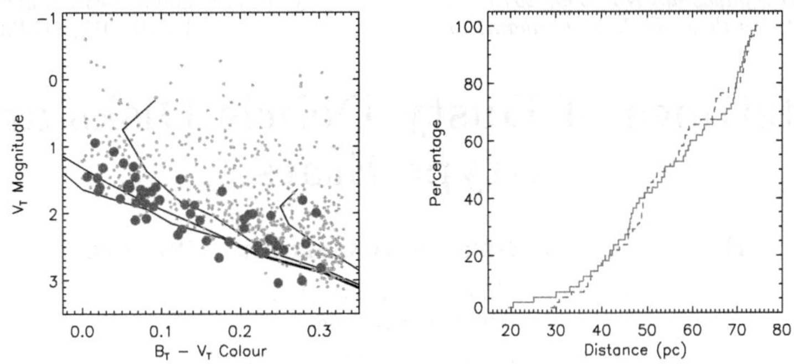

Figure 1. *(left panel)*: A color-magnitude diagram of the 55 excess stars identified within this study (large red points), with the full VAST sample (grey points), and 100, 200, 400, and 800 Myr isochrones (black lines; Siess *et al.* 2000) plotted for reference. *(right panel)*: Cumulative distance distributions for the 55-star excess sample (red histogram), and the 110-star control sample (blue dashed histogram). The control sample was selected such that the distance distributions of the two samples were statistically similar.

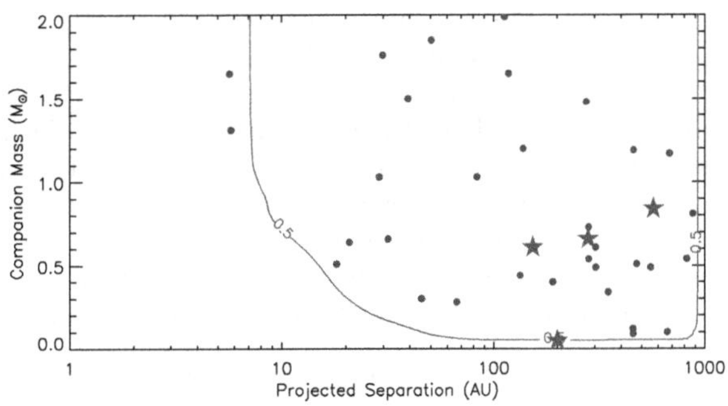

Figure 2. The separation and mass of the companions identified within the excess sample (red stars) and the control sample (blue points). The 50% sensitivity contour of the full VAST survey is plotted for reference (green curve).

10 and 100 AU, a range within which no stellar companions were resolved within the excess sample. A more detailed analysis of the sensitivity to stellar companions for both samples, as a function of projected separation, will be undertaken in order to determine the significance of this deficiency. The depressed multiplicity fraction of the excess sample is consistent with previous studies of debris disk hosting stars (e.g. Kraus *et al.* 2012, Rodriguez & Zuckerman 2012). Previous studies have also shown a higher excess fraction for stars with close (< 3 AU) binaries (Trilling *et al.* 2007), a range which future extensions of this study will cover.

References

De Rosa, R. J., *et al.* 2013, *MNRAS, submitted*
Hauschildt, P. H., Allard, F., & Baron, E. 1999, *ApJ*, 512, 377
Kraus, A. L., Ireland, M. J., Hillenbrand, L. A., & Martinache, F. 2012, *ApJ*, 745, 19
Rodriguez, D. R. & Zuckerman, B. 2012, *ApJ*, 745, 147
Siess, L., Dufour, E., & Forestini, M. 2000, *A&A*, 358, 593
Trilling, D. E., *et al.*, 2007, *ApJ*, 658, 1289

Exploring the Formation and Evolution of Planetary Systems
Proceedings IAU Symposium No. 299, 2013
M. Booth, B. C. Matthews & J. R. Graham, eds.

doi:10.1017/S1743921313008831

On Lifetimes of Dusty Debris Disks around A-type Stars

Ramiro de la Reza[1] **and Carolina Chavero**[2]

[1]Observatório Nacional, Ro de Janeiro, Brazil
email: delareza@on.br

[2]Observatorio Astronómico de Córdoba, Córdoba, Argentina
email: carolina@oac.uncor.edu

Abstract. Using a collection of 40 debris disks (DD) around dwarf A-type stars (from which 17 are spatially resolved) having well determined stellar ages (Vican 2012), we found that for a collisional self-stirring model of flux distributions (Kenyon & Bromley, 2008, 2010) dusty DD occupy for a large part the terminal age positions corresponding the end of the MS phase. This situation does not necessarily mean that the dust generation in these DD is finished at this stage. Dusty DD can continue their lives in the next Post-MS stages, as is the case of the recent first resolved dusty DD with planets around the K-type subgiant star κ CrB. This star is considered a "retired" A star (Bonsor *et al.* 2013). In order to explain the apparent lack of dusty DD in the giant phase, we propose here that this could be due, in part, to a dilution of any planetesimal belt existing around these stars, with the corresponding diminishing of collisions and subsequent dust generation. This dilution is provoked by the dynamical expansion of this belt produced by the important stellar mass loss during the RGB phase. This scenario is being studied by means of numerical simulations by de la Reza, Roig & Chavero 2013 (in preparation).

Keywords. planetary formation, debris disk, circumstellar disk

1. Motivation

The recent detection and spatial resolving of the first dusty debris disk (DD) with planets around the K-type subgiant star κ CrB (Bonsor *et al.* 2013) and the apparent absence of signatures of DD in giant stars, lead us to study the probable end of dust generation in DD. Early studies on evolution of DD have suffered from the lack of knowledge of well determined stellar ages. To study the end of dust production in DD stars around A-type stars, we use a recent published list of homogeneous age data for A-type DD stars, specially selected for studies with Herschel (Vican 2012). We also use the most complete collection of resolved (17) and non-resolved (23) A-type DD, totaling 40 objects which have been measured with *Spitzer* and *Hershel* observatories.

2. The collisional disks model

We consider the self-stirring collisional numerical models of DD and planet formation of Kenyon & Bromley (2008, 2010) hereafter (KB08, KB10). The emission fluxes of the new formed dust resulting from collisions are calculated (for different dimensions, strengths of the initial planetesimals, stellar masses and for *Spitzer* and *Herschel* selected wavelengths). Here, we use fluxes at 24 μm and 70 μm for which we dispose the most complete data. We also consider that radiation at 70 μm covers an important part of the disk (Sibthorpe *et al.* 2010, Booth *et al.* 2013). In the figures the fluxes ratios at 70 μm over 24 μm are presented in the form: $[70 - 24] = \log(24F_{70}) - \log(70F_{24})$, (Chavero

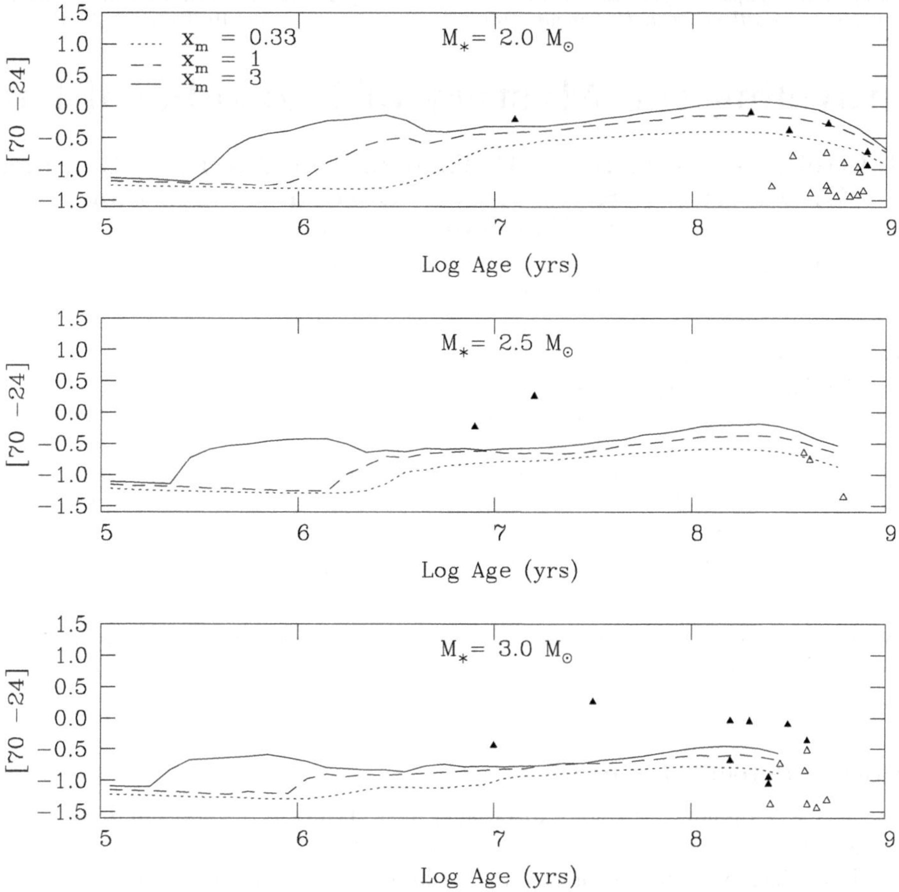

Figure 1. In this figure are superposed the observed flux ratios (in the form of [70 - 24] = $\log(24F_{70})$ - $\log(70F_{24})$) of the 40 DD stars at 70 μm and 24 μm on the distribution of dust radiation of the collisional model of Kenyon & Bromley (2008) in function of age. In this model the initial planetesimals are small with a radius of $\leqslant$ 1 km. Three stellar masses of 2.0 M$\odot$, 2.5 M$\odot$, and 3.0 M$\odot$ are considered representing the different A-type stars used here. Xm are model parameters with values from 3.0, 1.0 and 0.33 corresponding respectively to disks with smaller masses. Filled points correspond to resolved disks whereas open symbols to unresolved disks.

2009), where the fluxes are those calculated in KB08 for stellar masses of 2.0, 2.5 and 3.0 M $\odot$, masses that we consider representative for all types of A stars.

References

Bonsor, A. & Wyatt, M. C. 2010, *MNRAS*, 409, 163

Bonsor, A., kennedy, G. M., *et al.* 2013, *MNRAS*, 431, 4, 3025

Booth, M., *et al.* 2013, *MNRAS*, 248, 1263

Chavero, C. 2009, PhD thesis at Observatorio Nacional, Rio de Janeiro, Brazil

Kenyon, S. J. & Bromley, B. C. 2008, *ApJS* 179, 451

Kenyon, S. J. & Bromley, B. C. 2010, *ApJS*, 188, 242

Sibthorpe, B., & *et al.* 2010, *A&A*, 518.130

Vican, L. 2012, *AJ* 143, 135

Exploring the Formation and Evolution of Planetary Systems
Proceedings IAU Symposium No. 299, 2013
M. Booth, B. C. Matthews & J. R. Graham, eds.

doi:10.1017/S1743921313008843

Unraveling the Mystery of Exozodiacal Dust

S. Ertel[1], J.-C. Augereau[1], P. Thébault[2], O. Absil[3], A. Bonsor[1], D. Defrère[4], Q. Kral[2], J.-B. Le Bouquin[1], J. Lebreton[1], V. Coudé du Foresto[2,5]

[1]IPAG, UJF Grenoble; [2]LESIA, Observatoire de Paris; [3]Université de Liège; [4]Steward Observatory; [5]Bern University

Abstract. Exozodiacal dust clouds are thought to be the extrasolar analogs of the Solar System's zodiacal dust. Studying these systems provides insights in the architecture of the innermost regions of planetary systems, including the Habitable Zone. Furthermore, the mere presence of the dust may result in major obstacles for direct imaging of earth-like planets. Our EXOZODI project aims to detect and study exozodiacal dust and to explain its origin. We are carrying out the first large, near-infrared interferometric survey in the northern (*CHARA*/FLUOR) and southern (*VLTI*/PIONIER) hemispheres. Preliminary results suggest a detection rate of up to 30% around A to K type stars and interesting trends with spectral type and age. We focus here on presenting the observational work carried out by our team.

Keywords. infrared: stars, circumstellar matter, surveys, techniques: interferometric, techniques: high angular resolution

---

1. How to detect an exozodi

The thermal emission from hot exozodiacal dust (typically several hundreds of kelvin up to the sublimation temperature, e.g., Lebreton *et al.* 2007) results in a near-infrared excess above the stellar flux (typically 1% for known systems). This accuracy (at 3σ significance) is not reachable by simple photometry. It is necessary to spatially disentangle the dust and stellar emission. As the emission extends out to only a few hundred milliarcseconds for nearby stars, this is reachable only by interferometry. At short baselines ($\sim$10 to 40 m), considering fully extended (incoherent) dust emission around a nearly unresolved star, the emission results in a visibility drop compared to purely stellar visibilities that is equal to twice the disk/star flux ratio (Di Folco *et al.* 2007). Latest high precision near-infrared interferometers (*CHARA*/FLUOR, *VLTI*/PIONIER, Coudé du Foresto *et al.* 2003, Le Bouquin *et al.* 2011) allow us to detect this visibility drop.

2. An all-sky, near-IR survey for exozodiacal dust

We are carrying out the first large, near-infrared interferometric survey for exozodiacal dust (note the earlier results in the mid-infrared by Millan-Gabet *et al.* 2011). We are using the *CHARA*/FLUOR instrument in K band in the northern hemisphere and our *VLTI* visitor instrument PIONIER in H band in the southern hemisphere. A total of $\sim$ 200 stars are planned to be surveyed. The observing strategy and target selection is designed to form a large, statistical, and unbiased sample. A first sample of targets observed with CHARA consisting of 42 stars ($K < 4$ mag) has just been accepted for publication (Absil *et al.* 2013). The PIONIER sample (89 stars with $H < 5$ mag) has been observed and the data are under analysis (Ertel *et al.*, in prep.). Detection statistics for the CHARA sample are shown in Figure 1. Differences are most probably due to differences in sensitivity and lower excess in H band.

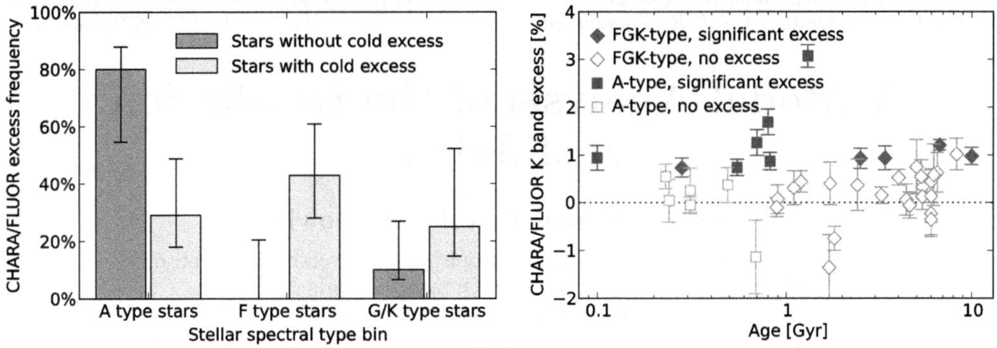

Figure 1. *Left: CHARA*/FLUOR *K* band detection rate of exozodiacal dust with respect to the spectral type of the host star and the presence or not of a cold, Edgeworth-Kuiper belt like debris disk. *Right: K* band excess as a function of stellar age.

Our magnitude limited sample allows for statistical investigation of correlations of the detection rate and excess levels with other parameters such as the stellar age, spectral type, or the presence of cold dust that could serve as a reservoir for replenishing the dust.

The most important results so far from the *CHARA*/FLUOR part of the survey (preliminary results from *VLTI*/PIONIER are largely consistent) are:

• The over all detection rate is ~ 30%, for sun-like stars ~ 20%.

• Stars with *and* without an outer reservoir harbor exozodiacal dust, but there *is* a correlation between the presence of cold and hot dust for sun-like stars.

• There is no significant correlation of the excess levels with the age of the star (Absil *et al.* 2013, Figure 1).

The latter would be expected in analogy to the well known trend in normal debris disks if the dust was produced in a steady state collisional process of larger bodies over the whole age of the system.

3. Future observational perspectives

The results from our survey as well as the state-of-the-art instruments and observing strategies allow for important and promising observational studies. Using *VLTI*/ PIONIER, and in the near future *VLTI*/GRAVITY and *VLTI*/MATISSE, the topics our team is currently working on include multi-wavelength observations constraining temperatures of known exozodis, the search for time variability of known exozodis on time scales of years, detailed studies of the hot dust in prominent debris disks, and the investigation of the connection between warm mid-IR and hot near-IR excesses.

References

Absil, O., Defrère, D., Coudé du Foresto, V., *et al.* 2013, *A&A*, 555A, 104A
Coudé du Foresto, V., Borde, P. J., Merand, A., *et al.* 2003, *SPIE*, 4838, 280C
Di Folco, E., Absil, O., Augereau, J.-C., *et al.* 2007, *A&A*, 475, 243D
Le Bouquin, J.-B., Berger, J.-P., Lazareff, B., *et al.* 2011, *A&A*, 535A, 67L
Lebreton, J., van Lieshout, R., Augereau, J.-C., *et al.* 2013, *A&A*, 555A, 146L
Millan-Gabet, R., Serabyn, E., Mennesson, B., *et al.* 2011, *ApJ*, 734, 67M

Exploring the Formation and Evolution of Planetary Systems
Proceedings IAU Symposium No. 299, 2013
M. Booth, B. C. Matthews & J. R. Graham, eds.

doi:10.1017/S1743921313008855

A Modal Analysis of the Irradiation Instability

Jeffrey Fung[1] **and Pawel Artymowicz**[1,2]

[1]Department of Astronomy and Astrophysics, University of Toronto,
50 St. George Street, Toronto, Ontario, Canada M5S3H4
email: fung@astro.utoronto.ca

[2]Department of Physical and Environmental Sciences, University of Toronto at Scarborough,
265 Military Trail, Scarborough, Ontario, Canada M1C1A4
email: pawel@utsc.utoronto.ca

Abstract. The irradiation instability is a disk instability caused by the radiation pressure cast by a central source onto an optically thick disk. The criterion for this instability depends on a sharp transition from an optically thin inner disk to an optically thick outer disk. The quickly diminishing radiation pressure in this transition region creates a radially compressing effect, which is in many ways similar to the effects of self-gravity. In this modal analysis, we demonstrate that a disk marginally stable to irradiation can develop global modes, with growth rates being of order the dynamical timescale of the disk. The non-linear evolution of the our model shows the formation of vortices near the transition region and spiral structures propagating into the optically thick region. Consequently the scale-height of our disk's inner edge becomes time-variable and can likely be observed as a variation in its infrared flux.

Keywords. instabilities, hydrodynamics, scattering, planetary systems: protoplanetary disks, accretion disks

---

Instabilities in accretion disks provide a mechanism for transferring angular momentum outward, resulting in accretion flows. Those that have been considered important to the evolution of accretion disks include the magnetorotational instability (MRI) (Balbus & Hawley 1998), gravitational instability (Lin & Pringle 1987, Gammie 2001), Papaloizou-Pringle instability (Papaloizou & Pringle 1984, 1985, 1987), and Rossby wave instability (RWI) (Lovelace *et al.* 1999). The list goes on if non-ideal MHD and the vertical shearing (Urpin & Brandenburg 1998) are also considered. Here we introduce a new form of instability associated with the radiation pressure exerted by a central object onto an optically thick disk which contains an inner edge. We term this instability the "irradiation instability".

The common parameter for describing the force of radiation pressure is β_0:

$$\beta_0 = \frac{\kappa_{\rm opa} L_*}{4\pi c G M_*}, \tag{0.1}$$

where L_* is the host's luminosity, M_* is the host's mass, and $\kappa_{\rm opa}$ is the opacity of the disk material; G and c are the gravitational constant and speed of light respectively. This parameter describes the fraction of gravity cancelled by radiation pressure, as both of them are central forces that diminish as r^{-2}. The force due to radiation pressure on dust grains is comparable to gravity, if the source is sufficiently luminous, or the grains are of an optimal size. For example, micron-sized grains under the illumination of an A type star have a β_0 well exceeding unity.

Radiation pressure can potentially destabilize disks. Consider the following: submerged in a disk where a porton of the illumination from the central source is blocked by layers of

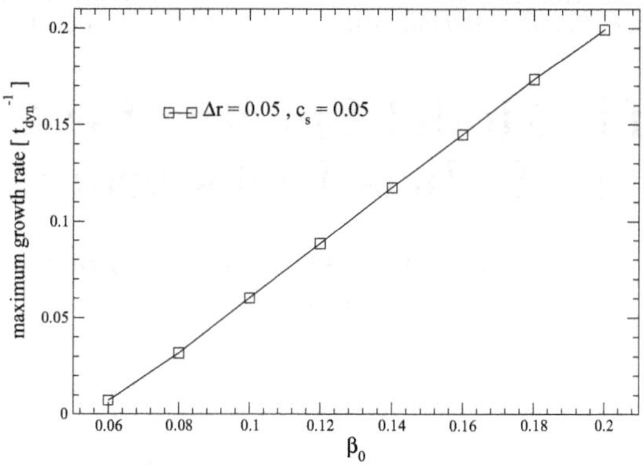

Figure 1. The growth rate of the asymmetric mode increases with β_0. There is no measurable modal growth for $\beta_0 < 0.06$. Δr and the sound speed c_s are non-dimensionalized, such that $r = 1$ and the Keplerian velocity $v_k = 1$ at the inner edge of the disk.

dust, an over-dense dust cloud slowly passes in front of a dust grain. As a result, a darker shadow is cast on to the grain. Consequently, the grain feels less push from radiation pressure, and it migrates towards the over-dense cloud, further increasing the density of the cloud. This process can potentially runaway and cause an instability if the shear in the disk fails to tear these clouds apart in time.

To capture this process, we derive a condition for local stability:

$$Q_{\rm rad} \equiv \left(\beta_0 e^{-\tau} \frac{\Omega^2}{\kappa^2} \frac{{\rm d}\tau}{{\rm d}\ln r} \right)^{-1} > 1, \tag{0.2}$$

where τ is the radial optical depth, κ is the epicyclic frequency, and Ω is the orbital frequency.

We perform a modal analysis where we solve the linearized set of Euler's equations and calculate the growth rates for individual azimuthal modes. We use a density profile that is constant, but with an inner edge where the density gradually falls to 1% of its value over a distance Δr. The τ profile is normalized such that $\tau = 2/3$ at the inner edge. Using this disk model, fig. 1 shows that the instability can grow as fast as a few dynamical timescales. While the growth rates plotted in fig. 1 do depend on our disk model, it is important to note that the disk flux is stable in the limit of a small β_0 demonstrating that radiation pressure is the driving force that destabilizes our disk.

We also investigate the fully non-linear regime of this instability through numerical simulations. We find that the asymmetric modes develop into vortices that remain at the inner edge of the disk. These features create a variable scale height at the inner edge and can result in an observable infrared variability (Flaherty & Muzerolle 2010).

References

Balbus, S. A. & Hawley, J. F. 1998, *Reviews of Modern Physics*, 70, 1
Flaherty, K. M. & Muzerolle, J. 2010, *ApJ*, 719, 1733
Gammie, C. F. 2001, *ApJ*, 553, 174
Lin, D. N. C. & Pringle, J. E. 1987, *MNRAS*, 225, 607
Lovelace, R. V. E., Li, H., Colgate, S. A., & Nelson, A. F. 1999, *ApJ*, 513, 805
Papaloizou, J. C. B. & Pringle, J. E. 1984, *MNRAS*, 208, 721
Papaloizou, J. C. B. & Pringle, J. E. 1985, *MNRAS*, 213, 799
Papaloizou, J. C. B. & Pringle, J. E. 1987, *MNRAS*, 225, 267
Urpin, V. & Brandenburg, A. 1998, *MNRAS*, 294, 399

Exploring the Formation and Evolution of Planetary Systems
Proceedings IAU Symposium No. 299, 2013
M. Booth, B. C. Matthews & J. R. Graham, eds.

doi:10.1017/S1743921313008867

Non-LTE Model Spectra for Gaseous Planetary Debris Disks around WDs

Stephan Hartmann, Thorsten Nagel, Thomas Rauch and Klaus Werner

Institute for Astronomy and Astrophysics, University of Tübingen, Germany
email: hartmann@astro.uni-tuebingen.de

Abstract. Metal-rich dust disks around white dwarfs are thought to be the debris of tidally disrupted rocky bodies. While normally the number of features to study the planetary material directly is very limited, ancillary gas disks around some of these white dwarfs provide the opportunity to do so. We used our Tübingen Accretion Disk code AcDc, assuming non-LTE conditions, to model the gaseous spectrum component. We investigated the chemical mixture as well as the surface density and effective temperature, and utilized the Ca II infrared triplet to determine the geometry of the disk.

Keywords. accretion disks, (stars:) white dwarfs, (stars:) planetary systems, line: profiles, radiative transfer

---

Analyses of ultraviolet and optical spectra of white dwarfs (WDs) revealed a large number of metal-rich single objects. The picture of the WDs' atmospheres getting polluted due to the accretion of planetary material, e.g. tidally shredded planetesimals, became quite successful in explaining these observations (Farihi, Jura & Zuckerman 2009). Usually the almost homogeneous H or He atmospheres are used as a detector for the spectral signatures of the pollution, as the debris disks themselves, consisting of cold, dusty material yield only few features to analyze the accreted planetary material. However, about 20% of these dusty disks are connected to an additional gaseous disk of similar radial extent (Melis *et al.* 2010). The hotter gas becomes noticeable through double peaked emission lines which provide the unique opportunity to directly analyze the debris material before it is accreted onto the WD. Hallmark to the gaseous disk component is the Ca II $\lambda\lambda$ 8498, 8542, 8662 Å infrared triplet (IRT), which often shows prominent asymmetric line profiles (Gänsicke *et al.* 2006)

The Tübingen Accretion Disk code AcDc (Nagel *et al.* 2004) calculates the spectrum of an axisymmetric, gaseous disk applying the well known α-disk method (Shakura & Sunyaev 1973). The disk is described in form of a set of discrete rings, for which vertical structure calculations can be performed independently from each other. Assuming non-local thermodynamic equilibrium conditions, the radiative transport, the equations of hydrostatic and energetic equilibrium, and the rate equations are solved simultaneously. During the final surface-integration step over the ring spectra, the Keplerian rotation is taken into account which results in double-peak line profiles.

To model the gaseous component of a metal-rich debris disk, we include H, C, O, Mg, Si, S, and Ca with abundances similar to a solar system asteroid-like mixture into our models (C = 2.82%, O = 48.08%, Mg = 15.33%, Si = 19.80%, S = 11.50%, and Ca = 2.48%, mass fractions). We calculated the disks in a non-stationary scheme using different surface density Σ and effective temperature T_{eff} values. The disk's radial extent can be varied by the number and radii of rings used to simulate the radial structure.

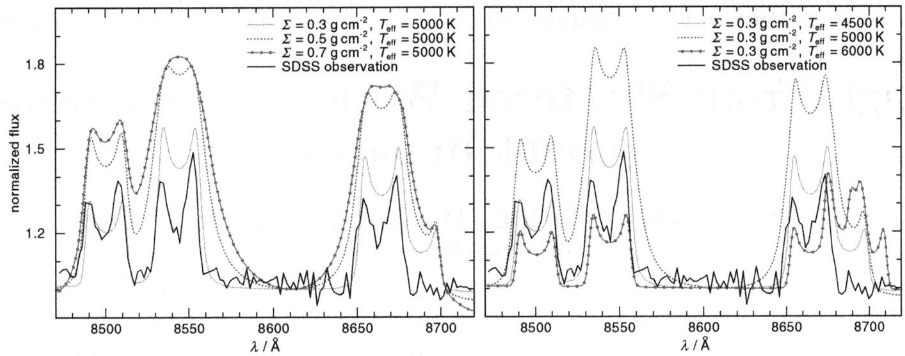

Figure 1. Ca II IRT of non-LTE disk models for three different values of Σ (left) and $T_{\rm eff}$ (right) compared with the observation of SDSS J122859.93+104032.9.

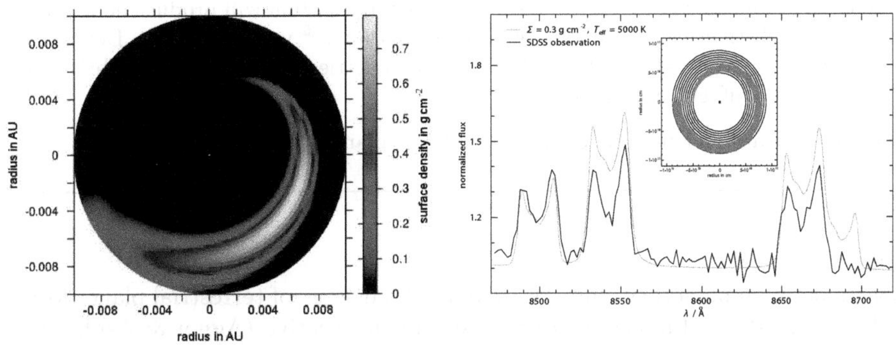

Figure 2. *Left:* A hydrodynamic calculation shows an asymmetric gas disk. *Right:* Modeled spectrum for a similar non-axisymmetric disk geometry (inlaid picture).

Disk spectra for radial constant values of Σ and $T_{\rm eff}$ were calculated to investigate the influence on the Ca II IRT (Fig. 1). Higher Σ values tend to broaden the lines and therefore reduce the dip between the lines' double peaks. Raising $T_{\rm eff}$ results in more similar line strengths of all three components. We compare the models to the observations of SDSS J122859.93+104032.9, the first WD hosting a gaseous debris disk reported (Gänsicke *et al.* 2006), achieving our best fit (green, solid line in all figures) for $\Sigma = 0.3\,{\rm g\,cm^{-2}}$ and $T_{\rm eff} = 5000\,{\rm K}$.

While for these models a circular disk is assumed, hydrodynamic simulations using FARGO (Masset 2000) suggest that a tidally disrupted rocky body would form a more asymmetric disk (Fig. 2, left panel). By modifying the surface-integration step of AcDc to use non-axisymmetric geometries, we are able to model the observed asymmetry of the Ca II IRT line profiles (Fig. 2, right panel). Further details of our analysis approach were described in Hartmann *et al.* (2011).

References

Farihi, J., Jura, M., & Zuckerman, B. 2009, *ApJ*, 694, 805

Gänsicke, B. T., Marsh, T. R., Southworth, J., *et al.* 2006, *Science*, 314, 1908

Hartmann, S., Nagel, T., Rauch, T., *et al.* 2011, *A&A*, 530, A7

Masset, F. 2000, *A&AS*, 141, 165

Melis, C., Jura M., Albert L., *et al.* 2010, *ApJ*, 722, 1078

Nagel, T., Dreizler, S., Rauch, T., *et al.* 2004, *A&A*, 428, 109

Shakura, N. I. & Sunyaev, R. A. 1973, *A&A*, 24, 337

Exploring the Formation and Evolution of Planetary Systems
Proceedings IAU Symposium No. 299, 2013
M. Booth, B. C. Matthews & J. R. Graham, eds.

doi:10.1017/S1743921313008879

Light from Shattered Worlds: Debris from Giant Impacts

Alan P. Jackson[1†], Mark C. Wyatt[1], William R. F. Dent[2] and Aki Roberge[3]

[1]Institute of Astronomy, University of Cambridge, Madingley Road, Cambridge, CB3 0HA, UK
[2]Joint ALMA Observatory, Alonso de Córdova 3107, Vitacura - Santiago, Chile
[3]Goddard Space Flight Centre, 8800 Greenbelt Road, Greenbelt, Maryland, 20771, USA
[†]email: `ajackson@ast.cam.ac.uk`

Abstract. Large impacts in the outer parts of a planetary system will produce debris discs that display a strong, distinctive, asymmetry, which will last for 10^5 year timescales. Debris resulting from a large impact may be able to explain the asymmetries in some known debris discs that have hitherto been difficult to understand.

Keywords. circumstellar matter, planetary systems, planetary systems: formation

1. Introduction

Impacts between planetary embryos drive the final stage of terrestrial planet formation (Kokubo & Ida 1998). Impacts can be erosive or destructive (Agnor & Asphaug 2004), and even efficiently accreting impacts produce debris equal to $\gtrsim$1 per cent the mass of the colliding bodies. Giant impacts occur not only in the terrestrial zone but anywhere in a planetary system; Pluto-Charon (Canup 2011) and Haumea (Brown *et al.* 2007) in the Kuiper belt are both believed to have experienced large impacts. Debris production through giant impacts is thus an implicit component of planetary system formation.

Jackson & Wyatt (2012) showed that debris produced in a Moon-forming like event around a nearby star should be detectable for tens of millions of years by typical *Spitzer* 24μm surveys. Detectable lifetime varies with the size of the largest fragments, but is over a million years even if these are only a few kilometres in size. The Moon-forming impact is a fairly low debris production event (Stewart & Leinhardt 2012), and thus this is a reasonable baseline level of expected debris from planetary scale giant impacts. In the outer reaches of a planetary system orbital timescales are much slower, and collision velocities generally lower, and as such the lifetime of a debris disc of constant mass will increase as its semi-major axis is increased. Debris discs resulting from giant impacts in the outer reaches of a planetary system may thus be very long lived.

2. Disc morphologies

Since all debris fragments in a giant impact debris disc are produced at the original impact point, the orbits of these fragments must, at least initially, pass through the impact point. As a result the disc will have a strong asymmetry, which lasts for typically around 1000 orbits before precession of the debris orbits smears it out. In the terrestrial zone this leads to asymmetry lifetimes of $\sim$1000 years, but in the outer regions the asymmetry can last for over 10^5 years and, around a nearby star, can be readily resolved by modern instruments.

In Fig. 1 we show an example of a debris disc produced by an impact at 85AU from the host star, both face-on and at several edge-on viewing angles, at 10^5yr after the impact. Dynamical evolution is followed using the MERCURY N-body package (Chambers 1999). The width of the debris ring is controlled by the velocity dispersion of the debris, in Fig. 1 we use an isotropic dispersion of 4 km s^{-1}; appropriate for roughly Mars size impactors. The progenitor is assumed to have been on a circular orbit. We also assume there are no massive (~Jupiter mass) planets nearby, but they may exist elsewhere in the system, e.g. the $7M_J$ β-Pic. b at 8.5 AU, without disrupting the disc. Bodies equivalent to the impactors can exist within the disc itself. These could be additional members of a primordial population, or the largest remnants of the impact.

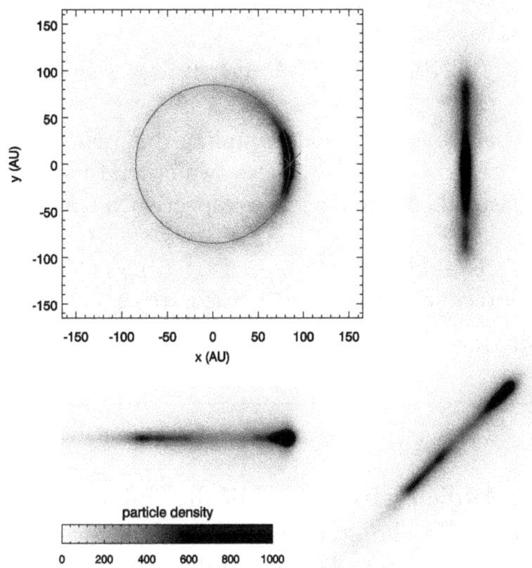

Figure 1. Images of a debris disk generated by a giant impact occuring at 85AU from the host star at 10^5yr after impact. Resolution is 10AU (0.5" at 20pc). At top left we show a face-on view of the disk, with impact-point and progenitor body orbit indicated. Arranged around this we show edge-on views of the disk. Clockwise from top right the lines of sight are at 0°, 45°, and 90° to the impact-point – star line. Spatial and colour scales are consistent between images, with the exception that the density in the face-on image is enhanced by a factor of 2 to aid visibility.

For impacts in regions where ambient temperature is below 50K and bodies contain significant quantities of CO ice, collisional grinding at the impact point may also produce a comet-like tail of CO gas that decays as it moves away from the impact point due to photo-dissociation, producing an even stronger asymmetry.

3. Conclusions

Dust from giant impacts is bright and detectable for ten million year timescales. Impacts in the outer reaches of a system will display distinctive asymmetries that last for 10^5 year timescales and can be resolved around nearby stars. A number of known debris discs, such as the well known disc of Beta Pictoris, display large asymmetries that may be explainable as the result of a large impact, and with new facilities such as ALMA the number of such discs is only likely to increase.

References

Agnor C. & Asphaug E., 2004, ApJL, 613, L157
Brown M. E., Barkume K. M., Ragozzine D., & Schaller E. L., 2007, Nature, 446, 294
Canup R. M., 2011, AJ, 141, 35
Chambers J. E., 1999, MNRAS, 304, 793
Jackson A. P. & Wyatt M. C., 2012, MNRAS, 425, 657
Kokubo E. & Ida S., 1998, Icarus, 131, 171
Stewart S. T. & Leinhardt Z. M., 2012, ApJ, 751, 32

Exploring the Formation and Evolution of Planetary Systems
Proceedings IAU Symposium No. 299, 2013
M. Booth, B. C. Matthews & J. R. Graham, eds.

doi:10.1017/S1743921313008880

LIDT-DD: A New Self-Consistent Debris Disc Model Including Radiation Pressure and Coupling Dynamical and Collisional Evolution

Q. Kral[1], P. Thebault[1] and S. Charnoz[2]

[1]LESIA-Observatoire de Paris, UPMC Univ. Paris 06, Univ. Paris-Diderot, France
email: quentin.kral@obspm.fr

[2]Laboratoire AIM, Universite Paris Diderot / CEA / CNRS, Institut Universitaire de France

Abstract. The first attempt at developing a fully self-consistent code coupling dynamics and collisions to study debris discs (Kral *et al.* 2013) is presented. So far, these two crucial mechanisms were studied separately, with N-body and statistical collisional codes respectively, because of stringent computational constraints.

We present a new model named LIDT-DD which is able to follow over long timescales the coupled evolution of dynamics (including radiation forces) and collisions in a self-consistent way.

Keywords. planetary systems: formation stars: circumstellar matter

---

1. Context & Objectives

Up to now, the evolution of debris discs was studied with two radically different approaches: 1) Dynamics (with N-body codes) to study the structures (e.g. Reche *et al.* (2008)) and 2) Collisions (with particle-in-a-box codes) to study the grinding process and the resulting collisional cascade in such a disc (e.g. Thébault & Augereau (2007)). While such separate studies can (and have) produce(d) important results, they suffer from unavoidable limitations. The absence of collisions in N-body codes can indeed strongly bias or even invalidate some results obtained in such approach: As an example, if collisional timescales are shorter than dynamical ones then collisions can hinder or even prevent the build-up of dynamical structures. For the purely statistical models these limitations are obviously the absence of or poor spatial resolution, but also the fact that dynamical processes might strongly affect impact rates and velocities and thus the collisional evolution.

The first attempts at partially coupling dynamics and collisions have been published recently (e.g., CGA and DyCoSS codes by Stark & Kuchner (2009) and Thebault *et al.* (2012)), most of them taking as a basis the N-body approach into which some collision-imposed properties are injected. But these codes are limited to steady state cases with only one perturber and no fragmentation.

Our new code LIDT-DD overcomes all these limitations and enables coupling dynamics (with radiation pressure) and collisions in a self-consistent way, and is able to follow complex interactions in discs over long timescales.

2. The LIDT-DD code: Principle & First Results...

LIDT-DD is based on the LIDT3D code developed by Charnoz *et al.* (2012) for protoplanetary discs, and strongly upgraded to account for the complexity of debris disc physics (high velocity collisions, radiation-pressure affected orbits, wide range of grains'

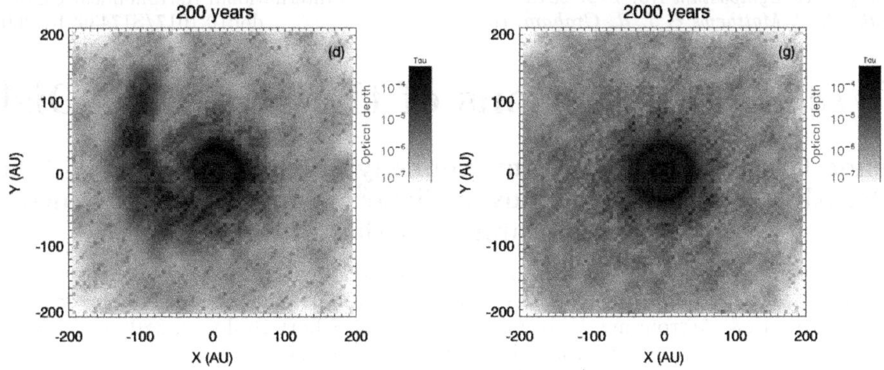

Figure 1. Total smoothed vertical geometrical optical depth evolution in the case of a massive planetesimal breakup within a debris disc.

dynamical behaviour, etc). In this 3D Lagrangian-Eulerian scheme, grains of a given size at a given location in a disc are grouped into "super-particles" (SPs), whose orbits are tracked with an N-body code and whose mutual collisions are treated using a particle-in-a-box scheme. To handle the complexity of grain dynamics, a sorting procedure, regrouping all SPs into dynamical families, has been implemented. A complex SP reassignment routine, looking for and reallocating all redundant SPs, prevents their number from diverging.

Our code has been tested for a set of simplified cases for which we reproduce well-known robust results. LIDT-DD has also been used on a first astrophysical case: the violent breakup of a massive planetesimal within a debris disc (see Fig 1). Dust due to the breakup is released within a debris disc and its dynamical and collisional evolution is followed. A spiral pattern develops and fades away in 2000 years. A secondary ring is created because of Keplerian shear and survives for millions of years.

3. Perspectives

We intend to use LIDT-DD to explore individual systems with "abnormal" flux excesses, such as HD172555 (e.g. Johnson *et al.* (2012)), and also to investigate how generic the massive-break-up scenario can be for explaining *all* bright debris discs with luminosities that cannot be explained by classical collisional cascades. Another area of interest for LIDT-DD are the planet-disc interactions and the extent to which planetary companions can sculpt debris discs. LIDT-DD enables the exploration of transient events, multi-planet systems as well as the feedback of the planetary perturbations on the collisional evolution. Another potential application of the code is the puzzling case of bright exozodiacal discs. Scenarios for explaining exozodis, such as massive impacts, planet scattering, falling evaporating bodies, can in principle also be investigated with LIDT-DD.

References

Charnoz, S. & Taillifet, E.,2012, ApJ, 753, 119
Johnson, B. C., Lisse, C. M., Chen, C. H., *et al.* 2012, ApJ, 761, 45
Kral, Q., Thebault, P., & Charnoz, S. 2013, A&A, 558, A121
Reche, R., Beust, H., Augereau, J.-C., & Absil, O. 2008, A&A, 480, 551
Stark, C. C. & Kuchner, M. J. 2009, ApJ, 707, 543
Thébault, P. & Augereau, J.-C. 2007, A&A, 472, 169
Thebault, P., Kral, Q., & Ertel, S. 2012, A&A, 547, A92

Exploring the Formation and Evolution of Planetary Systems
Proceedings IAU Symposium No. 299, 2013
M. Booth, B. C. Matthews & J. R. Graham, eds.

doi:10.1017/S1743921313008892

Herschel Observations of the HR 8799 Disk

Brenda C. Matthews[1,2], Grant Kennedy[3], Bruce Sibthorpe[4], Mark Booth[2,1], Hannah Broekhoven-Fiene[2,1], Mark Wyatt[3], Bruce Macintosh[5] and Christian Marois[1]

[1]National Research Council of Canada, Herzberg Astronomy & Astrophysics Program, 5071 West Saanich Road, Victoria, BC, V9E E7, Canada, email: brenda.matthews@nrc-cnrc.gc.ca
[2]Dept. of Physics & Astronomy, University of Victoria, Elliott Building, 3800 Finnerty Rd, Victoria, BC, V8P 5C2 Canada
[3]Institute of Astronomy, Madingley Rd, Cambridge CB3 0HA, UK
[4]SRON Netherlands Institute for Space Research, Groningen, The Netherlands
[5]Lawrence Livermore National Labs

Abstract. The HR 8799 four-planet host is known to host a multi-component disk from *Spitzer* observations. We have obtained *Herschel*† observations of the disk which provide increased sensitivity and resolution of its outer components: the planetesimal belt and halo. We find that the two components cannot be discerned from the spectral energy distribution alone, but require resolved images to independently identify them. In the resolved images, the halo stands out for its steep radial profile and large radial extent to 2000 AU, a factor of two larger than was estimated from *Spitzer* data.

Keywords. stars: individual (HR 8799), stars: circumstellar matter, planetary systems

---

1. Background

The excess emission indicating a debris disk was first detected around the A star HR 8799 by IRAS (Moshir *et al.* 1990). The disk was not resolved until after the detection of a directly imaged planetary system around HR 8799 (Marois *et al.* 2008). Using *Spitzer*, Su *et al.* (2009) identified three components present in the disk: a warm inner component and two colder outer components: the planetesimal belt extending from 100-300 AU and the halo extending to 1000 AU. Excepting that very extended halo, the architecture of the HR 8799 system is very similar to that of the Solar System, with a warm inner disk, four gas giants (Marois *et al.* 2010) and then an outer, cold planetesimal disk.

2. Herschel Observations

We have obtained deep maps of HR 8799 with *Herschel* using both PACS (Poglitsch *et al.* 2010) and SPIRE (Griffin *et al.* 2010) and thereby obtain flux measurements for the disk at 70, 100, 160, 250, 350 and 500 μm. The disk is resolved from 70 – 250 μm (Matthews *et al.* 2013). Figure 1 combines the *Herschel* data with earlier data from Moshir *et al.* (1990), Williams & Andrews (2006), Su *et al.* (2009), Ishihara *et al.* (2010) and Patience *et al.* (2011) to produce a well-sampled SED. We find a best fit model of two temperature components: 153 ± 15 K and 36 ± 1 K. The *Herschel* data are not sensitive to the warmer component, which is traced primarily by *Spitzer* MIPS and IRS data (Su *et al.* 2009).

† Herschel is an ESA space observatory with science instruments provided by European-led Principal Investigator consortia and with important participation from NASA.

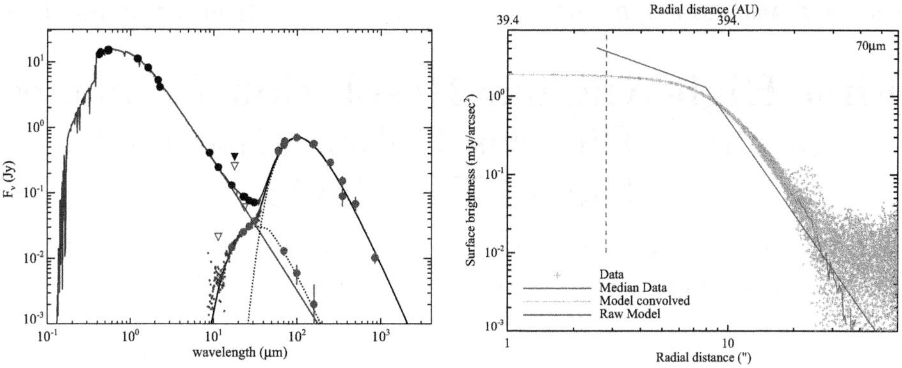

Figure 1. (left) The SED of the HR 8799 disk is shown, with independent curves for the warm blackbody component and the modified cold blackbody component. The stellar emission is fit with a T_{eff} of 7380 K and $L_* = 5.4L_\odot$, consistent with a late-A or early F-type star. The temperatures of the disk components are 153 ± 15 K (the warm unresolved component not detected by *Herschel*) and 36 ± 1 K (the cold resolved component). There is no temperature distinction between the halo and the planetesimal belt evident in the SED. (right) The radial profile of the disk at 70 μm. The HWHP is indicated by a dashed line. The break in the surface brightness profile around $10''$ is apparent and justifies the interpretation of a two-component model for the outer disk. The data constrain the surface brightness profile very well out to $20 - 30''$ ($800 - 1200$ AU). Figures are from Matthews *et al.* (2013), in preparation.

The *Herschel* images (not shown here) do not exhibit any azimuthal asymmetries. The radial profile produced by azimuthally averaging the data is shown in Figure 1. From this profile, it is clear that the cold component of the disk is comprised of material with two different distributions. The planetesimal belt exhibits a shallow profile from $90 - 310$ AU (the inner radius is constrained to within ± 10 AU) while the halo exhibits a much steeper profile. Separate fits to the resolved images show that the index of the radial profile of the planetesimal belt is ~ -1 (constraint of this parameter is degenerate with the unresolved central flux and the unresolved inner edge of the belt) while the better constrained halo profile index has values of -4.0 ± 0.3 and -3.5 ± 0.5 at 70 and 100 μm respectively. We are able to trace the halo emission to a radius of 2000 AU from the central star. We discuss the non-standard filtering applied to reconstruct the maps to capture the largest scale structures in Matthews *et al.* (2013).

These data illustrate that structural features in disks can be masked by the modeling of SEDs in the absence of resolved imaging. The fact that the halo of HR 8799 matches so closely the temperature of the planetesimal belt suggests that the grains are almost blackbodies. Non-blackbody grains would necessarily emit less efficiently than blackbodies and would therefore exhibit higher temperatures to maintain thermal equilibium.

References

Griffin, M. J., Abergel, A., Abreu, A., *et al.* 2010, A&A, 518, L3
Ishihara, D., Onaka, T., Kataza, H., *et al.* 2010, A&A, 514, A1
Poglitsch, A., Waelkens, C., Geis, N., *et al.* 2010, A&A, 518, L2
Matthews, B. C., *et al.* 2013, ApJ, in press
Marois, C., Macintosh, B., Barman, T., *et al.* 2008, Science, 322, 1348
Marois, C., *et al.* 2010, Nature, 468, 1080
Moshir, M., Kopan, G., Conrow, T., *et al.* 1990, Bull. AAS, 22, 1325
Patience, J., Bulger, J., King, R. R., *et al.* 2011, A&A, 531, L17
Su, K. Y. L., Rieke, G. H., Stapelfeldt, K. R., *et al.* 2009, ApJ, 705, 314
Williams, J. P. & Andrews, S. M. 2006, ApJ, 653, 1480

Exploring the Formation and Evolution of Planetary Systems
Proceedings IAU Symposium No. 299, 2013
M. Booth, B. C. Matthews & J. R. Graham, eds.

doi:10.1017/S1743921313008909

First High-Angular Resolution L' Images of the β Pictoris Debris Disc with the VLT / NaCo

Julien Milli[1,2], Dimitri Mawet[1], Olivier Absil[3], Anne-Marie Lagrange[2], David Mouillet[2], Julien H. Girard[1] and Jean-Charles Augereau[2]

[1]ESO, Alonso de Cordova 3107, Casilla 19001, Santiago 19, Chile, email: jmilli@eso.org

[2]IPAG, University Joseph Fourier, CNRS, BP 53, 38041, Grenoble, France

[3]AGO, Université de Liège, 17 Allée du Six Août, B-4000 Liège, Belgium

Abstract. Imaging debris discs in the L'-band (3.8 μm) is a difficult task. Quasi-static speckles from imperfect optics prevail below 1″whereas background emission is the dominant noise source beyond that separation and is much larger than at shorter wavelengths . We demonstrate here the potential of the newly commissioned AGPM coronograph on VLT/NaCo combined with advanced star and sky subtraction technique based on Principal Component Analysis, and we analyze the morphology of the β Pictoris disc.

Keywords. stars: individual (β Pictoris), techniques: high angular resolution, techniques: image processing

1. Introduction

Circumstellar discs are the birthplace of planets, especially the inner regions within a few tens of astronomical units. Imaging discs at that scale requires high-contrast and high-angular resolution techniques, as well as specific data reduction tools. Innovative methods such as Angular Differential Imaging, ADI (Marois *et al.* 2006) or Principal Components Analysis, PCA, (Soummer *et al.* 2012) were developed to detect and characterize faint companions. For discs, they are also used to reveal high spatial frequency features, e.g. Lagrange *et al.* (2012), but any extended structure is strongly biased (Milli *et al.* 2012). On the specific example of the β Pictoris system imaged with the VLT/NaCo, we present here a way to correct the biases induced by ADI and analyze the morphology of the disc in the L'-band.

2. Observations and data reduction specific to the disc geometry

The star β Pictoris was observed on January, 31 2013 with the AGPM coronograph that operates in the L'-band at 3.8 μm (see D. Mawet *et al.*, these proceedings). The star and background subtraction algorithm consisted in a PCA performed in four concentric annuli, retaining 5 components out of 90. The disc is detected from 0.5″ to 4.5″ at a mean 3σ level (Fig. 1). A low-frequency noise due to the strong and variable background emission prevents the detection beyond 4.5″. The warp seen at smaller wavelengths between 3 and 5″ (Lagrange *et al.* 2012) is marginally detected on the South-West (SW) side at 3″.

3. Morphological analysis and discussion

We estimated the position angle (PA) of each side of the disc by derotating the image to align each side horizontally so that the disc spine centers are zero on average between 1″ and 3″ (Lagrange *et al.* 2012). The measured PA is 28.3° ± 2° and 212.3° ± 2° for the NE and SW sides respectively. This is the first time the PA is measured in such a close separation range, the NE value is compatible with measurements at shorter wavelengths

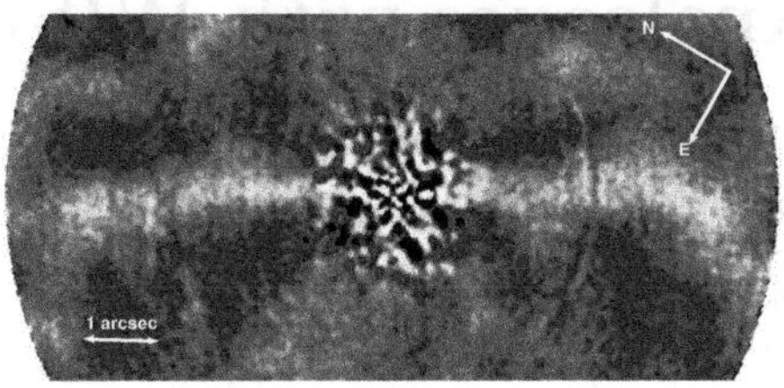

Figure 1. AGPM coronographic image. The colour scale is linear.

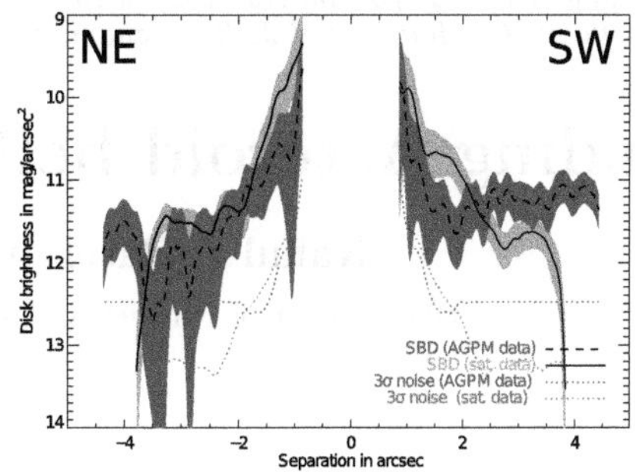

Figure 2. Disc SBD with error bars in colours.

and larger separations whereas the SW side is only marginally compatible. We interpret this as the superimposition of the warp component and preferential forward scattering on the SW side above the midplane. The surface brightness distribution (SBD) corrected from the flux losses related to ADI (see section 4) is shown in Fig. 2. We overplotted the same analysis performed on non-coronographic, saturated, NaCo archival data taken in the same band on September, 28 2010. Although the low-frequency background noise is lower, the observing strategy implied telescope offsets that limit the field of view to 3.5″ in radius and bright speckles and diffraction residuals from the telescope spiders prevent clear conclusions below 1″. Both images agree on a smooth SBD between 1″ and 3″ on the NE side, with a power-law index of –1.0. Golimowski *et al.* (2006) measured indices of –1.45, –1.43 and –1.38 for observations at 435 nm, 606 nm, and 814 nm respectively. Therefore these observations confirm that the SBD gets flatter at larger wavelengths. The brightness asymmetry detected in the coronographic image between 3″ and 4″ and seen in mid-infrared at 8.7 μm by Telesco *et al.* (2005) is not confirmed on the non-coronographic image.

4. Modelisation of the disc and correction of the ADI biases

A careful calibration of flux losses was performed using disc models generated with the GRaTeR code (Augereau *et al.* 1999). We assumed a radially symmetric distribution of optically thin dust with constant scattering cross-section. The radial distribution is described by two power-laws of exponent $\alpha_{in} = 2$ and $\alpha_{out} \leqslant 0$ with a peak density at the radius r_0. The disc model is subtracted from the initial data cube which is then re-reduced using PCA. We repeated this step iteratively by varying the free parameters of the disc model until the residuals are minimized. We performed this minimization for three discrete disc inclinations i (91°, 95°, and 100°) and both isotropic and anisotropic scattering. The best reduced χ^2 is obtained with an isotropic model of a disc having $i = 95°$, $PA = 30.3° \pm 0.8°$, $r_0 = 35.6^{+20}_{-5} AU$ and $\alpha_{out} = -0.7 \pm 0.2$. The peak density radius r_0 is poorly constrained. The best disc model was then used to calibrate the flux losses resulting from the PCA reduction (0.9mag/arcsec^2 at 2″) and to correct the SBD in Fig. 2.

References

Augereau, J. C., Lagrange, A. M., Mouillet, D., Papaloizou, J. C. B., & Grorod, P. A. 1999, A&A, 348, 557

Golimowski, D. A., Ardila, D. R., Krist, J. E., *et al.* 2006, AJ, 131, 3109

Lagrange, A.-M., Boccaletti, A., Milli, J., *et al.* 2012, A&A, 542, A40

Marois, C., Lafrenière, D., Doyon, R., Macintosh, B., & Nadeau, D. 2006, ApJ, 641, 556

Milli, J., Mouillet, D., Lagrange, A.-M., *et al.* 2012, A&A, 545, A111

Soummer, R., Pueyo, L., & Larkin, J. 2012, ApJ, 755, L28

Telesco, C. M., Fisher, R. S., Wyatt, M. C., *et al.* 2005, Nature, 433, 133

Exploring the Formation and Evolution of Planetary Systems
Proceedings IAU Symposium No. 299, 2013
M. Booth, B. C. Matthews & J. R. Graham, eds.

doi:10.1017/S1743921313008910

Finding Asteroid belt Analogues with WISE

Rahul I. Patel and Stanimir Metchev

Department of Physics & Astronomy, Stony Brook University,
100 Nicolls Rd,
Stony Brook, NY
11794-3800

Abstract. We present the detection of stars with infrared (IR) excesses attributed to circumstellar debris disks from the WISE All-Sky Survey at the WISE 12 and 22 μm bandpasses ($W3$ and $W4$, respectively). Excess flux at these wavelengths is significant because it traces material in the regions of terrestrial planet formation. We searched for debris disks by cross-matching Hipparcos main sequence stars with the All-Sky Data Release from WISE and seeking excess flux at $W3$ and $W4$. Our sample is confined to a volume of 75 pc around the sun, and outside the galactic plane ($|b| > 5°$). Debris disk-bearing stars were identified as 95%-confidence outliers in 2MASS/*WISE* color distributions, after checking for erroneous photometry and contamination from unrelated nearby objects.

Keywords. stars: circumstellar matter, statistics

1. Introduction and Cross-Match Overview

Numerous surveys have been conducted to search for dusty circumstellar debris disks around main-sequence stars in the last few decades by IRAS, the Spitzer Space Telescope, and ISO (e.g., Rhee *et al.* 2007, Bryden *et al.* 2009, Chen *et al.* 2006, Su *et al.* 2006, Habing *et al.* 2001). As of now, roughly 330 debris disks hosts are known within 75 pc, detected by the excess IR thermal emission produced by the dust in the system, above the expected photospheric flux of the star. The majority of these are long wavelength excesses ($> 30\mu m$), attributed to cold Kuiper-belt analogues.

Warm circumstellar dust (100–400 K) is more challenging to detect. Stars with this type of dust are interesting because the dust spans the frost line and habitable zone around a star. The Spitzer Space Telescope is responsible for detecting the majority of known warm dusty disks (e.g., Bryden *et al.* 2009, Su *et al.* 2009, Beichman *et al.* 2006), though like all pointed surveys, their limited coverage leads to biases in the overall completeness of the survey. The Wide-Field Infrared Survey Explorer (*WISE*; Wright *et al.* 2010) is particularly powerful in detecting these warm disks given its all-sky coverage at $W1, W2, W3$ and $W4$ bands (3.4, 4.6, 12 and 22 μm respectively). The contemporaneous coverage provides uniform 3–30μm photometry, which allows precise estimates of the stellar photospheres and hence detections of tenuous excesses from warm dust. Our study takes advantage of this, and we provide an unbiased and complete sample of stars with excess emission at $W3$ and $W4$ for main-sequence Hipparcos stars within 75 pc.

2. Excess Selection

We initially cross-matched all Hipparcos main-sequence stars within 120 pc and outside the galactic plane ($|b| > 5°$) to WISE sources. We used the larger 120 pc sample as a parent sample to accurately determine the photospheric behavior of stars in the *WISE* bands. We only used stars with a $B-V$ color between –0.1 to 1.3 mags and non-saturated

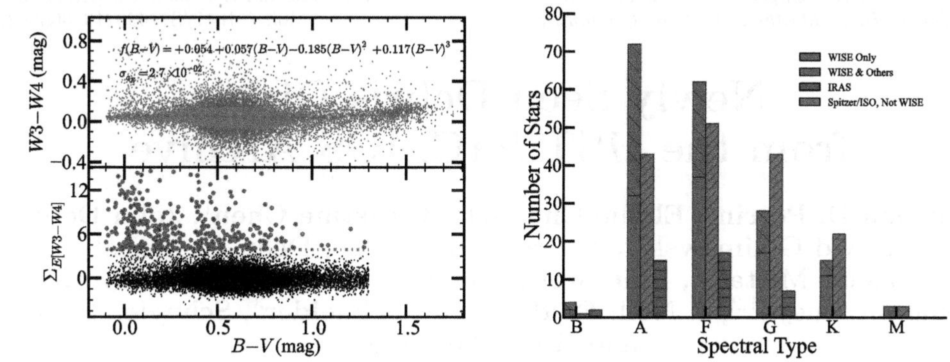

Figure 1. Left: An example *WISE* vs. $B-V$ color-color diagram for Hipparcos main-sequence stars within 120 pc, showing the correlation between IR and $B-V$ photospheric colors. We fit and subtract this dependence to eliminate photospheric biases in the excess. The residuals from the fit were used to assess the 95% confidence level above which stars were selected as excess candidates. **Right:** Spectral type distribution of stars with excesses between 10-30μm, comparing detections of past surveys to this study.

stars in the *WISE* $W2$, $W3$ and $W4$ bands. We also discarded objects with *WISE* flags indicating confusion, extended sources or variability. This ensured a clean sample of stars without any major systematic biases. We searched for excesses using $W2-W3$, $W2-W4$ and $W3-W4$ colors and also used 2MASS photometry for saturated stars in $W2$ (i.e., K_s-W3, K_s-W4). Stars with a 95% probability of having an excess were selected from a comparison to the intrinsic *WISE* or K_s-WISE color distribution as a function of stellar $B-V$ colors. The selection process can be seen in the left panel of Figure 1. Stars were selected this way for each of the considered color combinations and their 2MASS and *WISE* images were visually inspected for contamination from nearby objects.

3. Results

We detect 184 stars within 75 pc with IR excesses at 12 or 22 μm, 115 of which are new detections at these wavelengths. One hundred and two of the disks are new detections altogether and thus expand the number of known debris disks within 75 pc of the Sun by 36%. We find that 14% of A stars possess warm ($\gtrsim 100\ K$) excesses with $L_{dust}/L_* \gtrsim 10^{-4.5}$, while the incidence rate among solar type stars is 1%. A breakdown of the number of stars detected to have an IR excess can be seen in the right panel of Figure 1.

References

Beichman, C. A., *et al.* 2006a, ApJ, 652, 1674
Bryden, G., *et al.* 2009, ApJ, 705, 1226
Carpenter, J. M., *et al.* 2009, ApJS, 181, 197
Chen, C. H., *et al.* 2006, ApJS, 166, 351
Cutri, R. M., *et al.* 2012, Explanatory Supplement to the WISE All-Sky Data Release Products, Technical Report
Habing, H. J., *et al.* 2001, A&A, 365, 545
Rhee, J. H., Song, I., Zuckerman, B., & McElwain, M. 2007, ApJ, 660, 1556
Su, K. Y. L., *et al.* 2006, ApJ, 653, 675
Su, K. Y. L., *et al.* 2009, ApJ, 705, 314
Wright, E. L., *et al.* 2010, AJ, 140, 1868

Exploring the Formation and Evolution of Planetary Systems
Proceedings IAU Symposium No. 299, 2013
M. Booth, B. C. Matthews & J. R. Graham, eds.

doi:10.1017/S1743921313008922

Newly Seen Debris Disks from the HST NICMOS Archive

Marshall D. Perrin[1], Elodie Choquet[1], Christine Chen[1], John Debes[1], David Golimowski[1], J. Brendan Hagan[1], Dean C. Hines[1], Tushar Mittal[1,2], Margaret Moerchen[1], Mamadou N'Diaye[1], Laurent Pueyo[1,2], I. Neill Reid[1], Glenn Schneider[3], Schuyler Wolff[2] and Remi Soummer[1]

[1]Space Telescope Science Institute, 3700 San Martin Dr, Baltimore, MD, 21218 USA
[2] Johns Hopkins University, Baltimore MD, 21218 USA
[3] University of Arizona, Tucson AZ, 85721 USA

Abstract. By reprocessing the NICMOS coronagraphic archive using improved PSF subtraction methods, we have obtained new images of 5 debris disks, all previously unseen using classical PSF subtractions. Three of the disks are edge on and two appear to be ring like, one of which is extremely asymmetric.

Their stellar hosts are nearby, young F and G type stars (40-90 pc, 12–30 Myr), including one that is a close analog to the young sun at roughly the age at which terrestrial planets were assembling. This is a 25% increase in the sample of debris disks seen in scattered light. Analysis and modeling of the disk geometries is in process. Given these systems' youth, proximity, and brightness (V = 7.2 to 8.5), these will be superb targets for investigating planet formation, and are perfect targets for studies with GPI, SPHERE and JWST.

Keywords. circumstellar matter, planetary systems, techniques: image processing

1. Background

About 20 debris disks have previously been seen in scattered light, and a roughly similar number resolved in thermal emission. These imaged disks represent only a tiny portion of the thousands of nearby disks whose existence can be inferred from infrared excesses. But these imaged disks are tremendously valuable for advancing our understanding of nearby planetary systems. Due to the inherent degeneracies of SED-only modeling, imaging observations are essential to unambiguously determine the locations of debris and parent body belts and to ascertain the properties of the light-scattering dust grains.

2. Methods

By reprocessing the NICMOS coronagraphic archive using improved PSF subtraction methods, we have obtained new images of 5 debris disks, all previously unseen using classical PSF subtractions. The key elements of our reprocessing include:

(*a*) Careful recalibration and removal of NICMOS instrumental artifacts using optimal reference files for darks, flats, etc. ("LAPLACE" HST AR Program; PI Schneider)

(*b*) PSF subtraction using the KLIP algorithm (Soummer, Pueyo & Larkin 2012) applied to a large library of reference PSFs drawn from contemporaneous science programs. See the article on the ALICE pipeline by Choquet *et al.* in these proceedings.

(*c*) Disk images from multiple rolls were then merged using the iterative roll subtraction algorithm of Krist *et al.* 2010 to further reduce time-variable residuals.

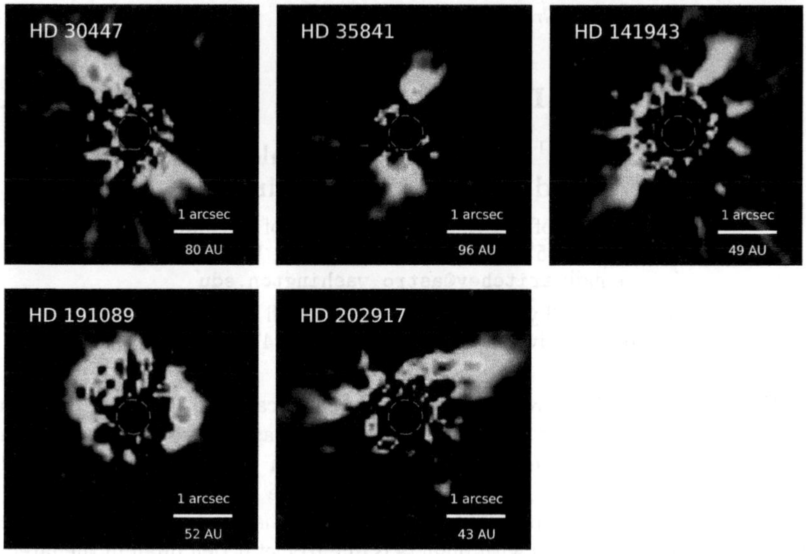

Figure 1. HST NICMOS F110W images of five debris disks, all on identical angular scale and oriented to have north up. These images show the combination of 2 rolls for each target; the disks are also seen consistently in the individual rolls but at lower S/N. Two of these disks have prior imaging detections at other wavelengths (HD 191089 at 18 μm, Churcher *et al.* 2011; HD 202917 in unpublished ACS GTO observations, Krist 2007) and in those cases properties such as disk position angles and inclinations are broadly consistent.

Compared to the prior existing sample of HST-imaged disks, these newly seen disks are preferentially young (<30 Myr) and near solar type (G7-F2). This is the result of selection biases in the NICMOS surveys from which we drew the data, which targeted young nearby systems for planet and disk searches.

3. Results

See Figure 1. **HD 30447**, a F3V star at 80 pc and member of the Columba Association, is seen to have a roughly edge-on disk, about 50% brighter on its northeast side. **HD 38541**, another F3V member of Columba, has the faintest and most compact disk in our sample, detected at $r < 1.5''$. **HD 141943**, a G2V star at 49 pc, is the first solar-type star < 50 Myr in age to have a debris disk imaged in scattered light. It has the lowest IR excess in our sample, $L_{IR}/L_\star \sim$1.2e-4. An edge on disk is seen from 0.7–2.6$''$ = 35–125 AU in projected separation. **HD 191089**, an F5V star at 52 pc and member of the β Pic moving group, shows an intermediate inclination ($i \sim 30°$) disk with one side significantly brighter than the other. **HD 202917**, a G7V star at 43 pc and member of the Tucana-Horologium association, shows a striking and dramatic asymmetric arc, suggestive of a partial ring seen near $i \sim 70°$, which is precisely aligned with the similar one-sided "needle" detected in prior ACS GTO observations (Krist *et al.* 2007).

Acknowledgements: Support for HST AR 12652 was provided by NASA through a grant from STScI, which is operated by AURA under NASA contract NAS5-26555.

References

Churcher, L., Wyatt, M., & Smith, R. 2011, *MNRAS*, 410, 2
Krist, J. E., Stapelfeldt, K. R., Bryden, G., *et al.* 2010, *AJ*, 140, 1051
Krist, J. 2007, Proceedings of the "Spirit of Lyot" conference.
Soummer, R., Pueyo, L. & Larkin, J. 2012, *ApJ*, 755, L28

Exploring the Formation and Evolution of Planetary Systems
Proceedings IAU Symposium No. 299, 2013
M. Booth, B. C. Matthews & J. R. Graham, eds.

doi:10.1017/S1743921313008934

Abundances in Stars with Debris Disks

Adam M. Ritchey[1], Guillermo Gonzalez[2], Myra Stone[1] and George Wallerstein[1]

[1]Department of Astronomy, University of Washington,
Box 351580, Seattle, WA 98195, USA
email: aritchey@astro.washington.edu

[2]Department of Physics and Astronomy, Ball State University,
2000 W. University Ave., Muncie, IN 47306, USA

Abstract. We present preliminary results of a detailed chemical abundance analysis for a sample of solar-type stars known to exhibit excess infrared emission associated with dusty debris disks. Our sample of 28 stars was selected based on results from the Formation and Evolution of Planetary Systems (FEPS) *Spitzer* Legacy Program, for the purpose of investigating whether the stellar atmospheres have been polluted with planetary material, which could indicate that the metallicity enhancement in stars with planets is due to metal-rich infall in the later stages of star and planet formation. The preliminary results presented here consist of precise abundances for 15 elements (C, O, Na, Mg, Al, Si, S, Ca, Sc, Ti, V, Cr, Fe, Co, and Ni) for half of the stars in our sample. We find that none of the stars investigated so far exhibit the expected trend of increasing elemental abundance with increasing condensation temperature, which would result from the stars having accreted planetary debris. Rather, the slopes of linear least-squares fits to the abundance data are either negative or consistent with zero. In both cases, our results may indicate that, like the Sun, the debris disk host stars are deficient in refractory elements, a possible signature of terrestrial and/or gas giant planet formation.

Keywords. stars: abundances, planetary systems

---

1. Sample Selection and Instrumentation

We initially selected 23 stars from those identified by Hillenbrand *et al.* (2008) as having "Tier 1" or "Tier 2" debris disks that had declinations above −30° and spectral types between F5 and K5. To this list, we added 12 additional stars with infrared excesses from Carpenter *et al.* (2009). Each star was observed with the echelle spectrograph of the 3.5 m telescope at Apache Point Observatory, which provides nearly complete optical wavelength coverage at a resolving power of $R \approx 31,500$. The exposure times were designed to yield signal-to-noise ratios of approximately 300:1. After reducing the data, we found that 7 stars displayed broad, blended absorption features, making them unsuitable for detailed chemical abundance analyses. Thus, these stars were dropped from the sample, leaving a total of 28 stars. Here, we present detailed results for 14 of the 28.

2. Detailed Chemical Abundance Analysis

Basic stellar parameters were determined spectroscopically through the usual excitation and ionization analysis and abundances were derived for 15 elements using the latest version of the Local Thermodynamic Equilibrium spectral analysis code MOOG. Our line list includes 53 Fe I lines, 6 Fe II lines, and 50 lines from other atomic species carefully chosen to be free from blends and telluric interference. Results of our spectroscopic analysis for the initial 14 stars are given in Table 1 along with the dust parameters listed in Hillenbrand *et al.* (2008).

Table 1. Derived Stellar Parameters, Slopes, and Dust Parameters from Hillenbrand *et al.*

Star	$T_{\rm eff}$ (K)	log g	[Fe/H]	Slope[1]	$T_{\rm dust}$ (K)	$R_{\rm dust}$ (AU)
HD 377	5701 ± 57	4.01 ± 0.06	-0.01 ± 0.05	-25.31 ± 3.43	58	23
HD 6963	5422 ± 53	4.23 ± 0.07	-0.19 ± 0.04	-20.08 ± 3.84	56	18
HD 8907	6301 ± 62	4.37 ± 0.06	$+0.00 \pm 0.05$	-10.42 ± 3.37	48	49
HD 17925	5068 ± 55	4.13 ± 0.11	$+0.06 \pm 0.03$	-12.92 ± 5.67	110	4
HD 25457	6260 ± 120	4.1 ± 0.2	$+0.11 \pm 0.09$	-8.66 ± 5.62	70	23
HD 70573	5761 ± 46	4.30 ± 0.08	-0.05 ± 0.03	-4.69 ± 2.81	41	35
HD 104860	5987 ± 94	4.35 ± 0.08	$+0.07 \pm 0.07$	-8.03 ± 3.32	46	42
HD 107146	5848 ± 62	4.40 ± 0.06	$+0.00 \pm 0.05$	-7.44 ± 2.93	52	30
HD 122652	6208 ± 42	4.46 ± 0.07	$+0.12 \pm 0.03$	-9.02 ± 2.06	55	31
HD 145229	5796 ± 57	4.02 ± 0.13	-0.18 ± 0.04	-13.97 ± 3.55	54	26
HD 150706	5857 ± 49	4.46 ± 0.06	-0.05 ± 0.04	-13.88 ± 3.01	58	23
HD 201219	5537 ± 51	4.33 ± 0.07	$+0.09 \pm 0.04$	-3.21 ± 3.66	53	23
HD 204277	6373 ± 88	4.6 ± 0.1	$+0.11 \pm 0.06$	-1.80 ± 3.53	<50	43
HD 206374	5520 ± 44	4.40 ± 0.08	-0.11 ± 0.03	-11.92 ± 2.89	<74	12

Note: [1]Slope of weighted linear least-squares fit to [X/H] versus $T_{\rm cond}$ (in units of 10^{-5} dex $\rm K^{-1}$).

3. Summary and Conclusions

We have completed a detailed chemical abundance analysis for 14 of the 28 debris disk stars in our sample, deriving basic stellar parameters and abundances for 15 elements.† All stars examined so far exhibit flat or negative slopes in plots of abundance ([X/H]) versus condensation temperature ($T_{\rm cond}$; taken from Lodders 2003), suggesting that the stellar atmospheres have not accreted refractory-rich planetary material. Since the Sun has been found to be deficient in refractory elements relative to solar twins (Meléndez *et al.* 2009), a result interpreted as a signature of terrestrial planet formation, and since these stars have compositions very similar to the Sun (and in some cases show even more of a deficiency in refractories), our results may indicate that planet formation has indeed occurred in these systems, and that the refractory elements are locked up in terrestrial planets or giant planet cores.

4. Future Work

Beyond completing the analysis for the remaining 14 stars in our sample, we will add abundance measurements for Zn since the derived slopes are heavily dependent on the abundances of the few elements with relatively low condensation temperatures (i.e., C, O, and S). We will also examine the impact of Non-Local Thermodynamic Equilibrium effects on the C and O abundances for the stars in our sample because, again, these effects could influence the slopes in some non-negligible way.

References

Carpenter, J. M., Bouwman, J., Mamajek, E. E., Meyer, M. R., Hillenbrand, L. A., Backman, D. E., Henning, T., Hines, D. C., Hollenbach, D., Kim, J. S., Moro-Martín, A., Pascucci, I., Silverstone, M. D., Stauffer, J. R., & Wolf, S. 2009, *ApJS*, 181, 197

Hillenbrand, L. A., Carpenter, J. M., Kim, J. S., Meyer, M. R., Backman, D. E., Moro-Martín, A., Hollenbach, D. J., Hines, D. C., Pascucci, I., & Bouwman, J. 2008, *ApJ*, 677, 630

Lodders, K. 2003, *ApJ*, 591, 1220

Meléndez, J., Asplund, M., Gustafsson, B., & Yong, D. 2009, *ApJ*, 704, L66

† Complete results for the full sample will be presented in Ritchey *et al.* (in preparation).

Exploring the Formation and Evolution of Planetary Systems
Proceedings IAU Symposium No. 299, 2013
M. Booth, B. C. Matthews & J. R. Graham, eds.

doi:10.1017/S1743921313008946

Planet signatures and Size Segregation in Debris Discs

Philippe Thébault[1]

[1]LESIA, Observatoire de Paris, F-92195, Meudon Principal Cedex, France

1. Context, Objective and Method

The response of a debris disc to a planetary perturber is the result of the complex interplay between gravitational effects, grain collisions and stellar radiation pressure (Stark & Kuchner 2009). We investigate to what extent this response can depart from the pure gravitational case when including grain collisional production and radiation pressure. We use the DyCoSS code (Thébault 2012), designed to study the coupled effect of collisions and dynamics for systems at steady state with one perturbing body. We focus on two outcomes: the 2D surface density profile of the disc+planet system, and the way the Particle Size Distribution (PSD) is spatially segregated within the disc. We consider two set-ups: 1) a narrow ring with an exterior "shepherding" planet, and 2) an extended disc in which a planet is embedded. For each case, the planet mass and orbit are explored as free parameters, and an unperturbed "no-planet" case is also considered. Another parameter is the disc's collisional activity, as parameterized by its optical depth τ.

2. Surface Density Profiles

We find that collisions always significantly damp planet-induced spatial structures. For the extended disc with embedded planet case, the planet's signature remains visible in face-on images if $M_{planet} \gtrsim M_{Saturn}$. If the system is seen edge-on, however, even a Jovian planet only leaves weak asymmetries in a collisionally active disc, even though some planet-induced signatures might be observable with high-resolution observations. For the narrow ring with shepherding planet case, we find that the planet cannot prevent the coupled effect of collisions and radiation pressure from injecting small dust grains in the dynamically "forbidden" region exterior to the ring (see Fig.1). For most cases, the radial density profile beyond the main ring closely mimics that in the absence of any planet, i.e., $\Sigma \propto r^{-1.5}$. Within the main ring itself, an exterior planet on a circular orbit leaves precessing structures that can be used to indirectly infer its presence. For an eccentric planet, the ring becomes elliptic and the pericentre glow effect is visible despite of collisions and radiation pressure, but detecting such features in observed discs is not an unambiguous indicator of the presence of an outer planet (Thébault *et al.*, 2012)

3. Spatial Segregation of the Size Distribution

We find that PSDs are almost always spatially segregated, even in the unperturbed no-planet case. Another surprising result is that, for most cases, the size dominating the disc's geometrical cross section is *not* that of the smallest bound grains. The only case where a standard PSD of $dN \propto s^{-3.5}ds$ is observed is for the unperturbed narrow ring case, but only within the ring itself. Beyond it, the size distribution is peaked around a size $s*$ given by $r_{ring} = (1 - 2\beta(s*)).r$ (β is the ratio of the radiation pressure to stellar

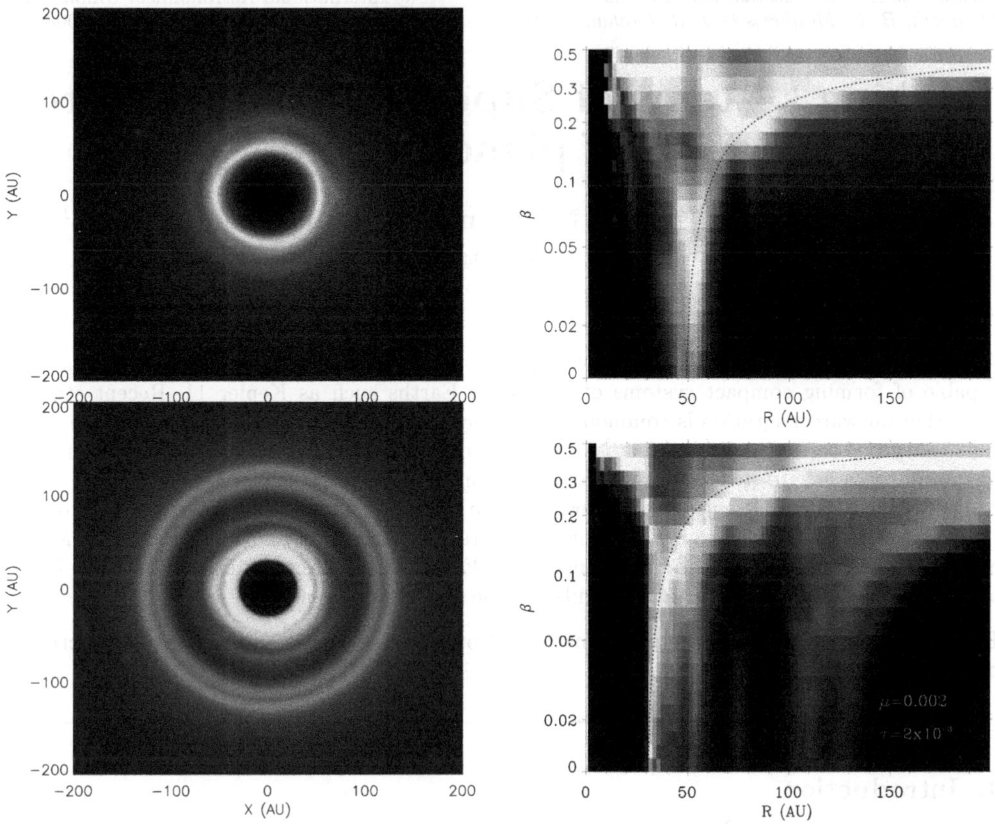

Figure 1. Steady state reached for (*upper panel*) the narrow ring with shepherding planet case and (*lower panel*) the extended disc with embedded planet case. For both cases: $\mu = M_p/M_{star} = 0.002$, $a_{pl} = 75\text{AU}$, $e_{pl} = 0$ and τ(optical depth)= 0.002. *Left panel*: normalized surface density profile viewed face-on. *Right panel*: Azimuthally averaged radial distribution of the geometrical cross-section as a function of grain sizes, as parametrized by their β value.

gravity). For an unperturbed wide disc, we find that, at radial location r, the PSD assumes a standard profile in $\sim dN \propto s^{-3.5}ds$ down to a size $s*$ given by $r_{inner-edge} = (1 - 2\beta(s*)).r$. For planet-perturbed discs, PSDs can be very complex. For instance, the L4 and L5 points are mostly populated by large grains, while dynamically unstable regions are populated with small, $\beta > 0.15$ grains (Thébault & Kral, 2013).

References

Stark, C., Kuchner, M. 2009, *ApJ*, 707, 543
Thébault, P. 2012, *A&A*, 537, 65
Thébault, P., Kral, Q., & Ertel, S. 2012, *A&A*, 547, 92
Thébault, P., Kral, Q., & Augereau, J.-C. 2013, *A&A*, in press

Exploring the Formation and Evolution of Planetary Systems
Proceedings IAU Symposium No. 299, 2013
M. Booth, B. C. Matthews & J. R. Graham, eds.

doi:10.1017/S1743921313008958

Making systems of Super Earths by inward migration of planetary embryos

Christophe Cossou[1,2], Sean N. Raymond[1,2] and Arnaud Pierens[1,2]

[1]Univ. Bordeaux, LAB, UMR 5804, F-33270, Floirac, France.
[2]CNRS, LAB, UMR 5804, F-33270, Floirac, France

Abstract. Using N-body simulations with planet-disk interactions, we present a mechanism capable of forming compact systems of hot super Earths such as Kepler 11. Recent studies show that outward migration is common in the inner parts of radiative disks. However we show that two processes naturally tip the balance in favor of inward migration. First the corotation torque is too weak to generate outward migration for planetary embryos less massive than $4\,M_{\oplus}$. Second, system of multiple embryos generate sustained non-zero eccentricities that damp the corotation torque and again favor inward migration. Migration and accretion of planetary embryos in realistic disks naturally produce super Earths in resonant chains near the disk inner edge. Their compact configuration is similar to the observed systems.

Keywords. Planets and satellites: formation, Protoplanetary disks, Planet-disk interactions, planetary systems, Methods: numerical

1. Introduction

Super Earths ($1-10\,M_{\oplus}$) are common (Mayor *et al.* 2011; Howard *et al.* 2012) even if none exist in our own solar system. They tend to be found in systems of multiple Super Earths in close-in, compact orbital configurations. For example, the Kepler 11 system contains five planets within the orbit of Mercury, each more massive than Earth (Lissauer *et al.* 2011).

Here we show how that gas disk-driven orbital migration naturally produces systems similar to the observed ones. We use N-body simulations that include torques from a 1-D gaseous disk to show how a system of planetary embryos starting at several AU embedded naturally migrates inward and accretes into a compact system of hot Super Earths. Some embryos can also grow and remain trapped on more distant orbits and presumably become giant planet cores.

2. Methods

We start from a 1D protoplanetary disk model, with parameters listed in Table 1. The key parameter is the surface density profile, defined as a power law in radial range $0.1 < R < 100\,\mathrm{AU}$:

$$\Sigma(R) = \Sigma_0 \left(\frac{R}{1\,\mathrm{AU}}\right)^{-d} \tag{2.1}$$

The inner edge of the disk is smoothed with a tanh function on a length scale of $H(R_{\rm in})$, where H is the scaleheight of the disk, so the density goes to zero at the inner edge. From the surface density profile, we self-consistently compute the temperature T, thermal diffusivity χ, scaleheight H and optical depth profiles τ. To calculate the temperature we use the following energy equation :

$$0 = -C_{\mathrm{BB}} + H_{\mathrm{en}} + H_{\mathrm{irr}} + H_{\mathrm{vis}} \tag{2.2}$$

with

$$C_{\mathrm{BB}} = 2\sigma \frac{T^4}{\frac{3}{8}\tau + \frac{\sqrt{3}}{4} + \frac{1}{4\tau}} \qquad H_{\mathrm{en}} = 2\sigma {T_{\mathrm{en}}}^4$$

$$H_{\mathrm{irr}} = 2\sigma {T_\star}^4 \frac{{R_\star}^2}{r^2}(1-\varepsilon) * \left[0.4\frac{R_\star}{r} + r\frac{\mathrm{d}}{\mathrm{d}r}\left(\frac{H}{r}\right)\right] \qquad H_{\mathrm{vis}} = \frac{9}{4}\nu\Sigma\Omega^2$$

C_{BB} is the disk's black body cooling and H_{en} is the black body heating by the disk envelope. H_{vis} is the viscous heating. H_{irr} is the heating from stellar irradiation (Chiang & Goldreich 1997; Menou & Goodman 2004), σ the Stephan-Boltzmann constant, ν the viscosity of the disk, $T_\star$ and $R_\star$ the temperature and radius of the central star respectively, Ω the angular speed in the disk at a given position and ε the disk's albedo. Starting at the outer edge of the disk, where we impose that $T = 10\,\mathrm{K}$, we retrieve the temperature profile by solving eq. (2.2) numerically.

We use the formulae of Paardekooper *et al.* (2011) to compute the torque exerted by our 1D disk on a planet of a given mass at a given orbital radius. The main differences with the model described in this paper are :

- The temperature profile is not a power law in our model, but instead a local power law between each point of a table of several hundred points spaced in orbital distance
- the scaleheight of the disk, defined as $H = \frac{1}{\Omega}\sqrt{\frac{k_B T}{\mu m_H}}$, is computed following the temperature profile, with k_B the Boltzmann constant, μ the mean molecular weight and m_H the mass of an hydrogen atom.

We implement type I eccentricity and inclination damping following Cresswell & Nelson (2008). We also include an eccentricity-migration feedback whereby the corotation torque is weakened for eccentric orbits (Bitsch & Kley 2010), using the equations from (Cossou *et al.* 2013).

$b/h = 0.6$	$\gamma = 7/5$	$\mu = 2.35$	$\alpha = 5 \cdot 10^{-3}$	$T_\star = 5700\,\mathrm{K}$	$R_\star = 4.65 \cdot 10^{-3}\,\mathrm{AU}$
Disk Albedo = 0.5	Disk : R $\in$ [0.1; 100] AU		$\Sigma(R) = 300 \cdot R^{-1/2}\,\mathrm{g/cm^2}$		

Table 1. Parameters of the disk. In addition to those parameters, note that the opacities were retrieved from the opacity table of Huré (2000). The viscosity is modeled via the α-prescription (Shakura & Sunyaev 1973).

Disk forces were added to the hybrid version of the `Mercury` integrator (Chambers 1999). Drag and migration forces are not applied to objects inside the inner cavity (inside 0.1 AU). Collisions were treated as inelastic mergers.

3. Mechanism

The right panel of Figure 1 shows the disk's migration map. Two zones of convergent migration exist. The first convergence zone causes all embryos between 0.2 and 0.8 AU to migrate toward 0.5 AU. The second causes embryos between 0.9 and ~100 AU to migrate toward ~ 15 AU. However, multiple planet systems do not actually migrate to convergence zones like isolated planets. Rather, they become trapped in chains of mean motion resonances. The resonant configurations sustain eccentricities large enough to attenuate the corotation torque (Bitsch & Kley 2010). The system stabilizes in an equilibrium

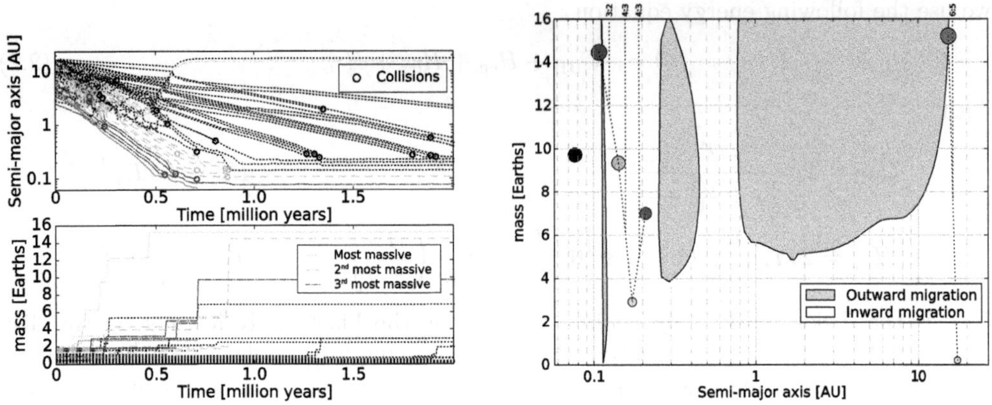

Figure 1. A simulation that forms a compact system of hot Super Earths. **Left**: The orbital evolution (top) and mass growth (bottom) of the embryos. **Right**: The final state of the system. Black lines represent zero-torque zones where isolated planets should stop migrating.

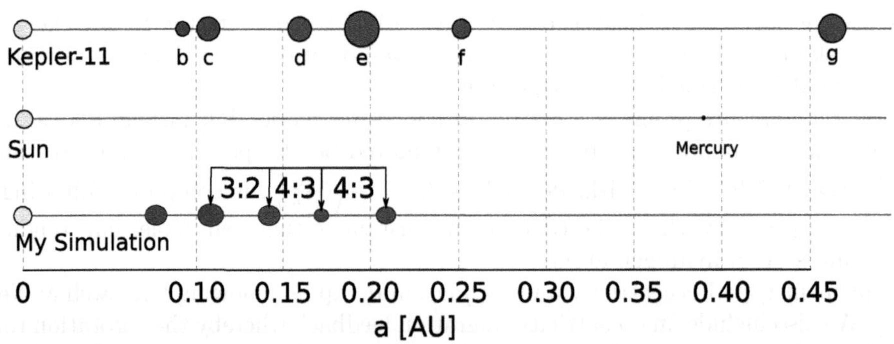

Figure 2. Final orbital configuration of the inner part of the simulation compared with the Solar System and Kepler 11.

position in the disk where the sum of all the torques felt by the planets cancels out (Cossou *et al.* 2013).

The left panel of Figure 1 shows the dynamics and accretion of a system of embryos embedded in the disk. The embryos comprised $60\,\mathrm{M}_\oplus$ in total, and started with masses randomly chosen from $0.1 - 2\,\mathrm{M}_\oplus$ spaced from $2 - 17\,\mathrm{AU}$, and the integration lasted for 10 Myr. At early times, embryos migrate inward because their masses are below the $\sim 4\,\mathrm{M}_\oplus$ threshold for outward migration (see right panel of Figure 1). Because of their different masses the embryos do not all migrate at the same rate. This leads to close encounters and collisions. Almost all of the embryos migrate inward and complete their growth close to the inner edge of the disk. The final configuration of planets is a compact resonant system. The most massive planet grows fast enough to reverse its migration and stabilizes at 15 AU. While migrating outward, it traps a lower mass planet in 6:5 resonance. The low-mass planet is pushed by the more massive one, and the two-planet system is effectively ruled by the zero-migration line of the more massive planet.

The embryos ended up at the inner edge of the disk for two reasons. First, objects less massive than $\sim 4\,\mathrm{M}_\oplus$ simply cannot migrate outward (see Figure 1). During its inward migration, an embryo must accrete quickly if it is to enter a zone of outward migration, as was the case for the outer massive planet. Second, embryos that do migrate outward may encounter other large embryos and become trapped in resonance, leading to sustained non-zero eccentricities that damp the corotation torque (Bitsch & Kley 2010) and tip the balance toward inward migration (Cossou *et al.* 2013). In this simulation the first mechanism dominated but both mechanisms can be important. The system was held up because a huge positive torque exists close to the inner edge due to the sudden decrease of the surface density profile (Masset *et al.* 2006). Our results are thus similar to those of Terquem & Papaloizou (2007) even though outward migration does occur in our modeled disk.

Figure 2 shows the inner simulated system compared to our Solar System and the Kepler-11 system. The simulated system has similar masses to Kepler-11 but its orbital configuration is even more compact, although there are no planets in the simulation between 0.25 and 15 AU. The four outer simulated planets are in a resonant chain but the innermost planet is not. This is because the innermost planet was pushed into the disk's inner gas-free cavity where it lacked the energy dissipation required for efficient resonant trapping.

4. Conclusion

In our disk model, a compact system of hot Super Earths is formed by migration and accretion of planetary embryos. Inward migration is favored. Low-mass embryos naturally migrate inward. Outward migration is stalled or even reversed by the eccentricities sustained through resonances (Cossou *et al.* 2013). In essence, the presence of multiple large embryos leads to non-zero eccentricities such that they no longer follow the migration map (right panel of Figure 1).

This formation mechanism for hot Super Earths is robust against a much wider range of disk parameters than the one presented here. We note that the survival of the simulated hot Super Earths may depend on the detailed structure of the inner edge of the gas disk, which provides the positive torque needed to balance the inward-driven system. Their survival also likely depends on how the disk evolves (Horn *et al.* 2012) and especially on how it dissipates. We plan to take these effects, and others such as stochastic forcing from turbulence, into account in future work.

References

Bitsch, B. & Kley, W. 2010, *A&A*, 523, A30
Chambers, J. E. 1999, *MNRAS*, 304, 793
Chiang, E. I. & Goldreich, P. 1997, *ApJ*, 490, 368
Cossou, C., Raymond, S. N., & Pierens, A. 2013, *A&A*, 553, L2
Cresswell, P. & Nelson, R. P. 2008, *A&A*, 482, 677
Horn, B., Lyra, W., Mac Low, M.-M., & Sándor, Z. 2012, *ApJ*, 750, 34
Howard, A. W., Marcy, G. W., Bryson, S. T., *et al.* 2012, *ApJS*, 201, 15
Huré, J.-M. 2000, *A&A*, 358, 378
Lissauer, J. J., Fabrycky, D. C., Ford, E. B., *et al.* 2011, *Nature*, 470, 53
Masset, F. S., Morbidelli, A., Crida, A., & Ferreira, J. 2006, *ApJ*, 642, 478
Mayor, M., Lovis, C., Pepe, F., Ségransan, D., & Udry, S. 2011, Astronomische Nachrichten, 332, 429
Menou, K. & Goodman, J. 2004, *ApJ*, 606, 520
Paardekooper, S.-J., Baruteau, C., & Kley, W. 2011, *MNRAS*, 410, 293
Shakura, N. I. & Sunyaev, R. A. 1973, *A&A*, 24, 337
Terquem, C. & Papaloizou, J. C. B. 2007, *ApJ*, 654, 1110

Discussion

SCHLICHTING: What determines the shape of the "whale"?

COSSOU: The shape of the whale (regions of the migration map where the torque is positive) are mainly determined by the non-saturated entropy related part of the corotation torque, i.e a torque that is hugely sensitive on the temperature profile. Thus, any change in temperature will be seen directly in the migration map. But the shape itself comes from whether or not you have a non saturated corotation torque or not. Inside the "whale", the torque is not saturated, the migration is positive, and the following inequations are fullfilled $t_{\text{U-turn}} < t_{\text{diff}} < t_{\text{libration}}$. In the case of the whale, we always have $t\text{visc} > t_{\text{rad}}$. The lower limit in mass of the whale is determined by the radiative diffusion timescale. When $t_{\text{U-turn}} > t_{\text{rad}}$, we are outside the whale because the corotation torque becomes linear. The upper mass limit of the whale comes when $t_{\text{lib}}/2 > t_{\text{visc}}$, thus the corotation torque saturates. But, the head of the whale does not exactly follow the equations $t_{\text{U-turn}} = t_{\text{rad}}$ and $t_{\text{lib}}/2 = t_{\text{visc}}$ because the commensurabilities between the main negative torque (Lindblad torque) and the main positive torque (horseshoe entropy related part of the corotation torque) changes due to transition from active to passive region of the disk. Past 4 AU, when the irradiation becomes dominant, Lindblad and corotation torques changes but not at the same rate. The final shape of the head of the whale is the line where the corotation torque (fully unsaturated or not) exactly balance the differential Lindblad torque.

Exploring the Formation and Evolution of Planetary Systems
Proceedings IAU Symposium No. 299, 2013
M. Booth, B. C. Matthews & J. R. Graham, eds.

doi:10.1017/S174392131300896X

Protostellar Disks, Planet Traps, and the Origins of Exoplanetary Systems

Ralph E. Pudritz[1,2] and Yasuhiro Hasegawa[3]

[1]Department of Physics and Astronomy, McMaster University,
Hamilton, ON L8S 4M1, Canada

[2]Origins Institute, McMaster University, Hamilton, ON L8S 4M1, Canada
email: pudritz@mcmaster.ca

[3]EACOA fellow, Institute of Astronomy and Astrophysics,
Academia Sinica (ASIAA), Taipei 10641, Taiwan
email: yasu@asiaa.sinica.edu.tw

Abstract. The mass-semimajor axis diagram for exoplanets is populated by at least three distinct planetary populations: hot Jupiters at small orbital radii, more massive Jovian planets gathered at about 1 AU, and a rapidly growing population of SuperEarths at short periods. Our work shows that low mass and rapidly migrating planetary cores get trapped at disk inhomogeneities, where strong density or thermal gradients exist (namely dead zone boundaries, ice lines, and disk heating transition regions). Planet growth and movement occur at rates dictated by planetary accretion, and the slow radial inward motion of the traps due to falling disk accretion rates during disk evolution. By combining the theory of traps in evolving disks with standard ideas about how protoplanets accrete, we develop evolutionary tracks of how planets evolve in the mass- semimajor axis diagram. Our models account for the planetary "pile-up" at 1AU, the origin of SuperEarths and hot Jupiters, and the relative scarcity of Jovian planets at large distances.

Keywords. accretion disks, (stars:) planetary systems: formation, (stars:) planetary systems: protoplanetary disks.

---

1. Introduction

Arguably the most important means that we have for exploring the origin of exoplanets and planetary systems is the mass-semimajor axis diagram. The discovery of large numbers of exoplanets by various kinds of surveys over the last decade reveals that several distinct kinds of planetary populations exist in this diagram (see Chiang & Laughlin 2013). The original discovery of the population of hot Jupiters at a relative orbit of Mercury and masses that are somewhat subJovian (denoted as Zone 1) has been augmented by several other populations: a region that is quite depleted of gas giants (Zone 2), more massive Jovian planets gathered around a semimajor orbital axis of 1 AU (Zone 3), an exciting new population of imaged young Jovian planets at large orbital radii $a > 10AU$ (Zone 4), and a strongly emerging population of SuperEarths of mass $1 - 10M_E$ and hot Neptunes close into their host stars (Zone 5). We highlight these populations in Figure 1 (Hasegawa & Pudritz (2013)). The location of planets in this diagram is largely a consequence of the formation and migration of planets in protostellar disks.

Theories of planet-disk interaction suggest that many parameters may control the migration and growth of planets in disks. In the context of the core-accretion model for planet formation, the population synthesis approach is often used to try to match populations seen in the mass-semimajor axis diagram. By varying a large number of disk parameters in this Monte Carlo approach, it was shown that Type I migration in

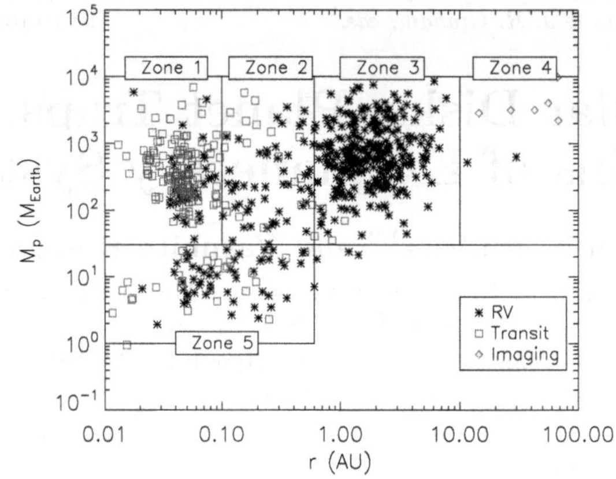

Figure 1. Mass - semimajor axis relation; data taken from exoplanets.eu. Planets observed by RV technique, stars; transits, squares; and direct imaging, diamonds. We denote Zone 1 - hot Jupiters, Zone 2 - exclusion of gas giants, Zone 3 - pile up of gas giants, Zone 4 - distant Imaged planets, and Zone 5 - SuperEarths and hot Neptunes. Adapted from Hasegawa & Pudritz (2013)

homogenous disks occurs far too rapidly ($\sim 10^5$ years for planets with $\sim 1M_\oplus$ at ~ 1 AU) to account for the data (eg. Ida & Lin 2008, Mordasini *et al.* 2009). A physical solution to this dilemma is that disks are quite inhomogenous structures and that rapidly migrating planetary cores become trapped in narrow regions with marked changes in density and temperature, - such as dead zone boundaries, ice lines, and heat transitions - that are regions of net zero torque (Masset *et al.* (2006), Hasegawa & Pudritz 2011). Planet traps, in turn, move to smaller disk radii as the disk accretion rate falls, which is on a several Myr timescale. The question then is - what kinds of evolutionary tracks connect progenitor protoplanetary cores, born and developed on their host planet traps, to their final positions in this diagram? We show that the combination of this slower radial inward movement of traps in evolving disks, coupled with the standard picture of accretion onto protoplanetary cores, provides an excellent explanation of these populations.

2. Evolutionary Tracks in the Mass-Semimajor Axis Diagram

The growth in mass of an accreting planetary core can be combined with the radial inward motion of its host planet trap in its evolving disk, to create a unique evolutionary track in the mass-semimajor axis diagram. Analytic solutions of the motion of the traps were worked out in Hasegawa & Pudritz 2012 (see also Hasegawa & Pudritz, these Proceedings). The three types of traps have different power law dependences on the mass accretion rate through the disk. The disk accretion rate is in turn driven by a combination of disk viscosity (cf. Lynden-Bell & Pringle 1974), which gradually clears out disk mass onto the star over some accretion time scale, coupled with the eventual rapid loss of the disk due to photoevaporation (Adams *et al.* 2004). We therefore model this process as depending upon two key parameters: the dimensionless disk accretion rate η_{acc} (in units of a fiducial TTS accretion rate $3 \times 10^8 M_\odot yr^{-1}$), and the dimensionless disk depletion time η_{dep} (in units of a fiducial disk lifetime, or depletion time, of $\simeq 10^6$ yr).

The increase in planetary mass due to accretion is well known to follow 3 stages (e.g. Pollack *et al.* 1996): the formation of rocky cores through runaway and oligarchic growth (Stage I) lasting $\simeq 10^5$ yr, the subsequent slow gas accretion of the cores and formation of envelopes surrounding them (Stage II) lasting $\simeq 10^6$ yr, and the rapid collapse of the envelopes and runaway gas accretion onto their cores once the core mass exceeds a critical

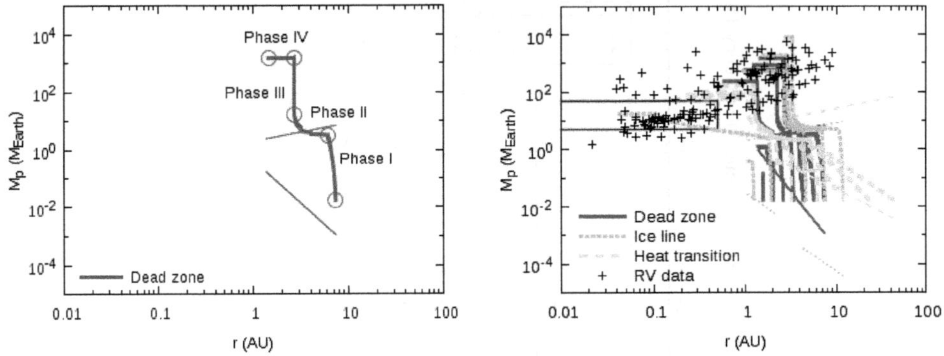

Figure 2. (a) Evolutionary track for an initial seed core being pulled along through the disk by a dead zone trap; (b) full set of tracks for the 3 types of planet traps; dead zones, the water ice line, and the heat transition. Adapted from Hasegawa & Pudritz 2012.

mass M_{cr} of the order of a few Earth masses (Stage III). In order to model these three physical processes, we adopt the formulation of Ida & Lin (2004). The whole accretion process may take a few Myr to complete.

In Figure 2(a) we show a typical evolutionary track of a small initial progenitor core (hundredth of an Earth mass) as it accretes and is pulled to smaller disk radii by a dead zone trap. The evolution undergoes 4 recognizable phases. In Phase I, runaway oligarchic growth of the core occurs on a much faster time scale than the disk, and trap, evolution time of 10^6 yr. Hence the track proceeds in a strongly vertical direction of increasing mass with only a slight inward shift of orbital radius. In phase II, slow gas accretion dominates over a long time span so that the track barely ascends in mass as the dead zone trap moves to smaller disk radii. In phase III, the core has reached its critical value and the planet accretes most of its mass very rapidly through runaway gas accretion - hence the extensive and nearly vertical motion in the diagram with little radial inward motion of the trap. At the end of this phase, the planet has achieved its gap opening mass and ceases to accrete. It now enters a slower, Type II migration phase wherein the planet migrates inwards with no gas accretion. This phase ends when the disk is finally dispersed by photoevaporation (Hasegawa & Pudritz (2012)). In Figure 2(b) we show a whole collection of tracks calculated for initial small cores distributed at different initial disk radii and carried along by the three respective traps. The tracks cover all of the populations in the earlier data set (based only on RV measurements), including a region (rectangle) that was claimed to be a "planetary desert" in the Ida and Lin picture.

3. Planet Formation Frequencies and Planetary Populations

We computed a large number of tracks (several hundred) for a population of disks that are controlled by just two parameters: the dimensionless disk accretion rate η_{acc} and disk depletion time η_{dep} (Hasegawa & Pudritz (2013). We define the planet formation frequency (PPF) for each zone as the fraction of the total tracks leading to zone i, integrated over all values of these two variables. Planets formed in different traps populate different types of target zones: Jovian planets around 1AU are largely fed by dead zone and ice line traps, whereas the closer in populations are fed by a combination of dead zone and heat transition traps. The PPFs of hot Jupiters are 1.6 %, those of gas giants around 1 AU are 24 % (the largest populations), and those of low-mass planets in tight orbits are 15 % . The SuperEarths are well explained as being failed cores in the longer

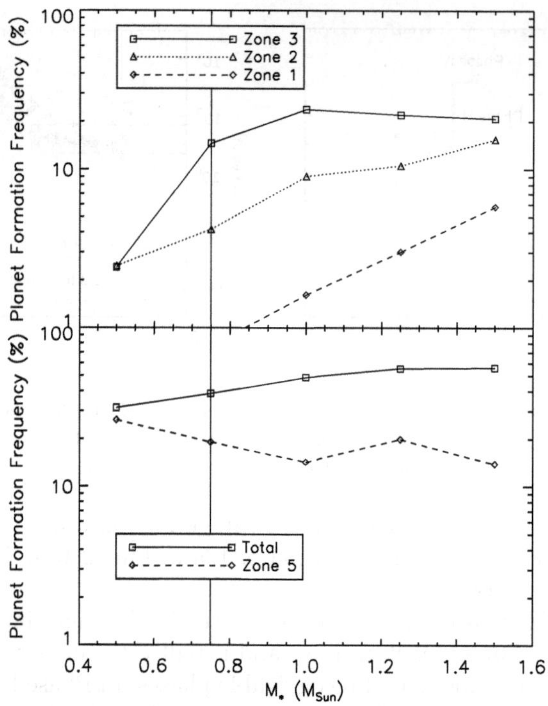

Figure 3. Planet Formation Frequencies (PPFs) - populations in the mass-semimajor axis diagram as a function of stellar mass. Adapted from Hasegawa & Pudritz 2013.

lived disk populations. Notably, there are only a few tracks that connect to giant planets at large radii, which agrees with the results of the NICI imaging surveys (see Vigan *et al.*, these proceedings) that find few objects beyond 10 AU. Finally, we show the PPFs for each zone, for stars of masses $0.5 - 1.5 M_\odot$ in Figure 3. The low-mass planets in tight orbits dominate the planetary population for low mass stars ($M \leqslant 0.7 M_\odot$), while massive Jupiters (zone 3) dominate for larger stellar masses, in good agreement with the observations.

References

Adams, F. C., Hollenbach, D., Laughlin, G., & Gorti, U.2004, *ApJ*, 611, 360
Chiang, E. & Laughlin, G.2013, *MNRAS*, 431, 3444
Hasegawa, Y. & Pudritz, R. E. 2011, *MNRAS*, 417, 1236
Hasegawa, Y. & Pudritz, R. E. 2012, *ApJ*, 760, 117
Hasegawa, Y. & Pudritz, R. E. 2013, *ApJ*, 778, 78
Ida, S. & Lin, D. N.C. 2004, *ApJ*, 604, 388
Ida, S. & Lin, D. N.C. 2008, *ApJ*, 673, 487
Lynden-Bell, D. & Pringle, J. E.1974, *MNRAS*, 168, 603
Masset, F. S., *et al.* 2006, *ApJ*, 642, 478
Mordasini, C., Alibert, Yl, & Benz, W.2009 *A&A*, 501, 1139
Pollack, J. B., Hubickyj, O., Bodenheimer, P. *et al.*,1996 *Icarus*,124, 62

Discussion

JACKSON: It looks like you may be forming to many planets in your 'exclusion zone' (~0.1-1 AU) compared to those at 1-10 AU, amd maybe compared to the hot Jupiters at <0.1 AU. Do you want to comment on that?

PUDRITZ: If you look carefully at our predictions for the total number of Zone 1 (hot Jupiters) vs. Zone 2 ("empty zone") there is a factor of 2-3 higher for Zone 1. This seems to agree with the observations.

TINNEY: How will planet multiplicity impact on your conclusions? It appears you could have situations where Super Earths are required to cross gas giant orbits in order to get to where we find them.

PUDRITZ: Planet traps tend to converge and sometimes intersect at late times in disk evolution. Therefore we would expect that under those circumstances, planet-planet interaction will become very important.

MACINTOSH: I should note that the statistical power of direct imaging surveys is much lower than Doppler, so the number of planets from say 10-50 AU could actually still be similar to 0.2-1 AU.

PUDRITZ: Interesting point. We will continue to run our models, but at the moment it still appears that it will be hard to populate the distant regions (>30 AU). We are eager to see the new data.

Exploring the Formation and Evolution of Planetary Systems
Proceedings IAU Symposium No. 299, 2013
M. Booth, B. C. Matthews & J. R. Graham, eds.

doi:10.1017/S1743921313008971

Collision parameters governing water delivery and water loss in early planetary systems

Thomas I. Maindl and Rudolf Dvorak

Universitätssternwarte, University of Vienna
A-1180, Vienna, Austria
email: thomas.maindl@univie.ac.at

Abstract. We investigate the distribution of encounter velocities and impact angles describing collisions in the habitable zone of the early planetary system. Here we present a catalogue of collision characteristics for a particular mass ratio of the colliding bodies and seven different planetesimal masses ranging from a tenth of Ceres' mass to 10 times the mass of the Moon. We show that there are virtually no collisions with impact speeds lower than the surface escape velocity and a similar velocity-impact angle distribution for different planetesimal masses if velocities are normalized using the escape velocity. An additional perturbing Jupiter-like object distorts the collision velocity and impact picture in the sense that grazing impacts at higher velocities are promoted if the perturber's orbit is close to the habitable zone whereas a more distant perturber has more the effect of a mere widening of the velocity dispersion.

Keywords. solar system: formation, celestial mechanics, methods: n-body simulations

---

1. Introduction

The goal of this study is to quantify and characterize the number of collisions between planetesimals in the early planetary system. Depending on the masses involved, we directly determine the velocity and the direction of the collisions between these bodies. In this approach we do not yet intend to accumulate them to larger ones; this is another task which is actually already in the stage of computations. The next step—this is the core of our overall project—is modeling the collisions in detail with our SPH code, which will allow us to track water content and explain water delivery processes by collisions and impacts (cf. Dvorak *et al.* 2012).

As dynamical model we use full n-body integrations of the Sun and a ring of planetesimals in the habitable zone. The bodies' initial orbital elements are uniformly distributed with $0.9 < a < 1.1$ AU (semi-major axis), $e < 0.1$ (eccentricity), and $i < 1^\circ$ (inclination), respectively. Additionally, we check the influence of a gas giant of $1\,\mathrm{M}_{♃}$ in different distances to the planetesimals which acts as a perturber in one particular scenario.

The scenarios differ in the chosen mass for the bodies which varies from $5\cdot10^{-11}\,\mathrm{M}_\odot \approx 1/10\,\mathrm{M}_{\mathrm{Ceres}}$ to $3\cdot10^{-7}\,\mathrm{M}_\odot \approx 10\,\mathrm{M}_{☾}$ (see Tab. 1). We use this relatively wide range of masses to see its influence on the collision velocities and impact angles of the two bodies. Each scenario is integrated for 10^6 years and includes 750 planetesimals.

In point mass-based n-body simulations we assume a collision to happen if two objects get as close as $R_{\mathrm{imp}} = R_P + R_T$ with R_P and R_T denoting the respective radii of spherical bodies represented by the point masses. In each scenario this distance is significantly smaller than the Hill radius (cf. Tab. 1).

Therefore—along with our interest in the statistical distribution of impact angles and velocities—we treat collisions as two-body problems of total mass $M_{\mathrm{tot}} = 2\,m = M_P + M_T$

Table 1. N-body simulation scenarios—m is the planetesimal mass, $R_{\rm imp}$ denotes the mutual distance of the barycenters upon impact for $C_{\rm P} = 0.3$, $C_{\rm T} = 0$, $M_{\rm P}/M_{\rm tot} = 0.1$, $r_{\rm Hill}$ the Hill radius at 1 AU and zero eccentricity, and $v_{\rm esc}$ the target's surface escape velocity (see text).

Scenario	m [$M_\odot$]	m [kg]	$R_{\rm imp}$ [10^6 m]	$r_{\rm Hill}$ [10^6 m]	$v_{\rm esc}$ [m/s]
Ce10	$5 \cdot 10^{-11}$	$9.95 \cdot 10^{19}$	0.3918	38	247
Ce	$5 \cdot 10^{-10}$	$9.95 \cdot 10^{20}$	0.8440	82	532
M10	$3 \cdot 10^{-9}$	$5.97 \cdot 10^{21}$	1.534	150	967
M3	$1 \cdot 10^{-8}$	$1.99 \cdot 10^{22}$	2.291	223	1444
M	$3 \cdot 10^{-8}$	$5.97 \cdot 10^{22}$	3.304	322	2083
3M	$1 \cdot 10^{-7}$	$1.99 \cdot 10^{23}$	4.936	481	3112
10M	$3 \cdot 10^{-7}$	$5.97 \cdot 10^{23}$	7.119	694	4488

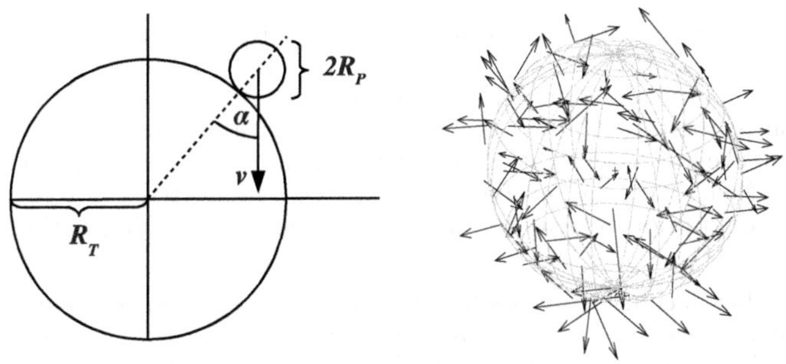

Figure 1. Left: impact geometry with impact angle α, impact velocity vector $\mathbf{v}$, and projectile and target radii R_P and R_T, respectively.
Right: schematic view of impact velocities in the 3M scenario. The arrows show the negative impact velocity for better visibility. The majority of impacts occurs from low inclination orbits.

with projectile and target masses M_P and M_T, respectively. For given M_P and given mean mean densities, $R_{\rm P}$, $R_{\rm T}$, and $R_{\rm imp}$ can be computed. As we are interested in water delivery by impacts we assume a solid basalt core with a layer or mantle of water ice, similar to the current models of Ceres (cf. Thomas *et al.* 2005).

We get the following relation for the radius of a body of mass $M_{\{P,T\}}$ consisting of basalt and a water ice shell (densities $\rho_{\rm b}$, $\rho_{\rm i}$) with a certain ice mass-fraction $C_{\{P,T\}}$:

$$R^3_{\{P,T\}} = \left[C_{\{P,T\}} + (1 - C_{\{P,T\}})\frac{\rho_{\rm i}}{\rho_{\rm b}}\right] M_{\{P,T\}} \frac{3}{4\pi}\frac{1}{\rho_{\rm i}}. \quad (1.1)$$

Keeping the future SPH-based simulations in mind we adopt the same values for the densities of ice and basalt as in Maindl *et al.* (2013): $\rho_{\rm i} = 917\,{\rm kg/m^3}$ and $\rho_{\rm b} = 2,700\,{\rm kg/m^3}$.

As a representative parameter set for collisions of interest for water delivery we choose a scenario with $M_{\rm P}/M_{\rm tot} = 0.1$ and respective water contents $C_{\rm P} = 30\,\%$ and $C_{\rm T} = 0$.

2. Results

In each scenario multiple collisions occur. As the initial "disk" of planetesimals is not considerably widened during the integration interval most of the impacts are close to the ecliptic (Fig. 1, right). As expected bigger masses correspond to bigger mutual perturbations and hence a larger number of collisions. The impact velocities show a relatively wide spread with a well defined lower boundary of about the target's surface escape velocity $v_{\rm esc}$ and an upper bound that varies between about 2.5 and 11.5 $v_{\rm esc}$. Fig. 2 shows scatter plots of velocities versus impact angles (Fig. 1, left defines the impact angle).

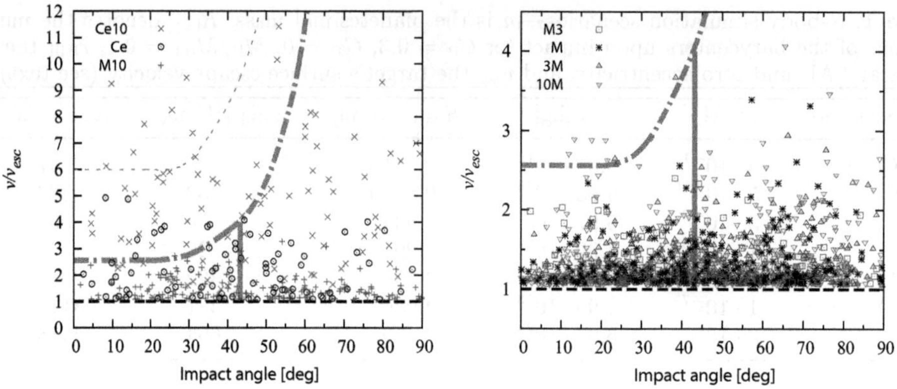

Figure 2. Collision velocities normalized by the target's surface escape velocities versus impact angle for the different scenarios. Note that high velocities only occur in the Ce10 scenario. The lines refer to the boundaries of different collision outcomes as given in Leinhardt & Stewart (2012): net erosion to the target above the dot-dashed curve, below it partial accretion to the left and hit-and-run events to the right of the thick vertical line. There is no collision in the area of perfect merging (below the dashed line, $v/v_{\rm esc} \lesssim 1$). The dotted curve in the left diagram denotes the critical disruption velocity for half the total mass remaining in the largest remnant.

Table 2. Total number of collision events $N_{\rm tot}$ for the different scenarios broken down to number of collisions N, impact velocities (mean $\bar{v}$ and median m_v), and their standard deviations σ for three impact angle intervals. An impact angle of 0° corresponds to a head-on collision. Velocity units are km/s.

		0° – 30°				30° – 60°				60° – 90°			
Scenario	$N_{\rm tot}$	N	$\bar{v}$	m_v	σ	N	$\bar{v}$	m_v	σ	N	$\bar{v}$	m_v	σ
Ce10	94	31	1.11	1.00	0.54	31	1.02	0.96	0.62	32	1.20	1.16	0.54
Ce	79	23	1.22	1.03	0.64	41	1.16	1.15	0.47	15	1.07	0.74	0.57
M10	158	45	1.51	1.42	0.44	61	1.42	1.28	0.44	52	1.34	1.28	0.34
M3	195	62	1.96	1.83	0.41	91	1.92	1.76	0.38	42	1.87	1.84	0.29
M	303	80	2.65	2.50	0.47	137	2.75	2.51	0.67	86	2.88	2.54	0.82
3M	449	126	3.93	3.64	0.78	197	4.11	3.84	0.94	126	4.26	3.76	1.13
10M	483	134	5.99	5.14	1.84	201	6.03	5.40	1.56	148	6.51	5.76	2.15

The results of all seven considered scenarios in absolute velocity units are summarized in Tab. 2. The standard deviations along with the relatively large deviations of the mean and median values confirm the wide spread of impact velocities as observed in the scatter plots, especially for large-mass scenarios.

Tab. 3 and the plot in Fig. 3 show how different perturbing "Jupiters" effect the collision velocities in the Ce scenario. Especially a gas giant very close to the habitable zone significantly increases the spread in the velocities due to highly perturbed orbits. Also, hit-and-run collisions happen more often in that case. The significantly larger spread in the velocities may also increase the number of destructive collisions—an effect that decreases towards a "real" Jupiter at larger distance $a_{\rm pert} = 5.2\,$AU.

3. Conclusions and further research

In our n-body calculations we confirm typical collision velocities in the early planetary system that range from the target's surface escape velocity $v_{\rm esc}$ up to a few times $v_{\rm esc}$ depending on the mass of the planetesimals. There is a clear tendency to collision velocities

Table 3. The effect of a perturbing body in the Ce scenario. The perturbing body's mass is $1\,M_{♃}$, its initial semi axis $a_{\rm pert}$. The other symbols and units are the same as in Tab. 2.

Scenario	$a_{\rm pert}$ [AU]	$N_{\rm tot}$	0° – 30° N	$\bar{v}$	m_v	σ	30° – 60° N	$\bar{v}$	m_v	σ	60° – 90° N	$\bar{v}$	m_v	σ
Ce	1.6	80	17	2.50	2.28	1.38	27	2.62	2.23	1.46	36	2.71	2.27	1.60
Ce	2.6	80	16	1.75	1.73	0.99	37	2.03	1.87	1.05	27	2.05	1.93	1.26
Ce	5.2	71	19	1.03	0.84	0.45	35	1.39	1.30	0.74	17	1.68	1.41	0.94

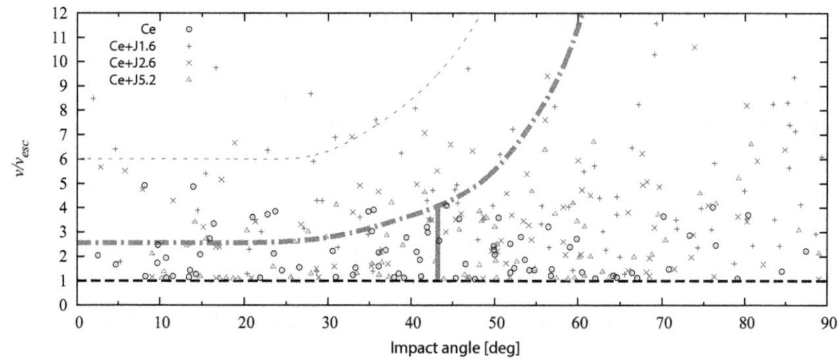

Figure 3. The Ce scenario without and with a perturbing gas giant of $1\,M_{♃}$ placed at 1.6, 2.6, and 5.2 AU from the Sun (denoted by Ce+J1.6, Ce+J2.6, and Ce+J5.2, respectively). The lines and symbols have the same meaning as in Fig. 2.

$v/v_{\rm esc}$ closer to 1 for heavier objects which may be explained by the two-body acceleration during the impact event dominating the initial velocity dispersion of the bodies for larger masses. The influence of perturbing bodies such as gas giants is significant in case of the perturber's orbit close to the planetesimals in the sense of a widened distribution of impact velocities and a tendency to promote hit-and-run collisions at high impact angles.

In the future, we will use collision velocities and angles corresponding to the partial accretion and hit-and-run scenarios in Figs. 2 and 3 as input to analyzing water transport via collisions. We will use our 3D SPH code that—among other features—includes elasto-plastic material modeling, brittle failure, self-gravity, and first order consistency fixes. It is introduced in Schäfer (2005) and Maindl *et al.* (2013) and is still developed further.

Acknowledgements

The authors wish to thank C. Schäfer and R. Speith for jump-starting us on collisions of solid bodies and SPH as well as for numerous fruitful discussions. We also thank Á. Süli and D. Bancelin for discussing and critically reviewing our n-body results. This research is produced as part of the FWF Austrian Science Fund project S 11603-N16.

References

Dvorak, R., Eggl, S., Süli, Á., Sándor, Z., Galiazzo, M., & Pilat-Lohinger, E. 2012, *AIP Conf. Proc.*, 1468, 137

Leinhardt, Z. M. & Stewart, S. T. 2012, *ApJ*, 745, 79

Maindl, T. I., Schäfer, C., Speith, R., Süli, A., Forgács-Dajka, E., & Dvorak, R. 2013, *AN*, 334, 9, 996

Schäfer, C. 2005, *Dissertation*, Eberhard-Karls-Universität Tübingen, Germany

Thomas, P. C., Parker, J. Wm., McFadden, L. A., Russell, C. T., Stern, S. A., Sykes, M. V., & Young, E. F. 2005, *Nature*, 437, 224

Exploring the Formation and Evolution of Planetary Systems
Proceedings IAU Symposium No. 299, 2013
M. Booth, B. C. Matthews & J. R. Graham, eds.

doi:10.1017/S1743921313008983

Protoplanetary Disk Evolution and Influence of the Host Star

Kévin Baillié and Sébastien Charnoz

Laboratoire AIM, Université Paris Diderot / CEA / CNRS
91191 Gif-sur-Yvette, France
email: kevin.baillie@cea.fr

Abstract. Based on a self-consistent coupling between protoplanetary disk thermodynamics, photosphere geometry and dynamics we designed a 1D-hydrodynamical numerical model for the spreading of the disks as a function of the star characteristics. We found that the evolution timescale increases for more massive or for a steeper surface density disk, and decreases for bigger stars or less turbulent disks. We found a strong dependency of the mass accretion rate versus the disk mass and a weaker dependency versus the star mass. Coupled with observed similar conclusions, we derived that the disk mass is scaling as $M_*^{1.6}$.

Keywords. planetary systems: protoplanetary disks, hydrodynamics, methods: numerical

1. Introduction

Recurrent observations (Andrews *et al.* 2009; Isella *et al.* 2009) in non-genetically related regions provide the following constraints:

- Surface-mass density : $\Sigma(r) \propto r^{-q}$ with $q \sim 1$
- Disk photosphere geometries are flared : $H_{\rm photo}(r) \propto r^{1.1-1.3}$
- Mass accretion rates : $\dot{M} \sim 10^{-8} - 10^{-6}\,{\rm M}_\odot/{\rm yr}$

We present our results of numerical simulations of the evolution of protoplanetary disks on large scales in order to explain the origins of these common properties.

2. Methods

We have designed a 1D-hydrodynamical code that takes into consideration

- a coupling between dynamics and thermodynamics through the turbulent viscosity $\nu = \alpha_{\rm visc} c_{\rm s} h_{\rm pr}$, where $\alpha_{\rm visc}$ is the viscosity prescription from Shakura & Sunyaev (1973), $c_{\rm s}$ is the local sound speed and $h_{\rm pr}$ is the pressure scale height.
- a coupling between the disk temperature and its photosphere geometry (considering both the star irradiation heating (Calvet *et al.* 1991) and the viscous heating.

3. Evolution and steady state

Protoplanetary disk evolution reaches a steady state (uniform mass flux) characterized by a surface-mass density profile in $\Sigma(r) \propto r^{-1}$. Inner regions evolve much faster and outer regions keep the memory of the initial conditions much longer (Fig. 1).

4. Results and discussion

These results are invariant with the initial disk : the steady state does not depend on the inital disk mass or distribution for a given star. However, the evolution timescales

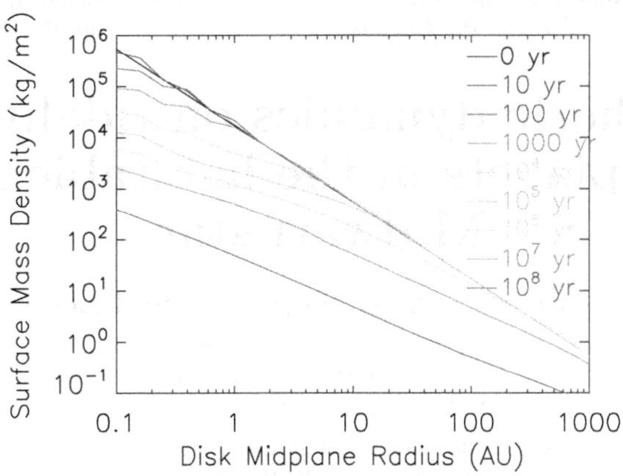

Figure 1. Evolution of the surface mass density profile.

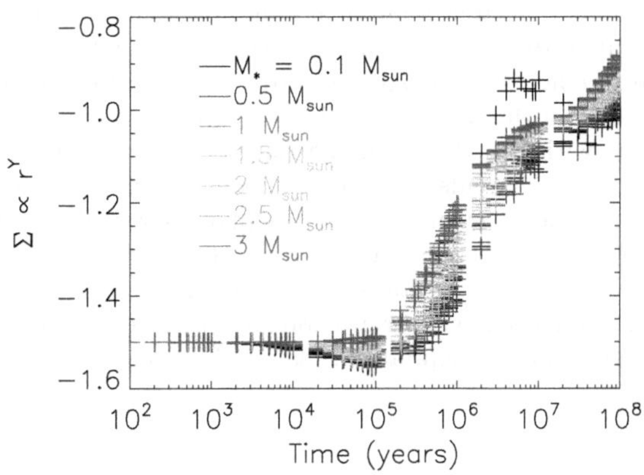

Figure 2. Temporal evolution of the surface mass density asymptotic trend for different star masses.

vary with the initial conditions : steeper or more massive initial disks will slow down the dynamical evolution and delay the reaching of the steady state while this steady state will be reached earlier for more massive stars (Fig. 2) and more turbulent disks (higher values of $\alpha_{\rm visc}$).

While varying independently the star mass and the disk mass, we were able to derive the following dependency for the mass accretion rate at 0.1 AU after 1 million years: $\dot{M} \propto M_{\rm disk} \cdot M_*^{0.22}$. Using observational results from Natta *et al.* (2006) along which $\dot{M} \propto M_*^{1.8}$, we predict a scaling of the disk mass along $M_{\rm disk} \propto M_*^{1.6}$.

References

Andrews, S. M., Wilner, D. J., Hughes, A. M., Qi, C., & Dullemond, C. P. 2009, *ApJ*, 700, 1502
Calvet, N., Patino, A., Magris, G. C., & DAlessio, P. 1991, *ApJ*, 380, 617
Isella, A., Carpenter, J. M., & Sargent, A. I. 2009, *ApJ*, 701, 260
Natta, A., Testi, L., & Randich, S. 2006, *A&A*, 452, 245
Shakura, N. I. & Sunyaev, R. A. 1973, *A&A*, 24,337

Exploring the Formation and Evolution of Planetary Systems
Proceedings IAU Symposium No. 299, 2013
M. Booth, B. C. Matthews & J. R. Graham, eds.

doi:10.1017/S1743921313008995

Atmospheric dynamics on tidally locked Earth-like planets in the habitable zone of an M dwarf star

Ludmila Carone[1], Rony Keppens[1] and Leen Decin[2]

[1]Centre for mathematical Plasma Astrophysics, Department of Mathematics, KU Leuven, Celestijnenlaan 200B, 3000 Leuven, Belgium
email: ludmila.carone@wis.kuleuven.be

[2]Instituut voor Sterrenkunde, KU Leuven, Celestijnenlaan 200D, 3000 Leuven, Belgium

Abstract. We investigated the large scale atmospheric circulation of Gl581g, a potentially habitable planet around an M dwarf star, using an idealized dry global circulation model (GCM) with simplified thermal forcing as a first step towards a systematic extended parameter study. The results are compared with the work of Joshi *et al.* (1997) who investigated a tidally-locked habitable Earth analogue with less than half the rotation period of Gl581g. The extent, form and strength of the atmospheric circulation in each model generally agree with each other, even though the models differ in key parameters such as planetary radius, surface gravity, forcing scheme and rotation period. The substellar point is associated with an uprising direct circulation-branch of a Hadley-like cell with return flow over the poles. It is compelling to assume that the substellar point of a tidally locked terrestrial exoplanet behaves dynamically like the Earth's tropic associated with clouds and precipitation, making it an ideal target for habitability.

Keywords. planetary systems

The 3D general circulation model MITgcm † was adapted to investigate the atmosphere of Gl581g, assuming an Earth-like composition (Tables 1 and 2).

Table 1. Planetary Parameters of Gl581g (Vogt *et al.* 2010)

Planetary mass and radius	$M_{pl} = 3.1\ M_{Earth}$ and $R_{Pl} = 1.45\ R_{Earth}$
Surface gravity	$g = 14.3\ m/s^2$
Planetary rotation and orbital period	$P_{rot} = P_{orb} = 36.5$ days.
surface pressure	$P_0 = 1$ bar
specific (dry) gas constant	$R = 287\ \frac{J}{kg K}$

Table 2. Model parameters

Horizontal resolution	128×64 using a C32 cubed sphere prescription (Adcroft 2001)
Vertical resolution	$20 \times \Delta P = 50$ mbar
Rayleigh friction at surface	$\tau_{drag} = 1$ d at surface, $\tau_{rad} = 0$ d for $P \leqslant 700$ mb (Held & Suarez, 1994)
Radiative relaxation time scale	$\tau_{rad} = 4$ d at surface, $\tau_{rad} = 40$ d for $P \leqslant 700$ mb (Held & Suarez, 1994)

Following the description of Held & Suarez (1994), temperatures are relaxed towards T_{force} via Newtonian cooling with $\tau_{rad} = 4 - 40$ days (Tab. 2). T_{force} is based on the prescription developed by Heng & Vogt (2011):

$$T_{force} = \left[T_0' + \Delta T_{EP}\cos(\Theta - 180°)\cos\Phi - \Delta T_z ln\left(\frac{P}{P_0}\right)\cos^2\Phi - T_{mod}\right]\left(\frac{P}{P_0}\right)^{2/7}$$

where Θ and Φ denote the longitude and latitude, and P and P_0 are the pressure and surface pressure, respectively. $T_0' = 268$ K is the surface temperature at the poles,

† http://mitgcm.org

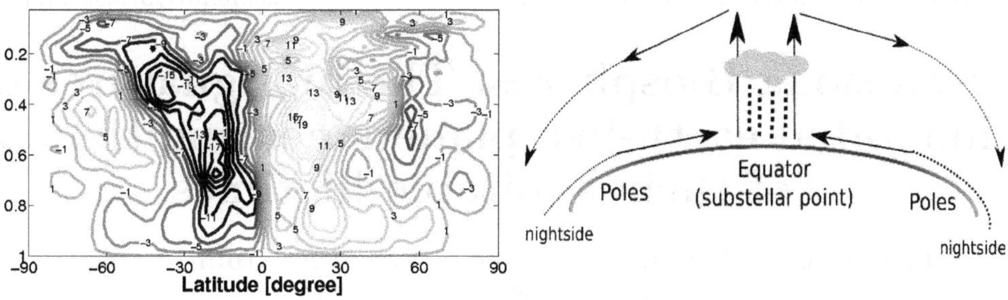

Figure 1. Left: Meridional mass overturning from our Gl581g model in units of 10^{10} kg/s agrees qualitatively with Joshi *et al.* (1997). Negative values (dark gray) indicate northward circulation, positive values (light gray) indicate southward circulation. **Right:** More schematic view of the circulation shown left with upwelling at the equator and return flow over the poles.

$\Delta T_{EP} = 10$ K is the temperature difference between the equator and the poles, and $\Delta T_z = 10$ K is a static stabilizing term along the equator. This model contains no stratosphere, yet. T_{mod} was added to suppress static instabilities at the substellar point by ensuring that the rate of decrease of temperature with respect to height,$-dT/dz$, is smaller than the dry adiabatic lapse rate.

$$T_{mod} = 10K\ ln\left(\frac{P}{P_0}\right) max(0, cos\Theta)$$

While horizontal flow patterns may differ between models e.g., Heng & Vogt (2011) and Joshi *et al.* (1997), the 3D atmospheric circulation observed in our model agrees generally with the results of Joshi *et al.* (1997), who investigated a tidally locked Earth analogue planet in the habitable zone of an M dwarf with $P_{rot} = P_{orb} = 16$ days.

In both models, the air mass transport is dominated by two large Hadley-like circulation cells, one fore each hemisphere, as highlighted in Fig. 1 left that shows a 'cross section' of meridional air mass circulation versus latitude and height in pressure. Fig. 1 right shows a simplified view of the two dominant circulation cells that transport heat from the surface of the hot substellar point at the equator towards the poles. The agreement in extent, form and strength of the circulation cells is compelling, even though the models differ in schemes used for thermal forcing and friction with the surface, and in planetary parameters like rotation, semi major axes, surface gravity and radius.

The upwelling at the substellar point makes this region an Earth tropic analogue, which could be associated with clouds and precipitation, if the temperatures at the surface allow for liquid water. This may be observed by 'probing' the substellar point of transiting terrestrial planets for water clouds by secondary transit observations or could be confirmed by increased albedo due to the clouds at the substellar point derived from high precision photometric light curves that the PLATO mission will provide†.

References

Adcroft, A., Scott, J. R., Marotzke, J. 2001, *Geophysical Research Letters*, 28, 1735

Held, I. M. & M. J. Suarez 1994, *Bulletin of the American Meteorological Society*, 75, 1825

Heng, K. & S. S. Vogt 2011, *MNRAS*, 415, 2145

Joshi, M. M., Haberle R. M., & Reynolds R. T. 1997, *Icarus*, 129, 450

Vogt, S. S., Butler, R. P., Rivera, E. J., Haghighipour N., Henry G. W., & Williamson, M. H. 2010, *ApJ*, 723, 954

† http://sci.esa.int/plato

Exploring the Formation and Evolution of Planetary Systems
Proceedings IAU Symposium No. 299, 2013
M. Booth, B. C. Matthews & J. R. Graham, eds.

doi:10.1017/S1743921313009009

Swansong Biospheres: The biosignatures of inhabited earth-like planets nearing the end of their habitable lifetimes

Jack T. O'Malley-James[1], Jane S. Greaves[1], John A. Raven[2] and Charles S. Cockell[3]

[1]School of Physics & Astronomy, University of St Andrews, St Andrews, UK
email: jto5@st-andrews.ac.uk

[2]Division of Plant Sciences, University of Dundee at TJHI, The James Hutton Institute, Invergowrie, Dundee, UK
email: j.a.raven@dundee.ac.uk

[3]UK Centre for Astrobiology, School of Physics & Astronomy, University of Edinburgh, Edinburgh, UK
email: c.s.cockell@ed.ac.uk

Abstract. The biosignatures of life on Earth are not fixed, but change with time as environmental conditions change and life living within those environments adapts to the new conditions. A latitude-based climate model, incorporating orbital parameter variations, was used to simulate conditions on the far-future Earth as the Sun enters the late main sequence. Over time, conditions increasingly favour a unicellular microbial biosphere, which can persist for a maximum of 2.8 Gyr from present. The biosignature changes associated with the likely biosphere changes are evaluated using a biosphere-atmosphere gas exchange model and their detectability is discussed. As future Earth-like exoplanet discoveries could be habitable planets nearing the end of their habitable lifetimes, this helps inform the search for the signatures of life beyond Earth

Keywords. Astrobiology, Sun: evolution, Earth.

1. Introduction

The increasing luminosity of the Sun as it ages will cause surface temperatures on Earth to increase, eventually leading to runaway ocean evaporation, making conditions more favourable for a unicellular microbial biosphere, similar to (but smaller than) that present for the first 2.5 Gyr of Earth's habitable lifetime (O'Malley-James *et al.* 2013a).

2. Temperature Model

The primary driver for habitability change is the solar luminosity increase over time, which can be modelled using the methods of Gough (1981), which, when combined with future orbital parameter changes, then determines surface insolation. This, coupled with greenhouse gas changes is used to calculate the flux of energy in and out ($F_{in/out}$) of the Earth climate system,

$$\frac{dT}{dt} = \frac{F_{in} - F_{out}}{C_p} \tag{2.1}$$

at different latitudes (where C_p is the heat capacity at constant pressure), giving a mean surface temperature profile for the planet. The temperature change with altitude can also be derived from this by calculating the adiabatic lapse rate.

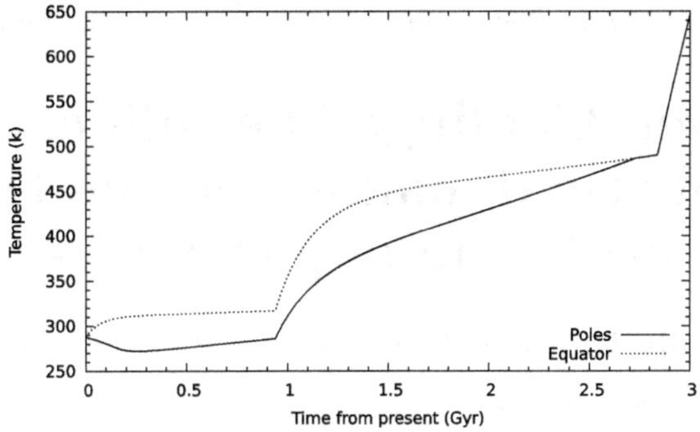

Figure 1. Change in global mean temperature over time with increasing solar luminosity. (O'Malley-James *et al.* 2013a).

3. Stages of biosphere death

Increased temperatures cause an increase in atmospheric water vapour, raising silicate weathering rates, which draws down more atmospheric CO_2. Decreasing CO_2 levels result in the gradual decline of plant species and the animal species they depend on (Calderia & Kasting, 1992; O'Malley-James *et al.* 2013a;b). This leaves behind a microbial biosphere.

Rapid ocean evaporation begins 1 Gyr from present. As water vapour is a greenhouse gas, there is a rapid increase in surface temperatures, reducing the extent of the biosphere by restricting life to remaining areas of liquid water. Such refuges include high altitude pools (for 2.2 Gyr from present), cold trap caves (for 2.8 Gyr from present) and the subsurface (O'Malley-James *et al.* 2013a).

The atmosphere-biosphere model is driven by the surface temperature model. For rapid extinction rates, the die-off of plants and animals may be indicated by elevated atmospheric CH_4 levels, isoprene and ethane (O'Malley-James *et al.* 2013b). Unlike the more productive microbial biospheres of the early Earth, the only potentially detectable biosignature of a far-future microbial biosphere is CH_4, although this may not reach remotely detectable levels in the atmosphere, essentially making a dying biosphere undetectable.

4. Summary and Conclusions

As Earth nears the end of its habitable lifetime it will only be capable of supporting microbial life adapted to multiple extreme conditions. This makes Earth a world predominantly inhabited by microorganisms for approximately two-thirds of its habitable lifetime, which, assuming a similar evolutionary history on Earth-like exoplanets, makes microbial biosignatures important in the search for life beyond the solar system.

References

Caldeira, C. & Kasting, J. F.: The lifespan of the biosphere revisited, Nature, Vol. 360, pp. 721–723, 1992.

Gough, D. O.: Solar interior structure and luminosity variations, Solar Phys., Vol. 74, pp. 21–34, 1981.

O'Malley-James, J. T., Cockell C. S., Greaves, J. S., & Raven, J. A.: Swansong Biospheres II: The final signs of life on terrestrial planets near the end of their habitable lifetimes, 2013b, arXiv: 1310.4841

O'Malley-James, J. T., Greaves, J. A., Raven, J. A., & Cockell, C. S.: Swansong Biospheres: refuges for life and novel microbial biospheres on terrestrial planets near the end of their habitable lifetimes, Int. J. Astrobiology, Vol. 12, pp. 99–112, 2013a.

Exploring the Formation and Evolution of Planetary Systems
Proceedings IAU Symposium No. 299, 2013
M. Booth, B. C. Matthews & J. R. Graham, eds.

doi:10.1017/S1743921313009010

Uneven Cooling: The influence of atmospheric dynamics on the thermal evolution of gas giants

Emily Rauscher[1] and Adam P. Showman[2]

[1]Department of Astrophysical Sciences, Princeton University,
4 Ivy Lane, Peyton Hall, Princeton, NJ 08544, U.S.A.
email: rauscher@astro.princeton.edu

[2]Lunar & Planetary Laboratory, University of Arizona,
1629 E University Blvd., Tucson, AZ 85721, U.S.A.
email: showman@lpl.arizona.edu

Abstract. Planets cool and contract as they age, with a cooling rate that depends on the efficiency with which they can transport heat out to space, first through the convective interior and then radiatively out through the atmosphere. The bottleneck for this cooling is the radiative-convective boundary (RCB), where the heat transport is the least efficient. Due to differential heating and atmospheric dynamics, the depth of the RCB can vary with latitude and longitude, meaning that the actual global cooling rate may differ from what would be calculated assuming a spherically symmetric RCB, as in 1D evolutionary models. Here we present models of the deep atmosphere of a generic hot Jupiter, calculate inhomogeneity in the RCB, and determine the resulting effect on the global thermal evolution. Although this issue can apply to any differentially heated gas giant, we focus on the hot Jupiter class of planet because: 1) the thick radiative zones above their deep RCBs can have a stronger influence on deforming the surface of the RCB than would generally be the case for a less-irradiated planet, and 2) an uneven RCB should increase the cooling rate, potentially exacerbating the mismatch between the large radii measured for some hot Jupiters and the smaller radii expected from evolutionary models.

Keywords. hydrodynamics, radiative transfer, planets: general, stars: evolution

---

Context. The evolutionary cooling rate of a planet is set by the flux at the boundary between its inner, convective region and its outer, radiative atmosphere (the radiative-convective boundary, RCB) and is easy to calculate if the temperature, pressure, and opacity at the RCB are known (Arras & Bildsten 2006). Hot Jupiters are a class of gas giants on close-in, circular orbits around their host stars and are assumed to be tidally locked into synchronous rotation, such that they experience strongly asymmetric heating. It has been proposed that the depth of the RCB should vary between their permanent day and night sides, allowing a hot Jupiter to cool more efficiently through its night side (Guillot & Showman 2002; Budaj *et al.* 2012; Spiegel & Burrows 2013). However, our 3D atmospheric circulation models predict that winds can efficiently homogenize day-night temperature differences in the deep atmosphere, near the RCB, leading to a temperature structure varying primarily in latitude but not longitude (e.g., Showman *et al.* 2009; Rauscher & Menou 2012). Since at deep pressures we still see a non-uniform temperature structure (maintained by the circulation pattern), uneven cooling could occur.

Model description. Informed by our 3D circulation models, we use a 2D model for the deep atmosphere, with pressure as the vertical coordinate and latitude as the horizontal, under the assumption that there is no significant longitudinal variation in the temperature structure. We built our model with standard radiative transfer (two-stream double-gray,

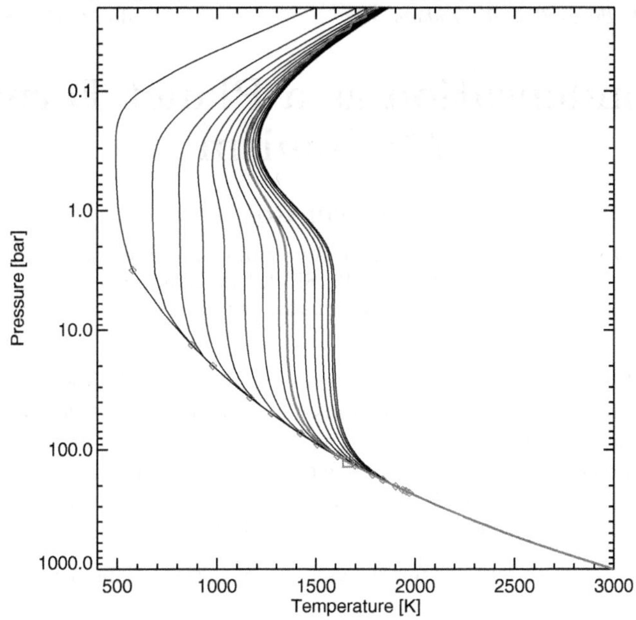

Figure 1. Temperature-pressure profiles from our model atmosphere. The black lines are at latitudes ranging from near the equator (hottest profile) to near the pole (coldest profile), with a pink diamond marking the location of the radiative-convective boundary. The thick blue line is calculated by our code, assuming a globally averaged stellar flux. The RCB for this 1D model is marked by the blue square. (The temperature increase high in the atmosphere is a result of our choice to have the infrared absorption coefficient scale linearly with pressure while the optical coefficient is constant. This region effectively acts only as an upper boundary condition.)

transitioning to flux-limited diffusion at high optical depth), dynamical heat transport based on local conditions, a standard adjustment of convectively unstable regions to the adiabatic lapse rate, and a matrix solution method to reach local flux equilibrium (radiative and dynamic). We then self-consistently calculate the global cooling luminosity, by integrating over the uneven flux escaping from the non-uniform RCB.

Initial results. We are testing the code without any dynamical heat transport before including the influence of atmospheric circulation on the temperature structure. This places an upper limit on how much the cooling luminosity through an uneven RCB could differ from what would be calculated by a 1D, uniform model. In the particular example shown in Fig. 1, the cooling rate through the uneven RCB is 30% higher than through the uniform RCB, implying faster evolutionary cooling and a smaller radius than would otherwise be predicted, and further exacerbating the problem of bloated hot Jupiter radii.

References

Arras, P. & Bildsten, L. 2006, *ApJ*, 650, 394

Budaj, J., Hubeny, I., & Burrows, A. 2012, *A&A*, 537, A115

Guillot, T. & Showman, A. P. 2002, *A&A*, 385, 156

Rauscher, E. & Menou, K. 2012, *ApJ*, 750, 96

Showman, A. P., Fortney, J. J., Lian, Y., Marley, M. S., Freedman, R. S., Knutson, H. A., & Charbonneau, D. 2009, *ApJ*, 699, 564

Spiegel, D. S. & Burrows, A. 2013, arXiv:1303.0293

Exploring the Formation and Evolution of Planetary Systems
Proceedings IAU Symposium No. 299, 2013
M. Booth, B. C. Matthews & J. R. Graham, eds.

doi:10.1017/S1743921313009022

Ice Condensation as a Planet Formation Mechanism

Katrin Ros

Lund Observatory, Department of Astronomy and Theoretical physics,
Lund University, Box 43, SE-221 00 Lund, Sweden
email: katrin.ros@astro.lu.se

Abstract. Particles in protoplanetary discs grow rapidly to millimetre-sizes via coagulation, but further growth to centimetre-sized pebbles is not yet completely understood. We investigate particle growth by ice condensation in a model where we take the dynamical behaviour of vapour and ice particles into account, as well as the size evolution due to condensation and sublimation. Our results show that efficient growth from dust to pebbles is possible close to the water ice line at ~3 AU, with particles growing from millimetres to decimetres on a time scale of 10 000 yr.

Keywords. Methods: numerical, planetary systems: formation, planetary systems: protoplanetary discs

1. Introduction

Millimetre-sized particles in protoplanetary discs form by collisions in which dust grains stick together via contact forces. For larger particles, however, experiments show that collisions tend to lead to fragmentation or bouncing rather than to sticking (Blum & Wurm 2008). This is problematic since it is known from observations that pebbles, approximately centimetre-sized particles, do exist in protoplanetary discs (Testi *et al.* 2003). These pebbles are needed for continued growth towards planetesimals and planets, both for further coagulation and for clumping via streaming instabilities, in vortices or in pressure bumps (Johansen *et al.* 2007; Barge & Sommeria 1995; Johansen *et al.* 2009). Accretion of icy pebbles is also a way to efficiently form the cores of ice giants and gas giants (Lambrechts & Johansen 2012). We have investigated the formation of pebbles via condensation, close to the water ice line at ~3 AU (Lecar *et al.* 2006). Outwards of ice lines solid densities are enhanced as material condenses out, leading to larger particles sizes. Additionally, collisions between water ice particles are less destructive than corresponding ones for silicates (Wada *et al.* 2009). Particles outwards of the ice line can thus profit from both growth by condensation and of an enhanced coagulation probability.

2. Model

We have developed an idealized condensation model, in which we investigate particle size evolution due to condensation and sublimation in a dynamical model. We take into account the coupling of vapour and particles to the turbulent gas, and model the resulting turbulent diffusion as a random walk. Sedimentation to the midplane as particles decouple from the turbulent eddies is also accounted for, as well as radial drift due to the headwind particles experience from the pressure-supported gas. Condensation and sublimation are modelled in a Monte Carlo scheme, whereas particle collisions are ignored. For a more detailed description of the model, see Ros & Johansen 2013.

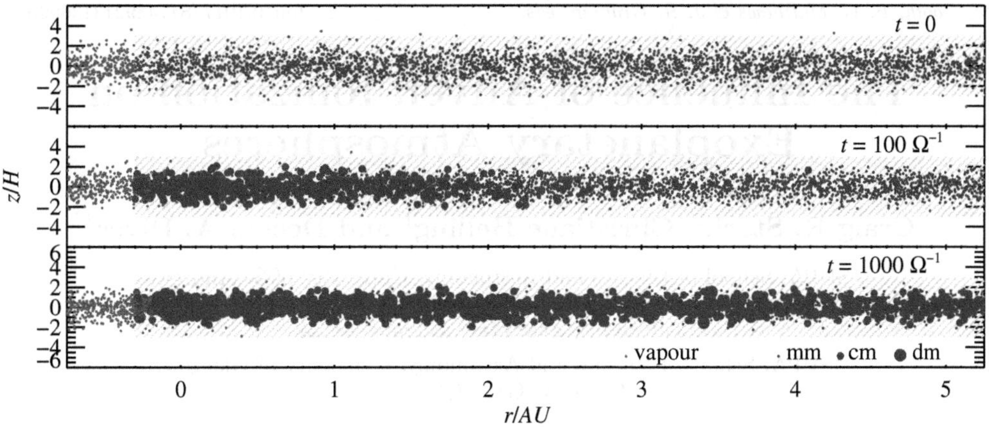

Figure 1. Distribution and sizes of particles at $t = 0$, $t = 100\,\Omega^{-1}$ and $t = 1000\,\Omega^{-1}$, where $1\Omega^{-1} \approx 1\,\mathrm{yr}$ at 3 AU, where the ice line is located. Ice particles (shown as blue filled circles) grow and sediment to the midplane as vapour (shown as red dots) condenses onto them. The region marked by grey dashed lines is the condensation region, which consists of the cold midplane region ($|z| \lesssim 3\,H$, where H is scale heights) outwards of the radial ice line ($r \gtrsim 3\,AU$). The number of particles is inversely scaled with size for visibility.

3. Conclusions

We find that ice condensation is an efficient growth mechanism for particles up to sizes of large pebbles. In our simulations, particles grow from millimetres to decimetres on a time scale of 10 000 yr in a turbulent disc ($\alpha = 10^{-2}$, where the strength of the turbulence is parameterized by the dimensionless constant α introduced by Shakura & Sunyaev 1973). These particles are large enough to explain the observed pebbles in protoplanetary discs and to enable further growth into planetesimals via instabilities and gravitational collapse. Figure 1 shows the distribution and sizes of ice particles at three different times.

We find that radial mixing due to turbulent diffusion plays an important role. Large particles form preferentially close to the radial ice line at ~3 AU, but are radially transported outwards throughout the entire simulation domain. This suggests that the enhanced particle growth close to ice lines has implications for the particle size distribution also farther out in the disc.

References

Barge, P. & Sommeria, J. 1995, *A&A*, 295, L1
Blum, J. & Wurm, G. 2008, *ARAA*, 46, 21
Johansen, A., Oishi, J. S., Mac Low, M.-M., *et al.* 2007, *Nature*, 448, 1022
Johansen, A., Youdin, A., & Klahr, H. 2009, *ApJ*, 697, 1269
Lambrechts, M. & Johansen, A. 2012, *A&A*, 544, A32
Lecar, M., Podolak, M., Sasselov, D., & Chiang, E. 2006, *ApJ*, 640, 1115
Ros, K. & Johansen, A. 2013, *A&A*, 552, A137
Shakura, N. I. & Sunyaev, R. A. 1973, *A&A*, 24, 337
Testi, L., Natta, A., Shepherd, D. S., & Wilner, D. J. 2003, *A&A*, 403, 323
Wada, K., Tanaka, H., Suyama, T., Kimura, H., & Yamamoto, T. 2009, *ApJ*, 702, 1490

Exploring the Formation and Evolution of Planetary Systems
Proceedings IAU Symposium No. 299, 2013
M. Booth, B. C. Matthews & J. R. Graham, eds.

doi:10.1017/S1743921313009034

The Influence of Alfvén Ionization on Exoplanetary Atmospheres

Craig R. Stark[1], Christiane Helling[1] and Declan A. Diver[2]

[1]SUPA, School of Physics and Astronomy, University of St Andrews,
St Andrews, KY16 9SS, UK
email: craig.stark@st-andrews.ac.uk

[2]SUPA, School of Physics and Astronomy, University of Glasgow,
Glasgow, G12 8QQ, UK

Abstract. Observations of radio emission from exoplanets and brown dwarfs suggest that such objects harbour atmospheric magnetized plasmas. However, the degree of thermal ionization for such substellar objects is insufficient to qualify the ionized component as a plasma, posing the question: what ionization processes can efficiently produce the required plasma? We propose Alfvén ionization as a mechanism for producing localized pockets of ionized gas in the atmosphere, having sufficient degrees of ionization ($\gtrsim 10^{-7}$) that they constitute plasmas. We outline the criteria required for Alfvén ionization to occur and justify its applicability in substellar atmospheres. For the model atmospheres considered here, our results show that degrees of ionization ranging from $10^{-6} - 1$ can be obtained.

Keywords. plasmas, stars: atmospheres, stars: low-mass, brown dwarfs

1. Alfvén Ionization

Ultracool dwarfs (low-mass objects with spectral type later than M7; such as Brown Dwarfs, BDs, and gas giant planets, GPs) are observed to be strong sources of coherent radio emission (e.g. see Route & Wolszczan 2012 and Hallinan *et al.* 2007), inferring the presence of atmospheric plasmas. We propose Alfvén ionization as a mechanism to produce the required plasma that is the source of the radiation.

In Alfvén ionization a constant stream of neutral gas impinges on a low-density magnetized plasma. The inflowing neutral atoms collide with and displace the plasma ions, leaving behind a significant charge imbalance that accelerates electrons to energies sufficient to ionize the local gas via electron-neutral impact ionization. Alfvén ionization requires: (i) an initial, low-density magnetized seed plasma; and (ii) a neutral gas flow that reaches a critical threshold speed (Alfvén 1960). The critical speed is obtained by equating the kinetic energy of the flow to the electrostatic potential energy of an electron in an electric potential equal to the ionization potential of a neutral gas particle ϕ_I: $\frac{1}{2}m_{\rm gas}v_c^2 = e\phi_I$. Therefore, the critical neutral gas flow speed is $v_c = (2e\phi_I/m_{\rm gas})^{1/2}$. Species such as Potassium ($v_c = 4.63$ kms^{-1}) have the smallest critical speeds whereas Hydrogen ($v_c = 51.02$ kms^{-1}) and Helium (34.43 kms^{-1}) have the largest speeds. In substellar atmospheres the required seed plasma can be easily generated from local discharge events (lightning) in mineral clouds (Helling *et al.* 2011, 2013) and cosmic ray bombardment (Rimmer *et al.* 2013). Furthermore, typical magnetic flux densities of GPs ($\approx O(10$ G$)$) can ensure that the seed plasma is sufficiently magnetized.

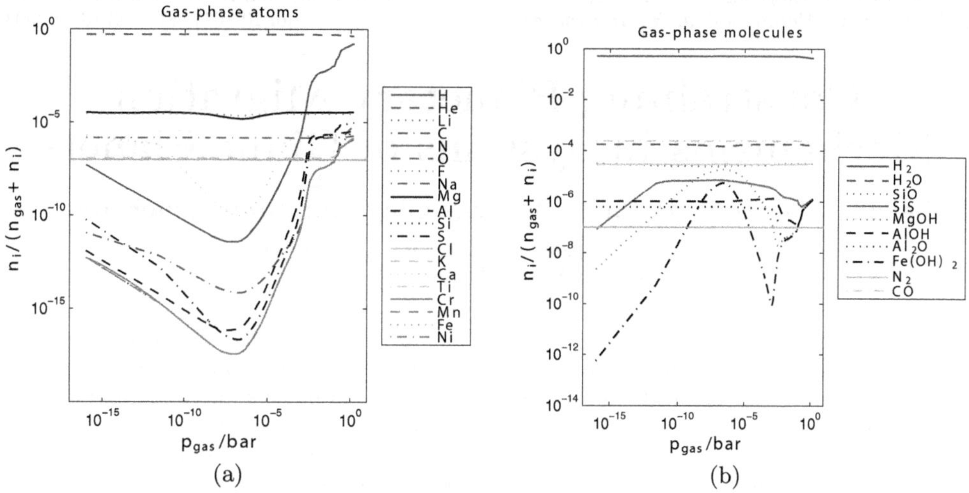

Figure 1. The degree of Alfvén ionization for a GP atmosphere. The grey horizontal line signifies the degree of ionization (10^{-7}) required to constitute a plasma.

2. Degree of Alfvén Ionization in exoplanetary atmospheres

We are interested in Alfvén ionization in GP and BD atmospheres. GPs are similar objects to BDs but form differently and have a weaker magnetic field. Without loss of generality, we consider an example GP atmosphere using DRIFT-PHOENIX which describes non-equilibrium cloud formation in substellar envelopes (Woitke & Helling 2003, Helling *et al.* 2008, Helling *et al.* 2008b, Witte *et al.* 2009, Witte *et al.* 2011) characterized by $\log g = 3.0$, $T_{\rm eff} = 1500$ K, for solar metallicity. Assuming conditions (i) and (ii) can be met, Alfvén ionization can ionize the entirety of the gas in a localized volume, leaving a plasma with an electron number density equal to the gas component number density (assuming 100% ionization) plus the initial seed magnetized plasma number density (Stark *et al.* 2013). Figs. 1(a) and 1(b) show the resulting degree of Alfvén ionization, if specific individual species constituting the atmospheric gas are entirely ionized (on their own) in a localized volume. In general, if in a localized atmospheric pocket a particular species can be 100% ionized, then the species with the greatest number density will yield the highest degree of ionization. To summarise: if entirely ionized on their own He, Fe, Mg, Na, H_2, CO, H_2O, N_2 and SiO all consistently increase the degree of ionization beyond 10^{-7} throughout the model atmosphere considered here.

References

Alfvén, H. 1960, *Rev. Mod. Phys.*, 32, 4, 710
Hallinan, G., *et al.* 2007, *ApJ*, 663, L25-L28
Helling, Ch., Jardine, M., Witte, S., & Diver, D. A. 2011, *ApJ*, 727, 4
Helling Ch., Jardine, M., Stark, C. R., & Diver, D. A. 2013, *ApJ*, 767, 136.
Helling, Ch., *et al.* 2008, *MNRAS*, 391, 1854
Helling, Ch., Woitke, P., & Thi, W.-F 2008b, *A&A*, 485, 547
Rimmer, P. B. & Helling, Ch. 2013, *ApJ*, 774, 108
Route, M. & Wolszczan, A. 2012, *ApJ*, 747, L22
Stark, C. R., *et al.* 2013, *ApJ*, 776, 11
Witte, S., Helling, Ch., Hauschildt, P. H. 2009, *A&A*, 506, 1367.
Witte, S., Helling, Ch., Barman, T., Heidrich, N., Hauschildt, & P. H. 2011, *A&A*, 529, A44.
Woitke, P. & Helling, Ch. 2003, *A&A*, 399, 297

Exploring the Formation and Evolution of Planetary Systems
Proceedings IAU Symposium No. 299, 2013
M. Booth, B. C. Matthews & J. R. Graham, eds.

doi:10.1017/S1743921313009046

Constraining Planetary Migration Mechanisms in Systems of Giant Planets

Rebekah I. Dawson, Ruth A. Murray-Clay and John Asher Johnson

Harvard-Smithsonian Center for Astrophysics
60 Garden St, MS-10, Cambridge, MA 02138

Abstract. It was once widely believed that planets formed peacefully in situ in their proto-planetary disks and subsequently remain in place. Instead, growing evidence suggests that many giant planets undergo dynamical rearrangement that results in planets migrating inward in the disk, far from their birthplaces. However, it remains debated whether this migration is caused by smooth planet-disk interactions or violent multi-body interactions. Both classes of model can produce Jupiter-mass planets orbiting within 0.1 AU of their host stars, also known as hot Jupiters. In the latter class of model, another planet or star in the system perturbs the Jupiter onto a highly eccentric orbit, which tidal dissipation subsequently shrinks and circularizes during close passages to the star. We assess the prevalence of smooth vs. violent migration through two studies. First, motivated by the predictions of Socrates *et al.* (2012), we search for super-eccentric hot Jupiter progenitors by using the "photoeccentric effect" to measure the eccentricities of Kepler giant planet candidates from their transit light curves. We find a significant lack of super- eccentric proto-hot Jupiters compared to the number expected, allowing us to place an upper limit on the fraction of hot Jupiters created by stellar binaries. Second, if both planet-disk and multi-body interactions commonly cause giant planet migration, physical properties of the proto-planetary environment may determine which is triggered. We identify three trends in which giant planets orbiting metal rich stars show signatures of planet-planet interactions: (1) gas giants orbiting within 1 AU of metal-rich stars have a range of eccentricities, whereas those orbiting metal- poor stars are restricted to lower eccentricities; (2) metal-rich stars host most eccentric proto-hot Jupiters undergoing tidal circularization; and (3) the pile-up of short-period giant planets, missing in the Kepler sample, is a feature of metal-rich stars and is largely recovered for giants orbiting metal-rich Kepler host stars. These two studies suggest that both disk migration and planet-planet interactions may be widespread, with the latter occurring primarily in metal-rich planetary systems where multiple giant planets can form. Funded by NSF-GRFP DGE-1144152.

Keywords. Planetary systems, planets and satellites: general, celestial mechanics

---

1. Introduction

Approximately 1% of stars host hot Jupiters, ousted from their birthplaces to short-period orbits (Wright *et al.* 2012) via mechanisms that remain debated. Proposed theories fall into two classes: smooth disk migration (e.g. Goldreich & Tremaine 1980), and migration via gravitational perturbations, either by stars (e.g. stellar binary Kozai, Wu & Murray 2003) or sibling planets (including planetary Kozai, e.g. Naoz *et al.* 2011; scattering, e.g. Rasio & Ford 1996; and secular chaos, e.g. Wu & Lithwick 2011). We assess the prevalence of these two classes through two studies. In Section 2, we search for super-eccentric hot Jupiter progenitors in the *Kepler* sample, expected if hot Jupiters migrated via gravitational perturbations. If both planet-disk and multi-body interactions commonly cause giant planet migration, physical properties of the proto-planetary environment may determine which is triggered. In Section 3, we identify three trends in which giant planets orbiting metal rich stars show signatures of planet-planet interactions.

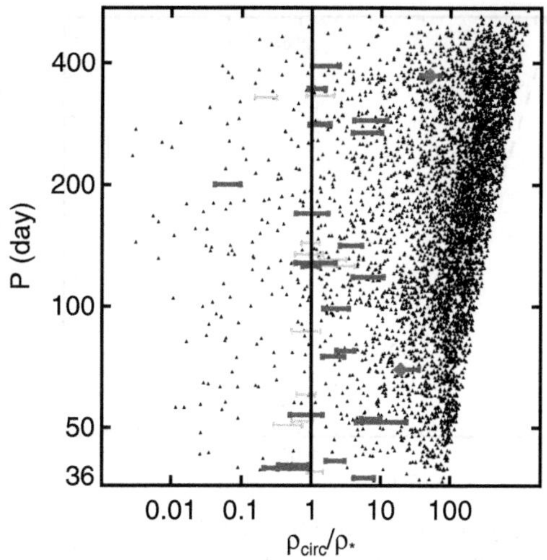

Figure 1. Expected 2D posterior for orbital period P vs. $\rho_{\rm circ}/\rho_\star$. The values we measured for *Kepler* giant planet candidates (Batalha *et al.* 2013) are overplotted. Thin, light bars: candidates with companions in their systems. Thick, dark bars: candidates with no known companions. We do not see the expected half a dozen candidates in the region of high-probability density. Diamonds: KOI-1474.01 (bottom), KOI-211.01 (top).

2. A paucity of super-eccentric Jupiters

One or more companions can create a hot Jupiter by perturbing a cold Jupiter onto an eccentric orbit, which tidal forces shrink and circularize during close passages to the star. Socrates *et al.* (2012, S12 hereafter) predicted that if this "high-eccentricity migration" (HEM) is the primary channel for producing hot Jupiters, the *Kepler* candidate collection should harbor a population of super-eccentric Jupiter-sized planets that are in the midst of tidal circularization. We have been using what we term the "photoeccentric effect" to measure individual eccentricities of Jupiter-sized planets from their transit light curves (Dawson & Johnson 2012). Dawson *et al.* (2012) identified KOI-1474.01 as a transiting planet candidate with a long orbital period (69.7 days), a large eccentricity ($e = 0.81 \pm 0.10$), and transit timing variations caused by a massive outer companion. However, uncertainty in the candidate's eccentricity made it ambiguous whether KOI-1474.01 is one of the proto-hot Jupiters predicted by S12 or, alternatively, a failed-hot Jupiter beyond the reach of tidal circularization over its host star's lifetime.

Here we examine the entire sample of *Kepler* Jupiters to assess whether the planets expected from HEM are present. We perform a Monte Carlo simulation to predict the signature in the transit light curve observable $\rho_{\rm circ}/\rho_\star$ expected from the super-eccentric proto-hot Jupiters. Following the procedure described in Dawson *et al.* (2013), we generate a two-dimensional (2D) probability distribution in $(P, \rho_{\rm circ}/\rho_\star)$ (Figure 1, black dots), where P is the orbital period, $\rho_\star$ the bulk stellar density, $\rho_\star = M_\star/(\frac{4}{3}\pi R_\star^3)$, and $\rho_{\rm circ}$ is the stellar density measured from the light curve under the assumption of a circular orbit. Next, we fit stellar evolution models to the host stars and transit models to the light curves of *Kepler* giant planet candidates (see Dawson *et al.* 2012, Dawson *et al.* 2013). Finally, we combine the $\rho_{\rm circ}$ and $\rho_\star$ posteriors into a posterior of $\rho_{\rm circ}/\rho_\star$ for each candidate, marginalized over all other parameters. None of the candidates fall in the high-probability area of the prediction. Applying the statistical procedure described in Dawson *et al.* (2013), we find that, with 95.8% confidence, we detected too few super-eccentric proto-hot Jupiters to be consistent with the prediction. We place a

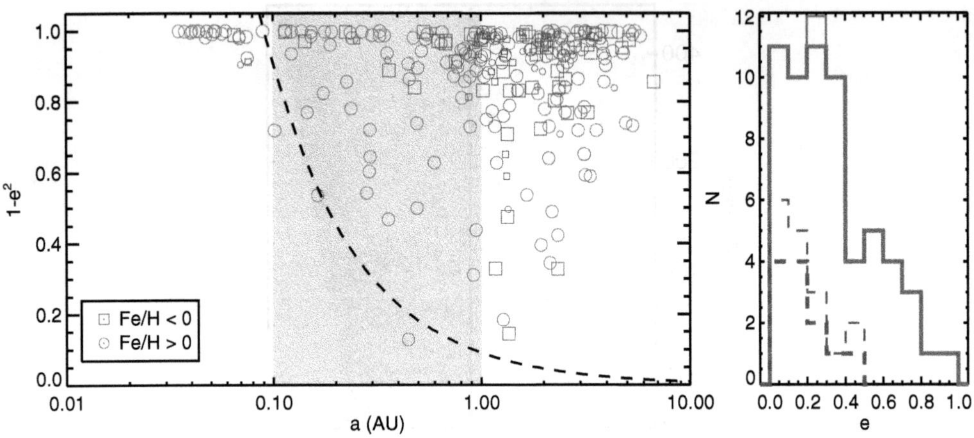

Figure 2. Left: "Valley" (gray region) giant planets orbiting metal-rich stars ([Fe/H]$\geqslant$0, circles) have a range of eccentricities; those orbiting metal-poor stars ([Fe/H]<0, squares) are confined to low eccentricities. Small symbols represent stars with $\log g < 4$. For reference, above the dashed line (a tidal circularization track ending at 0.1 AU) planets are unlikely to experience significant tidal circularization. We plot the quantity $1 - e^2$ to emphasize high-eccentricity planets. Right: Eccentricity distributions of Valley planets orbiting metal-rich (solid) and metal-poor (dashed) stars. The bold distributions omit stars with $\log g < 4$.

two-sigma upper limit of 33% on the fraction of moderately-eccentric proto-hot Jupiters ($0.2 < e < 0.6$) that began beyond the ice line with $e \to 1$. Since this is the only pathway open to hot Jupiters produced by stellar binary Kozai oscillations, the upper limit is also on the fraction of hot Jupiters created by stellar binaries.

3. Giant planets orbiting metal-rich stars show signatures of planet-planet interactions

We find three ways in which the properties of hot Jupiters and Valley giants depend on host star metallicity (see Dawson & Murray-Clay 2013 for a quantative, statistical assessment):

(*a*) Gas giants with $a <$ 1AU orbiting metal-rich stars have a range of eccentricities, whereas those orbiting metal-poor stars are restricted to lower eccentricities (Figure 2; see also Taylor 2012).

(*b*) Metal-rich stars host most eccentric proto-hot Jupiters undergoing tidal circularization (Dawson & Murray-Clay 2013, Figure 2).

(*c*) The pile-up of short-period giant planets, missing in the *Kepler* sample, is a feature of metal-rich stars and is largely recovered for giants orbiting metal-rich *Kepler* host stars (Figure 3; Dawson & Murray-Clay 2013, Figure 5) .

Hot Jupiters and Valley giants (those with semi-major axes beyond 0.1 AU but interior to the pile-up at 1 AU) are both thought to have been displaced from their birthplaces. Therefore these metallicity trends can be understood if smooth disk migration and planet-planet scattering both contribute to the early evolution of systems of giant planets. We expect disk migration could occur in any system, but only systems packed with giant planets – which most easily form around metal-rich stars – can scatter giant planets inward to large eccentricities (Trend 1). Some of these tides shrink and circularize (Trend 2), creating a pile-up of short-period giants (Trend 3). Moreover, these trends support planet-planet interactions (e.g. scattering, secular chaos, or Kozai) as the dynamical

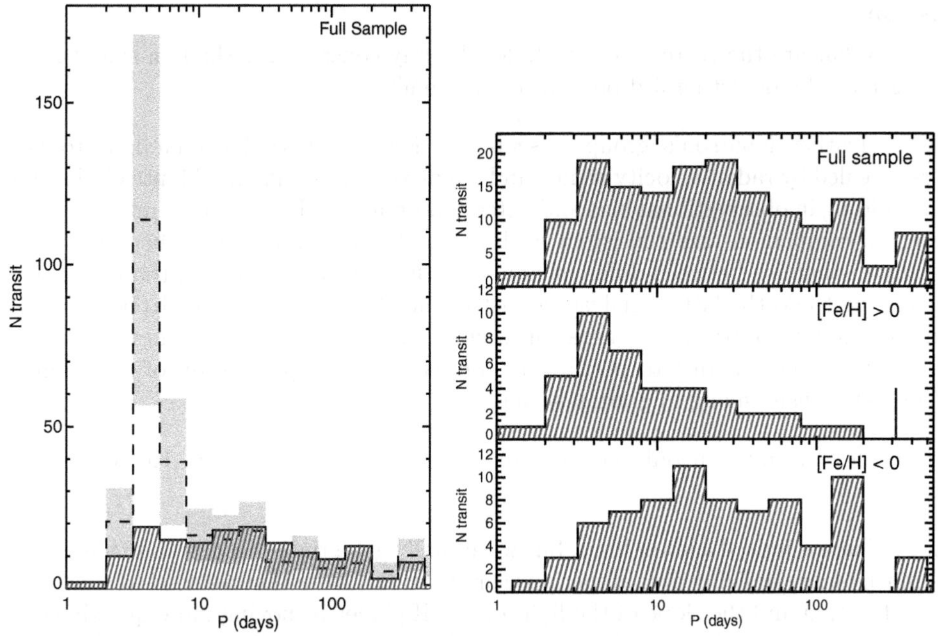

Figure 3. Left: Striped striped: number of transiting giant planets detected by *Kepler*. Black dashed: expected number based on the RV-discovered (i.e. excluding planets discovered by transit surveys) sample. The gray error bars are from uncertainties in $C_{\rm norm}$, not the Poisson uncertainties of each individual bin. The two distributions are consistent at long periods, but the *Kepler* sample lacks a short period pile-up. Right: Number of transiting giant planets observed by *Kepler* without a stellar metallically cut (top), with [Fe/H]$\geqslant$0 (middle), and with [Fe/H]<0 (bottom). In the metal-rich sample (middle), we recover the shape of the short-period pile-up seen in the RV sample (black-dashed line, Figure 2). In contrast, the metal-poor sample (bottom) is depleted in short-period giants.

migration mechanism for delivering close-in giant planets, rather than stellar Kozai. This is consistent with the conclusion of Section 2 that stellar Kozai does not produce most hot Jupiters, based on the lack of super-eccentric proto-hot Jupiters.

References

Batalha, Rowe, Bryson, *et al.* 2013, *ApJS*, 204, 24
Dawson, R. I. & Johnson, J. A. 2012, *ApJ*, 756, 122
Dawson, R. I., Johnson, J. A., Morton, *et al.* 2012, *ApJ*, 761, 163
Dawson, R. I., & Murray-Clay, R. A. 2013, *Ap. Lett.*, 767, L24
Dawson, R. I., Murray-Clay, R. A., & Johnson, J. A. 2013, ArXiv e-prints 1211.0554
Goldreich, P. & Tremaine, S. 1980, *ApJ*, 241, 425
Guillochon, J., Ramirez-Ruiz, E., & Lin, D. 2011, *ApJ*, 732, 74
Naoz, S., Farr, W. M., Lithwick, Y., *et al.* 2011, *Nature*, 473, 187
Rasio, F. A. & Ford, E. B. 1996, Science, 274, 954
Socrates, A., Katz, B., Dong, S., & Tremaine, S. 2012, *ApJ*, 750, 106
Steffen, J. H., Ragozzine, D., Fabrycky, D. C., *et al.* 2012, Proceedings of the NAS, 109, 7982
Taylor, S. F. 2012, arXiv:1211.1984
Wright, J. T., Marcy, G. W., Howard, A. W., *et al.* 2012, *ApJ*, 753, 160
Wu, Y. & Lithwick, Y. 2011, *ApJ*, 735, 109
Wu, Y. & Murray, N. 2003, *ApJ*, 589, 605

Discussion

RAFIKOV: What are the current prospects for directly constraining the non-smooth migration scenario by detecting 3rd body in the system?

DAWSON: Heather Knutson's group is searching for long period companions to hot Jupiters revealed by radial-velocity trends and adaptive-optics imaging. Monte Carlo simulations, folding in observational bias and detection limits, could be used to infer whether the detected companions are consistent with being the third body perturbers and even potentially distinguish between different non-smooth migration scenarios, though it may be challenging to do the latter analysis robustly. Failed hot Jupiters, like KOI-1474, are also important for constraining migration scenarios, and some of them have detected transit-timing variations due to perturbing companions, the population of which may also help distinguish among migration scenarios.

WIKTOROWICZ: Don't you only get a lower limit to eccentricity from Kepler transit durations?

DAWSON: There is indeed a lower limit, but we actually get a tighter constraint than just a lower limit by marginalizing over all periapse orientations. Note that we need to constrain both the duration and the slope of the light curve. If planet is moving more quickly than if it were on a circular orbit, a wide range of periapse orientations produce a light curve similar to that if the planet were transiting at periapse, so the inferred eccentricity probability distribution is mostly concentrated close to the minimum eccentricity.

Exploring the Formation and Evolution of Planetary Systems
Proceedings IAU Symposium No. 299, 2013
M. Booth, B. C. Matthews & J. R. Graham, eds.

doi:10.1017/S1743921313009058

A Hot Jupiter in a Nearly Polar Orbit

Brett C. Addison[1,2], Chris G. Tinney[1,2], Duncan J. Wright[1,2], Daniel Bayliss[3], George Zhou[3], Joel D. Hartman[4,5], Gáspár Á. Bakos[4,5] and Brian Schmidt[3]

[1]Exoplanetary Science Group, School of Physics, University of New South Wales, 2052, Australia

[2]Australian Centre of Astrobiology, University of New South Wales, 2052, Australia

[3]Research School of Astronomy and Astrophysics, Australian National University, Canberra, ACT 2611, Australia

[4]Department of Astrophysical Sciences, Princeton University, NJ 08544, USA

[5]Harvard-Smithsonian Center for Astrophysics, Cambridge, MA 02138, USA

Abstract. We measured the spin-orbit misalignment for WASP-79b, a transiting hot Jupiter from the WASP survey. Using the Rossiter-McLaughlin effect during the transit event, we determined the sky-projected obliquity to be $\lambda = -106^{+10}_{-8}{}^{\circ}$. This result indicates that the planet is in a nearly polar orbit.

Keywords. planets and satellites: formation — stars: individual (WASP-79) — techniques: radial velocities

---

1. Introduction

Key additional insights into the formation, migration, and evolution of the more than 900 exoplanets known (as at June 2013), can be obtained from the measurement of the sky-projected spin-orbit angle of exoplanetary systems through spectroscopic measurements of the Rossiter-McLaughlin effect (Winn *et al.* 2005).

We have carried out spectroscopic follow-up measurements of the Rossiter-McLaughlin effect for WASP-79b, a recently discovered hot Jupiter from the WASP Southern Hemisphere transit survey (Smalley *et al.* 2012). WASP-79b has a mass M_P of $0.90 \pm 0.09\, M_J$, radius R_P of $1.70 \pm 0.11\, R_J$, and orbits an F5 star with $T_{eff} = 6600 \pm 100$K (Smalley *et al.* 2012). We detect the radial velocity anomaly due to the Rossiter-McLaughlin effect and determine that it is in a nearly polar orbit (Addison *et al.* 2013).

2. Rossiter-McLaughlin Effect

The Rossiter-McLaughlin effect is an anomaly in the radial velocity curve of a host star that is seen during an exoplanetary transit caused as the planet occults the stellar disk (Rossiter 1924; McLaughlin 1924; Ohta *et al.* 2005). Measurement of this anomaly provides us with an estimate of the projected rotational velocity of the stellar disk ($v \sin i_\star$) and the projected spin-orbit angle (λ) between the planetary orbital axis and the stellar spin axis (Winn *et al.* 2005).

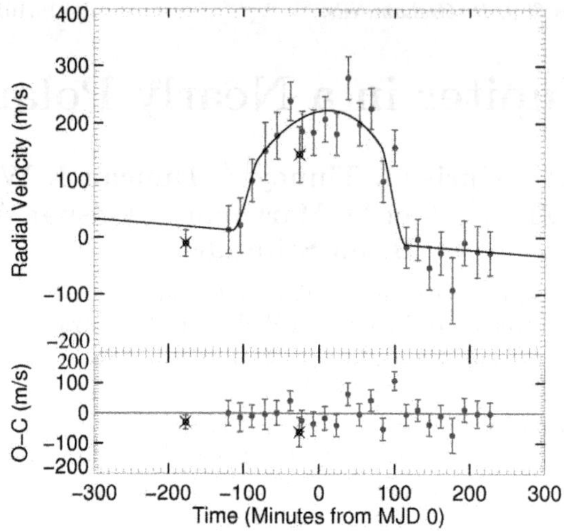

Figure 1. Spectroscopic radial velocities of WASP-79 plotted as a function of time along with the best fitting main-sequence model and corresponding residuals. The filled circles are our data, while the two open cross-circles are velocities by Smalley *et al.* (2012).

3. Observations, Analysis, & Results

WASP-79 was observed using the CYCLOPS2 (see Horton *et al.* 2012) optical-fiber bundle feeding the UCLES echelle spectrograph on the 3.9 m Anglo-Australian Telescope. The data was reduced and wavelength calibrated using custom MATLAB routines (Wright and Tinney, in prep. & see Addison *et al.* 2013). Radial velocities were computed using the IRAF task, *fxcor*, by cross-correlation with a spectrum of a bright template star (HD86264) of similar spectral type.

We have developed a modeling system (the Exoplanetary Orbital Simulation and Analysis Model, or ExOSAM) to determine λ and $v \sin i_\star$ from the planetary orbit, a calculated in-transit lightcurve, and a velocity anomaly Hirano *et al.* (2010).

Smalley *et al.* (2012) derived two preferred solutions for WASP-79 – one with it on the main sequence ($R_\star = 1.64 \pm 0.08\ R_\odot$) and one with it evolved just off the main sequence ($R_\star = 1.91 \pm 0.09\ R_\odot$). Using parameters from both solutions, we found $\lambda = -106^{+10}_{-8}$ ° and $v \sin i_\star = 17.5^{+1.3}_{-1.4}$ kms^{-1} for the main sequence case and $\lambda = -85^{+13}_{-33}$ ° and $v \sin i_\star = 16.0^{+1.3}_{-1.3}$ kms^{-1} for the non-main sequence case (Addison *et al.* 2013). Figure 1 shows the observed positive Rossiter-McLaughlin velocity anomaly, indicating that WASP-79b is in a nearly polar orbit. Only a handful of planetary systems display such extreme orbits. Additional samples of spin-orbit measurements will help to provide clearer insights into the mechanisms driving planetary migration and spin-orbit misalignments.

References

Addison, B. C., Tinney, C. G., Wright, D. J., *et al.* 2013, *ApJL*, 774, 9
Hirano, T., Suto, Y., Taruya, A., *et al.* 2010, *ApJ*, 709, 458
Horton, A., Tinney, C. G., Case, S., *et al.* 2012, *In SPIE Conf. Ser.*, 8446
McLaughlin, D. B. 1924, *ApJ*, 60, 22
Ohta, Y., Taruya, A., & Suto, Y. 2005, *ApJ*, 622, 1118
Rossiter, R. A. 1924, *ApJ*, 60, 15
Smalley, B., Anderson, D. R., Collier-Cameron, A., *et al.* 2012, *A&A*, 547, A61
Winn, J. N., Noyes, R. W., Holman, M. J., *et al.* 2005, *ApJ*, 631, 1215

Exploring the Formation and Evolution of Planetary Systems
Proceedings IAU Symposium No. 299, 2013
M. Booth, B. C. Matthews & J. R. Graham, eds.

doi:10.1017/S174392131300906X

Tidal Dissipation and Eccentricity Pumping: Implications for the Depth of the Secondary Eclipse of 55 Cnc e

E. Bolmont[1,2], F. Selsis[1,2], S. N. Raymond[1,2], J. Leconte[3], F. Hersant[1,2], A.-S. Maurin[1,2] and J. Pericaud[1]

[1]Univ. Bordeaux, LAB, UMR 5804, F-33270, Floirac, France
email: emeline.bolmont@obs.u-bordeaux1.fr

[2] CNRS, LAB, UMR 5804, F-33270, Floirac, France

[3] Laboratoire de Météorologie Dynamique, Institut Pierre Simon Laplace, Paris, France

Abstract. We use the super-Earth 55 Cnc e (Fischer *et al.* 2008) as a case-study to address an observable effect of tidal heating. This transiting short-period planet belongs to a compact multiple system with massive planets. We investigate whether planet-planet interactions can force the eccentricity of this planet to a level that would affect the eclipse depth observed with Spitzer.

Keywords. Planets and satellites: fundamental parameters – Planets and satellites: dynamical evolution and stability – Planet-star interactions

---

1. Introduction

The super Earth 55 Cnc e is a transiting short-period planet that belongs to a compact multiple system with massive planets. Its eccentricity is excited by the presence of the other planets and it is damped by tidal forces. The resulting tidal heating will modify the flux of the planet and thus affect the eclipse depth observed with Spitzer.

2. Method

Using the constant time-lag model (Hut 1981), we first calculated the observed planet flux as a function of albedo and eccentricity for different tidal dissipation constants and for two extreme cases: a planet without heat redistribution and a planet with full heat redistribution. We derived the values of albedo and eccentricity that match the observed transit depth (Demory *et al.* 2012: D12).

We then performed N-body simulations of the planetary system including tides and General Relativity to follow the evolution of the eccentricity of planet e. As 55 Cnc e is part of a five-planet system, the eccentricity of 55 Cnc e depends on its tidal dissipation factor σ_p, and on the eccentricities of the other disturbing planets. In order to study the dynamical and tidal evolution of this system we used a code (Bolmont, Raymond & Leconte 2013) based on Chambers (1999)'s N-body code Mercury.

We then compared the range of eccentricities given by the simulations with the eccentricities required to alter the eclipse depth (Bolmont *et al.* 2013).

3. Results

We investigate the cases for which tides contribute to more than 10% of the planet flux (*condition 1*) constrained by two observations: the modeled secondary transit depth

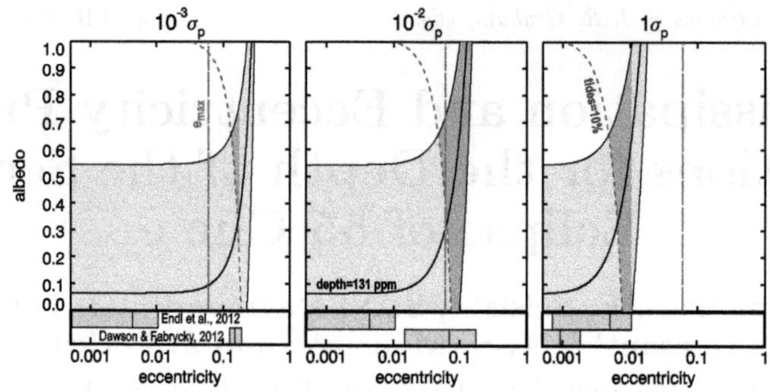

Figure 1. Albedo vs eccentricity for three different dissipation factors. The green areas satisfy *condition 2*, the blue areas satisfy *condition 1*, *condition 2* and agree with the eccentricity range from simulations.

must agree with the observations (*condition 2*) and the eccentricity must be below the upper limit of 0.06 (D12). The higher 55 Cnc e's tidal dissipation factor, the smaller its eccentricity. The higher its tidal dissipation factor, the lower the eccentricity needed to alter the secondary depth. For the orbital configuration from Dawson & Fabrycky (2010), the tidal flux is potentially observable for dissipation rates around $10^{-2} \times \sigma_p$†. For Endl *et al.* (2012), the tidal flux is potentially observable for dissipation rates around $1 \times \sigma_p$.

Fig. 1 shows the dependance of the inferred albedo on eccentricity when taking into account tides for three different dissipation factors for the case of no heat redistribution.

Without tides, the albedo must be below 0.55. With tides and for dissipation factors of $10^{-3} \times \sigma_p$ and $10^{-2} \times \sigma_p$, tidal heating is observable assuming the orbital configuration of Dawson & Fabrycky (2010) with an albedo between 0.5 and 0.7. For $1 \times \sigma_p$ the agreement is met for the Endl *et al.* (2012) configuration with an albedo up to 0.8.

4. Conclusion

Transiting exoplanets in multiple planet systems – like 55 Cancri – are exceptional targets for testing tidal models because their tidal luminosity may be observable. Future multi-wavelength observations of eclipse depth and phase curves (for instance with EChO and JWST) may allow us to better resolve the temperature map of these planets and break the degeneracy between albedo and tidal heating that remains for single-band observations. In addition, an accurate determination of the eccentricity will make it possible to constrain the dissipation rate of the planet and probe its internal structure.

References

Fischer, D. A., and 11 colleagues 2008, *ApJ* 675, 790-801.
Hut, P. 1981, *A&A*, 99, 126
Bolmont, E., Selsis, F., Raymond, S. N., Leconte, J., Hersant, F., Maurin, A.-S., & Pericaud, J. 2013, A & A, 556, A17.
Demory, B.-O., Gillon, M., Seager, S., *et al.* 2012, *ApJ*, 751, L28
Leconte, J., Chabrier, G., Baraffe, I., & Levrard, B. 2010, *A&A*, 516, A64+
Bolmont, E., Raymond, S. N., Leconte, J. in prep
Dawson, R. I. & Fabricky, D. C. 2010, *ApJ*, 722, 937
Endl, M., Robertson, P., Cochran, W. D., MacQueen, P. J., Brugamyer, E. J., Caldwell, C., Wittenmyer, R. A., Barnes, S. I., & Gullikson, K. 2012, *ApJ*, 759, 19.

† $\sigma_p \propto k_{2,\oplus}\Delta t_\oplus$, where $k_{2,\oplus}$ is Earth's Love number of degree two and $\Delta t_\oplus$ its time lag.

Exploring the Formation and Evolution of Planetary Systems
Proceedings IAU Symposium No. 299, 2013
M. Booth, B. C. Matthews & J. R. Graham, eds.

doi:10.1017/S1743921313009071

CARMENES: Blue Planets Orbiting Red Dwarfs

A. Quirrenbach[1], P. J. Amado[2], J. A. Caballero[3], H. Mandel[1], R. Mundt[4], A. Reiners[5], I. Ribas[6], M. A. Sánchez Carrasco[2], W. Seifert[1], M. Azzaro[7], D. Galadí[7] and the CARMENES Consortium[1,2,3,4,5,6,7,8,9,10,11]

[1]Landessternwarte, Zentrum für Astronomie der Universität Heidelberg, Königstuhl 12, D-69117 Heidelberg, Germany

[2]Instituto de Astrofísica de Andalucía (CSIC), Glorieta de la Astronomía s/n, E-18008 Granada, Spain

[3]Centro de Astrobiología (CSIC-INTA), Campus ESAC, PO Box 78, E-28691 Villanueva de la Cañada, Madrid, Spain

[4]Max-Planck-Institut für Astronomie, Königstuhl 17, D-69117 Heidelberg, Germany

[5]Institut für Astrophysik (IAG), Friedrich-Hund-Platz 1, D-37077 Göttingen, Germany

[6]Institut de Ciències de l'Espai (CSIC-IEEC), Campus UAB, Facultat Ciències, Torre C5 - parell - 2a planta, E-08193 Bellaterra, Barcelona, Spain

[7]Calar Alto Observatory (MPG-CSIC), Centro Astronómico Hispano-Alemán, Jesús Durbán Remón, 2-2, E-04004 Almería, Spain

[8]Departamento de Astrofísica, Facultad de Física, Universidad Complutense de Madrid, E-28040 Madrid, Spain

[9]Thüringer Landessternwarte Tautenburg, Sternwarte 5, D-07778 Tautenburg, Germany

[10]Instituto de Astrofísica de Canarias, Vía Láctea s/n, E-38205 La Laguna, Tenerife, Spain, and Dept. Astrofísica, Universidad de La Laguna (ULL), E-38206 La Laguna, Tenerife, Spain

[11]Hamburger Sternwarte, Gojenbergsweg 112, D-21029 Hamburg, Germany

Abstract. CARMENES (**C**alar **A**lto high-**R**esolution search for **M** dwarfs with **E**xo-earths with **N**ear-infrared and optical **E**chelle **S**pectrographs) will conduct a radial-velocity survey of ~ 300 M dwarfs with the 3.5 m telescope at the Calar Alto Observatory. The CARMENES instrument is currently under construction; it consists of two independent échelle spectrographs, which together cover the wavelength range $0.55 - 1.7\,\mu$m at a spectral resolution of $R = 82,000$. The spectrographs and the fiber input are designed with a goal of 1 m/s radial velocity precision using simultaneous calibration with emission-line lamps.

Keywords. instrumentation: spectrographs, techniques: radial velocities, stars: late-type, (stars:) planetary systems

1. Introduction

M-type stars span the mass range of $0.1 - 0.6\,M_{\odot}$, and therefore obtaining statistics of planetary system occurrence and architecture for these stars is of great importance for understanding the physics of planet formation and evolution, and its dependence on stellar host mass. Radial-velocity searches for planets around M dwarfs benefit from a large signal and a short orbital period of planets in the habitable zone, which provides an opportunity for finding habitable planets in the Solar neighborhood. However, the current number of planet detections with M-star hosts is still low, due to the faintness of the targets in the visible wavelength range and the intrinsic stellar jitter. The CARMENES project (see also Quirrenbach *et al.* 2010, 2012) is aimed at filling this gap, by constructing a radial-velocity instrument optimized for planet searches of mid- to late-type M dwarfs.

2. The CARMENES Instrument

For mid-M to late-M spectral types, the wavelength range around 1000 nm (Y band) is the most important region for RV work. Therefore, the efficiency of CARMENES will be optimized in this range. Since CCDs do not provide high enough efficiency above 900 nm, a near-IR detector is required. It is thus natural to adopt an instrument concept with two spectrographs, one equipped with a CCD for the range 550 – 1050 nm, and one with HgCdTe detectors for the range from 0.9 – 1.7 μm.

Each spectrograph will be coupled to the 3.5 m telescope with its own optical fiber. The front end will contain a dichroic beam splitter and an atmospheric dispersion corrector, to feed the light into the fibers leading to the spectrographs. Additional fibers are available for simultaneous injection of light from emission line lamps for RV calibration.

The spectrographs are mounted on benches inside vacuum tanks, which are located in climatic chambers inside the coudé laboratory of the 3.5 m dome. The near-IR vacuum tank is equipped with a temperature stabilization system; the thermal design of both spectrographs provides the capability of keeping the temperature of the optical bench constant to within ± 0.01 K over 24 h. The visible-light spectrograph will be operated near room temperature, the NIR spectrograph will be cooled to ~ 140 K.

3. The CARMENES Survey

The main scientific driver for CARMENES is the search for very low-mass planets (i.e., Earth-analogs and "super-Earths") around mid- to late-type M dwarfs. A long-term RV precision of 1 m s^{-1} is sufficient to detect a $2\,M_\oplus$ planet in the middle of the HZ of an M5 star. For stars later than ~M4 ($M < 0.25 M_\odot$), such precision will yield detections of super-Earths of $5M_\oplus$ and smaller inside the entire width of the HZ.

Our survey strategy is to intensively monitor a well-characterized sample of ~ 300 M dwarfs. If we assume that about 30–40 % of the M dwarfs have low-mass planets and a high chance of some being in the habitable zone, we may expect to find between 50 and 100 suitable planets. Even if the real frequency is much lower, still a substantial number of detections will be available, and CARMENES will put tight constraints on the abundance and orbital parameters of low-mass planets in the investigated mass regime.

The expected CARMENES planet harvest is, at any rate, sufficient to carry out a reliable statistical analysis of the planet population and shed light on the architecture of planetary systems and on their formation mechanisms. The CARMENES survey will also provide valuable constraints on $\eta_\oplus$ for M dwarfs. Given the transit probability and the favorable selection bias, there is a good chance of finding 1 or 2 transiting low-mass planets in the habitable zone. These hold extraordinary value for future investigations.

References

Quirrenbach, A., Amado, P. J., Mandel, H., *et al.* (2010). *CARMENES: Calar Alto high-Resolution search for M dwarfs with Exo-earths with Near-infrared and optical Echelle Spectrographs.* In *Ground-based and airborne instrumentation for astronomy III.* Eds. McLean, I. S., Ramsay, S. K., & Takami, H., SPIE 773513, p. 1–14

Quirrenbach, A., Amado, P. J., Seifert, W., *et al.* (2012). *CARMENES. I: Instrument and survey overview.* In *Ground-based and airborne instrumentation for astronomy IV.* Eds. McLean, I. S., Ramsay, S. K., & Takami, H., SPIE 84460R, p. 1–13

Exploring the Formation and Evolution of Planetary Systems
Proceedings IAU Symposium No. 299, 2013
M. Booth, B. Matthews & J. R. Graham & eds.

doi:10.1017/S1743921313009083

Eccentricity Dependence on Iron Abundance

Stuart F. Taylor

Participation Worldscope/Global Telescope Science
Sedona, Arizona, U.S., H. Kong
email: astrostuart@gmail.com

Abstract. The occurrence and eccentricity distribution of planets as a function of period is significantly different for iron-rich and iron-poor planet systems. We find that iron-poor stars with planets having periods between 525 and 600 days have higher eccentricity than such systems outside this range. If whole planet pollution causes the correlation of giant planet eccentricity with stellar iron abundance, then this cluster could be due to a paucity of pollution in this period range. Newly reported patterns of planet occurrence must result from planet system architectural features such as the snow line, followed by subsequent migration. Different results favor pollution or higher initial iron abundance causing the higher occurrence fraction of giant planets hosted by iron-rich stars, but the two explanations could be complementary. Relations between planet and stellar parameters are a major product of planet-finding, which promise further insights into star-planet system formation and evolution. Collaborators are sought to study these patterns. We expect a spirited debate over the relative contributions of initial abundances, disk accretion, and whole planet accretion.

Keywords. stars: abundances, planetary systems: dynamical evolution, tidal interactions, binaries (including multiple): close, planetary systems: formation, stars: fundamental parameters

1. Introduction

There has been a long debate over whether the correlation between iron abundance in stars ($[Fe/H]_*$) and giant planet occurrence is due to higher formation rates around iron-rich stars, or if the stars were "polluted" by accretion of iron-rich disk material. Gonzalez (1997) cited the lack of a correlation of $[Fe/H]_*$ with the mass of the convective zone (CZ) to reject the pollution hypothesis, but Murray & Chaboyer (2002) showed that the then observed metallicities could be explained by the pollution of $\sim 5\ M_\oplus$ into the convective zone (CZ) in combination with higher instrinsic metallicity and selection effects. Fischer & Valenti (2005, hereafter FV05) ruled out that the occurrence-metallicity correlation is due to disk accretion by showing that there is no abundance variation of Na, Si, Ti, or Ni, etc. by condensation temperature as would be expected if dust grains were infalling.

Newly found correlations between $[Fe/H]_*$ and planet properties are presented as evidence that pollution from planets merging with stars makes a contribution to $[Fe/H]_*$. Such "whole planet" pollution would not be affected by stellar radiation, so would not vary by condensation temperature. Unexpected detail in planet numbers and eccentricity presented in T13b and here favor an explanation that whole planets pollute stars. The correlation of high iron abundance in binary SWPs (BSWPs) (Taylor 2013b; hereafter T13b, and presentation at conference) suggests that in the presence of a binary companion, only in systems with higher initial iron can planet formation overcome the inhibiting effects of gravitational perturbations and irradiation.

It has been suggested (Dawson & Murray-Clay 2013; hereafter DM13) that the correlation of iron-rich stars hosting planets with more highly eccentric orbits (T13b, Taylor 2012, 2013a; hereafter T12b, T13a) and with the presence or absence of the "hot Jupiter pileup" (DM13), could be due to planet-planet scattering being more frequent in the more crowded planetary systems formed from iron-rich systems (DM13). which could

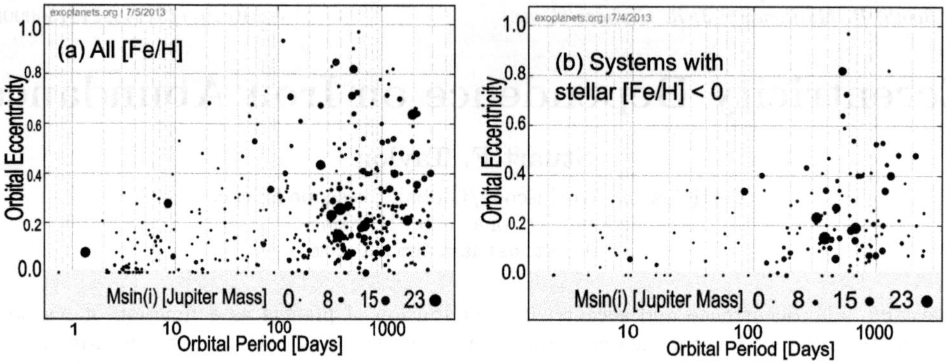

Figure 1. Eccentricity as function of period for systems found by radial velocity. This comparison of all systems (a) with low $[Fe/H]_*$ systems (b) shows that for most period ranges, high-e systems are correlated with above-solar metallicity (a), but the low iron abundance systems (b) have a different pattern. We find a spike of five anomalously high-e systems confined within a narrow-period range, (b).

be due to higher iron formation rates, whole planet pollution, or both. The correlation of higher $[Fe/H]_*$ with binary SWPs (BSWPs) could be due to planet formation in the presence of a stellar companion requiring a higher initial iron abundance to overcome the inhibiting effects of gravitational perturbations and irradiation from the stellar companion. However, the eccentricity-period ($e - P$) distribution of T13b varies with the period of the planet in a manner that likely reveals details of planet migration. To explain all of these, whole planet pollution could be further increasing the $[Fe/H]_*$ of systems crowded due to being preferentially initially metal-rich. If observed $[Fe/H]_*$ is due to both initial iron abundance and to pollution, it is important to determine the relative contributions.

2. Unexpected dependence of $e - P$ on $[Fe/H]_*$ and binarity

The eccentricity-period relation, shown in Fig. 1, can be seen to be very different for systems with $[Fe/H]_*$ below and above that of the sun. T13b and Taylor (2013c) also show that planets of single stars and binaries are separate populations. The eccentricity distribution as a function of period is described in T13b and Taylor (2013c), which show that iron-rich and poor populations below 500 days and above 600 days are different populations, with 99% and 96.5% probability respectively. Work using fewer statistics had not seen this, and said no correlation was evidence against pollution (e.g. FV05). There is a "spike" of iron-poor high eccentricity systems with periods from 520 to 600 days which are presented as evidence of pollution by T13b and Taylor (2013c).

References

Dawson, Rebekah I. & Murray-Clay, Ruth A., 2013, *ApJL*, 767, L24.
Fischer, Debra A. & Valenti, Jeff, 2005, *ApJ*, 622, 1102-1117.
Murray, N. & Chaboyer, B., 2002, *ApJ*, 566, 442-451.
Gonzalez, Guillermo, 1997, *MNRAS*, 285, 403-412.
Taylor, S. F. 2012, arXiv:astro-ph/1211.1984.
Taylor, S. F. 2013a, arXiv:astro-ph/1301.4229.
Taylor, S. F. 2013b, arXiv:astro-ph/1305.5197.
Taylor, S. F. 2013c, extended version, to be submitted, arXiv:astro-ph.

Exploring the Formation and Evolution of Planetary Systems
Proceedings IAU Symposium No. 299, 2013
M. Booth, B. C. Matthews & J. R. Graham, eds.

doi:10.1017/S174392131300759X

Spin-Orbit Angles as a Probe to Orbital Evolution

A. H. M. J. Triaud[1], A. Collier Cameron[2], D. Queloz[3], D. R. Anderson[4], D. J. A. Brown[2], B. Smalley[2], F. Bouchy[3], M. Lendl[3] and M. Gillon[5]

[1]Department of Physics, Kavli Institute of Astrophysics & Space Research, Massachusetts Institute of Technology, Cambridge, MA 02139, USA

[2]SUPA, School of Physics and Astronomy, University of St Andrews, North Haugh, St Andrews, Fife KY16 9SS, UK

[3]Observatoire Astronomique de l'Université de Genève, 51 Chemin des Maillettes, CH-1290 Sauverny, Switzerland

[4]Astrophysics Group, School of Physical and Geographical Sciences, Lennard-Jones Building, Keele University, Staffordshire ST5 5BG, UK

[5]Institut d'Astrophysique et de Geéophysique, Université de Liège, Allée du 6 Août, 17 (Bât. B5C) Sart Tilman, B-4000 Liége, Belgium

Abstract. We observed with HARPS, the Rossiter-McLaughlin effect for 40 of the 75 transiting hot Jupiters discovered in the Southern Hemisphere by WASP. Our observations reveal a wide distribution in orbital inclinations indicative of past dynamical interactions. Our data also demonstrate the important effect produced by tidal interactions in shaping the spin–orbit (β) angle distribution. We briefly present and interpret the data we collected in a series of graphs.

Keywords. binaries: eclipsing, planetary systems, techniques: radial velocities

---

The Rossiter-McLaughlin effect is observed using radial velocities and occurs when a planet transits its host star. From its shape we can measure the projected angle onto the sky, called β (or sometimes λ in literature), between the stellar spin axis and the planet's orbital axis (Gaudi & Winn, 2007).

Fig. 1a & b show that the fraction of aligned to misaligned objects is larger when planets orbit stars that have a massive outer convective layer (red triangles) corresponding to $T_{\rm eff}$ <6150 K. Tides are dissipated more effectively when there is an outer convective layer (Zahn 1977). Winn *et al.* (2010) have postulated that tides have changed the distribution in spin-orbit angles of planets orbiting stars <6150K, but left the distribution intact for systems >6350K. It is still unclear whether a putative misaligned population of planets orbiting the colder stars could tidally realign or would instead tidally infall into the star, and only leave the mostly aligned population that is observed today.

Fig. 2a only displays planets orbiting the "cold star" sub-sample against a proxy for the strength of tidal interactions: $a/R_\star$. We observe that the weaker the tides the more likely misaligned orbits occur. As a recent addition, we point towards WASP-80b, who orbits a K7-M0 star. Although the coldest in our sample, it is observed on an inclined, circular orbit just like hot Jupiters orbiting mid F stars (Triaud *et al.* 2013). This indicates hot Jupiters may have been more frequently misaligned in the past. No pattern with $a/R_\star$ is visible for the sample orbiting hotter stars. This is expected: tides are weakened due to the lack of an outer convective layer. But stars $>1.2M_\odot$ are known to cool as they age on the main sequence. As they cool they develop an outer convective layer. We should therefore expect a change of spin–orbit angle (β) distribution, as one planet, originally

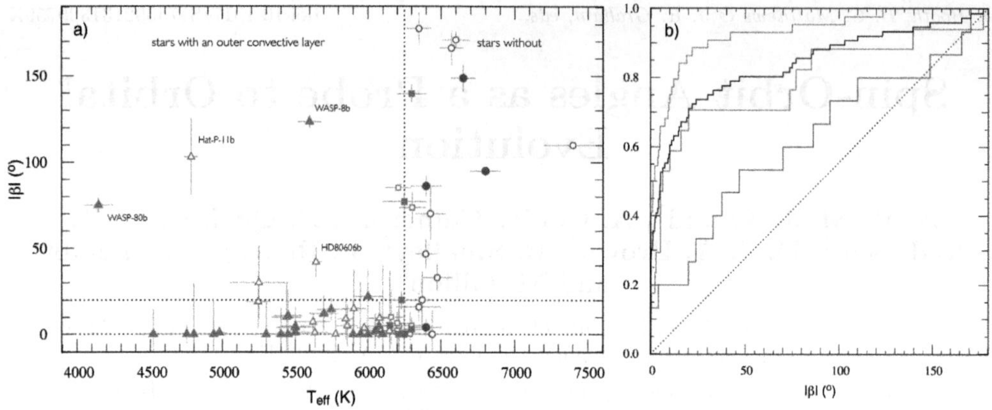

Figure 1. a) updated from Winn *et al.* (2010). The spin–orbit angle β as a function of effective temperature of the host star. Symbols are shaped and colour-coded as a function of $T_{\rm eff}$. Filled symbols represent our HARPS results. Open diagrams come from other observations efforts. b) show cumulative distribution for each of the different symbol shapes. Top curve is for $T_{\rm eff} <$ 6150 K (red triangles), middle for $6150 \leqslant T_{\rm eff} < 6350$ K (purple squares), bottom for $T_{\rm eff} \geqslant$ 6350 K (blue discs), with the black cumulative representing the total population. The variety of angles indicate the presence of a process leading some planets onto highly inclined orbits.

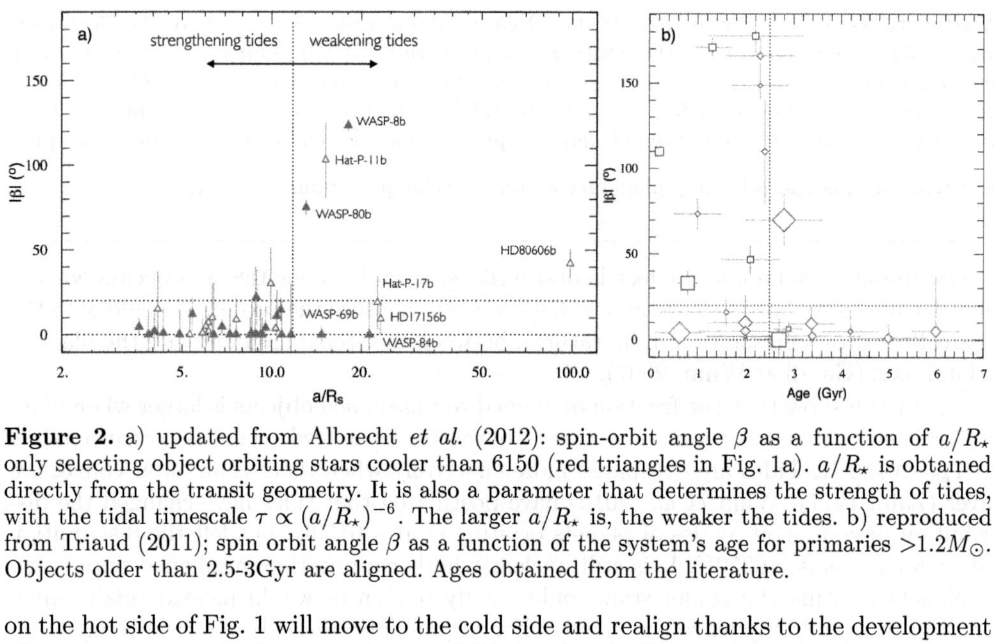

Figure 2. a) updated from Albrecht *et al.* (2012): spin-orbit angle β as a function of $a/R_\star$ only selecting object orbiting stars cooler than 6150 (red triangles in Fig. 1a). $a/R_\star$ is obtained directly from the transit geometry. It is also a parameter that determines the strength of tides, with the tidal timescale $\tau \propto (a/R_\star)^{-6}$. The larger $a/R_\star$ is, the weaker the tides. b) reproduced from Triaud (2011); spin orbit angle β as a function of the system's age for primaries $>1.2 M_\odot$. Objects older than 2.5-3Gyr are aligned. Ages obtained from the literature.

on the hot side of Fig. 1 will move to the cold side and realign thanks to the development of an outer convective layer on the star. This is what is displayed in Fig. 2b.

References

Albrecht, S., Winn, J. N., Johnson, J. A., *et al.* 2012, *ApJ* 757, 18
Gaudi, B. S. & Winn J. N. 2007, *ApJ*, 655, 550
Triaud, A. H.M. J. 2011, *A&A* 534, L6
Triaud, A. H. M. J., Anderson, D. R., Doyle, A. P., *et al.* 2013, *A&A* 551, A80
Winn, J. N., Fabrycky, D., Albrecht S. & Johnson, J. A. 2010, *ApJL* 718, L145
Zahn, J.-P. 1977, *A&A* 57, 383

Author Index